Jan Nijman

Professor of Geography, University of Amsterdam
Professor Emeritus of Geography and Regional Studies,
University of Miami

Peter O. Muller

Senior Professor of Geography and
Regional Studies, University of Miami

H.J. de Blij

The late John A. Hannah Professor of Geography,
Michigan State University

THE WORLD TODAY

CONCEPTS AND REGIONS IN GEOGRAPHY

SEVENTH EDITION

WITH

Richard Grant

Professor of Geography and Regional Studies, University of Miami

You-tien Hsing

Professor of Geography, University of California, Berkeley

Michael Shin

Associate Professor of Geography, University of California, Los Angeles

WILEY

VICE PRESIDENT & DIRECTOR	Petra Recter
EXECUTIVE EDITOR	Jessica Fiorillo
DEVELOPMENT EDITOR	Jennifer Yee
SENIOR MARKET SOLUTIONS ASSISTANT	Kathryn Hancox
EXECUTIVE MARKETING MANAGER	Christine Kushner
ASSOCIATE DIRECTOR, CONTENT PRODUCTION SERVICES	Kevin Holm
PRODUCTION EDITOR	John Du Val
PRODUCT DESIGN MANAGER	Geraldine Osnato
SENIOR PHOTO EDITOR	Billy Ray
SENIOR DESIGNER	Wendy Lai
COVER PHOTO	© Jan Nijman

This book was set in 10.5 Adobe Garamond Pro Regular by Cenveo® Publisher Services and printed and bound by Quad Graphics. This book is printed on acid-free paper.

Main text binder version: 9781119116363
Evaluation ISBN: 9781119236481

Printed in the United States of America

10 9 8 7 6 5 4 3 2 1

Harm de Blij, 1935–2014

This Seventh Edition of The World Today *is dedicated to Harm de Blij, our long-time co-author, mentor, and friend. Harm revolutionized Geography textbook writing in the United States, and he was without equal in communicating geographic knowledge and awareness to wider audiences. His legacy will live on in our books.*

PREFACE

This is the Seventh Edition of a book that, since its first appearance in 2003, has undergone several significant changes. We first wrote this book because so many of you, teachers and students alike, expressed the need for a text that provided an introduction to concepts and regions in world geography—but that did so in a concise manner and in the context of current world affairs. That is still the thrust of the book today, and this Seventh Edition, we believe, meets that challenge better than ever.

We continue our established approach to the integrated treatment of regions and concepts, but the book contains several innovative features as well as some major revisions in the coverage of selected world geographic realms. Some of the most profound changes in the world during the mid-2010s involve the Russian/Central Asian, East Asian, and North African/Southwest Asian realms, all with global ramifications. Elsewhere, as always, geographies are being reshaped by geopolitical, environmental, economic, and/or social forces—from the Arctic to Myanmar and from the Panama Canal to Afghanistan.

As always, we rely in large part on our own field observations: much of our time and research for this book involves field visits and research in many of the world's realms. This is expressed, most concretely, in the numerous field notes and photographs supplied by the authors throughout this book. For the geographer, there is just no substitute for being there.

GEOGRAPHIC KNOWLEDGE, GLOBAL AWARENESS

The book before you is an information highway to geographic knowledge and global awareness. Universities and colleges prepare their students to become critical thinkers, creative problem solvers, and, increasingly, global citizens. If we are now living in a "global village," it means that students must get to know the world like they do their own local areas. But while the influence and interests of the United States reach across the globe, it is no secret that geographic illiteracy at home remains widespread. Much of what students encounter in this book is of immediate, practical value to them—as citizens, as consumers, as travelers, as voters, as job-seekers. North America is a geographic realm with steadily intensifying global connections and relationships. Those involvements require countless, often instantaneous decisions, whether they relate to international business, media, affairs of state, disaster relief, and many other spheres. Such decisions must be based on the best possible knowledge of the world beyond our continent. That understanding can be gained by studying the layout of our world, its environments, societies, traditions, resources, policies, development strategies, and other properties—in short, its **regional geography**.

THE NEW AUTHOR TEAM

The World Today carries the lifelong imprint of Harm de Blij, our founding author, mentor, faculty colleague, and cherished friend. Moreover, this book always will reflect his perspective—even as the world continues to change and the discipline of Geography evolves. The author team is changing as well, and up to a few weeks before his passing Harm de Blij was heavily involved in the discussions and highly supportive of the impending changes in this Seventh Edition. Jan Nijman and Peter Muller, Harm's long-time coauthors, are the main authors of this revision, and Richard Grant and You-tien Hsing, respectively, have contributed the revised chapters on Subsaharan Africa and East Asia. Richard Grant is Professor of Geography and Regional Studies as well as the Director of the Urban Studies Program at the University of Miami. He is one of the world's foremost experts on contemporary Subsaharan Africa and has more than two decades of field research experience in Southern and West Africa. That work continues, and he also has affiliations with universities in Ghana and South Africa. You-tien Hsing is both Professor of Geography and holder of the Pamela P. Fong Family Distinguished Chair in China Studies at the University of California, Berkeley. She not only brings a wealth of knowledge on rapidly transforming East Asia, but is also a leading authority on the emerging geographies of China. And she, too, draws on an enormous range of field experience all across China as well as in many other parts of East Asia.

The author team is also fortunate to have been joined by Michael Shin, who provided indispensable expertise and support on the use of digital geographic data and cutting-edge methods of interactive learning in our discipline. Michael Shin is Associate Professor of Geography at the University of California, Los Angeles, a highly regarded teacher of GIS at that UCLA Department, and possesses the rare ability to combine high-level technical skills with in-depth thematic expertise (in political geography). Michael's contribution is focused in the digital version of this Seventh Edition. We are delighted that Richard, You-tien, and Michael have joined us to enrich the book with their knowledge and insights.

THE WORLD TODAY: PRINT OR DIGITAL?

The print version of the book is also available in digital form and we—both authors and publisher—have invested substantially in both formats. We are now well underway in preparing for a future in which users will increasingly opt for the digital version because it facilitates greater interactive learning and incorporates an array of online connections to monitor real-time changes in the world around us. At this time, we want to be sure that you have a choice and that both products, print and digital, are at the leading edge of world regional geography. This revision of the print book has, in fact, benefitted notably from our work on the digital version—the new *Map Analysis* feature (elaborated below) is a direct outcome of these efforts. We know you will be pleased with this new edition, and we also encourage you to explore the digital version to see which option best suits your particular needs and those of your students.

THE NEW *MAP ANALYSIS* FEATURE

Our book is well known for the quality and currency of its maps, an indispensable tool in the *understanding* of geographic questions. We have now gone a step further and have added a new feature in each chapter that actively draws students 'into' the map, challenging them to use and develop their analytical skills in map interpretation. These new maps typically present the spatial distribution of one or more variables tied to concepts highlighted in that chapter of the text; the accompanying caption provides pertinent information and asks one or two open-ended analytical questions, usually related to the mapped spatial patterns and their possible explanations. Since the questions are open-ended and we do not supply any answers, they are well suited to class discussion or short writing assignments.

For example, the South Asia chapter discusses the concept of megacities, with the *Map Analysis* focused on comparing the shapes of the urbanized areas centered by Dhaka, Bangladesh, and Los Angeles (see Fig. 8-9). The questions posed to student readers concern their thoughts on why the areal sizes are so different (L.A. is about 17 times larger than Dhaka), even though their population sizes are roughly equal (ca. 15 million). The answers involve notions of density that are, in turn, related to factors such as mobility, modes of transportation, housing conditions, and income level.

We know how challenging it can be to engage students with maps in a meaningful way, especially in getting them to *read* maps. We believe the new *Map Analysis* feature provides a most useful instrument to help achieve one of the central learning objectives in any (regional) geography course. If you are interested in exploring the digital version of this book, be sure to check out the new *Geographic Information Analysis* modules: these are the online counterparts of the *Map Analysis* feature in this print book. They are specifically designed to heighten geographical knowledge, to increase spatial awareness, and to engage students with interactive, GIS-based, spatial analyses. A set of questions—both multiple choice and discussion-oriented—accompanies each module to facilitate student learning, progress, and assessment.

A NEW ORGANIZATIONAL STRUCTURE

Organizationally, this Seventh Edition differs from its predecessor in four distinct ways. First, it is both shorter and more concise. We are keenly aware of the persistent challenge to instructors and students in World Regional Geography trying to cover all of the world's realms in a single course. We have judiciously condensed the text by more than 10 percent without compromising explanation of the most essential topics and concepts. Moreover, the emphasis in this book, more than ever, is on the world *today*—on illuminating the role of geography in the global affairs of the late 2010s.

The second difference involves the reordering of the chapters that follow the Introduction. Previous editions began with Europe, followed by Russia, then turned to the Americas, and so on. Of course, there is no self-evident sequence for the presentation of the world's realms; users are always free to choose their own sequence, and we have maintained the flexibility of the book to fit every possible ordering scheme. Nonetheless, we decided to make a major change and begin the regional chapters in this edition with North America—because that is where most of us are based; because, overall, it presents the most familiar terrain; and because it is from here that we view and develop perspectives on the rest of the world. Starting with North America, then, is not to privilege that realm but rather to acknowledge our own positionality for this survey of world regions. Following North America (Chapter 1), the succession of chapters is largely based on proximity: the second regional chapter covers Middle America, and the third South America; from there we move to Europe (Chapter 4), followed by Russia/Central Asia (5); then on to North Africa/Southwest Asia (6), that is followed by Subsaharan Africa (7). Chapters 8 through 12—South Asia (8), East Asia (9), Southeast Asia (10), Austral Realm (11), and Pacific Realm/Polar Futures (12)—follow the same sequence used in the Sixth Edition. It might therefore be useful to examine this new organization more closely for the purposes of course planning.

Third, the world's ongoing geographic transformations have led us to reconsider some critical boundaries that separate certain realms. In this edition we pay more attention to *transition zones* between realms in general—but we also saw the need to redraw a few boundaries. The chapter on the Russian realm now includes the additional region of Central Asia, which, as Turkestan, was previously allocated to the North Africa/Southwest Asia realm. At the same time, we have shifted Afghanistan to the South Asian realm from its previous positioning within the North Africa/Southwest Asia realm. The reasons for these changes are elaborated in the appropriate chapters, but they are quite clearly related to the changing regional dynamics of the 2010s. Whereas the world's geography is still anchored to a relatively stable set of realms, at a finer scale, often on the edges of realms, pieces of the regional puzzle are shifting.

Finally, the Seventh Edition again consists of single chapters for each world geographic realm. Thus we have moved away from subdividing realms into A and B chapters, a result of our streamlining and making the text more concise (in which separate A and B chapters became unnecessary because we felt they imposed too rigid a structure). Internally, each regional chapter preserves our practice of focusing first on large-scale, realmwide geographies, and then turning to the more detailed coverage of each realm's regions, often breaking them down further into subregions and/or individual countries.

HIGHLIGHTS OF NEW MATERIAL

Since the Sixth Edition appeared in 2013, the world has undergone some momentous changes, all of which have geographic ramifications. Occasionally, those changes are so rapid and transformative that they force us to rethink the fundamentals of particular realms, regions, and countries, or even of the global system that binds them together. In the mid-2010s, Russia's revanchist policies in its Near Abroad, China's relentlessly expanding global presence, and the spiraling conflicts in North Africa/Southwest Asia are cogent examples. This Seventh Edition not only contains myriad valuable updates but also, where necessary, provides reappraisals of the fundamental nature of the world's ever-changing geography.

No summary can adequately encapsulate all that this new edition contains in terms of text, field notes, maps, or photos, but here are some chapter-by-chapter highlights:

The chapter-length **Introduction**, as usual, provides the foundation and sets the stage for the regional chapters, discussing a broad range of basic geographic concepts along with a series of cornerstone world maps. It includes a new section on concepts of sustainable development and inclusive development. The newly featured *Map Analysis* focuses on the critical relationship between economic growth and sustainable development.

Chapter 1, North America, has a new section and map on the U.S.-Mexican border that, simultaneously, is a sharp boundary *and* a transition zone. Also look for new text on high-technology clusters and new maps on U.S. agriculture as well as Miami's world-city connections.

Chapter 2, Middle America, directs attention to the beginnings of a new era in United States-Cuba relations, the implications of the widened Panama Canal that opens in 2016, and the apparent diversification—from exclusively Mexico to incorporating the Caribbean—of U.S.-bound routes for smuggled cocaine and other illicit drugs.

Chapter 3, South America, contains a new section on the slowdown of the commodity boom that affects so much of this realm, and another section on Colombia's recovery from years of civil strife and economic near-paralysis (including the remarkable revival of the city of Medellín). The *Map Analysis* in this chapter compares national incomes to the variation in retail prices of iPhones.

Chapter 4, Europe, discusses the increasing pressures on the European Union concerning Greece, even as the EU continued to expand in recent years (Croatia joined as the 28th member-state in 2013, and Latvia and Lithuania had joined the eurozone by 2015). The Ukrainian crisis receives considerable attention and is the focus of the *Map Analysis* of this chapter. Also look for new material on Islamic jihadist terrorism in France, as well as the staggering human drama of the refugee crisis caused by the tide of undocumented immigration occurring across the central Mediterranean Basin.

Chapter 5, Russia/Central Asia, emphasizes the increasingly revanchist policies of the Putin regime, not only their consequences for the 'Near Abroad' (extending from Ukraine to the Transcaucasian Transition Zone to Central Asia) but also in the context of Russia's rapidly fading democratic prospects. There is a new section (and map) on Russia's power as the major oil and gas supplier of Europe, and a major new section on Central Asia—where Russia and China compete for influence, and the Chinese are busily planning their 'New Silk Road' connection to Europe through Russia's 'backyard.'

Chapter 6, North Africa/Southwest Asia (NASWA), takes a close look at the anticlimactic aftermath of the 'Arab Spring' movement and the dislocation and turmoil it triggered in various regions and countries ranging from Syria to Yemen, from Libya to Somalia. There are several new maps: one details the trans-Mediterranean tidal flow of undocumented immigrants from NASWA to Europe and the refugee crisis it spawned in Southern Europe; another shows the territorial expansion of ISIS (Islamic State) across Syria and Iraq through mid-2015. Note, too, that this chapter now concludes with the regional discussion of the crucial African Transition Zone (previously part of the chapter on Subsaharan Africa).

Chapter 7, Subsaharan Africa, bears the imprint of our new contributing author, Africa specialist Richard Grant. It contains several new field notes and documents the Ebola crisis of 2014–2015. The *Map Analysis* feature in this chapter addresses the challenges of communication and infrastructure in this enormous postcolonial realm. There are new maps on patterns of economic growth and the spatial distribution of Chinese investments across the realm, both linked to an analysis of the heavy reliance on raw materials.

Chapter 8, South Asia, now includes Afghanistan as a transition zone on the realm's western flank. That new section includes a map of the reconfigured West region of South Asia, highlighting the critical connections between Afghanistan and Pakistan—which in turn, are crucial to relations between Pakistan and India. Also look for added material on Indian Ocean geopolitics, gender relations, population dynamics, and megacities.

Chapter 9, East Asia, is enriched throughout by our new collaborating author, You-tien Hsing. There are several new field notes and maps to reflect China's westward march of both economic development and urbanization (also the topic of the *Map Analysis* in this chapter). Look for an entirely rewritten section on Korea and for an updated discussion of geopolitical controversies in the East China Sea. This edition pays special attention to Tibet, which is undergoing rapid change but where culture has thus far proven resilient (see both the front and back covers).

Chapter 10, Southeast Asia, contains newly-added sections and maps on the Mekong River Basin and the astonishing proliferation of dams along the trunk stream and its tributaries; on the regional impacts of Myanmar's liberalization (vis-à-vis India, Thailand, and China); on the coming geopolitical confrontation over Chinese territorial claims in the South China Sea; on China's singular dominance in this realm's international trade; and on the intensifying persecution of Myanmar's Muslim minority—the Rohingyas—thousands of whom have fled the country in desperate search of safer havens in Malaysia and elsewhere in the realm.

Chapter 11, The Austral Realm, includes an expanded section on Australia's environmental challenges, especially concerning climate change and growing water shortages. Look for a new map that shows changes in precipitation and the distribution of desalination plants along parts of the country's coastline. The *Map Analysis* in this chapter addresses the geography of Australia's Aboriginal population.

Chapter 12, The Pacific Realm and Polar Futures, in addition to new field notes and photos, contains an updated section on the accelerated melting of the polar ice cap and its consequences for competing territorial claims in the Arctic. The *Map Analysis* centers on rival claims by Russia and Denmark, and on the implications of the shrinking Arctic ice cap for the navigational potential of both the Northeast and Northwest Passages.

ANCILLARIES

A broad spectrum of print and electronic ancillaries are available to support instructors:

Test Bank—Prepared by Dennis Geffert, of Cuyahoga Community College Western. Includes over 1200 multiple-choice, true/false, and fill-in-the-blank short-answer questions.

PowerPoint Lecture Slides—Prepared by Rhonda Reagan of Blinn College. These slides highlight key chapter topics to help reinforce students' grasp of essential concepts.

PowerPoint Slides with Text Images—Images, maps, and figures from the text are available in PowerPoint format.

Instructors may use these images to customize their presentations and to provide additional visual support for quizzes and exams.

Clicker Questions—A bank of questions is available for instructors who are using personal-response-system technology in their courses.

Instructor's Media Guide—Includes information about the media resources (videos, animations, interactive maps) available for use with this text and offers tips and suggestions on how to use these resources in your course, in conjunction with WileyPLUS Learning Space and Orion.

To see a complete listing of these ancillaries, or to gain access to them upon adoption and purchase, please visit: **http://www.wiley.com/college/sc/nijman**.

ACKNOWLEDGMENTS

Over the 45 years since publication of the First Edition of *Geography: Realms, Regions, and Concepts* (and joined over the past dozen years by the appearance of this book), we have been fortunate to receive advice and assistance from literally thousands of people. One of the rewards associated with the publication of a book of this kind is the unremitting stream of correspondence and other feedback it generates. Geographers, economists, political scientists, education specialists, and others have written us, often with fascinating enclosures. We make it a point to respond personally to every such letter, and our editors have communicated with many of our correspondents as well. Moreover, we have considered every suggestion made, and several who wrote or transmitted their reactions through other channels will see their recommendations in print in this edition.

STUDENT RESPONSE

A major part of the correspondence we receive comes from student readers. We would like to take this opportunity to extend our deep appreciation to the several million students around the world who have studied from our books. In particular, we thank the students from more than 150 different colleges across the United States who took the time to send us their opinions. Students told us they found the maps and graphics attractive and functional. We have not only enhanced the map program with exhaustive updating but have added a number of new maps to this Seventh Edition, as well as making significant changes in many others. Generally, students have told us that they found the pedagogical devices quite useful. We have kept the study aids the students cited as effective: a boxed list of each chapter's key concepts, ideas, and terms (numbered for quick reference in the text itself); a box that summarizes each realm's major geographic qualities; and a Glossary of Terms at the end of the book.

FACULTY FEEDBACK

In assembling the Seventh Edition, we are indebted to the following people for advising us on several matters:

Thomas Bell, *University of Tennessee, Knoxville*
Carmen Brysch, *Auburn University (Alabama)*
Edmar Da-Silva, *Broward College, North Campus (Florida)*
Chuck Fahrer, *Georgia College & State University*
Dennis Geffert, *Cuyahoga Community College Western*
Daniel J. Grafton, *South Plains College (Texas)*
Jason B. Greenberg, *Sullivan University (Kentucky)*
Margaret Gripshover, *Western Kentucky University*
Joshua Kegan, *University of New Hampshire*
Chi Kin Leung, *California State University, Fresno*
Sainan Lin, *University of Massachusetts, Amherst*
Max Lu, *Kansas State University*
Ian MacLachlan, *University of Lethbridge (Alberta)*
Veronica Mormino, *William Rainey Harper College (Illinois)*
Thomas R. Mueller, *California University of Pennsylvania*
Clifton W. Pannell, *University of Georgia*
Jessica J. Patton, *Texas Wesleyan University*
Bimal Paul, *Kansas State University*
Sylvia Powell, *Kennesaw State University (Georgia)*
Timothy Pruett, *University of New Hampshire*
Rhonda Reagan, *Blinn College (Texas)*

In addition, several faculty colleagues from around the world assisted us with earlier editions, and their contributions continue to grace the pages of this book. Among them are:

James P. Allen, *California State University, Northridge*
Stephen S. Birdsall, *University of North Carolina*
J. Douglas Eyre, *late of University of North Carolina*
Fang Yong-Ming, *Shanghai, China*
Edward J. Fernald, *Florida State University*
Richard C. Jones, *University of Texas at San Antonio*
Gil Latz, *Indiana University-Purdue University Indianapolis*
Melinda S. Meade, *late of University of North Carolina*
Henry N. Michael, *late of Temple University (Pennsylvania)*
J. R. Victor Prescott, *University of Melbourne (Australia)*
John D. Stephens, *University of Washington*
Canute Vander Meer, *University of Vermont*

Faculty members from a large number of North American colleges and universities (as well as a number of independent scholars) continue to supply us with vital feedback and much appreciated advice. Our publisher arranged several feedback sessions, and we are most grateful to the following for showing us where and how the text could be strengthened and made more precise:

Martin Arford, *Saginaw Valley State University (Michigan)*
Donna Arkowski, *Pikes Peak Community College (Colorado)*
Greg Atkinson, *Tarleton State University (Texas)*
Christopher Badurek, *Appalachian State University (North Carolina)*
Michele Barnaby, *Pittsburg State University (Kansas)*
Denis Bekaert, *Middle Tennessee State University*
Donald J. Berg, *South Dakota State University*

Jill (Alice) Black, *Missouri State University*
Patricia Boudinot, *George Mason University (Virginia)*
Kathleen Braden, *Seattle Pacific University*
Sean Chenoweth, *University of Louisiana, Monroe*
David Cochran, *University of Southern Mississippi*
Joseph Cook, *Wayne County Community College (Michigan)*
Deborah Corcoran, *Southwest Missouri State University*
Marcelo Cruz, *University of Wisconsin at Green Bay*
William V. Davidson, *Louisiana State University*
Larry Scott Deaner, *Kansas State University*
Jeff De Grave, *University of Wisconsin, Eau Claire*
Jason Dittmer, *Georgia Southern University*
James Doerner, *University of Northern Colorado*
Steven Driever, *University of Missouri, Kansas City*
Elizabeth Dudley-Murphy, *University of Utah*
Dennis Ehrhardt, *University of Louisiana, Lafayette*
Bryant Evans, *Houston Community College*
William Flynn, *Oklahoma State University*
William Forbes, *Stephen F. Austin State University (Texas)*
Bill Foreman, *Oklahoma City Community College*
Eric Fournier, *Samford University (Alabama)*
Gary A. Fuller, *University of Hawai'i*
Randy Gabrys-Alexson, *University of Wisconsin, Superior*
William Garbarino, *Community College of Allegheny County (Pennsylvania)*
Hari Garbharran, *Middle Tennessee State University*
Chad Garick, *Jones County Junior College*
Jon Goss, *University of Hawai'i*
Debra Graham, *Messiah College*
Ellen Hansen, *Emporia State University (Kansas)*
Sara Harris, *Neosho County Community College (Kansas)*
John Havir, *Ashland Community & Technical College*
John Hickey, *Inver Hills Community College (Minnesota)*
Shirlena Huang, *National University of Singapore*
Edward Jackiewicz, *California State University, Northridge*
Ingrid Johnson, *Towson State University (Maryland)*
Kris Jones, *Saddleback College (California)*
Uwe Kackstaetter, *Front Range Community College, Westminster (Colorado)*
Cub Kahn, *Oregon State University*
Thomas Karwoski, *Anne Arundel Community College (Maryland)*
Robert Kerr, *University of Central Oklahoma*
Eric Keys, *University of Florida*
Jack Kinworthy, *Concordia University (Nebraska)*
Marti Klein, *Saddleback College (California)*
Christopher Laingen, *Kansas State University*
Heidi Lannon, *Santa Fe College (Florida)*
Unna Lassiter, *Stephen F. Austin State University (Texas)*
James Leonard, *Marshall University (West Virginia)*
Richard Lisichenko, *Fort Hays State University (Kansas)*
Catherine Lockwood, *Chadron State College (Nebraska)*
George Lonberger, *Georgia Perimeter College*
Claude Major, *Stratford Career*
Steve Matchak, *Salem State College (Massachusetts)*
Alan Marcus, *Towson University (Maryland)*
Chris Mayda, *Eastern Michigan University*

Brian McCabe, *Hutchinson Community College (Kansas)*
Trina Medley, *Oklahoma City Community College*
Cristian Melo, *Florida International University*
Dalton W. Miller Jr., *Mississippi State University*
Ernando F. Minghine, *Wayne State University (Michigan)*
Irene Naesse, *Orange Coast College (California)*
William Nieter, *St. Johns College (New York)*
Valiant C. Norman, *Lexington Community College (Kentucky)*
Richard Olmo, *University of Guam*
Pai Yung-Feng, *New York City*
J. L. Pasztor, *Delta College (Michigan)*
Iwona Petruczynik, *Mercyhurst College (Pennsylvania)*
Paul E. Phillips, *Fort Hays State University (Kansas)*
Rosann Poltrone, *Arapahoe Community College (Colorado)*
Jeff Popke, *East Carolina University*
David Privette, *Central Piedmont Community College (North
    Carolina)*
Joel Quam, *College of DuPage (Illinois)*
Craig Revels, *Central Washington University*
Joelia Robinson, *Houston Community College*
Paul Rollinson, *Missouri State University*
Rinku Roy Chowdhury, *Clark University (Massachusetts)*
A. L. Rydant, *Keene State College (New Hampshire)*
Justin Scheidt, *Ferris State University (Michigan)*
Kathleen Schroeder, *Appalachian State University (North Carolina)*
Betty Shimshak, *Towson University (Maryland)*
Nancy Shirley, *Southern Connecticut State University*
Dmitrii Sidorov, *California State University, Long Beach*
Dean Sinclair, *Northwestern State University (Louisiana)*
Richard Slease, *Oakland, North Carolina*
Susan Slowey, *Blinn College (Texas)*
Dean B. Stone, *Scott Community College (Iowa)*
Jamie Strickland, *University of North Carolina at Charlotte*
Ruthine Tidwell, *Florida Community College at Jacksonville*
Irina Vakulenko, *Collin County Community College (Texas)*
Cathy Weidman, *Austin, Texas*
Kirk White, *York College of Pennsylvania*
Thomas Whitmore, *University of North Carolina*
Leon Yacher, *Southern Connecticut State University*
Keith Yearman, *College of DuPage (Illinois)*
Laura Zeeman, *Red Rocks Community College (Colorado)*
Yu Zho, *Vassar College (New York)*

PERSONAL APPRECIATION

We want to express our personal appreciation to those who have been especially helpful in our revisions of this book in the present and/or earlier editions. The map of Russian physiography was updated thanks to the recommendations of Mika Roinila of the State University of New York, College at New Paltz. Special thanks also go to Charles Pirtle, Professor of Geography at Georgetown University's School of Foreign Service for his advice on the Middle America chapter; to Joshua Kegan and his instructor, Timothy Pruett (both of the University of New Hampshire), for suggesting improvements in our map of Europe's regions; to Bilal Butt (postdoctoral fellow, University of Wisconsin-Madison) for his help with the Subsaharan Africa chapter; and to C. Frederick Lohrengel II of Southern Utah University for his advice on the Southeast Asia chapter. We are also grateful to Beth Weisenborn, Virtual Course Coordinator for the Department of Geography at Michigan State University, and all the instructors of GEO 204 (World Regional Geography) for their most valuable input. And we salute Chuck Fahrer of Georgia College and State University for his many useful suggestions.

We also record our appreciation to a number of geographer colleagues who generously provided advice in their areas of expertise. In particular, at the University of Miami's Department of Geography and Regional Studies, this includes faculty colleagues Richard Grant, Miguel Kanai, Elvira Restrepo, Shouraseni Sen Roy, Ira Sheskin, Justin Stoler, and Diana Ter-Ghazaryan as well as GIS Lab Manager Chris Hanson.

We are privileged to work with a team of professionals at John Wiley & Sons that is unsurpassed in the college textbook publishing industry. As authors we are acutely aware of these talents on a daily basis during the pivotal production stage, especially the coordination and leadership skills of Development Editor Jennifer Yee, whose experience and wide range of talents enabled her to take on this position at the last minute. Our new Executive Editor, Jessica Fiorillo, also hit the ground running and helped guide this project to fruition at a time of considerable change. We have benefitted from the skills and savvy of our Senior Marketing Manager, Christine Kushner, and from the diligent assistance of Kathryn Hancox, our Market Solutions Assistant. We also want to express our admiration for the superb contributions of Jeanine Furino, one of the most gifted production editors we have ever worked with; and we are grateful to Wiley Media Specialist John Du Val's help in coordinating the production process. Our book is known for its splendid maps, and a good deal of credit goes to our cartographer-in-chief, Don Larson (ably assisted by Terry Bush) of Mapping Specialists, Ltd. in Fitchburg, Wisconsin. We are also thankful to Senior Photo Editor Billy Ray and Senior Designer Wendy Lai for giving this book its good looks. Last but not least, we are grateful to Petra Recter, Vice President and Director, for her strong support of one of the longest-running success stories in college textbook history.

Finally, we express our deep gratitude to our families for this latest round of seeing us through the demanding schedule of creating this Seventh Edition of *The World Today*. And we wish to acknowledge the inspiration we received from Bonnie de Blij in the difficult months following Harm's passing in 2014. We want her to know that Harm's legacy will live on in this book and those to follow.

Jan Nijman
Coral Gables, Florida

Peter O. Muller
Coral Gables, Florida

July 10, 2015

WileyPLUS Learning Space

Includes **ORION** Adaptive Practice

An easy way to help your students learn, **collaborate**, and **grow**.

Diagnose Early

Educators assess the real-time proficiency of each student to inform teaching decisions. Students always know what they need to work on.

Facilitate Engagement

Educators can quickly organize learning activities, manage student collaboration, and customize their course. Students can collaborate and have meaningful discussions on concepts they are learning.

Measure Outcomes

With visual reports, it's easy for both educators and students to gauge problem areas and act on what's most important.

Instructor Benefits

- Assign activities and add your own materials
- Guide students through what's important in the interactive e-textbook by easily assigning specific content
- Set up and monitor collaborative learning groups
- Assess learner engagement
- Gain immediate insights to help inform teaching

Student Benefits

- Instantly know what you need to work on
- Create a personal study plan
- Assess progress along the way
- Participate in class discussions
- Remember what you have learned because you have made deeper connections to the content

www.wileypluslearningspace.com

WILEY

BRIEF CONTENTS

To gain access to Appendix B, please visit:
http://www.wiley.com/college/sc/nijman.

CONTENTS

To gain access to Appendix B, please visit:
http://www.wiley.com/college/sc/nijman.

THE
WORLD
TODAY

CONCEPTS AND REGIONS IN GEOGRAPHY

SEVENTH EDITION

Nature and culture, entwined in geography. The Tuscany region in the north-western corner of peninsular Italy contains some of the most elegant cultural landscapes in the world with its rolling hills, contoured grain fields, olive groves, and vineyards.

INTRODUCTION
WORLD
REGIONAL
GEOGRAPHY:
A GLOBAL PERSPECTIVE

FIGURE G-1

Elevation (m)

Ice cap

6000
3000
1500
600
0
-150
-1500
-3000
-6000

Goode Homolosine Projection
Scale 1:145,105,000

What are your expectations as you open this book? You have signed up for a course that will take you around the world to try to understand how it functions today. You will discover how interesting and challenging the discipline of geography is. We hope that this course, and this book, will open new vistas, bring new perspectives, raise your awareness of place, and help you navigate our increasingly complex and often daunting world.

You could not have chosen a better time to study geography. The world is changing on many fronts and is doing so ever faster in response to the rapid advancement of communication technologies. The world truly is 'shrinking,' and the likelihood has never been greater that your professional career will be taking you to places far from home. As globalization steadily progresses, geography is becoming ever more important. Moreover, the United States has the world's most globalized economy, with political influence to match. Intentionally or not, the United States is affecting nations and peoples, lives and livelihoods, from pole to pole. That power confers on Americans a responsibility to learn as much as they can about those places, nations, and livelihoods, so that the decisions of their government representatives are well informed. But in this respect, the United States is no superpower. Geographic literacy is a measure of international comprehension and awareness, and Americans' geographic literacy ranks low. That is a liability, both for the United States and the rest of the world because such geographic fogginess afflicts not only voters but also the representatives they elect, from local school boards to the U. S. Congress.

A WORLD ON MAPS

Just a casual glance at the pages that follow reveals a difference between this and other textbooks: there are almost as many maps as there are pages. Geography is more closely identified with maps than any other discipline, and we urge you to give as much (or more!) attention to the maps in this book as you do to the text. It is often said that a picture is worth a thousand words, and the same applies to maps. When we write "see Figure XX," we really mean it . . . and we hope that you will get into the habit. We humans are territorial creatures, and the boundaries that fence off our 200 or so countries reflect our divisive ways. Other, less visible borders—between religions, languages, wealth, and poverty—partition our planet as well. When political and cultural boundaries are at odds, there is nothing like a map to summarize the circumstances. Just look, for example, at the map of the African Transition Zone in Chapter 6: this corridor's turbulence and challenges are steeped in geography.

Maps in Our Minds

All of us carry in our minds maps of what psychologists call our *activity space*: the apartment building or house we live in, the streets nearby, the way to school or workplace, the general layout of our hometown or city. You will know what lane to use when you turn into a shopping mall, or where to park at the movie theater. You can probably draw from memory a pretty good map of your hometown. These **mental maps** **1** allow you to navigate your activity space with efficiency, predictability, and safety. When you arrived as a first-year student on a college or university campus, a new mental map would have started forming. At first you needed a GPS, online, or hard-copy map to find your way around, but soon you dispensed with that because your mental map

became sufficient. And it will continue to improve as your activity space expands.

If a well-formed mental map is useful for decisions in daily life, then an adequate mental map is surely indispensable when it comes to decision making in the wider world. You can give yourself an interesting test. Choose some part of the world, beyond North America in which you have an interest or about which you have a strong opinion—for example, Israel, Iran, Russia, North Korea, or China. On a blank piece of paper, draw a map that reflects your impression of the regional layout there: the country, its neighbors, its internal divisions, major cities, water bodies, and so forth. That is your mental map of the place. Put it away for future reference, and try it again at the end of this course. You will have proof of your improved mental-map inventory.

Mapping Revolutions

The maps in this book show larger and smaller parts of the world in various contexts. Some depict political configurations; others display ethnic, cultural, economic, or environmental features. *Cartography* (the making of maps) has undergone a dramatic and continuing technological revolution. Earth-orbiting satellites equipped with remote-sensing technology (special onboard sensors and imaging instruments) transmit remotely sensed information to computers on the surface, recording the expansion of deserts, the shrinking of glaciers, the depletion of forests, the growth of cities, and myriad other geographic phenomena. Earthbound computers possess ever-expanding capabilities not only to organize this information but also to display it graphically. This allows geographers to develop a *geographic information system (GIS)*, bringing geospatial data to a monitor's screen that would have taken months to assemble just three decades ago.

A parallel map revolution is embodied in the astounding proliferation of navigation systems in cars and on mobile phones. Smartphones allow the use of maps on the go, and many of us, in the developed world at least, have already become dependent on them to traverse cities, find a store or restaurant, even to move around shopping malls. Whereas the personalized maps on our smartphones allow us to navigate more efficiently, the maps in this book are aimed at better understanding the world and its constituent parts.

Satellites—even spy satellites—cannot record everything that occurs on the Earth's surface. Sometimes the borders between ethnic groups or cultural sectors can be discerned by satellites—for example, in changing house types or religious shrines—but this kind of information tends to require on-the-ground verification through field research and reporting. No satellite view of Iraq could show you the distribution of Sunni and Shi'ite Muslim adherents. Many of the boundaries you see on the maps in this book cannot be observed from space because long stretches are not even marked on the ground. So the maps you will be "reading" here have their continued uses: they summarize complex situations and allow us to begin forming durable mental maps of the areas they represent.

There is one other point we should make that is especially important when it comes to world maps: never forget that the world is a sphere, and to project it onto a two-dimensional flat surface must necessarily entail some very significant distortions. Try peeling an orange and flattening the entire peel on a surface—you

will have to tear it up and try to stretch it in places to get the job done. Take a look at Figure G-1 and note how the Atlantic Ocean and other segments of the planetary surface are interrupted. You can produce a map like this in many different ways, but you will always end up distorting things. When studying world maps, there is nothing like having a globe at hand to remind you of our three-dimensional reality.

GEOGRAPHY'S PERSPECTIVE

Geography is sometimes described as the most interdisciplinary of disciplines. That is a testimonial to geography's historic linkages to many other fields, ranging from geology to economics and from sociology to environmental science. And, as has been the case so often in the past, geography is in the lead because interdisciplinary studies and research are now more prevalent than ever.

A Spatial Viewpoint

Most disciplines focus on one key theme: economics is about money; political science is about power; psychology is about the mind; biology is about lifeforms. Geography is about the explanation of space on the Earth's surface. More specifically, geographers are concerned with the organization of *terrestrial space*.

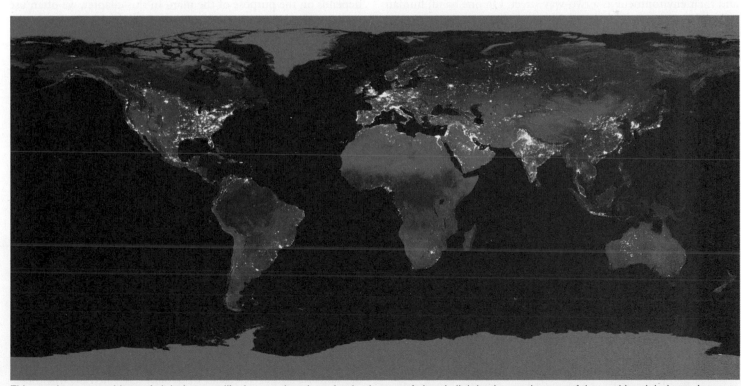

NASA Earth Observatory / NOAA NGDC

This map is an assemblage of nighttime satellite images that show the dominance of electric lighting in certain parts of the world and darkness in others. What does this spatial distribution suggest about the geography of development?

Social space (cities, buildings, political boundaries, etc.) as well as natural space (climates, terrain, vegetation, etc.) are not randomly configured. Instead, there generally prevails a particular order, regularity, even predictability about the ways in which space is organized. Sometimes it is the deliberate work of human beings, and sometimes it is the work of nature, but both produce specific patterns. Geographers consider these spatial patterns and processes as not only interesting but also crucial to how we live and how we organize our societies. The **spatial perspective** **2** has defined geography from its beginning.

Environment and Society

There is another connection that binds geography and has done so for centuries: an interest in the relationships between human societies and the natural environment. Geography lies at the intersection of the social and natural sciences and integrates perspectives from both, being the only discipline to do so explicitly. This perspective comes into play frequently: environmental modification is in the news on a daily basis in the form of worldwide climate change, but this current surge of global warming is only the latest phase of endless atmospheric and ecological fluctuation. Geographers are involved in understanding current environmental issues not only by considering climate change in the context of the past, but also by looking carefully at the implications of global warming for human societies. Geographers are acutely aware that human beings will always be part of nature, no matter how far technology advances.

More generally, think of this relationship between humans and their environment as a two-way street. On one hand, human beings have always had a transformative effect on their natural surroundings, from the burning of forests to the creation of settlements. On the other hand, humans have always been heavily dependent on the natural environment, their individual and collective behaviors very much a product of it. There are so many examples that it is hard to know where to begin or when to end: we eat what nature provides, and traditional diets vary regionally; rivers allow us to navigate and connect with other peoples—or they serve as natural boundaries like the Rio Grande; wars are fought over access to water or seaports; landlocked countries seem to have different cultures from those of islands; and so on.

Spatial Patterns

Geographers, therefore, need to be conversant with the location and distribution of salient features on the Earth's surface. This includes the natural (physical) world, simplified in Figure G-1, as well as the human world, and our inquiry will view these in temporal (historical) as well as spatial perspective. The spatial structure of cities, the layout of farms and fields, the networks of transportation, the configurations of rivers, the patterns of climate—all these form part of our investigation. As you will find, geography employs a comprehensive spatial vocabulary with meaningful terms such as area, distance, direction, clustering, proximity, accessibility, and

many others we will encounter in the pages ahead. For geographers, some of these terms have more specific definitions than is generally assumed. There is a difference, for example, between *area* (surface) and *region*, between *boundary* and *frontier*, and between *place* and *location*. Sometimes, what at first may seem to be simple ideas turn out to be rather complex concepts.

Scale and Scope

One prominent term in the geographic vocabulary is **scale** **3**. Whenever a map is created, it represents all or part of the Earth's surface at a certain level of detail. Obviously, Figure G-1 displays a very low level of detail; it is little more than a general impression of the distribution of land and water as well as lower and higher elevations on our planet's surface. A limited number of prominent features such as the Himalayas and the Sahara are named, but not the Pyrenees Mountains or the Nile Delta. At the bottom of the map you can see that one inch at this scale must represent about 1650 miles of the real world, leaving the cartographer little scope to insert information.

A map such as Figure G-1 is called a *small-scale* map because the ratio between map distance and real-world distance, expressed as a fraction, is very small at 1:103,750,000. Increase that fraction (i.e., zoom in), and you can represent less territory—but also enhance the amount of detail the map can exhibit. In Figure G-2, note how the fraction increases from the smallest (1:103,000,000) to the largest (1:1,000,000). Montreal, Canada, is just a dot on Map A but an urban area on Map D. Does this mean that world maps like Figure G-1 are less useful than larger-scale maps? It all depends on the purpose of the map. In this chapter, we often use world maps to show global distributions as we set the stage for the more detailed discussions to follow. In later chapters, the scale tends to become larger as we focus on smaller areas, even on individual countries and cities. But whenever you read a map, be aware of the scale because it is a guide to its utility.

The importance of the scale concept is not confined to maps. Scale plays a fundamental role in geographic research and in the ways we think about geographic problems—scale in terms of *level of analysis*. This is sometimes referred to as *operational scale*, the scale at which social or natural processes operate or play out. For instance, if you want to investigate the geographic concentration of wealth in the United States, you can do so at a range of scales: within a neighborhood, a city, a county, a State,* or at the national level. You choose the scale that is the most appropriate for your purpose, but it is not always that straightforward. Suppose you had to study patterns of ethnic segregation: what do you think would be the most relevant scale(s)?

---

*Throughout this book we will capitalize State when this term refers to an administrative division of a country: for example, the U.S. State of Ohio or the Australian State of New South Wales. Since this term is also synonymous with country (e.g., the state of Brazil), we use the lower case when referring to such a national state.

EFFECT OF SCALE

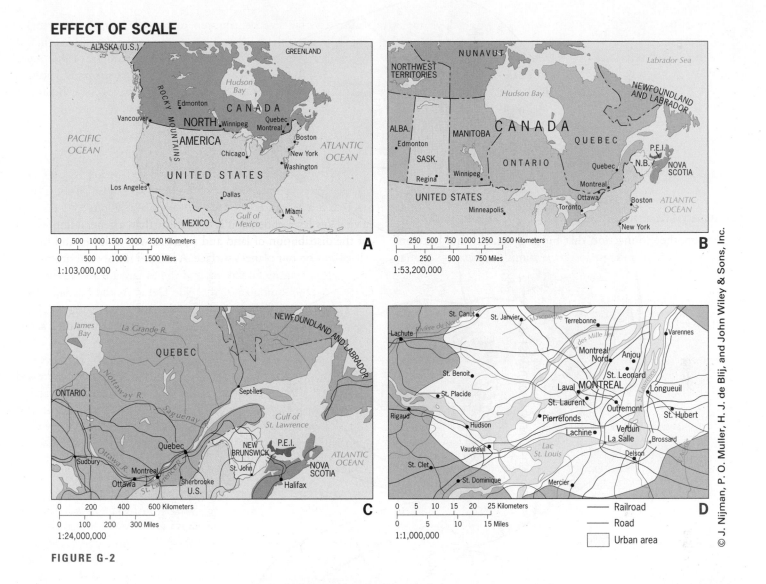

© J. Nijman, P. O. Muller, H. J. de Blij, and John Wiley & Sons, Inc.

FIGURE G-2

In this book, our main purpose is to understand the geography of the world at large and how it works, and so, inevitably, we must deal with broad spatial entities. Our focus is on the world's realms and on the main regions within those realms, and in most cases we will have to forego analyses at a finer scale. For our purposes, it is the big picture that matters most.

WORLD GEOGRAPHIC REALMS

Ours is a globalized, interconnected world, a world of international trade and travel, migration and movement, tourism and television, financial flows and Internet traffic. It is a world that, in some contexts, has taken on the properties of a "global village"—but that village still has its neighborhoods. Their names are Europe, South America, Southeast Asia, and others familiar to us all. Like the neighborhoods of a city or town, these global neighborhoods may not have sharply defined borders, but their persistence, after

tens of thousands of years of human dispersal, is beyond doubt. We call such global neighborhoods **geographic realms** **4** . Each of the realms possesses a particular combination of environmental, cultural, and organizational properties.

These blended, characteristic qualities are imprinted on the landscape, giving each realm its own traditional attributes and social settings. As we come to understand the human and environmental makeup of these geographic realms, we learn not only where they are located but also why they are located where they are (a central question in geography), how they are constituted, and what their future is likely to be in our fast-changing world. Figure G-3, therefore, forms the overall framework for our investigation in this book.

Criteria for Geographic Realms

The existence and identification of world geographic realms depends on a combination of factors. Our world offers a highly complex and variable environment of large and small continents,

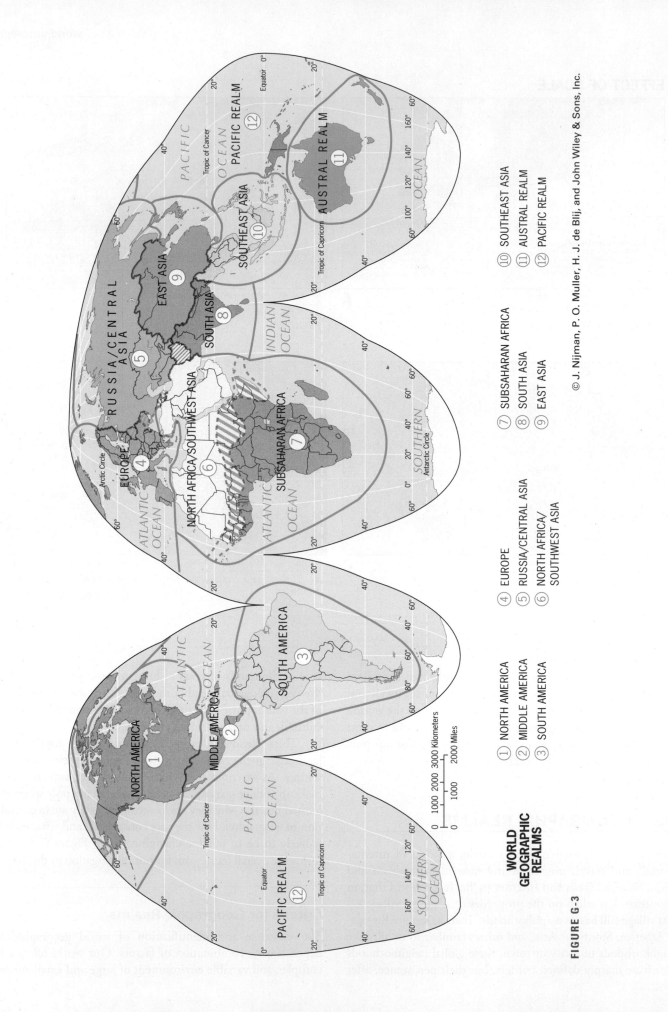

WORLD
GEOGRAPHIC
REALMS

① NORTH AMERICA
② MIDDLE AMERICA
③ SOUTH AMERICA

④ EUROPE
⑤ RUSSIA/CENTRAL ASIA
⑥ NORTH AFRICA/
   SOUTHWEST ASIA

⑦ SUBSAHARAN AFRICA
⑧ SOUTH ASIA
⑨ EAST ASIA

⑩ SOUTHEAST ASIA
⑪ AUSTRAL REALM
⑫ PACIFIC REALM

FIGURE G-3

© J. Nijman, P. O. Muller, H. J. de Blij, and John Wiley & Sons, Inc.

8

enormous oceans and countless waterways, innumerable islands, diverse habitats and cultures, and intricate political geographies. What constitutes a realm depends on the circumstances, but we can still identify three main sets of criteria that apply to all realms:

- *Physical and Human* Geographic realms are based on sets of spatial criteria. They are the largest units into which the inhabited world can be divided. The criteria on which such a broad regionalization is based include both physical (that is, natural) and human (or social) yardsticks. For example, South America is a geographic realm because physically it is a continent and culturally it comprises comparable societies. The realm called South Asia, on the other hand, lies on a Eurasian landmass shared by several other geographic realms; high mountains, wide deserts, and dense forests combine with a distinctive social fabric to create this well-defined realm centered on India.

- *Functional* Geographic realms are the result of the interaction of human societies and natural environments, a *functional* interaction revealed by farms, mines, fishing ports, transport routes, dams, bridges, villages, and countless other features that mark the landscape. According to this criterion, Antarctica is a continent but not a geographic realm.

- *Historical* Geographic realms represent the most comprehensive and encompassing definition of the great clusters of humankind in the world today. China lies at the heart of such a cluster, as does India. Most of Africa constitutes a geographic realm from the southern margin of the Sahara (an Arabic word for desert) to the Cape of Good Hope and from its Atlantic to its Indian Ocean shores. Geographic realms are generally defined on the basis of physical features, but they are also the product of historical evolution.

Figure G-3 displays the 12 world geographic realms based on these criteria. As we will show in greater detail later, waters, deserts, and mountains as well as cultural and political shifts mark the borders of these realms. We shall discuss the positioning of these boundaries as we examine each realm.

Delineating Realms: Boundaries and Transition Zones

Oceans and seas are the most common natural boundaries of the world's realms, such as the South Atlantic to Subsaharan Africa's west or the North Atlantic to North America's east. But where two geographic realms meet, **transition zones** 5 , not sharp boundaries, often mark their contacts.

We need only remind ourselves of the border zone between the geographic realm in which most of us live, North America, and the adjacent realm of Middle America. The line in Figure G-3 coincides with the boundary between Mexico and the United States, crosses the Gulf of Mexico, and then separates Florida from Cuba and the Bahamas. But Hispanic influences are strong in North America north of this boundary, and the U.S. economic influence is strong south of it. The line, therefore,

represents an ever-changing zone of regional interaction. Again, there are many ties between South Florida and the Bahamas, but the Bahamas resemble a Caribbean more than a North American society. Metropolitan Miami has so many ethnic Cuban residents that it is sometimes referred to as the second-largest Cuban city after Havana.

In Africa, the transition zone from Subsaharan to North Africa is so wide and well defined that we have put it on the world map; elsewhere, transition zones tend to be narrower and less easily represented. In the second decade of this century, such countries as Belarus (between Europe and Russia/Central Asia) and Afghanistan (between Southwest Asia and South Asia) lie in inter-realm transition zones. Remember, over much (though not all) of their length, borders between realms are zones of regional change.

Transition zones are fascinating spaces: it is almost as if they rebel against a clear ordering of the world's geography. They remind us that the world is a restless and contested place with shifting boundaries and changing geographic fortunes. They challenge the geographer and, above all, they underscore just how complex the study of geography is.

Two Types of Realms

The world's geographic realms can be divided into two categories. The first are *monocentric* realms that are dominated by a single major political entity, in terms of territory and/or population. North America (United States), Middle America (Mexico), East Asia (China), South Asia (India), Russia/Central Asia, and the Austral Realm (Australia) are all monocentric realms. They are, in their entirety, heavily influenced by the presence of that one country. It is as if the realm is organized around them.

The second type of realm is *polycentric* in nature. In these, the appearance, functioning, and organization of the realm are dispersed among a number of more or less equally influential regions or countries. Europe, North Africa/Southwest Asia, Subsaharan Africa, and the Pacific Realm all fall into this category. Polycentric realms can be more volatile in some ways, their development determined by the sum of many different parts.

Two of the world's realms are a bit more difficult to categorize. Southeast Asia is a dynamic realm that contains almost a dozen countries, some of them regarded as emerging economies. Arguably, Indonesia is becoming the most influential power, but it would be premature to label this a monocentric realm. The other realm that seems to fall in between is South America. Here it is Brazil that has the biggest population and increasingly the largest and most influential economy. South America, more emphatically than Southeast Asia, may be moving toward a monocentric spatial organization.

Of course, some of the dominant powers in the monocentric realms influence events beyond their realm and demonstrate a truly global reach. The United States has dominated world events (though not unchallenged) since World War II, but in recent decades it has had to make way for newly emergent powers such as Japan and Germany. Nowadays, China exercises a major influence not just across East and Southeast Asia but around the globe. Our discussion of the various realms will give due consideration to the influence of global trends and outside powers.

REGIONS WITHIN REALMS

The compartmentalization of the world into geographic realms establishes a broad global framework, but for our purposes a more refined level of spatial classification is needed. This brings us to an important organizing concept in geography: the **region 6**. To establish regions within geographic realms, we need more specific criteria.

Let us use the North American realm to demonstrate the regional idea. When we refer to a part of the United States or Canada (e.g., the South, the Midwest, or the Prairie Provinces), we employ a regional concept—not scientifically but as part of everyday communication. We reveal our perception of local or distant space as well as our mental image of the region we are describing.

But what exactly is, for example, the Midwest? How would you draw this region on the North American map? Regions seem easy to imagine and describe, but they can be difficult to outline with precision on a map. You might suggest that, broadly, the Midwest lies between the Ohio River and the Rocky Mountains. Or you might use the borders of States: certain States are part of this region, others are not. Alternatively, you could use agriculture as the principal criterion: the Midwest is where corn and/or soybeans occupy a certain percentage of the farmland. Each method results in a different delimitation; a Midwest based on physical boundaries is different from States from a Midwest based on State borders or farm production. Therein lies an important principle: regions are devices that allow us to make spatial generalizations, and they are based on criteria to help us construct them. If you were studying the geography of presidential-election politics, then a Midwest region defined by State boundaries would make sense. If you were studying agricultural distributions, you would need a different definition.

Criteria for Regions

Given these different dimensions of the same region, we can identify properties that all regions have in common:

- **Area** To begin with, all regions have *area*. This observation would seem obvious, but there is more to this idea than meets the eye. Regions may be intellectual constructs, but they are not abstractions: they exist in the real world, and they occupy space on the Earth's surface.

- **Boundaries** It follows that regions have *boundaries*. Occasionally, nature itself draws sharp dividing lines, for instance, along the crest of a mountain range or along a coast. More often, regional boundaries are not self-evident, and we must

determine them using criteria that we establish for that purpose. For example, to define a cultural region, we may decide that only areas where more than 50 percent of the population belong to a particular religion or speak a certain language qualify to be part of that region.

- **Location** All regions also possess *location*. Often the name of a region contains a locational clue, as in Amazon Basin or Indochina (a region of Southeast Asia lying between India and China). Geographers refer to the **absolute location 7** of a place or region by providing the latitudinal and longitudinal extent of the region with respect to the Earth's grid coordinates. A more useful measure is a region's **relative location 8**, that is, its location with reference to other regions. Again, the names of certain regions reveal aspects of their relative locations, as in *Mainland* Southeast Asia and *Equatorial* Africa.

- **Homogeneity** Many regions are marked by a certain *homogeneity* or sameness. Homogeneity may lie in a region's cultural properties, its physical characteristics, or both. Siberia, a vast region of northeastern Russia, is marked by a sparse human population that resides in widely scattered, small settlements of similar form, frigid climates, extensive areas of permafrost (permanently frozen subsoil), and cold-adapted vegetation. This dominant uniformity makes it one of Russia's natural and cultural regions, extending from the Ural Mountains in the west to the Pacific Ocean in the east. When regions display a measurable and often visible internal homogeneity, they are called **formal regions 9**. But not all formal regions are visibly uniform. For instance, a region may be delimited by the area in which, say, 90 percent or more of the people speak a particular language. This cannot be seen in the landscape, but the region is a reality, and we can use that criterion to draw its boundaries accurately. It, too, is a formal region.

- **Regions as Systems** Other regions are marked not by their internal sameness but by their functional integration—that is, by the way they work. These regions are defined as **spatial systems 10** and are formed by the areal extent of the activities that define them. Take the case of a large city with its surrounding zone of suburbs, urban-fringe countryside, satellite towns, and farms. The city supplies goods and services to this encircling zone, and it buys farm products and other commodities from it. The city is the heart, the *core* of this region, and we call the surrounding zone of interaction the city's **hinterland 11**. But the city's influence wanes along the outer *periphery* of that hinterland, and there lies the boundary of the functional region of which the city is the focus. A **functional region 12**, therefore, is usually forged by a structured, urban-centered system of interaction. It has a core and a periphery.

Interconnections

Even if we can easily demarcate a particular region and even if its boundaries are sharp, that does not mean it is isolated from other parts of the realm or even the world. All human-geographic regions are more or less interconnected, being linked to other regions. As we shall see, globalization is forging ongoing integration and connections among regions around the world. Trade, migration, education, television, computer linkages, and other interactions sometimes blur regional identities. Interestingly, globalization tends to have a seemingly paradoxical effect: in some ways, regions and places become more alike, more homogeneous (think of certain consumption patterns), but in other respects the contrasts can become stronger (for example, a reassertion of ethnic or religious identities).

THE PHYSICAL SETTING

Natural Landscapes

The landmasses of Planet Earth present a jumble of **natural landscapes** **13** ranging from rugged mountain chains to smooth coastal plains (Fig. G-1). Certain continents are readily linked with a dominant physical feature—for instance, North America and its Rocky Mountains, South America with its Andes and Amazon Basin, Europe with its Alps, Asia with its Himalaya Mountains and massive river basins, and Africa with its Sahara and Rift Valley complex. Physical features have long influenced human activity and movement. Mountain ranges form barriers to movement but have also channeled the spread of agricultural and technological innovations around them. Large deserts similarly form barriers as do rivers, although rivers also facilitate accessibility and connectivity between people. River basins in Asia still contain several of the planet's largest population concentrations: the

advantages of fertile soils and ample water supplies that first enabled clustered human settlement now sustain hundreds of millions in crowded South and East Asia.

As we study each of the world's geographic realms, we will find that physical landscapes continue to play significant roles in this modern world. That is one reason why the study of world regional geography is so important: it puts the human map in environmental as well as regional perspective.

Geology and Natural Hazards

Our planet may be 4.5 billion years old, but it is far from placid. As you read this chapter, Earth tremors are shaking the still-thin crust on which we live, volcanoes are erupting, storms are raging. Even the continents themselves are moving measurably, pulling apart in some areas, colliding in others. Hundreds of thousands of human lives are lost to natural calamities of this sort in almost every decade.

More than a century ago, the German geographer Alfred Wegener used spatial analysis to explain something that is obvious even from a small-scale map like Figure G-1: the apparent jigsaw-like fit of the landmasses, especially across the South Atlantic Ocean. His theory of **continental drift** **14** held that the landmasses on the map were actually pieces of a supercontinent, ***Pangaea***, that existed hundreds of millions of years ago. Today we know that the continents are "rafts" of relatively light rock that rest on slabs of heavier rock known as **tectonic plates** **15** whose movement is propelled by giant circulation cells in the red-hot magma below (when this molten magma reaches the surface through volcanic vents, it is called lava).

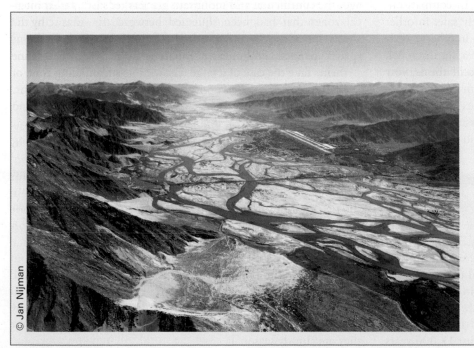

© Jan Nijman

@ FROM THE FIELD NOTES...

"On the descent into Tibet's Lhasa Gongga Airport, I have a great view of the Yarlung Zangbo Valley, its braided stream channels gently flowing toward the distant east. The Yarlung Zangbo is the highest major river on Earth, running from the Tibetan Plateau into northeastern India where it joins the mighty Brahmaputra River that continues on to Bangladesh where it empties into the Indian Ocean. It was mid-October and the water levels were low. The landing strip of the airport can be seen in the center-right of the photo, on the south bank. The airport is quite far from Lhasa, the Tibetan capital, located about 62 kilometers (40 mi) to its southwest. Despite major road and tunnel construction, it is still more than an hour's drive. The airport had to be built away from the city and in this widest part of the valley because it allows the easiest landings and takeoffs in this especially rugged terrain. It lies at 3700 meters above sea level (12,100 ft), one of the highest airports in the world."

FIGURE G-4

When moving tectonic plates collide, earthquakes and volcanic eruptions result, and the physical landscape is thrown into spectacular relief. Figure G-4 shows the outlines of the tectonic plates and the coincidence of earthquakes and volcanic eruptions. The Earth's largest ocean is almost completely encircled by active volcanoes and earthquake epicenters. Appropriately, this geologic frame is known as the **Pacific Ring of Fire** **16**. In comparison, Russia, Europe, Africa, and Australia are relatively safe; in other realms, the risks are far greater in one sector than in others (western as opposed to eastern North and South America, for instance). As we shall discover, for certain parts of the world the activity mapped in Figure G-4 presents a clear and present danger. Some of the world's largest cities, such as Tokyo and Mexico City, lie in zones that are highly vulnerable to sudden disaster—as indeed occurred with Japan's huge 2011 earthquake and tsunamis not very far north of Tokyo.

Climate

The prevailing **climate** **17** constitutes a key factor in the geography of realms and regions (in fact, some regions are essentially defined by climate). But climates change: those dominating in certain regions today may not have prevailed there several thousand years ago. Thus any map of climate, including the maps in this chapter, is but a still-picture of our constantly changing world.

Climatic conditions have swung back and forth for as long as the Earth has had an atmosphere. Imagine this: just 18,000 years

ago, great ice sheets had spread all the way south to the Ohio River Valley, covering most of the Midwest; this was the peak of a glaciation that had lasted about 100,000 years (Fig. G-5). The Antarctic Icesheet was bigger than ever, and even in the tropics, great mountain glaciers pushed down valleys and onto plateaus. But then the **Holocene** **18** warming began, about 10,000 years ago, the continental and mountain glaciers receded, and ecological zones that had been squeezed between the advancing ice sheets spread north and south. In Europe, particularly, living space expanded and human numbers grew.

Global Climate Change

Today we are living in an era of **global climate change** **19**, particularly natural global warming that has been accelerated by anthropogenic (human-source) causes. Since the Industrial Revolution, we have been emitting gases that have enhanced nature's *greenhouse effect* whereby the sun's radiation becomes trapped in the Earth's atmosphere. This is leading to a series of atmospheric changes, especially the overall warming of the planet. One important international organization of experts, the Intergovernmental Panel on Climate Change (IPCC), predicts an increase of about 2°C (3.6°F) in the course of the twenty-first century for the world at large, but with significant regional variability (e.g., more at higher latitudes, less at lower latitudes). Precipitation patterns are predicted to become more variable, particularly in regions where they are already seasonal.

FIGURE G-5

This change in temperature may seem small but is expected to have significant impacts on global climate patterns, agricultural zones, and the quality of human lives. The full ramifications are not known, but scenarios are being modeled so that societies can confront the changes that are coming. Leaders of some countries are more skeptical than others, and some have already made greater adjustments than others. One of the most significant consequences of global climate change is that the icecap atop the Arctic Ocean is melting faster than even recent models predicted, with environmental and geopolitical implications. We pick up this issue in Chapters 5 and 12.

Climate Regions

We have just learned how variable climate can be, but in a human lifetime most of us see relatively little evidence of this variability—although this may change as a result of global warming. We talk about the weather (the immediate state of the atmosphere) in a certain place at a given time, but as a technical term *climate* defines the aggregate, total record of weather conditions at a place, or in a region, over the entire period during which records have been kept. Figure G-6 may appear rather complicated, but even a glance at this map shows its utility. Devised long ago by Wladimir Köppen and later modified by Rudolf Geiger, it represents climatic regions

through a combination of colors and letter symbols. In the legend, note that the **A** climates (rose, gold, and peach) are equatorial and tropical; the **B** climates (tan, yellow) are dry; the **C** climates (shades of green) are temperate, that is, moderate and neither hot nor cold; the **D** climates (shades of purple) are cold; the **E** climates (blue) are frigid; and the **H** climates (gray) prevail in highlands like the Himalayas/Tibetan Plateau and the Andes.

A good way to get a sense of how this map works is to start with a climate with which you are familiar. If you live in the southeastern United States (the large green **C** area extending from near the Great Lakes to Florida), you would feel at home—climatically—in the green area in southeastern South America, in southeastern Australia, and in southeastern China. If you live in the **B** zones of the U.S. Southwest, either in the tan (Texas) or the yellow (Arizona) area, conditions in much of Australia, southwestern Africa, or the Middle East would be familiar to you. This can even be taken to the city scale. If you enjoy the climate of San Francisco, you will rediscover it in Santiago (Chile), Cape Town (South Africa), Athens (Greece), and Perth (Australia). Details will differ, of course—higher wind incidence or maybe lower overall humidity—but in general, the similarities will be greater than such differences. And if you live in Canada, you would be prepared for much of Russia and northeastern China.

Even at this relatively small geographic scale, Figure G-6 can tell us a great deal about entire climate regions. For instance,

© Jan Nijman

"What a sight! Mount Kilimanjaro in northern Tanzania, just south of the equator (see Fig. G-1), is the highest peak on the African continent and the highest freestanding mountain on Earth; at 5895 meters (19,340 ft), it actually consists of three closely-grouped volcanoes. To get a sense of the scale of this photo, the rim of the outer crater is about one kilometer (0.6 mi) wide. In the local language of the Maasai people, Kilimanjaro means "source of water," attributable to the streams fed by rain and snow at higher elevations—and, importantly, the glaciers that have existed here for tens of thousands of years. Nonetheless, on this 23rd day of November 2014, it is frightening to observe how little of these glaciers remain. Scientists estimate that this shrunken glacial surface is now down to only about two square kilometers (0.8 sq mi). Global warming affects places and people all over the world, and Tanzania's Maasai are not exempt."

among the equatorial/tropical (*Af*) climates, the areas with the darkest (rose) shade get the most rain and it falls year-round, so that this is where tropical rainforests still stand. But in the peach-colored (*Aw*) areas, the rainfall regime is subject to dry seasons, and the vegetation reflects this by its very name: savanna. Much more open space and more widely scattered stands of trees prevail in these tropical grasslands. And a third kind of regime is signified by the *m* after the *A*: for *monsoon*, the annual copious rainy season on which the lives of hundreds of millions of people still depend.

Here is an important qualification to keep in mind when studying Figure G-6: this map is also a still-picture of a changing scene, a single frame from an unspooling film. Global climate never stops changing, and less than a century from now those changes may compel revisions to this map. If current predictions about rising sea levels turn out to be accurate, we may even have to start redrawing some familiar coastlines.

You will find larger-scale maps of climate in several of the regional chapters that follow, but it is always useful to refer back to this Köppen-Geiger map whenever the historical or economic geography of a region or country is under discussion. This world climatic map reflects agricultural opportunities and limitations as well as climatic regimes, and as such helps explain some enduring patterns of human distribution on our planet. We turn next to this crucial topic.

REALMS OF POPULATION

Earlier we noted that population numbers by themselves do not define geographic realms or regions. Population distributions, and the functioning society that gives them common ground, are more

significant criteria. That is why we can identify one geographic realm (the Austral) with less than 30 million people and another (South Asia) with more than 1.7 billion inhabitants. Neither population numbers nor territorial size alone can delimit a geographic realm. Nevertheless, the map of global population distribution shows some major clusters that are part of certain realms (Fig. G-7).

Before we examine these clusters in some detail, remember that the world's human population now rounds off at 7.1 billion (see Appendix B, available online, for a detailed breakdown)—confined to the landmasses that constitute less than 30 percent of our planet's surface, much of which is arid desert, inhospitable mountain terrain, or frigid tundra. (And keep in mind that Fig. G-7 is yet another still-picture of a continually changing scene as the rapid growth of humankind continues.) After thousands of years of slow growth, world population during the nineteenth and twentieth centuries grew at an increasing rate. That rate has recently been slowing down, even imploding in some parts of the world. But consider this: it took about 17 centuries following the birth of Christ for the world to add 250 million people to its numbers; now we are adding 250 million about every three years. While the *rate* of population growth has declined in some parts of the world, in absolute terms the global population continues to grow apace and is expected to reach around 9.5 billion by 2050.

This raises the important question as to whether there are limits to the Earth's carrying capacity—will there be enough food to go around? That question has become more and more pressing over the past decade due to rapidly rising food prices resulting from increased demand in China and India, a general dietary shift from grains to meat and vegetables, and the use of agricultural resources for the production of biofuels. The actual increase of

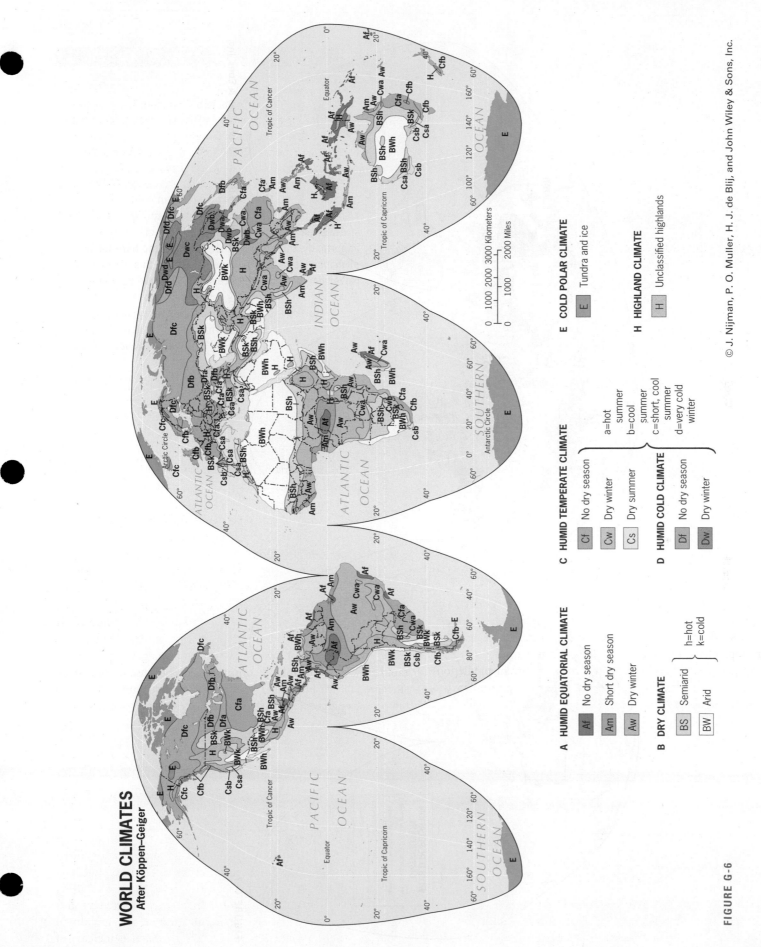

WORLD CLIMATES
After Köppen–Geiger

A HUMID EQUATORIAL CLIMATE

Af	No dry season
Am	Short dry season
Aw	Dry winter

B DRY CLIMATE

| BS | Semiarid | h=hot |
| BW | Arid | k=cold |

C HUMID TEMPERATE CLIMATE

Cf	No dry season	a=hot summer
Cw	Dry winter	b=cool summer
Cs	Dry summer	c=short, cool summer

D HUMID COLD CLIMATE

| Df | No dry season | d=very cold winter |
| Dw | Dry winter |

E COLD POLAR CLIMATE

| E | Tundra and ice |

H HIGHLAND CLIMATE

| H | Unclassified highlands |

0   1000   2000   3000 Kilometers
0   1000   2000 Miles

© J. Nijman, P. O. Muller, H. J. de Blij, and John Wiley & Sons, Inc.

FIGURE G-6

15

WORLD POPULATION DISTRIBUTION

One dot represents 100,000 people

INDIVIDUAL COUNTRY TOTALS ARE SHOWN
IN THE DATA TABLE IN APPENDIX B.

0   1000   2000   3000 Kilometers
0   1000   2000 Miles

© J. Nijman, P. O. Muller, H. J. de Blij, and John Wiley & Sons, Inc.

population is only part of the problem; our growing appetite for certain products is another. And we are not just talking about food—think drinking water, fossil fuels, and minerals as well. Thus, it seems inconceivable that almost 10 billion people by mid-century could be consuming the way we do today in the developed world.

Major Population Clusters

One way to present an overview of the location of people on the planet is to create a map of population distribution [20] (Fig. G-7). As you can see in the map's legend, each dot represents 100,000 people, and the clustering of large numbers of them in certain areas as well as the near-emptiness of others is readily evident. By the way, there is a technical difference between **population distribution** and *population density*, which is another way of showing where people live. Density maps reveal the number of persons per unit area.

- *South Asia* The *South Asia* population cluster lies centered on India and includes its populous neighbors, Pakistan and Bangladesh. This huge agglomeration of humanity focuses on the wide plain of the Ganges River (**A** in Fig. G-7). South Asia recently became the world's largest population cluster, overtaking East Asia in 2010.

- *East Asia* The *East Asia* population cluster is centered on China and includes the Pacific-facing Asian coastal zone from the Korean Peninsula to Vietnam. Not long ago, we would have reported this as a dominantly rural, farming population, but rapid economic growth and burgeoning urbanization have altered the picture. In China's interior river basins of the Huang (Yellow) and Chang/Yangzi (**B** and **C** on the map), and in the Sichuan Basin between these two letters, most of the people remain farmers. But the booming cities of coastal, and increasingly interior, China are attracting millions of new inhabitants, and in 2011 the Chinese urban population surpassed the 50 percent milestone.

- *Europe* The third-ranking population cluster, *Europe*, also lies on the Eurasian landmass but at the opposite end from China. The European cluster, including western Russia, counts more than 700 million inhabitants, which puts it in a class with the two larger Eurasian concentrations—but there the similarity ends. In Europe, the key to the linear, east-west orientation of the axis of population (**D** in Fig. G-7) is not a fertile river basin but a zone of raw materials for industry. Europe is among the world's most highly urbanized and industrialized realms, its human agglomeration sustained by factories and offices rather than paddies and pastures.

The three world population concentrations just discussed (South Asia, East Asia, and Europe) account for about 4 billion of the world's 7.1 billion people. No other cluster comes close to these numbers. The next-ranking cluster, eastern North America, is only about one-quarter the size of the smallest Eurasian concentrations.

As in Europe, the population in this zone is concentrated in major metropolitan complexes; the rural areas are now relatively sparsely settled.

Toward an Urban World

Geographic realms and regions display varying levels of urbanization [21], the proportion of the total population residing in cities and towns (which globally stands at 55 percent today). Some regions are urbanizing far more rapidly than others, a phenomenon we will investigate as we examine each realm. For now, it is important to be aware of the distinction between urbanization level and urban growth rates. The urbanization level refers to the proportion of a country's or region's total population living in cities (however defined by national governments). The urban growth rate, on the other hand, concerns the growth of the urban population over time, usually on an annual basis. So, for example, the U.S. urbanization level is 81 percent, whereas its yearly urban growth rate is around 0.2 percent; for Tanzania, the urbanization level is far less, about 31 percent, but the urban growth rate is a much higher 2.3 percent. It makes sense that urbanization levels and growth rates are often inversely related: once the percentage of the population already living in cities reaches a certain level, urban growth rates will subside, as has been the case for the United States. Alternatively, when urbanization levels are low, there are ample opportunities for rapid urban growth.

Thus the fastest urban growth rates tend to occur in realms and regions in which a large proportion of the population resides in rural areas, as in Subsaharan Africa as well as South and East Asia. It is also worth noting—in these realms where urbanization levels remain low and urban growth rates are high—that a large number of megacities [22] have emerged over the past three decades. These huge urban agglomerations are defined as containing populations of greater than 10 million, and in 2014 28 of them could be found on the world map. Among the latter, only two are in North America (New York and Los Angeles), whereas 5 are located in South Asia and no less than 6 in China alone.

REALMS OF CULTURE

Imagine yourself aboard a riverboat on Africa's White Nile, heading upstream (south) from Khartoum, Sudan. The desert sky is blue, the heat is searing. You pass by villages that look much the same: low, square, or rectangular dwellings, some recently whitewashed, others gray, with flat roofs, wooden doors, and small windows. The minaret of a modest mosque may rise above the houses, and there may be a small central square. There is very little vegetation; here and there a hardy palm tree stands in a courtyard. People on the paths wear long white or colored robes and headgear. A few goats lie in the shade, and children are playing on the dusty ground.

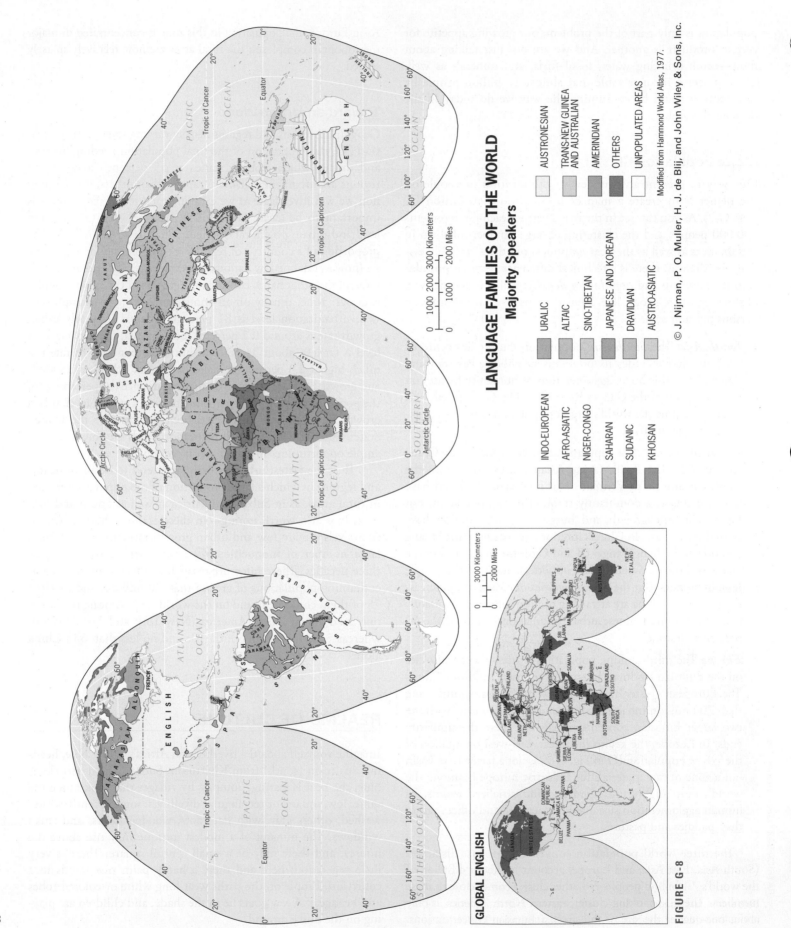

LANGUAGE FAMILIES OF THE WORLD
Majority Speakers

INDO-EUROPEAN		URALIC		AUSTRONESIAN
AFRO-ASIATIC		ALTAIC		TRANS-NEW GUINEA AND AUSTRALIAN
NIGER-CONGO		SINO-TIBETAN		AMERINDIAN
SAHARAN		JAPANESE AND KOREAN		OTHERS
SUDANIC		DRAVIDIAN		UNPOPULATED AREAS
KHOISAN		AUSTRO-ASIATIC		

Modified from Hammond World Atlas, 1977.

© J. Nijman, P. O. Muller, H. J. de Blij, and John Wiley & Sons, Inc.

GLOBAL ENGLISH

FIGURE G-8

18

"After a chat with the manager of a high-rise hotel in Mumbai (formerly Bombay), Zach Woodward and I are allowed access to the roof in early evening to take shots of the city's mushrooming skyline. Zach is an undergraduate photography major who accompanied me on this field research trip and he has a sharp eye for both depth and scenic drama. In the foreground, waterfront-lining Marine Drive, also known as "The Queen's Necklace," faces the Arabian Sea as the setting sun shines faintly on some of the low-rise Art Deco apartment blocks that date back to the 1930s. We are at the southern tip of the Mumbai Peninsula, from where the British ruled their South Asian empire during colonial times. Today, the urban region has expanded far beyond this slender peninsula onto the surrounding mainland, and the metropolitan-area population exceeds 21 million. Looking northward, Mumbai's unrelenting densification presents itself as the proverbial concrete jungle."

© Jan Nijman

All of this is part of Sudan's rural **cultural landscape** **23**, the distinctive attributes of a society imprinted on its portion of the world's physical stage. The cultural landscape concept was initially articulated in the 1920s by a University of California geographer named Carl Sauer, who stated that "a cultural landscape is fashioned from a natural landscape by a culture group" and that "culture is the agent; the natural environment the medium." What this means is that people, starting with their physical environment and using their culture as their agency, construct a landscape layered with forms such as buildings, gardens, and roads, and is also distinguished by modes of dress, aromas of food, and sounds of music.

Continue your journey southward on the Nile, and you will soon witness a remarkable transition. Quite suddenly, the square, solid-walled, flat-roofed houses of Sudan give way to the round, wattle-and-thatch, conical-roofed dwellings of South Sudan. You may note that clouds have appeared in the sky: it rains more here, and flat roofs will not do. The desert has yielded to green. Vegetation, natural as well as planted, grows between houses, flanking even the narrow paths. The villages seem less orderly, more varied. You'll see women, often in colorful dresses, carrying water in jars on their heads from a nearby well or the river to their huts. You have traveled from one cultural landscape into another, from

PATRICK BAZ / Getty Images, Inc.

Cultura Travel/Philip Lee Harvey / Getty Images, Inc.

Settlement landscapes in Sudan (left, the town of Umm Dawban just north of the capital, Khartoum) sharply contrast with those in South Sudan (right, the village of Nyaro just south of the Sudan-South Sudan border).

Arabized, Islamic Africa to animist/Christian Africa. More broadly, you have crossed the boundary between two geographic realms (see the accompanying photo pair).

Although no geographic realm, not even the Austral Realm, is marked by only a single cultural landscape, these landscapes can help define both realms and regions. The cultural landscape of the high-rise North American city with its far-reaching suburbs differs from that of Europe; the organized terraced paddies of Southeast Asia are unlike anything to be found in the rural cultural landscape of neighboring Australia. Important variations of cultural landscapes also occur within geographic realms, such as between highly urbanized and dominantly rural (and more traditional) areas, and they can be quite useful in the process of delineating several of the world's regions as well.

The Geography of Language

Language is the essence of culture. People tend to feel passionately about their mother tongue, especially when they believe it is threatened in some way. In the United States today, the English Only movement reflects many people's fears that the primacy of English as the national language is under threat as a result of immigration. As we will see in later chapters, some governments try to suppress the languages (and thereby the cultures) of minorities in mistaken attempts to enforce national unity, provoking violent reactions.

In fact, many languages emerge, thrive, and die out over time, and linguists estimate that the number of lost languages is in the tens of thousands—a process that continues. One year from the day you read this, about 25 more languages will have become extinct, leaving no trace. Just in North America, more than 100 native languages have been lost over the past half-century. Some major languages of the past, such as Sumerian and Etruscan, have left fragments in later languages. Others, like Sanskrit and Latin, live on in their modern successors. At present, about 6900 languages remain, half of them classified by linguists as endangered ("hot spots" include the Amazon Basin, the Andes to its west, Siberia, and northern Australia). By the end of this century, the bulk of the world's population will be speaking just a few hundred languages, which means that many millions will no longer be able to speak their ancestral mother tongues.

Scholars have tried for many years to unravel the historic roots and branches of the global "language tree," and their debates continue. Geographers trying to map the outcome of this research keep having to modify the pattern, so you should take Figure G-8 as a work in progress, not the final product. At minimum, there are some 15 so-called *language families*, groups of languages with a shared but usually distant origin. The most widely distributed language family, the Indo-European (shown in yellow on the map), includes English, French, Spanish, Russian, Persian, and Hindi. This encompasses the languages of European colonizers that were carried and implanted worldwide, English most of all. Figure G-9 shows that there are still thousands of languages spoken, but by

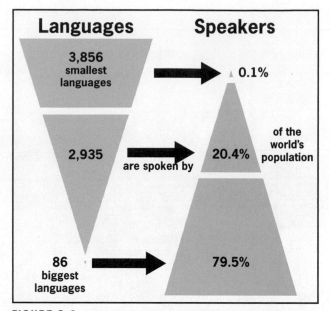

FIGURE G-9   After *Why Languages Die* by Harrison (2008), Fig. 1.5, p. 14. By permission of Oxford University Press, USA.

very small numbers of people; their continued existence is therefore under threat.

Today, English serves as the national or official language of many countries and outposts, and remains the *lingua franca* (common second language) of government, commerce, and higher education in many multicultural societies (see Fig. G-8 inset map). In the postcolonial era, English became the chief medium of still another wave of ascendancy now in progress: globalization. But even English may eventually go the way of Latin, morphing into versions you already can hear as you travel to different parts of the world. For example, in Hong Kong, Chinese and English are producing a local "Chinglish" you may hear in the first taxi you enter. In Lagos, Nigeria, where most of the people are culturally and ethnically Yoruba, a language called "Yorlish" is emerging. It is hard to keep up with the constant evolution of language.

Landscapes of Religion

Religion played a crucial role in the emergence of ancient civilizations and has shaped the course of world history. Hinduism, for instance, was one of the earliest religions that helped shape an entire realm (South Asia). Later, Buddhism, Christianity, and Islam emerged as major belief systems, often splitting up into various branches stretching across realms and regions. Figure G-10 shows the current distribution of world religions. Our world has become more complicated in recent times, and its patterns of religion are increasingly diffuse and dynamic. But today, still, we find that geographic realms are often dominated by a single religion or family of religions: Christianity in Europe and the Americas;

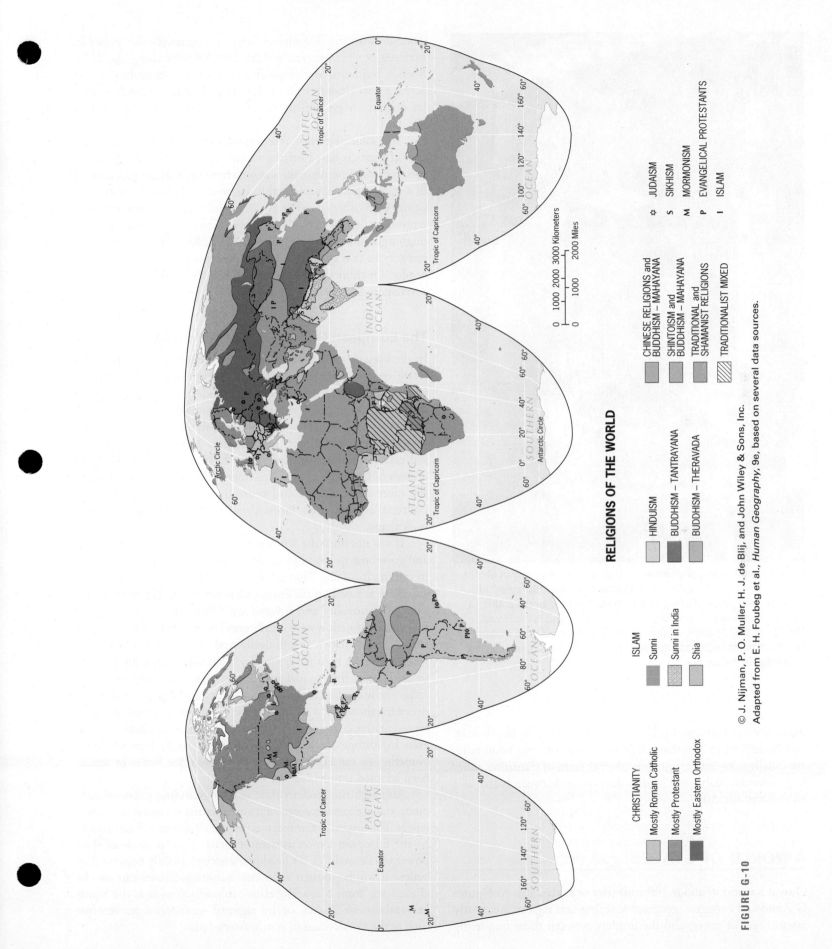

RELIGIONS OF THE WORLD

CHRISTIANITY
- Mostly Roman Catholic
- Mostly Protestant
- Mostly Eastern Orthodox

ISLAM
- Sunni
- Sunni in India
- Shia

- HINDUISM
- BUDDHISM – TANTRAYANA
- BUDDHISM – THERAVADA

- CHINESE RELIGIONS and BUDDHISM – MAHAYANA
- SHINTOISM and BUDDHISM – MAHAYANA
- TRADITIONAL and SHAMANIST RELIGIONS
- TRADITIONALIST MIXED

- ✡ JUDAISM
- S SIKHISM
- M MORMONISM
- P EVANGELICAL PROTESTANTS
- I ISLAM

0 1000 2000 3000 Kilometers
0 1000 2000 Miles

© J. Nijman, P. O. Muller, H. J. de Blij, and John Wiley & Sons, Inc.

Adapted from E. H. Foubeg et al., *Human Geography*, 9e, based on several data sources.

FIGURE G-10

Mustafa Ozer/AFP/Getty Images, Inc.

The *hajj* is the yearly pilgrimage of Muslims to the holy city of Mecca in Saudi Arabia. The pilgrimage is referred to as the fifth "pillar" of Islam, the obligation of every able-bodied Muslim to worship Allah in this holiest of sites at least once in their lifetime. This is the Grand Mosque of Mecca on November 17, 2010, as more than two million Muslim pilgrims launched into the final rituals of this largest religious pilgrimage in the world.

Islam in North Africa/Southwest Asia; Hinduism in South Asia; and Buddhism in mainland Southeast Asia. But the boundaries are usually fuzzy and frequently take the form of transition zones (e.g., between North and Middle America or between North and Subsaharan Africa).

A WORLD OF STATES

Ours is a world of about 200 countries or **states 24**. As Figures G-7 and G-11 suggest, geographic realms and regions are mostly assemblages of states, and the borders between them frequently

coincide with the boundaries between countries—for example, between North America and Middle America along the U.S.-Mexico border. It is also possible for a realm boundary to cut across a state (Indonesia is one), so that state and realm boundaries do not always concur (see the African Transition Zone in Fig. G-3).

The political territorial organization of the world within a system of states is based on the notion of **sovereignty 25**. It is a concept from international law, which means that the government of a state reigns supreme within its borders. Normally, states recognize each other's sovereignty, but this becomes a matter of contention in times of conflict and war. Think, for example, of the ongoing conflict in which Russia is challenging Ukraine's sovereignty in that country's eastern provinces next to the Russian border. Or think of China's refusal to even consider the possible sovereignty of Taiwan.

The European Origins of the Modern State

In the long course of human history, the modern state is a relatively recent invention, and so is the international system of which it forms the cornerstone. The modern state emerged from other varieties of politico-territorial organization that date back to the earliest complex civilizations. In the study of ancient history, scholars sometimes use the term *polity* or *proto-state* to indicate the difference. Ever since farm surpluses enabled the growth of large and prosperous towns, this was accompanied by the more sophisticated and centralized exercise of power and political organization. From these origins, the earliest states took shape.

It was not until the seventeenth century that European rulers and governments began to negotiate treaties that defined the state in international law. That is why the modern state is quite often described as based on the **European state model 26**, with definitions of nationality and sovereignty. Often, the model assumed that state and nation were ideally conterminous, so that a *nation-state* would enclose an ethnically and culturally homogeneous people within a national boundary. That was never truly the case (even France, the "model of models," had its minorities), and today the ideal state is defined as a clearly and legally defined territory inhabited by a citizenry governed from a capital city by a representative government. As we shall discover in Chapter 4, not even in Europe itself are all governments truly representative. Nonetheless, the European state model has, for better or worse, been adopted throughout the world.

Although the modern state is a longstanding phenomenon, many states experience external and internal pressures and challenges. For instance, European states are challenged "from above" by the European Union as member-states transfer some of their power to "Brussels" (EU's headquarters city), mainly because they believe it will be to their economic advantage. States can also be challenged "from below" by ethnic minorities, such as the Kurds in southeastern Turkey or by regional secessionist movements such as the Catalonians in northeastern Spain.

Power and Geopolitics

As the term **geopolitics** **27** implies, geography plays a major role in the politics of the state, in the international relations among states within realms and regions, and even in warfare and terrorism. Geography does not, of course, explain everything, but it is certainly an essential dimension in the understanding of world politics. A state's power and influence are related to its size, access to natural resources, location vis-à-vis other states, trading routes, access to waterways, and the like. More than ever today, news reports about conflicts in any part of the world—whether on television, the website and hard-copy pages of leading newspapers, or in newsmagazines—are accompanied by maps crucial to the explanation of these clashes.

Even though the role of geography in political affairs can be regarded as fundamental, it certainly does not remain static. Relative geopolitical location itself is subject to change, especially in response to new technologies. For example, states and regions once deemed impregnable because they could not easily be reached by enemies (such as the United States prior to the age of aviation), can swiftly see their invulnerability disappear. It is not hard to imagine how the arrival of long-range missiles, state-of-the-art spy-satellite technology, advances in geospatial and mapping systems, and other high-tech military innovations such as drones can hugely impact a country's strategic planning and positioning.

The territorial functioning and integrity of many states in this new century has also been challenged by the emergence, proliferation, and growing sophistication of terrorist groups. Today's most heavily contested zones of conflict are found in the North Africa/Southwest Asian realm (e.g., Syria, Iraq, Libya) where longstanding sectarian disputes have been complicated by the rise of violent, radical organizations such as Al Qaeda and ISIS. Increasingly, these threats are also being felt far beyond that realm, in places such as western Europe, Russia, and West Africa. You will learn about the geographic dimensions of these and other key conflicts

in the chapters ahead. For now, it is important to realize that the world map of states is far from stable in certain realms and regions, and that states are not the only major geopolitical actors on the global scene.

GEOGRAPHIES OF DEVELOPMENT

Finally, as we prepare for our study of world regional geography, it is all too clear that realms, regions, and states do not enjoy the same level of prosperity. The field of *economic geography* focuses on spatial aspects of the ways people make their living, and deals with patterns of production, distribution, and consumption of goods and services. As with all else in this world, these patterns reveal significant variation. Individual states report the nature and value of their imports and exports, farm and factory output, and many other economic data to the United Nations and other international agencies. From such information, economic geographers can measure the comparative well-being of the world's countries. The concept of **development** **28** is used to gauge a population's economic, social, and institutional growth and overall well-being.

As the definition above indicates, the meaning of development is not confined to economics but can be expressed, for example, in access to schools or medical care, democratic institutions, environmental stewardship, social inclusiveness, political freedom, and the quality of governance. Development is, therefore, a broad theme that not only attracts attention from economic geographers but also from geographers that take a more comprehensive approach. In recent years, the dual notions of **sustainable development** **29** and **inclusive development** **30** have emerged as particularly important themes. The first focuses on the long-term viability of economic patterns in relation to the physical environment and natural

© Jan Nijman

@ FROM THE FIELD NOTES...

"As I sit enjoying breathtaking views of the spectacular Andean highlands of southern Peru, a farmer slowly comes up the side of the hill with her donkeys. She tells me that her name is Carmen and that the donkeys are carrying hay grass to market in the nearby town of Chinchero. The hay will feed guinea pigs that many rural and small-town Peruvian families keep in their home . . . to eat. Guinea pigs are said to be quite tasty, and they are much cheaper than beef as well as a good source of protein. Carmen's family owns a small plot of land on one of these hillsides and the donkeys are her main means of transportation. Although incomes for small farmers here are low and life is simple, she exudes an air of confidence and pride; she is at home here, and I am on her turf. I'm now thinking about trying guinea pig before I depart . . . "

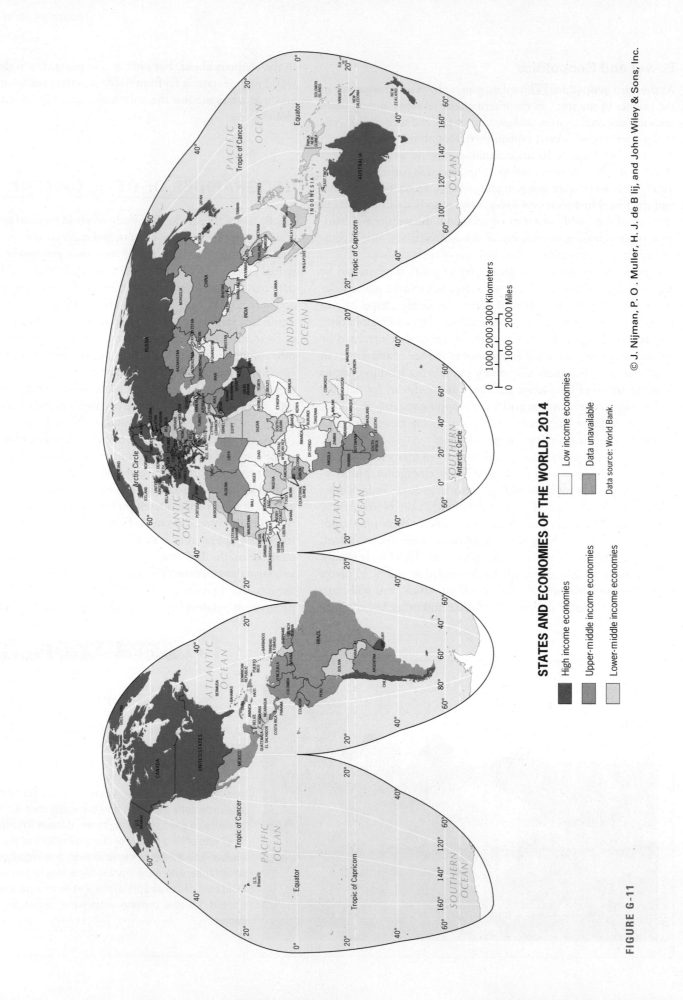

STATES AND ECONOMIES OF THE WORLD, 2014

■ High income economies
■ Upper-middle income economies
□ Lower-middle income economies
□ Low income economies
□ Data unavailable

Data source: World Bank.

FIGURE G-11

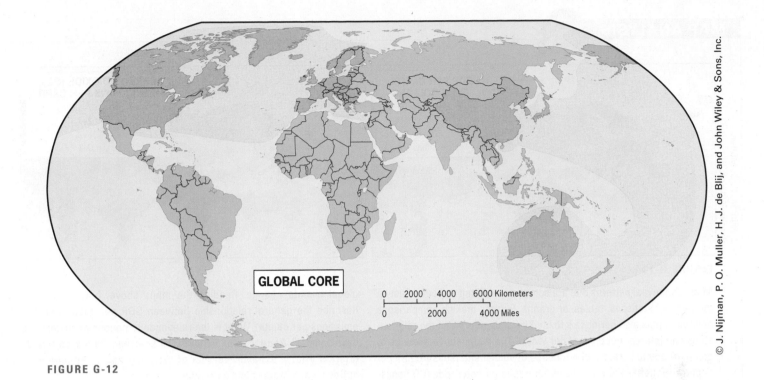

GLOBAL CORE

FIGURE G-12

resources (think of pollution; think of finite resources like fossil fuels). The second emphasizes the extent of equal economic (and social) opportunities for different population groups, especially minorities and the poor.

Development in Spatial Perspective

Various schemes to group the world's states into economic-geographic categories have come and gone, and others will probably arise in the future. For our purposes, the classification scheme used by the World Bank (one of the agencies that monitor economic conditions across the globe) is the most effective. It sorts countries into four categories based on the success of their economies: (1) high-income, (2) upper-middle income, (3) lower-middle income, and (4) low-income. These categories, when mapped, display important regional patterns (see Fig. G-11). Compare this map to our global framework (Fig. G-3), and you can see the role of economic geography in the layout of the world's geographic realms. Also evident are regional contrasts within realms—for instance, between Brazil and its western neighbors, between South Africa and most of the rest of Subsaharan Africa, and between west and east in Europe. Thus the geographic scale at which we consider development is of critical importance.

A Complex, Core-Periphery World

The economic success of human societies on the Earth's surface tends to be concentrated in certain areas while bypassing others. This was true in ancient and more recent historical times, and it is true today. The contemporary world economy—an integrated,

international spatial system—is structured within the framework of a dominant **global core 31** and a subservient **global periphery 32**. During the nineteenth century, this core more or less coincided with western Europe, controlling as it did vast areas of the world through its empires. In the twentieth century, the core expanded first to North America and then grew to include Japan, Australia, and New Zealand. By the 1980s, Hong Kong, Singapore, Taiwan, and South Korea had become part of the core as well. And as the twenty-first century opened, the newest entrant stepped forward: Pacific-fronting China. Today's map of this still evolving global core is mapped in tan in Figure G-12.

It is also important to realize that **core-periphery relationships 33** are not limited to the global scale. Countries themselves can and do exhibit such patterns as well. China again is a case in point: its coastal provinces form the core, while the interior and westernmost reaches of the country are part of China's periphery. Except for a few special cases, all countries contain **core areas 34**. These national cores are most often anchored by the country's capital and/or largest city: Paris (France), Tokyo (Japan), Buenos Aires (Argentina), and Bangkok (Thailand) are just a few notable examples. Larger countries may have more than one core area, such as Australia with its eastern and western coastal cores and intervening periphery. And uneven development also exists at the local scale, especially in the complex socioeconomic landscapes of urban areas.

The world continues to exhibit major differences in productivity and well-being. One of the most intriguing economic-spatial outcomes of globalization is that a growing number of countries have accelerated their development—but this growth is often confined to specific city-regions, whereas the rest of the country remains quite poor. Figure G-12, therefore, is provisional: it

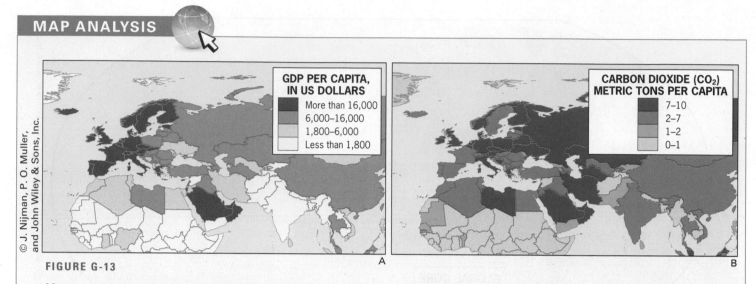

MAP ANALYSIS

FIGURE G-13

Maps can reveal patterns that are sometimes difficult to discern just on the basis of statistics, tables, or graphs. Further, making comparisons between maps also permits us to identify and understand the relationships and interconnections of our world. The maps above allow us to compare golbal patterns of economic development, measured as per capita GDP (left), with patterns of per capita carbon dixoide (CO_2) emissions (right). CO_2 emissions are linked to the greenhouse effect and global climate change. Reading the maps above, how would you describe the general relationship between GDP per capita and CO_2 emissions per capita? What is the relevance to notions of sustainable development? **Access your WileyPLUS Learning Space course to interact with a dynamic version of this map and to engage with online map exercises and questions.**

conveys a broad impression of the global core, but we realize that it is not really a contiguous geographic space (and neither is the global periphery). The core-periphery concept is most useful in thinking about the organization of space as well as the geographies of power and dependence. But it is difficult to map precisely, not least because it operates simultaneously at multiple scales. For instance, the United States constitutes part of the global core; its northeastern seaboard anchors part of the national core area; and New York's Midtown and Lower Manhattan form the core of its largest city.

We also need to add a cautionary note about spatially aggregated statistics. The concept of development, as measured by data that reflect totals and averages for entire national populations, entails certain pitfalls that we should be aware of. When a state's economy is growing as a whole, and even when it is "booming" relative to other states, this does not automatically mean that every citizen is better off and the income of every worker is rising. Averages have a way of concealing regional variability and local stagnation. In very large states such as India and China, it is useful to assess regional, provincial, and even local economic data to discover to what extent the whole country is sharing in "development." In the case of India, we should know that the State of Maharashtra (containing the megacity of Mumbai) is far ahead of most others States when it comes to sharing in the national economy. In China, the coastal provinces of the Pacific Rim still outstrip those of the west. Thus national-level statistics can conceal as much as they reveal, and this is why *inclusive development* is becoming a leading policy goal around the world.

GLOBALIZATION

Globalization 35 is essentially a geographical process in which spatial relations—economic, cultural, political—shift to the macroscale. What this means is that what happens in one place has repercussions in places ever more distant—thereby integrating the entire world into a "global village," linked ever more tightly by recent advances in communications and transport technologies. Globalization now comes into our homes via television, computers, and smartphones; news today has never traveled faster, and sometimes even government leaders turn to the Internet on their personal electronic devices to get the latest reports on international events.

Globalization, however, is not all that new. The second half of the nineteenth century, for instance, also witnessed major advances in the intensification of global interdependence. It was particularly affected by such new technologies as the steamship, the railway, and the telegraph, which subsequently were followed by first-generation motor vehicles and airplanes. But with today's newest technologies, the global village is achieving an unprecedented level of long-distance interconnectivity. Thus geography and our knowledge of the world's realms and regions have become more important than ever for understanding and navigating this transforming macrospatial reality.

Globalization and Environmental Challenges

Globalization plays out in various spheres, from the environmental to the cultural to the economic. One of today's most pressing

"Would you like to float in an 'infinity pool' atop a 57-story skyscraper that provides amazing views? You can in Singapore. During a 2012 visit, I had spotted this futuristic-looking tower across the water from downtown and thought it would be worth a visit. It turned out to be the massive Marina Bay Sands Hotel (developed by the Sands Corporation of Las Vegas) that opened for business in 2010. Singapore is one of the best examples of a city propelled into prominence by the forces of globalization. Located strategically on one of the world's busiest shipping lanes (the Strait of Malacca) and possessing a fine harbor, this minuscule city-state was bound to benefit from expanding world trade. It now boasts the busiest transshipment port in the world, and the city has also become a shopping magnet for elites all across the Southeast Asian realm. Like most 'world-cities,' Singapore is a hub of both production and consumption."

© Jan Nijman

environmental issues is global warming, which threatens every corner of our planet. It is clear that the problem must be addressed internationally, yet it is far from easy to agree on strategies. Some countries are bigger polluters than others, some have more resources than others, and some are more developed than others. How to divide the burdens? Countries from around the world have been meeting for many years now at the United Nations Climate Change Conference, the most recent one in Paris in late 2015. The main purpose is to reach agreement on collective reductions in the emission of greenhouse gases so as to limit global warming—but progress has been slow. Figure G-13 provides some insight into the relationship between economic growth and environmental impact. Often, national economic growth comes at a price, but in the end all of us pay.

Globalization and Population Movements

Culturally, too, the world is coming closer together. This is apparent in the global spreading of new fashions, music, foods, and other innovations. In a more problematic way, it is also expressed in global migration flows. In past times, **migration** **36** was relatively uncommon because most people were rooted in their home environment, where they lived out their entire lives. When residential relocation did occur, it used to be one-way, with people migrating from one place to another and then staying put. But over the past quarter-century, migration flows have intensified, in part because people increasingly possess better knowledge about opportunities elsewhere. Moreover, it is now much easier to travel back and forth, which allows migrants to maintain close ties with their countries of origin. Not surprisingly, as the number of highly mobile *transnational migrants* has risen, these transients have become major players in the spreading of cultures around the world. Examples include Algerians in Paris, Haitians in Montreal, Cubans in Miami, Mexicans in Los Angeles, Indians in Singapore, and Indonesians in Sydney.

Yet it is also important to keep in mind that people's mobility is often constrained because certain parts of this highly uneven world are so much better off than others. High-income countries exert the strongest pull on migrants—but all too often they cannot gain access. Millions of workers aspire to leave the poverty of the global periphery and seek a better life somewhere in the core. Trying to get there, every year many of them perish in the waters of the Mediterranean, the Caribbean, and the Atlantic. Others risk

ITALIAN NAVY/DPA/Corbis Images

Refugees on a boat on the Mediterranean Sea between North Africa and Sicily who are being rescued by the Italian navy on June 29, 2014. By the end of that year alone, Italy's navy and coast guard had rescued more than 130,000 desperate immigrants.

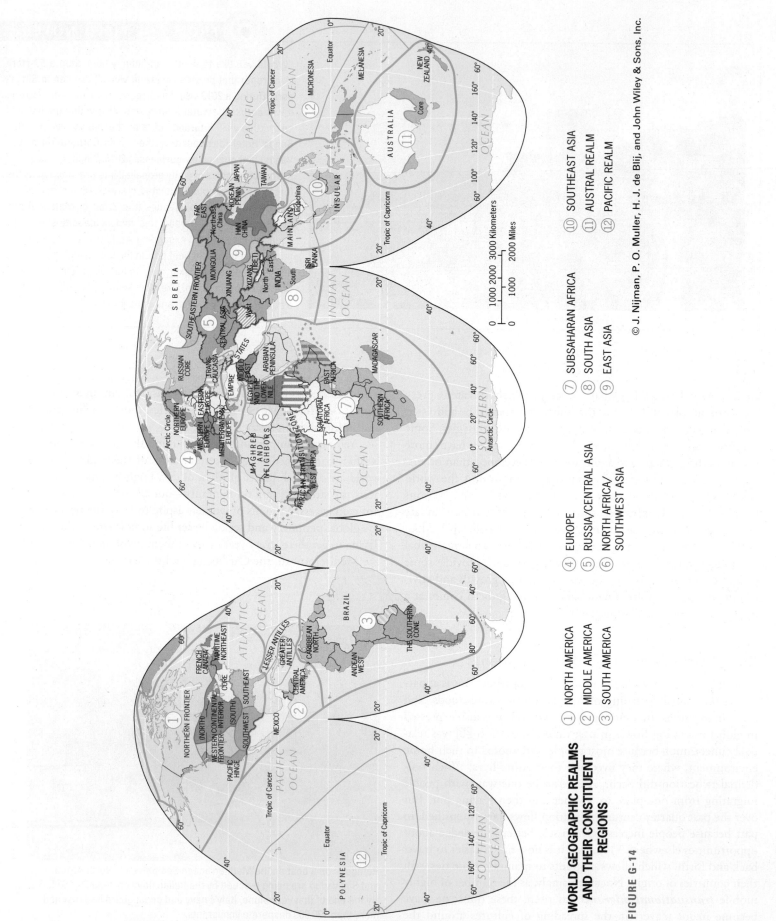

**WORLD GEOGRAPHIC REALMS
AND THEIR CONSTITUENT
REGIONS**

① NORTH AMERICA
② MIDDLE AMERICA
③ SOUTH AMERICA

④ EUROPE
⑤ RUSSIA/CENTRAL ASIA
⑥ NORTH AFRICA/
   SOUTHWEST ASIA

⑦ SUBSAHARAN AFRICA
⑧ SOUTH ASIA
⑨ EAST ASIA

⑩ SOUTHEAST ASIA
⑪ AUSTRAL REALM
⑫ PACIFIC REALM

© J. Nijman, P. O. Muller, H. J. de Blij, and John Wiley & Sons, Inc.

FIGURE G-14

28

their lives at the barriers that encircle the global core as if it were a gated community—from the "security fence" between Mexico and the United States to the walls that guard Israel's safety to the razor wire that encircles Spain's outposts on North Africa's shore.

Economic Globalization, Growth, and Inequality

Most economists, politicians, and businesspeople have a favorable view of economic globalization because it facilitates the expansion of international capitalism, open markets, and free trade. In theory, such globalization stimulates commerce, breaks down barriers to foreign trade, brings jobs to remote places, and promotes social, cultural, and political exchanges. High-technology workers in India are employed by computer firms based in California. Japanese cars are assembled in Thailand. American footwear is made in China. Fast-food restaurant chains spread standards of service and hygiene as well as familiar (and standardized) menus from Tokyo to Tel Aviv to Tijuana. If wages and standards of employment are lower in peripheral countries than in the global core, production will shift there and the development gap will shrink. Everybody wins. In general, the ***gross national income (GNI)***[†] of countries that increase their foreign trade (and thereby become more "globalized") rises, whereas the GNI of those with diminished trading activity actually declines.

But there is another, more complicated issue: the liberalization and opening up of national economies to global markets is often accompanied by widening inequalities. In other words, uneven development within countries becomes more pronounced. As noted earlier, this imbalance is obvious in a country like China, the world's fastest-growing economy over the past two decades; much of this growth occurred in its Pacific coastal zone, not in the interior of the country, and income differentials became ever wider. The same is true in India and most other ***emerging markets***. Moreover, this phenomenon is not confined to developing economies. In his widely read 2013 book entitled *Capital in the Twenty-First Century*—now translated into more than 30 languages worldwide (globalization!)—the French economist Thomas Piketty shows that inequality is growing in the United States and Europe as well. Inequality in the United States is discussed in Chapter 1; that discussion underscores why a regional approach is vital to interpreting this process at both national and global scales.

REGIONAL AND SYSTEMATIC GEOGRAPHIES

At the beginning of this chapter, we introduced a map of the great geographic realms of the world (Fig. G-3). We then addressed the

---

[†]Gross national income (GNI) is the total income earned from all goods and services produced by the citizens of a country, within or outside of its borders, during a calendar year. *Per capita GNI* is a widely used indicator of the variation of spendable income around the globe, and is most useful for comparing countries to one another.

task of dividing these realms into regions, and we used criteria ranging from physical geography to economic geography. The result is Figure G-14. On this map, note that we display not only the world geographic realms but also the regions into which they subdivide. The numbers in the legend reveal the order in which the realms and regions are discussed, starting with North America (1) and ending with the Pacific Realm (12).

As this introductory chapter demonstrates, this book does not merely offer a series of descriptions of places and areas. We combine the study of realms and regions with a look at geography's ideas and concepts—the notions, generalizations, and basic theories that make the discipline what it is. We continue this approach in the chapters ahead so that we will become better acquainted with the world and with geography as a science. By now you are aware that geography is a wide-ranging, multifaceted discipline. It is often described as a social science, but that is only half the story: geography, in fact, straddles the divide between the social and the physical sciences. Many of the ideas and concepts you will encounter involve the multiple interactions between human societies and natural environments.

Regional geography allows us to view the world in an all-encompassing way. As we have seen, regional geography integrates information from many sources to create an overall image of our divided world. Those sources are not random. They represent topical or ***systematic geography***. Research in the systematic fields of geography makes our world-scale generalizations possible. As Figure G-15 shows, these systematic fields connect closely to those of other disciplines. Cultural geography, for example, is allied with anthropology; it is the spatial perspective that distinguishes

FIGURE G-15

cultural geography. Economic geography focuses on the spatial dimensions of economic activity; political geography concentrates on the spatial imprints of political behavior. Other systematic fields include historical, medical, behavioral, environmental, and urban geography. We will also draw on information from biogeography, marine geography, population geography, geomorphology, and climatology (as we did earlier in this chapter).

These systematic fields of geography are so named because their approach is thematic, not regional. Take the geographic study of cities, urban geography. Urbanization is a worldwide process, and urban geographers can identify certain human activities that all cities in the world exhibit in one form or another. But cities also display regional properties. The typical American city is quite distinct from, say, the European or the Chinese city. Regional geography, therefore, borrows from the systematic field of urban geography, but it injects this regional perspective. In the following chapters we call upon these systematic fields to provide you with a better understanding of the world's realms and regions. As a result, you will gain insights into the discipline of geography as well as the regions we investigate.

We will begin our investigation of the world's realms and regions in North America. This is, after all, where most of you were born and raised—and it is from this geographic realm that your expanding worldview has evolved.

POINTS TO PONDER ?

- According to some measures, China in 2015 surpassed the United States and became the world's biggest national economy.

- The global human population in 2016 is 7.1 billion and is predicted to reach 9.5 billion by 2050. Will there be enough food and water to sustain everyone?

- Global warming is expected to cause a global rise in sea level of least 10 centimeters (4 inches) by the end of the century. Which cities are most vulnerable?

- Fast-forwarding globalization: the number of smartphones in the world surpassed the 2 billion mark in 2016; most of these devices have map navigation capabilities.

Austin, Texas—highly regarded for its burgeoning technology-led economy—is one of the fastest growing American cities in the 2010s. As this photo of the 2014 Austin City Limits Music Festival shows, the Lone Star State's capital also occupies a prominent position on the U.S. popular culture scene. Austin is a product of the postindustrial Sunbelt boom of the past half-century, an anchor of the new high-tech industrial corridor that stretches from the Dallas-Fort Worth Metroplex in the north to Greater San Antonio in South Texas. In addition to its function as a far-reaching employment magnet for young people, Austin is considered to be one of the most 'livable' cities in the nation.

IN THIS CHAPTER

- North America's high-tech regions
- Changing geographies of inequality
- Bilingualism in Canada
- Detroit's bankruptcy
- U.S.-Mexico borderland dynamics
- China's impact on the realm's Pacific coast

CONCEPTS, IDEAS, AND TERMS

Physiographic region	1	Gentrification	11
Continentality	2	Sunbelt	12
Rain shadow effect	3	Migration	13
Federation	4	Melting pot	14
Fossil fuel	5	First Nations	15
Urban system	6	World-City	16
American Manufacturing Belt	7	Technopole	17
Outer city	8	Pacific Rim	18
Deindustrialization	9	Tar sands	19
Central business district (CBD)	10	Boreal forest	20

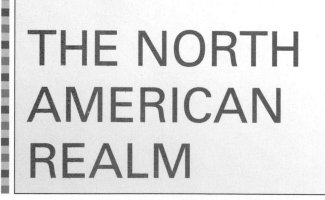

1

THE NORTH AMERICAN REALM

FIGURE 1-1

In our survey of the world's geography, we first turn to the Western Hemisphere. It contains the two great interconnected continents that separate the Atlantic and Pacific oceans and extend, very nearly, from pole to pole, flanked by numerous islands large and small, indented by gulfs and bays of historic and economic import, and endowed with an enormous range of natural resources. Two continents—North and South—form the Americas, but three geographic realms blanket them (North, Middle, and South). In the context of physical (natural) geography, North America—from Canada's Ellesmere Island in the far north to the Isthmus of Panama in the south—is a continent. In terms of modern human geography, the northern continent is subdivided into the North American and Middle American realms along a transition zone marked by a political as well as a physical boundary between the United States and Mexico (Fig. 1-1). In Texas, from the Gulf of Mexico to El Paso/Ciudad Juárez, the Rio Grande forms this border. From El Paso westward to the Pacific Ocean, straight-line boundaries, hardened by fences and walls, separate the North from the Middle. We begin in North America.

DEFINING THE REALM

North America is constituted by two of the world's most highly advanced countries in virtually every measure of social and economic development. Blessed by an almost unlimited range of natural resources and bonded by trade as well as culture, Canada and the United States are locked in a mutually productive embrace that is reflected by the economic statistics: in an average recent year, about two-thirds of Canadian exports went to the United States, and just over half of Canada's imports came from its southern neighbor. For the United States, Canada is not quite as dominant, but it is still its leading export market and the second-biggest importer.

This realm is also defined on the basis of broad cultural traits, ranging from urban landscapes and a penchant for mobility to religious beliefs, language, and political persuasions. Both countries rank among the world's most highly urbanized: nothing symbolizes the North American city as strongly as the skyscrapered panoramas of New York, Chicago, and Toronto—or the vast, beltway-connected suburban expanses of Los Angeles, Washington-Baltimore, and Houston. North Americans are also the most mobile people on Earth. Each year about one out of every eight individuals in the United States changes residence, a proportion that has declined lately but still leads the world.

MAJOR GEOGRAPHIC QUALITIES

NORTH AMERICA

1. North America encompasses two of the world's biggest states territorially (Canada is the second-largest in size; the United States is third).

2. Both Canada and the United States are federal states. Canada's is divided into ten provinces and three territories. The United States consists of 50 States, the Commonwealth of Puerto Rico, and a number of island territories under U.S. jurisdiction in the Caribbean Sea and the Pacific Ocean.

3. Both Canada and the United States are plural societies. Canada's pluralism is most strongly expressed in regional bilingualism. In the United States, divisions occur largely along ethnic, racial, and income lines.

4. North America's population, not large by international standards, is one of the world's most highly urbanized and mobile, its enduring growth largely propelled by continuing waves of immigration.

5. By world standards, this is a rich realm where high incomes and high rates of consumption prevail. North America possesses a highly diversified resource base, but nonrenewable fuel and mineral deposits are consumed prodigiously.

6. North America has long been home to one of the world's great manufacturing complexes that generated an intricately connected, mature urban system. Over the past few decades, North America has come to rely increasingly on a highly advanced information economy and high-technology industries.

7. The two countries heavily depend on each other for supplies of critical raw materials (e.g., Canada is a leading source of U.S. energy imports) and have long been each other's primary trading partners.

8. The U.S.-Mexico border separates the North American realm from Middle America; it is a sharp boundary in some respects but in other ways a broadening transition zone.

9. Continued immigration and high transnational mobility make for an exceptionally diverse multicultural realm.

THE U.S-MEXICO BORDERLAND: BOUNDARY OR TRANSITION ZONE?

A stretch of the fortified fence along the Mexican border, suggesting a sharp boundary between the United States and Mexico. This is a ground-level view of the improvised wall that slashes westward, between Tijuana, Mexico, and the southernmost suburbs of San Diego, California, as it makes its way toward the Pacific shore on the horizon. The poverty-stricken cityscape of this eastern part of Tijuana extends almost to the rusty barrier itself, while development on the U.S. side has kept its distance from this well-patrolled segment of the border.

Most of what is now the southwestern United States used to belong to Mexico. In the so-called Treaty of Guadalupe Hidalgo (1848), Mexico ceded the territory to the United States. Today, that history is still reflected here in the presence of substantial numbers of Mexicans, Mexican Americans, and Americans of Mexican descent. Therefore, in terms of cultural geography, the U.S.-Mexico borderland should be viewed as a broad transition zone marked by a decidedly Mexican imprint on the landscape on both sides of the actual boundary. This duality extends into other spheres as well. The infusion of Mexican ethnicity profoundly shapes the demography of this quadrant of the country (see Fig. 1-2). And in the twenty-first century, the economies of the United States and Mexico are becoming ever more closely linked as their longstanding trade partnership intensifies—fully two-thirds of Mexican exports are now being shipped to its northern neighbor.

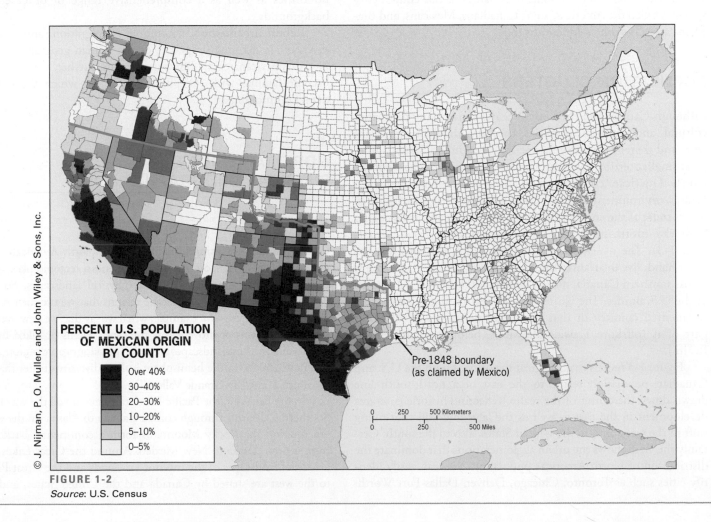

PERCENT U.S. POPULATION OF MEXICAN ORIGIN BY COUNTY

- Over 40%
- 30–40%
- 20–30%
- 10–20%
- 5–10%
- 0–5%

Pre-1848 boundary (as claimed by Mexico)

0   250   500 Kilometers
0   250   500 Miles

© J. Nijman, P. O. Muller, and John Wiley & Sons, Inc.

FIGURE 1-2

Source: U.S. Census

Despite noteworthy multilingualism, in most areas English is the *lingua franca* of both countries. And notwithstanding all the religions followed by minorities, the great majority of both Canadians and Americans are Christians (whose U.S. total has dropped sharply since 2007). Moreover, both states are stable democracies, and both have federal systems of government.

Driving from Michigan across southern Ontario to upstate New York does not present a sharp contrast in cultural landscapes. Americans visiting Canadian cities (and vice versa) find themselves in mostly familiar settings. Canadian cities have fewer impoverished neighborhoods, no ethnic ghettos (although low-income ethnic districts do exist), lower crime rates, and, in general, better public transportation; but rush hour in Toronto very much resembles rush hour in Chicago. The boundary between the two states remains porous, even after the events of 9/11. Americans living near the border shop for lower-priced medicines in Canada; Canadians who can afford it seek medical treatment in the United States.

None of this is to suggest that Canadians and Americans don't have their differences—they do. Most of these differences may seem subtle to people from outside the realm, but they are important nonetheless, particularly for Canadians who at times feel dominated by Americans whose political or cultural values they do not always share. You can even argue over names: if this entire hemisphere is called the Americas, aren't Canadians, Mexicans, and Brazilians as well as others Americans too?

POPULATION CLUSTERS

Although Canada and the United States share many historical, cultural, and economic qualities, they also differ in significant ways and thereby diversify this realm. The United States, somewhat smaller territorially than Canada, occupies the heart of the North American continent and, as a result, encompasses a greater environmental range. The U.S. population is dispersed across most of the country, forming major concentrations along both the (north-south trending) Atlantic and Pacific coasts (Fig. 1-3). The overwhelming majority of Canadians, on the other hand, live in an interrupted east-west corridor that extends across southern Canada, mostly within 300 kilometers (200 mi) of the U.S. border. The United States, again unlike Canada, is a fragmented country in that the broad peninsula of Alaska is part of it (offshore Hawai'i, however, belongs in the Pacific Realm).

Figure 1-3 reveals that the great majority of both the U.S. and Canadian population reside to the east of a north-south line drawn down the center of the realm, reflecting historic core-area development in the east as well as the later and still-continuing shift to the west, and, in the United States, also to the south. Certainly this map shows the urban agglomerations that dominate the distribution of North America's population: you can easily identify cities such as Toronto, Chicago, Denver, Dallas-Fort Worth,

San Francisco, and Vancouver. Overall, nearly 80 percent of the realm's population is concentrated in cities and towns, a higher proportion even than Europe's.

The population of the United States, which reached 316.1 million in 2015, is growing 25 percent faster than Canada's, and the 400-million mark may be surpassed as soon as 2050. This is an unusually high rate of growth for a high-income economy, resulting from a combination of natural increase (which leads today) and substantial immigration (on a pace to lead after 2030).

Although Canada's overall growth rate is significantly lower than that of the United States, immigration contributes proportionately even more to this increase than in the United States. With 35.2 million residents, Canada, like the United States, has been relatively open to legal immigration, and as a result both societies exhibit a high degree of cultural diversity in ancestral and traditional backgrounds. Indeed, Canada recognizes two official languages, English and French (the United States does not even designate English as such); and by virtue of its membership in the British Commonwealth, East and South Asians form a larger sector of Canada's population than Asians and Pacific Islanders do in the United States. In contrast, cultural diversity in the United States reflects large Hispanic (17.4 percent), African American (13.2 percent), and Asian (6.3 percent) minorities as well as a comprehensive range of other ethnic backgrounds.

Robust urbanization, substantial immigration, and cultural diversity are defining properties of the human geography of the North American realm. But before going further, we must also become familiar with the physical stage on which the human drama is unfolding.

NORTH AMERICA'S PHYSICAL GEOGRAPHY

Physiographic Regions

One of the distinguishing properties of the North American landmass, all of which lies on the North American tectonic plate, is its remarkable variegation into regional physical landscapes. So well defined are many of these landscape regions that we use their names in everyday parlance—for example, when we say that we flew over the Rocky Mountains or drove across the Great Plains or hiked in the Appalachians. These landscapes are called **physiographic regions 1**, and nowhere else in the hemisphere is their diversity greater than to the north of the Rio Grande Valley (Fig. 1-4).

In the Far West, the Pacific Mountains extend all the way from Southern California through coastal Canada to Alaska. In the western interior, the Rocky Mountains form a continental backbone from central Alaska to New Mexico. Around the Great Lakes, the low-relief landscapes of the Interior Lowlands and the Great Plains to the west are shared by Canada and the United States, and the

FIGURE 1-3

international boundary even divides the Great Lakes. In the east, the Appalachian Mountains (as the cross-sectional inset shows, no match for the Rockies) form a corridor of ridges, valleys, and plateaus that represent a familiar topography from Alabama and Georgia to Nova Scotia and Newfoundland. If there is a major physiographic province that belongs to only one of the realm's two countries, it is the Canadian Shield, scoured bare by the Pleistocene glaciers that deposited their pulverized rocks as fertile soil in the U.S. Midwest, that is, in the Interior Lowlands.

Climate

This diversity of landscape is matched by a variety of climates. Take another look at Figure G-6, and you can see the lineaments of

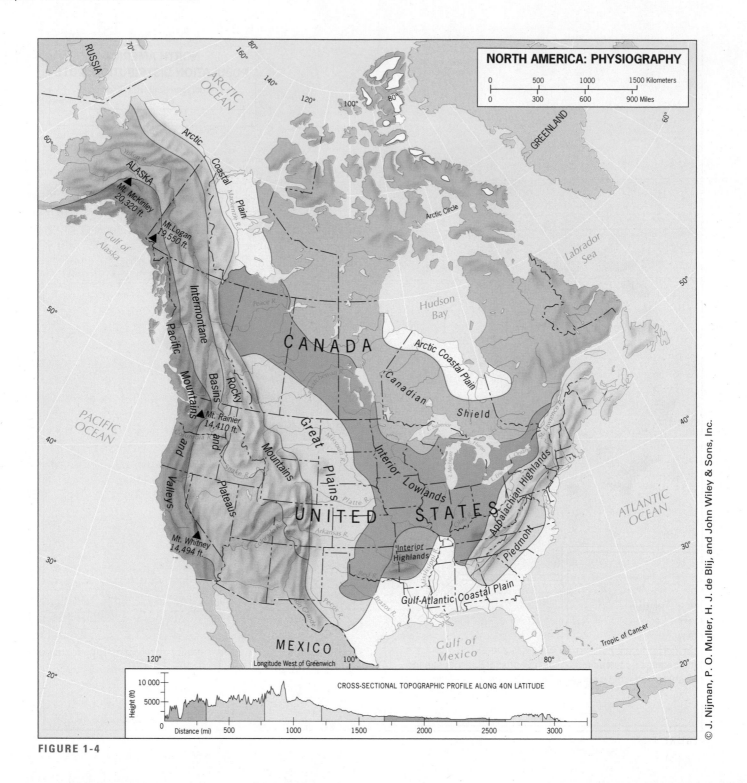

NORTH AMERICA: PHYSIOGRAPHY

CROSS-SECTIONAL TOPOGRAPHIC PROFILE ALONG 40N LATITUDE

FIGURE 1-4

landscape mirrored in the contrasts of climate. North America may not have it all—there are no areas of true tropical environment in the North American realm to speak of except at the southern tip of Florida—but it does exhibit a great deal of variation. North America contains moist coastal zones and arid interiors, well-watered plains, and even bone-dry deserts. On the world map, *Cf* and *Df* climates are especially good for commercial farming; note how large North America's share of these environments is.

Temperature-wise, the map leaves no doubt: the farther north you go, the colder it gets. And even though coastal areas derive moderation from warmer offshore waters, the rigors of **continentality 2** (inland climate environment remote from ameliorating maritime influences) set in not far from the coast. Hot summers, frigid winters, and limited precipitation make these higher-latitude continental interiors difficult places to make it on the land. Figure G-6 has much to do with the southerly concentration of

Superstorm Sandy, the powerful remnant of a Caribbean hurricane of the same name, will forever be remembered along the Mid-Atlantic seaboard of the northeastern United States (and far beyond as the second-costliest natural disaster in American history, through mid-2015). It struck with a vengeance at high tide on the night of October 28–29, 2012, unleashing sustained winds of up to 130 kilometers per hour (80 mph) and a 2.6-meter (9-ft) storm surge of seawater. The brunt of Sandy's landfall was absorbed by the low-lying northern New Jersey shore. The heaviest wind and flooding damage occurred on the narrow offshore barrier islands that 'guard' and parallel the mainland. This aerial view encompasses the oceanside resort town of Mantoloking on the well-known barrier island that terminates a few miles south at the Jersey Shore's popular Island Beach State Park. The inset photo was taken a few months prior to Sandy, and the most dramatic landscape change is the new inlet at the end of the bridge (center of main photo) that breached the island between the Atlantic Ocean (right) and Barnegat Bay (left), the lagoon that separates Mantoloking from Brick Township on the mainland.

the Canadian people as well as the lower population densities that mark the realm's interior.

In the west, especially in the United States, you can see the effects of the Pacific Mountains on inland areas. Moisture-laden air arrives from the ocean, the mountain wall forces the air upward, cools it, condenses the moisture in it, and produces rain—the rain for which Seattle and Portland as well as other cities of the Pacific

Northwest are (in)famous. By the time the air crosses the mountains and descends on the landward side, most of the moisture has been extracted from it, and the forests of the ocean-facing side quickly yield to scrub and brush. This **rain shadow effect** **3** extends all the way across the Great Plains; North America does not turn moist again until the Gulf of Mexico sends humid tropical air northward into the eastern interior via the Mississippi Basin.

In a very general way, therefore, and not including the coastal strips along the Pacific, nature divides North America into an arid west and a humid east. Again the population map reveals more than a hint of this: draw a line approximately from Lake Winnipeg to the mouth of the Rio Grande, and look at the contrast between the (comparatively) humid east and drier west when it comes to population density. Water is a critical part of this story.

Between the Rocky Mountains and the Appalachians, North America lies open to air masses from the frigid north and tropical south. In winter, southward-plunging polar fronts send frosty, bone-dry air masses deep into the heart of the realm, turning even places like Memphis and Atlanta into iceboxes; in summer, hot and humid tropical air surges northward from the Gulf of Mexico, giving Chicago and Toronto a taste of the tropics. Such air masses clash in low-pressure systems along weather fronts loaded with lightning, thunder, and, frequently, dangerously destructive tornadoes. And the summer heat brings an additional threat to the Gulf-Atlantic Coastal Plain (Fig. 1-4): hurricanes capable of inflicting catastrophic devastation on low-lying areas (see the accompanying photo). These tropical cyclones also prune natural vegetation, replenish groundwater reservoirs, fill natural lakes, and flush coastal channels.

Great Lakes and Great Rivers

Two great drainage systems lie between the Rockies and the Appalachians: (1) the five Great Lakes that drain into the St. Lawrence River and the Atlantic Ocean, and (2) the mighty Mississippi-Missouri river system that carries water from a vast interior watershed to the Gulf of Mexico, where the Mississippi forms one of the world's major *deltas*. Both natural systems have been modified by human engineering. In the case of the St. Lawrence Seaway, a series of locks and canals has created a direct shipping route, via the Great Lakes and their outlet, from the Midwest to the Atlantic. The Mississippi has been fortified by artificial levees that, while failing to contain the worst of flooding, have enabled farmers to cultivate the most fertile of American soils.

NATIVE AMERICANS AND EUROPEAN SETTLEMENT

When the first Europeans set foot on North American soil, the continent was inhabited by millions of people whose ancestors had reached the Americas from Asia via Alaska, and possibly also across the Pacific, more than 14,000 years earlier (and perhaps as long as 30,000 years ago). In search of Asia, the Europeans misnamed them Indians. These Native Americans or **First Nations**—as they are now called, respectively, in the United States and Canada—were organized into hundreds of nations with a rich mosaic of languages and a great diversity of cultures.

The eastern nations were the first to bear the brunt of the European invasion. By the end of the eighteenth century, ruthless and land-hungry settlers had driven most of the native peoples living along the Atlantic and Gulf coasts from their homes and lands, initiating a westward push that was to devastate indigenous society. Today, less than 3 million Native Americans remain in the United States, and about one-third that number in Canada. In the United States they are left with only about 4 percent of the national territory in the form of mostly impoverished reservations. Aboriginal peoples today are better off territorially in Canada where they hold titles to large tracts of land, especially in the northern sectors of British Columbia and Quebec (but keep in mind that northern Canada is nearly empty to begin with and that environmental conditions there can be quite harsh).

The current population geography of North America is rooted in the colonial era of the seventeenth and eighteenth centuries that was dominated by Britain and France. The French sought mainly to organize a lucrative fur-trading network, while the British established settlements along the coast of what is now the eastern seaboard of the United States (the oldest, Jamestown, was founded in tidewater Virginia more than four centuries ago).

CULTURAL FOUNDATIONS

The modern American creed, if one can be identified, is sometimes idealistically characterized as exhibiting an adventurous drive, a liking for things new, a willingness to move, a sense of individualism, an aggressive pursuit of goals and ambition, a need for societal acceptance, and a firm sense of destiny. These qualities are not unique to modern U.S. culture, of course, but in combination they seem to have created a particular and pervasive mindset that is reflected in many ways in this geographic realm. Generally speaking, these tendencies tend to translate into an intense pursuit of educational and other goals to fulfill high expectations concerning upward socioeconomic mobility. Thus the *American Dream*, whether a genuine prospect or an ideological fabrication, is very much an expression of this set of cultural values, and it is closely entwined with the character of North America as a land of immigrants.

Facilitating these aspirations is *language*. None of these goals would be within reach for so many were it not for the use of English throughout most of the realm. In Europe, by contrast, language inhibits the mobility so routinely practiced by Americans: workers moving from Poland to Ireland, for example, found themselves at a disadvantage in competition with immigrants from Ghana or Sri Lanka. In North America, a worker from

Arkansas would not even consider language to be an issue when applying for a job in Toronto.

Also reflecting the cultural values cited above is the role of *religion*, which sets American (more so than Canadian) culture apart from much of the rest of the Western world. The overwhelming majority of North Americans express a belief in God, and a (declining) majority regularly attend church, in contrast to much of Europe. Three out of four people in the United States say they adhere to a religion, and four out of ten say they attend a religious service once a week (more than in Iran!). Religious observance is a virtual litmus test for political leaders; no other developed country prints 'In God We Trust' on its currency.

Figure 1-5 shows the mosaic of Christian faiths that blankets the realm, but a map at this scale can only suggest the broadest outlines of what is a much more intricate pattern. Protestant denominations are estimated to number in the tens of thousands in North America, and no map can capture them all. Southern (and other) Baptists form the majority across the U.S. Southeast from Texas to Virginia, Lutherans in the Upper Midwest and northern Great Plains, Methodists in a belt across the Lower Midwest, and Mormons in the interior West centered on Utah. Roman Catholicism prevails in most of Canada as well as the U.S. Northeast and Southwest, where ethnic Irish and Italian adherents form majorities in the Northeast and Hispanics in the Southwest. Behind these patterns lie histories of proselytism, migration, conflict, and competition, but tolerance of diverse religious (even nonreligious) views and practices is a hallmark of this realm.

THE FEDERAL MAP OF NORTH AMERICA

The two states that constitute North America may have arrived at their administrative frameworks with different motives and at different rates of speed, but the result is unmistakably similar: their internal political geographies are dominated by straight-line boundaries of administrative convenience (Fig. 1-6). In comparatively few places, physical features such as rivers or the crests of mountain ranges mark internal boundaries, but by far the greater length of boundaries is ruler-straight. Even the international boundary between Canada and the United States west of the Great Lakes mostly coincides with a parallel of latitude, 49°N.

The reasons are all too obvious: the framework was laid out before, in some areas long before, significant white settlement occurred. There was something arbitrary about these delineations, but by delimiting the internal administrative units early and clearly, governments precluded later disputes over territory and resources. In any case, neither Canada nor the United States was to become a **unitary (centralized) state**: both countries are federations **4**. In a federal state, power is shared between a country's central government and its highest-ranking political subdivisions. The Canadians call these first-order subdivisions provinces, whereas in the United States they are, as the country's name implies, States.

NORTH AMERICA: PREDOMINANT RELIGIONS
DENOMINATION

- Anglican
- Baptist
- Christian/Disciples
- Pentecostal
- Lutheran
- Methodist
- Mormon
- Roman Catholic
- United Church
- No Majority
- No reported denominations

FIGURE 1-5

It is no trivial matter to take a careful look at some of the provinces and States that compartmentalize this realm because some are of greater consequence than others. We have already noted that Canada's pair of official languages includes French as well as English. Although French-speakers are in the minority, they are heavily concentrated in and around the Province of Quebec, which occupies a unique position in federal Canada. In many ways the pivotal Canadian province is Ontario, containing the country's largest and most globalized city (Toronto) as well as the capital (Ottawa) on its eastern river border with Quebec. In the United

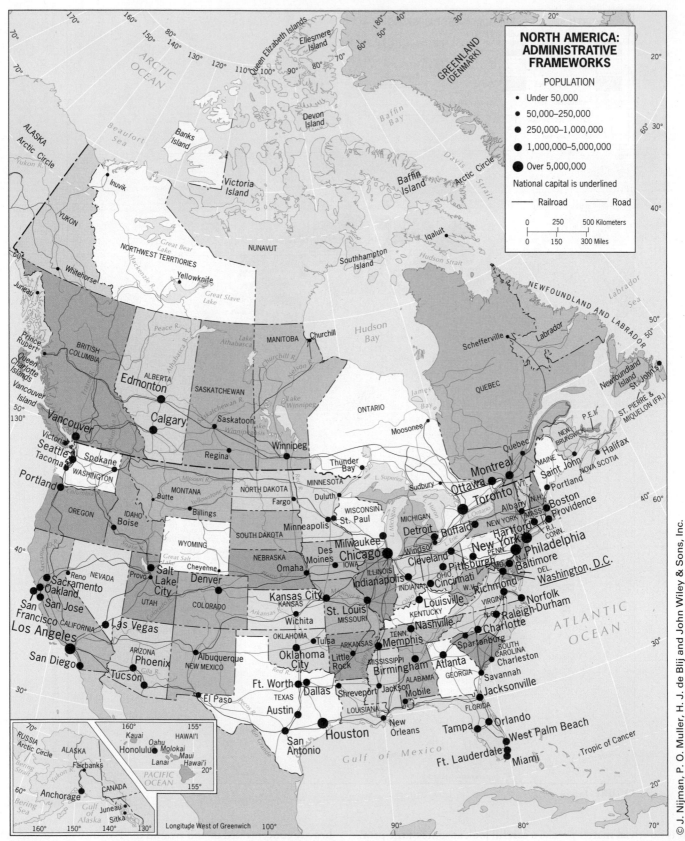

FIGURE 1-6

States, key States in the political and economic geography of the federation (indeed, the four most populated, in descending order) are California facing the Pacific, Texas bordering Mexico, Florida pointing toward the Caribbean and South America beyond, and New York with its window on the northern Atlantic Ocean.

The rectangular layout of the realm seems to symbolize North America's modernity and its stability, whereas the federal structure itself (with considerable autonomy at the State or provincial level) reflects the culture of freedom and independence. If federalism works within these high-income economies, shouldn't the rest of the world emulate it? Federalism does have its assets, but it also carries some significant liabilities.

THE DISTRIBUTION OF NATURAL RESOURCES

One of these liabilities has to do with the variable allotment of natural resources among the States and provinces. In Canada, Alberta is favored with massive energy reserves (and even more in the offing), but its provincial government is not always eager to see the federal administration take that wealth to assist less affluent provinces. In the United States, when oil prices are high, Texas benefits while California's budget suffers. So it is important to compare the map of natural resources (Fig. 1-7) with the map of States and provinces.

FIGURE 1-7

© J. Nijman, P. O. Muller, H. J. de Blij, and John Wiley & Sons, Inc.

Water certainly is a natural resource, and North America as a realm is comparatively well supplied with it despite concerns over long-term prospects in the American Southwest and the Far West, where some States depend on sources in other States. Another concern focuses on lowering the water tables in some of North America's most crucial *aquifers* (underground water reserves) in which a combination of overuse and decreasing replenishment foretells the rise of supply problems.

North America is endowed with abundant reserves of *minerals* that are mainly found in three zones: the Canadian Shield north of the Great Lakes, the Appalachian Mountains in the East, and the mountain ranges of the West. The Canadian Shield contains substantial iron ore, nickel, copper, gold, uranium, and diamonds. The Appalachians yield lead, zinc, and iron ore. And the western mountain zone has significant deposits of copper, lead, zinc, molybdenum, uranium, silver, and gold.

In terms of **fossil fuel** `5` energy resources (oil, natural gas, and coal), North America is also quite well endowed, though the voracious demand of the United States, the highest in the world, cannot be met by domestic supplies alone and generates significant imports. Figure 1-7 displays the distribution of oil, natural gas, and coal.

The leading *oil*-production areas lie (1) along and offshore from the Gulf Coast, where the floor of the Gulf of Mexico is yielding a growing share of the output; (2) in the Midcontinent District, from western Texas to eastern Kansas; and (3) along Alaska's North Slope facing the Arctic Ocean. An important development is taking place in Canada's northeastern Alberta, where oil is being drawn from deposits of *tar sands* in the vicinity of Fort McMurray. The process is expensive and can reward investors only when the price of oil is comparatively high, but the reserves of oil estimated to be contained here may exceed those of Saudi Arabia.

The distribution of *natural gas* reserves resembles that of oilfields because petroleum and natural gas tend to be found in similar geological formations (i.e., the floors of ancient shallow seas). What the map cannot reveal is the volume of production, in which this realm leads the world (Russia and Iran lead in proven reserves). That output has risen significantly in recent decades as natural gas has become the fuel of choice for electricity generation in North America. Driving this expansion has been the widespread application of hydraulic fracturing (*fracking*) technology. By injecting pressurized fluids into deeply buried shale rocks to create fractures, vast quantities of trapped gas can be extracted, resulting in surging supplies and dropping prices.

The *coal* reserves of North America, perhaps the largest on the planet, are found in Appalachia, beneath the Great Plains of the United States as well as Canada, and in the southern Midwest among other places. These reserves guarantee an adequate supply for centuries to come, although coal has become a less desirable fuel owing to its polluting emissions that contribute to global warming.

Given the relative scarcity of oil and the geopolitical disadvantages of having to import more than one-quarter of what is needed

Wind 'farms' are spreading to many more places in North America as the clean-energy campaign to generate electricity from sources other than fossil fuels accelerates. This cluster of modern wind turbines on a cattle ranch in southern Wyoming (the livestock provide a sense of scale) exploits a flat-floored windstream corridor between areas of higher relief.

in the United States (27 percent in 2014), you might expect the use of nuclear energy to be more widespread, especially because it is relatively cheap to produce. The United States has just over 60 nuclear power plants that provide about one-fifth of all electrical energy (Canada has 5 plants that produce less than 2 percent of the country's electricity). But ever since the notorious leak and near-calamity at Pennsylvania's Three Mile Island in 1979, the fear of nuclear accidents has been pervasive. In fact, the federal government did not approve a new nuclear plant again until 2012, a facility that is slated to open in Georgia in 2017.

URBANIZATION AND THE SPATIAL ECONOMY

Industrial Cities

When the Industrial Revolution (which began in England around 1800) crossed the Atlantic to America in the 1870s, it took hold so successfully and advanced so robustly that only 50 years later North America was surpassing Europe as the world's mightiest industrial stronghold. Industrialization progressed in tandem with urbanization as manufacturing plants in need of large labor supplies were built in and near cities. This, in turn, propelled rural-urban migration (including the migration of African Americans from the South to northern cities during the first half of the twentieth century) and the growth of individual cities. At a broader scale, a *system* of new cities emerged that specialized in the collection and processing of raw materials and the distribution of manufactured products. This **urban system** `6` was interconnected through a steadily expanding network of railroads, themselves a product of the Industrial Revolution.

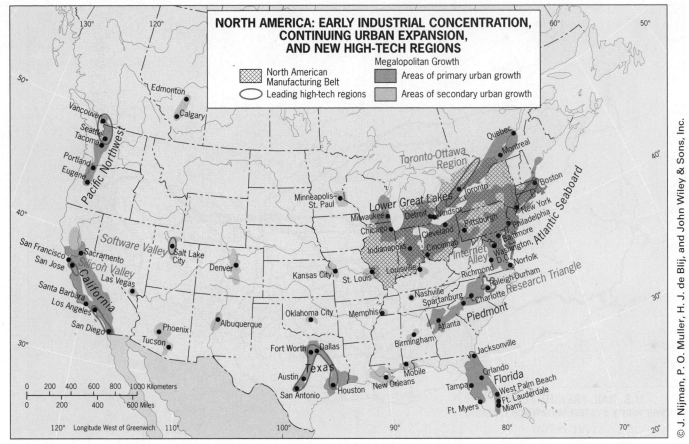

FIGURE 1-8

As new technologies and innovations emerged, and as specializations such as Detroit's automobile industry strengthened, an **American Manufacturing Belt** **7** (that extended into southern Ontario) evolved into the foundation of a North American Core (Fig. 1-8). This core area, on the way to becoming the world's most productive and important, contained the majority of the realm's industrial activity and leading cities, including New York, Chicago, Toronto, and Pittsburgh (the 'steel city').

In the course of interlinked urbanization and industrialization, the spatial economy of North America underwent profound changes. The ***primary*** sector, involving the extraction of raw materials from nature (agriculture, mining, fishing), was rapidly mechanized, and the workforce in this sector shrank considerably. Employment in the ***secondary*** sector, using the input of raw materials and manufacturing them into finished products, grew rapidly. The ***tertiary*** sector, entailing all kinds of services to support production and consumption (banking, retail, transport), expanded as well. Both the secondary and tertiary sectors were overwhelmingly concentrated in cities.

Realm of Railroads

The United States first became a single, integrated, continental-scale economy when long-distance railroads were built during the middle decades of the nineteenth century. This rail network was launched in the Northeast and soon expanded into and across the Midwest, the realm's industrial heartland (Fig. 1-9). The realm's flourishing network grew denser in the first half of the twentieth century, but then rapidly declined between 1960 and 1980. Railroading subsequently entered a new phase of development in the 1980s, marked by government deregulation and declining shipping costs. Between 1981 and 2012, those costs dropped by more than 50 percent, making the movement of goods by rail in the United States and Canada among the cheapest in the world, half of what it costs in Europe or in Japan. Presently, rail freight in North America accounts for more than one-third of the entire freight market (that also includes trucking, shipping, and air transport), which ranks it among the highest in the world's wealthy countries.

This far-flung rail infrastructure not only requires constant maintenance but also frequent updating (which rail companies mostly pay for). One example is the new Alameda Corridor, a major project involving a 20-mile-long rail cargo express line that links the ports of Long Beach and Los Angeles to the Southern California mainlines near downtown Los Angeles (which from there connects to the national rail system). It consists of a series of bridges and underpasses that allow unimpeded transport, separated from all other traffic. Later in this decade, even more rail capacity will be needed with the 2016 opening of the enlarged Panama Canal. That is bound to translate into a major increase of

FIGURE 1-9

trans-Pacific cargo headed for U.S. ports on the Gulf of Mexico and the Atlantic seaboard, requiring a commensurate increase in overland transfer by rail to get these goods to their destinations.

Deindustrialization and Suburbanization

Cities continued to evolve, along with the structure of the economy and the development of new transportation and communication technologies. The mass introduction of automobiles and the accompanying construction of a large-scale highway system, especially after World War II, had two major effects on cities. First, it resulted in much higher interconnectedness and mobility among cities. Second, it propelled a process of *suburbanization*—the transformation from compact city to widely dispersed metropolis through the evolution of residential suburbia into a complete **outer city** 8 with its own clusters of office businesses and industries, sports and entertainment complexes, and myriad other amenities. As the newly urbanized suburbs increasingly captured major economic activities, many central cities saw their comparative status diminish.

Importantly, this transformation coincided with the **deindustrialization** 9 that devastated American cities after 1970, manifested in the loss of manufacturing jobs as a result of automation and the relocation of production to countries with lower wages. The suburban outer cities now became the primary destination of intrametropolitan migrants and of economic activity and employment in the service sector. Meanwhile, the **central business district (CBD)** 10 of the central city lost its role as the dominant metropolitan core, with many of its functions now reduced to serving the less affluent populations of the central city's close-in neighborhoods. The fate of Detroit (see photo) is often cited as exemplifying this sequence of events, but to a certain degree all major U.S. cities suffered.

The Information Economy and City Regions

Since the 1980s, a number of North American cities have rebounded from the destructive effects of deindustrialization and unemployment in conjunction with the arrival of the *information economy*.

An abandoned house in Detroit's Brush Park neighborhood, barely two miles north of downtown, sits by itself during the summer of 2014. The poster child of a city ravaged by deindustrialization, the Motor City has followed a downward trajectory in population, workforce, and tax base since the 1960s. In 2013, Detroit was forced to file for bankruptcy, by far the biggest American city ever to do so. Some say that Brush Park now possesses 'regrowth potential' thanks to its low real estate values, but to date few developers have taken notice.

and accounting), finance, research and development, and the like. The geography of the information economy is quite different from that of the old manufacturing industries and is still not fully understood. Some of it concentrates in established CBDs such as Midtown Manhattan; other economic activities center around important hubs of the Internet infrastructure to benefit from maximum bandwidth (e.g., North Dallas); and still other activities cluster in suburban areas containing large pools of highly skilled workers, notably Silicon Valley, metropolitan Seattle, and North Carolina's Research Triangle. The largest and best known high-technology clusters in North America are mapped in Figure 1-8.

Northern California's *Silicon Valley*—the world's leading hub of computer research and development as well as the headquarters of the U.S. microprocessor industry—exquisitely illustrates the locational dynamics of this sector of the spatial economy. Proximity to Stanford, a world-class research university; adjacent to cosmopolitan San Francisco; the major local concentration of highly educated and skilled workers; a strong, thriving business culture; ample local investment capital; high-quality housing; and a scenic environment with good weather—all combine to make Silicon Valley the prototype for similar developments now springing up around the globe. Under different names (e.g., *technopolis* is used in Brazil, France, and Japan; *science park* in China, Taiwan, and South Korea), these ultra-modern, campus-like complexes symbolize the digital economy just as the smoke-belching factory embodied the industrial age of the past.

Many northern cities had lost population and experienced 'hollowing out' for nearly two decades, but for a few lucky ones the tide had turned by the end of the twentieth century. Employment in the tertiary sector started to grow rapidly, especially in high technology, producer services (such as consulting, advertising,

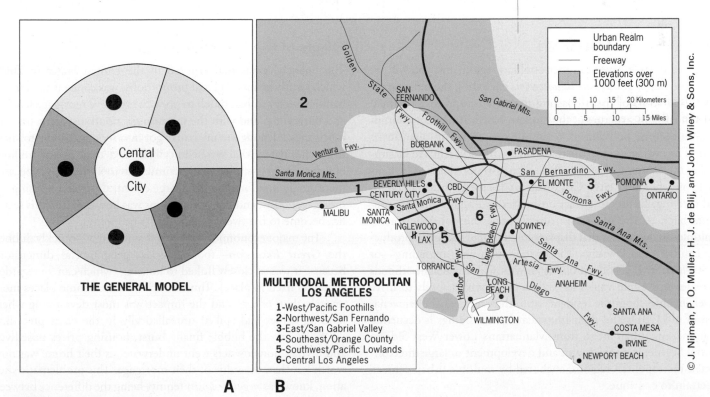

THE GENERAL MODEL

MULTINODAL METROPOLITAN LOS ANGELES

1-West/Pacific Foothills
2-Northwest/San Fernando
3-East/San Gabriel Valley
4-Southeast/Orange County
5-Southwest/Pacific Lowlands
6-Central Los Angeles

A   **B**

FIGURE 1-10

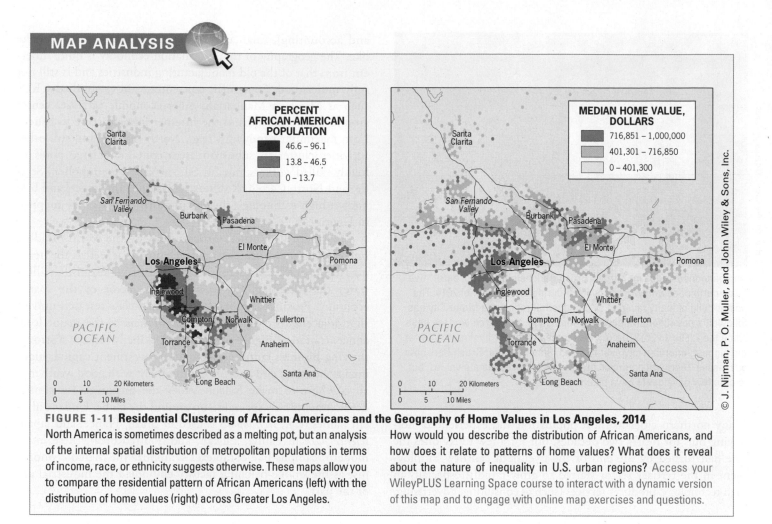

MAP ANALYSIS

FIGURE 1-11 Residential Clustering of African Americans and the Geography of Home Values in Los Angeles, 2014

North America is sometimes described as a melting pot, but an analysis of the internal spatial distribution of metropolitan populations in terms of income, race, or ethnicity suggests otherwise. These maps allow you to compare the residential pattern of African Americans (left) with the distribution of home values (right) across Greater Los Angeles.

How would you describe the distribution of African Americans, and how does it relate to patterns of home values? What does it reveal about the nature of inequality in U.S. urban regions? Access your WileyPLUS Learning Space course to interact with a dynamic version of this map and to engage with online map exercises and questions.

The Polycentric Metropolis

Fly into any large American metropolitan region, and you will see the high-rises of **suburban downtowns** encircling the old central-city CBD, some of them boasting their own impressive skylines. Thus the overall structure of the modern North American metropolis has become polycentric, resembling a pepperoni pizza in its general form (Fig. 1-10). The traditional CBD still tends to be situated at the center, much of its former cross-traffic diverted by beltways; but the outer city's CBD-scale nodes are both ultramodern and thriving. Efforts to attract businesses and higher-income residents back to the old CBD often entail the construction of multiple-use high-rises that displace low-income residents, producing conflicts and lawsuits. The revitalization and upgrading—or gentrification **11**—of crumbling downtown-area neighborhoods increases real estate values as well as taxes, which tends to drive lower-income, longtime residents from their homes. The rejuvenation of old CBDs and their immediate surroundings is occurring in numerous urban areas, from Manhattan's Lower West Side to central Seattle; but the growth and development of large interconnected metropolitan regions, anchored by multiple urban centers, is certain to continue.

Effects of the Great Recession

The financial crisis that erupted in the United States in 2008, which swiftly assumed global proportions, was concentrated in the banking sector; it had much to do with excessive mortgage lending and borrowing, and with the increasingly nontransparent trade in mortgages. Although many banks got into serious trouble for having accumulated what were clinically labeled 'toxic assets,' millions of American households were victimized as their mortgages became increasingly unaffordable. To a large extent, this was a housing (or mortgage) crisis, and one of its lessons was how essential this sector has become to the twenty-first-century economy.

The major economic downturn that resulted—quickly dubbed the **Great Recession**—took on clear geographical dimensions because it was so closely linked to housing (which can be regarded as capital fixed in place). Thus the crisis affected some places more severely than others, and the impact was most devastating where housing values had spiked unrealistically in the years preceding 2008. When the bubble finally burst, housing prices nosedived and myriad homeowners went underwater as their home was now worth less than the value of their mortgage. This problematic situation, known as *negative equity* (equity being the difference between market value and mortgage balance), rendered homeowners less affluent—now they could not afford to sell because paying off the bank would result in a significant loss.

Changing Geographies of Economic Opportunity

Two major legacies of the Great Recession are the exacerbation of growing economic inequality and the reshaping of the map of economic opportunity. In the late 2010s, the longstanding gap in personal income continues to widen at an accelerating rate as the distance between affluence and poverty grows while the ranks of the middle class between them steadily shrinks. In the United States today, the number of families living in middle-income neighborhoods is more than one-third less than it was in 1970; during that same half-century, residential areas housing the wealthy and the poor have doubled in size.

Over the past two decades across North America, the most lucrative employment opportunities have shifted from manufacturing, construction, and other eroding middle-class job sectors into the burgeoning health-care, technology, informatics, and energy sectors that require higher levels of education and training. As a result, the pay-scale spectrum is beginning to resemble the bimodal pattern of the income-based residential fabric—expanding lower and upper extremes at the expense of the contracting middle.

The States with the greatest inequality tend to be those with a strong presence of high-tech and information-based activities such as New York, Connecticut, Massachusetts, Texas, and California. Certain southern States still exhibit historical trends of inequality that date back to the days of the antebellum plantation economy and slavery. And accelerating inequality can be observed at the urban scale, where patterns of residential segregation are shaped not only by income and social class, but also by ethnicity (see Fig. 1-11).

THE MAKING OF A MULTICULTURAL REALM

The Virtues of Mobility and Immigration

Given the mobility of North America's population, the immigrant streams that continue to diversify it, the economic changes affecting it, and the forces acting upon it, the population distribution map (Fig. 1-3) should be viewed as the latest still in a motion picture that has been unreeling for four centuries. Slowly at first, then with accelerating speed after 1800, North Americans pushed their settlement frontier westward to the Pacific. Even today, such shifts continue. Not only does the center of gravity of population continue to move westward, but within the United States it is also shifting southward—the latter a drift that has gained momentum since the 1960s as the advent of universal air conditioning made the so-called **Sunbelt 12** States of the U.S. southern tier ever more attractive to internal migrants.

The current population map is the still-changing product of numerous forces. For centuries, North America has attracted a pulsating influx of immigrants who, in the faster-growing United States, were rapidly assimilated into the societal mainstream. Throughout the realm, people have sorted themselves geographically to maximize their proximity to evolving economic opportunities, and they have shown little resistance to relocating as these opportunities successively favored new locales.

During the past century, such transforming forces have generated a number of major **migrations 13**, of which the still-continuing shift to the West and South is only the latest. Five others were: (1) the persistent growth of metropolitan areas, first triggered by the impact of the late-nineteenth-century Industrial Revolution in North America; (2) the large-scale movement of African Americans from the rural South to the urban North during the latter stages of the industrial era; (3) the shift of tens of millions of urban residents from central cities to suburbs, and subsequently to exurbs even farther away from the urban core; (4) the return migration of millions of African Americans from the deindustrializing North back to the growing opportunities in the South (in metropolises such as Atlanta and Charlotte); and (5) the strong and steady influx of immigrants from outside North America including, in recent times, Mexicans, Cubans, and other Latinos;

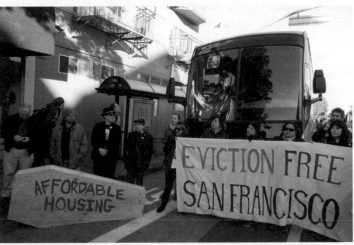

© Steve Rhodes/Demotix / Corbis Images

Protesters in the city of San Francisco during December 2013 block a private luxury bus (illegally using a public bus stop) that ferries affluent Apple employees to their jobs in suburban Silicon Valley. Throughout 2013 and 2014, various groups of residents protested the 'invasion' of their neighborhoods by highly paid professionals from Apple, Google, and other blue-chip technology companies. This swarm of new homebuyers and renters had triggered an especially rapid episode of gentrification, resulting in skyrocketing real estate values and rents—and more than a few evictions. Swift turnovers like these can transform the character of a neighborhood almost overnight—for instance, here generating a demand for a specialized bus service that did not exist a few months earlier. Central San Francisco and New York's Manhattan south of 59th Street are examples of 'hot' downtown residential zones that have prompted urban geographers to predict that considerable wealth will soon be a prerequisite for dwelling in such high-cachet urban locales.

South Asians from India and Pakistan; and East and Southeast Asians from Hong Kong, Vietnam, and the Philippines.

The Challenge of Multiculturalism

Growing diversity, especially at a time of globalization and *transnationalism*, comes with challenges. The sheer dimensions of immigration into the United States, from virtually all corners of the world, creates an increasingly complex ethnic and cultural mosaic that tests **melting pot** 14 assertions at every geographic scale. Today there are more people of African descent living in the United States than in Kenya. The number of Hispanic residents in America is nearing half the population of Mexico. Miami is the second-biggest Cuban city after Havana, and Montreal is the largest French-speaking city in the world after Paris. In short, there are sufficient immigrant numbers in the United States for them to create durable societies within the overall national fabric. The challenge is to ensure that the great majority of the immigrants become full participants in that larger society.

Such are the numbers, and so diverse are the cultural traditions, that the melting pot that America once was is today morphing into something else. Spatially, the United States is now completing its transformation into a *mosaic culture*, an increasingly heterogeneous complex of separate, more or less uniform 'tiles' whose residents spend less time than ever interacting and 'melting.' This new pattern applies not only to new immigrant groups but also to existing communities, underscored by the proliferation of walled-off, gated housing complexes in metropolitan areas across the realm.

In Canada, the melting-pot notion is put to the test with large numbers of (mostly affluent) Chinese immigrants arriving in Vancouver and elsewhere who purchase and renovate (or frequently replace) traditional homes in long-stable Vancouver neighborhoods, arousing the ire of numerous locals. Across the realm, multicultural tendencies now vary on a regional basis, depending on the subtle interplay of economics, existing cultural patterns, immigration, and local politics.

REGIONALISM IN CANADA: DIVISIVE FORCES

Canada's Spatial Structure

Territorially the second-largest state in the world after Russia, Canada is administratively divided into only 13 subnational entities—10 provinces and 3 territories (the United States has 50). The provinces—where 99.7 percent of all Canadians live—range in dimensions from tiny, Delaware-sized Prince Edward Island to Quebec, nearly twice the size of Texas (Fig. 1-12).

As in the United States, the smallest provinces lie in the northeast. Canadians call these four the Atlantic Provinces (or simply Atlantic Canada): Prince Edward Island, Nova Scotia, New Brunswick, and the mainland-island province named Newfoundland

and Labrador. Gigantic, mainly *Francophone* (French-speaking) Quebec and populous, heavily urbanized Ontario, both flanking Hudson Bay, form the heart of the country. Most of western Canada (which is what Canadians call everything west of Ontario) is organized into the three Prairie Provinces: Manitoba, Saskatchewan, and Alberta. In the Far West, beyond the Canadian Rockies and extending to and along the Pacific Ocean, lies the tenth province, British Columbia.

Of the three territories in Canada's Arctic North, the Yukon is the smallest—but nothing is small in Arctic Canada. With the Northwest Territories and recently created Nunavut, the three territories cover almost 40 percent of Canada's total area. The population, however, is extremely small: only about 115,000 people inhabit this vast, frigid frontier zone.

The case of Nunavut merits special mention. Created in 1999, this newest Territory is the outcome of a major aboriginal land claim agreement between the *Inuit* people (formerly called Eskimos) and the federal government, and encompasses all of Canada's eastern Arctic as far north as Ellesmere Island (Fig. 1-12), an area far larger than any other province or territory. With around one-fifth of Canada's total area, Nunavut—which means 'our land'—contains only about 37,000 residents, of whom 80 percent are Inuit.

Canada's population is clustered in a discontinuous ribbon, roughly 300 kilometers (200 mi) wide, nestled against the U.S. border and along ocean shores, whereas most of the country's energy reserves lie far to the north of this corridor. As such, the map of Canada's human settlement (Figs. 1-3 and 1-12) bears out this generalization and emphasizes the significance of environment. People choose to live in the more agreeable areas to the south, but territorial control of abundant northern resources is highly important. This population pattern creates cross-border affinities with major American cities in several places (Toronto-Buffalo, Windsor-Detroit, Vancouver-Seattle), although very little of this occurs in the three Prairie Provinces of the Canadian West, which for the most part adjoin sparsely settled North Dakota and Montana.

Cultural Contrasts

Canada's capital, Ottawa, is symbolically located astride the Ottawa River, which marks the boundary between English-speaking Ontario and French-speaking Quebec. Though comparatively small, Canada's population is markedly divided by culture and tradition, and this division has a pronounced regional expression. Linguistically, 75 percent of Canada's citizens speak English as their mother tongue, and 23 percent speak French (18 percent of the population claim to be bilingual); the remaining 2 percent speak neither English nor French. The spatial concentration of more than 90 percent of the country's Francophones in Quebec accentuates this division (Fig. 1-12), and the province overwhelmingly continues to function as the historic, traditional, and emotional focus of French culture in Canada.

Over the past half-century, a strong nationalist movement has emerged in Quebec, and at times it has demanded outright

FIGURE 1-12

separation from the rest of Canada. This ethnolinguistic division continues to be the litmus test for Canada's federal system, and the issue continues to pose a latent threat to the country's future national unity. The historical roots of the matter are related to French-British competition since the seventeenth century in which Britain gained the upper hand. To this day, the British monarch remains the official head of state (as Canada remains part of the British Commonwealth), represented by the governor general. It is one of the issues that riles Quebec nationalists.

There does seem to be an overall decline in support for Quebec's independence, however. This is partly due to a number of laws that have been enacted to reassure the primacy of Québécois culture in the province (a 2006 federal parliamentary resolution effectively recognizes the Québécois as a *distinct nation*). In addition, there appears to be a growing realization among the French-speaking people that continued integration within Canada combined with a high degree of autonomy is the best possible option. But the struggle for a maximum degree of autonomy continues, along with accompanying political and ethnic tensions.

The Ascendancy of Indigenous Peoples

On the wider Canadian scene, the culture-based Quebec struggle has stirred ethnic consciousness among the country's 1.4 million native peoples—dominated by **First Nations** **15**, but also including Inuit and Métis. They, too, have received a sympathetic hearing in Ottawa, and in recent years breakthroughs have been achieved with the creation of Nunavut as well as local treaties for limited self-government in northern British Columbia.

A foremost concern among these peoples—who constitute 3.9 percent of Canada's population—is that their aboriginal rights be protected by the federal government against the provinces of which they are a part. This is especially true for the largest First Nation of Quebec, the Cree, whose historic domain covers the northern half of the province. As Figure 1-12 shows, the territory of the Cree (north of the red dashed line) is no unproductive wilderness: it contains the James Bay Hydroelectric Project, a massive scheme of dikes and dams that has transformed much of northwestern Quebec and generates hydroelectric power for a huge market within and outside the province.

DISTRIBUTION OF MAJOR RACIAL/ETHNIC GROUPS, CONTERMINOUS UNITED STATES, 2010

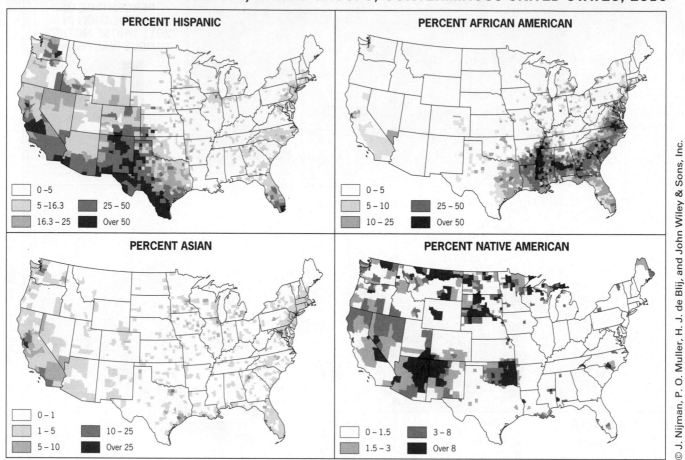

FIGURE 1-13

REGIONALISM AND ETHNICITY IN THE UNITED STATES

In Canada, indigenous peoples were able to hold on to sparsely populated areas in the north. In contrast, in the United States, the relentless push westward by European settlers and the U.S. government in effect dispossessed—and decimated—the original inhabitants. In the end, they were confined to some 300 reservations located mainly in the western half of the country (Fig. 1-13, lower-right map). As noted earlier, there are fewer than 3 million Native Americans left today, with the Navajo and Cherokee being the largest surviving nations. Whereas in Canada First Nations hold significant territorial power, in the United States Native Americans are in a far weaker position to challenge the federal government.

Within U.S. society, more so than in Canada, the social fabric and ethnic relations are determined by past and present waves of immigration. In certain cases, immigration has helped forge a particular regional geography. This was especially true in the case of the hundreds of thousands of Africans who were forcibly brought to the United States in bondage from the seventeenth to the mid-nineteenth centuries and put to work as slaves on plantations in the American South.

Slavery was abolished in 1865, but racial segregation persisted in a variety of legal forms until the civil rights movement crested in the 1960s. In violation of the Constitution, a wide range of State and local Jim Crow laws were enacted to codify the repression of African Americans. This legacy is still visible in the present-day geography of race in the United States, with blacks regionally concentrated in the Southeast, from eastern Texas to Maryland (Fig. 1-13, upper-right map). We noted earlier that substantial numbers of blacks migrated north to work in the industrial centers of the American Manufacturing Belt during the second quarter of the twentieth century; major African American population clusters persist in those northern cities, but they are not clearly apparent at the scale of this national map.

Race continues to be a sensitive issue in the U.S., more so than in Canada, because of this terrible history, enduring rifts between blacks and whites, continuing deprivation among African Americans, and the recurrent racial injustices impeding the nation's evolution into a truly postracial society. Emblematic of this challenged transition were the violent 2014–2015 confrontations that erupted first in the St. Louis suburb of Ferguson after the fatal shooting of an unarmed black teenager by a white police officer (who was later exonerated), and then inner-city Baltimore where another unarmed black died violently while in police custody.

Another highly visible regional expression of ethnic migration is the large concentration of Hispanics in the southwestern United States (Fig. 1-13, upper-left map; see also Fig. 1-2). Since 1980, the number of Hispanics in the United States has more than tripled to surpass 55 million, forming the nation's biggest ethnic minority—and more than one-sixth of the country's total population. For some Anglophone Americans who fear the erosion of the status of English in their communities, language remains an emotional issue. Indeed, Hispanics are now dispersing and increasing their presence in many more parts of the United States, ranging from the rural areas of the Southeast and Midwest to thousands of urban communities from Massachusetts to Washington State to the small towns that still remain on the Great Plains.

Finally, the Asian immigrant population, whose numbers have risen over the past few decades, is geographically the most agglomerated of the leading ethnic minorities (Fig. 1-13, lower-left map). Ethnic Chinese remain the most numerous among the nearly 20 million Americans who identify themselves as Asian, but this diverse minority also includes Japanese, Filipinos, Koreans, and Vietnamese as well as a sizeable cohort of South Asian Indians. Whatever their origins, the great majority have remained in California, where many have achieved economic success and upward mobility.

REGIONS OF NORTH AMERICA

REGIONS OF NORTH AMERICA

Mountain ranges, coastal zones, deserts, low-lying plains, and a range of climatic environments all demand different kinds of human adaptation. Successive waves of immigrants brought with them a variety of cultures; sometimes the differences were subtle, but sometimes they were quite stark. As Figure 1-14 shows, a majority of this realm's regions include both countries regardless of the international boundary. We can regionalize North America's geography by combining functional and formal principles. The resulting map of nine regions is by no means the only solution, but for our general purposes these boundaries are the most pertinent, and the regional distinctions they facilitate will be helpful in understanding the overall fabric of the realm.

THE NORTH AMERICAN CORE (1)

The historic Core region of the North American realm integrates what have long been the most prominent parts of the United States and Canada (Fig. 1-14). They contain the largest cities and the federal capitals of both countries, the leading financial markets, the largest number of corporate headquarters, dominant media centers, prestigious universities, cutting-edge research complexes, and the busiest airports and intercity expressways. Moreover, both the U.S. and Canadian portions of the North American Core still contain roughly one-third of their respective national populations. Political and business decisions, investments, and many other commitments made here affect not just North America but the world at large.

Nonetheless, the dominance of the historic Core has been declining. As was true of the American Manufacturing Belt—the region with which the Core coincides—it once commanded unmatched supremacy. But deindustrialization and the subsequent emergence of the information economy have eroded that dominance as competitors to the south and west continue to siphon away some of its key functions. An example is the State of Michigan, which is experiencing long-term economic decline as jobs in the automobile industry have steadily disappeared over the past few decades. In fact, household incomes in Michigan fell by more than 20 percent just between 2000 and 2009, the most precipitous drop in the nation. Second and third on this unenviable list were two other States of the old Manufacturing Belt: Michigan's southern neighbors, Indiana and Ohio.

THE MARITIME NORTHEAST (2)

When you travel north from Boston into New Hampshire, Vermont, and Maine, you move across one of North America's historic culture hearths whose identity has remained strong for four centuries. Even though Massachusetts, Connecticut, and Rhode Island—traditional New England States—share these qualities, they have become part of the Core, so that on a functional as well as formal basis, the Maritime Northeast extends from the northern border of Massachusetts to Newfoundland, thereby incorporating all four Canadian Atlantic Provinces (Fig. 1-14).

Small fishing ports still dot the New England coast: Chatham, Massachusetts, at the elbow of Cape Cod.

© H. J. de Blij

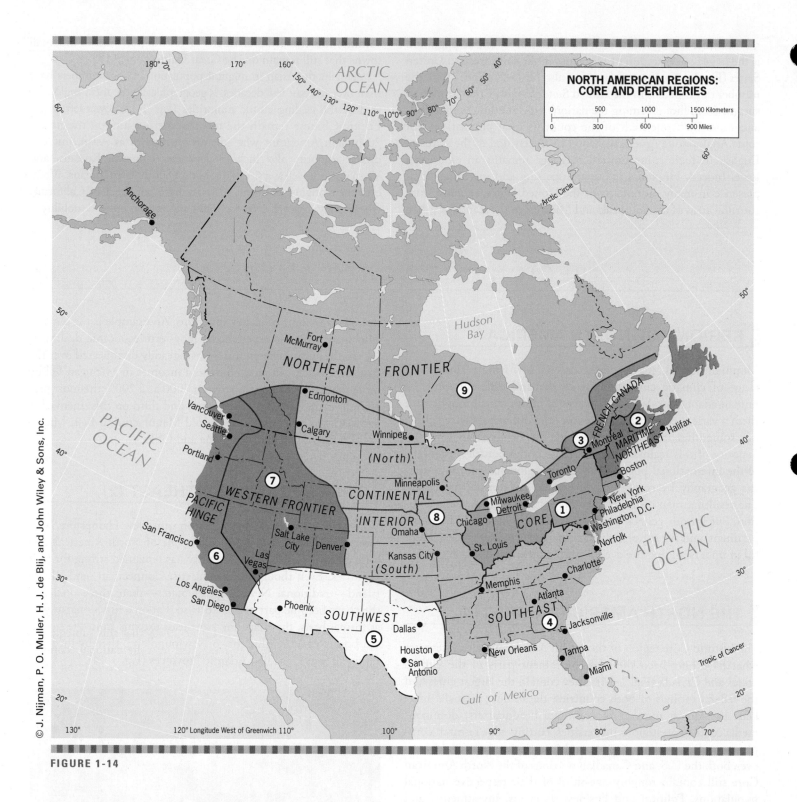

FIGURE 1-14

Difficult environments, a maritime orientation, limited resources, and a persistent rural character have combined to slow economic development here. Primary industries—fishing, logging, and some farming—are mainstays, although recreation and tourism have boosted the regional economy in recent decades. Upper New England experiences the spillover effect of metropolitan Boston's prosperity to some degree, but not enough to ensure steady and sustained economic growth. Atlantic Canada has endured economic hard times as well in the recent past. For example, overfishing has seriously depleted fish stocks. Alternate opportunities focus on this subregion's spectacular scenery and the tourism it attracts. Economic prospects have also been boosted in Canada's poorest province, Newfoundland and Labrador, by the discovery of significant offshore oil reserves. Production is now well underway,

An unmistakably French cultural landscape in North America: the Rue Saint-Louis in Quebec City, the capital of Quebec Province. Not an English sign in sight in this hearth of French/Québécois culture. The green spires in the background belong to Le Chateau Frontenac, a grand hotel built on the site of old Fort St. Louis (the Count of Frontenac was a prominent governor of what was once known as New France). French-speaking, Roman Catholic Quebec forms the heartland of French Canada.

and in the 2010s additional energy deposits have been discovered both on and around Newfoundland Island.

FRENCH CANADA (3)

Francophone Canada constitutes the inhabited southern portion of Quebec, focused on the St. Lawrence Valley from near Montreal northeastward to the river's mouth (Fig. 1-14). It also includes the French-speaking Acadians who reside in neighboring New Brunswick (Fig. 1-12). The French cultural imprint on the region's cities and towns is matched in farming zones by the narrow, rectangular *long lots* perpendicular to the river, also of French origin.

The economy of French Canada was historically less industrial and more rural than the adjacent Core region (and Toronto far outranks Montreal here). Nevertheless, Montreal's economic role has advanced since 2000, with information technology, telecommunications, and biopharmaceuticals all growing in importance. In some ways, French Canada is a richly endowed part of North America. Northern Quebec produces so much hydroelectricity that it exports the surplus to Ontario and to the northeastern United States. Most of this power is generated via the James Bay Project, a series of dams and power stations on the eastern side of Hudson Bay along the La Grande River (Fig. 1-12). In fact, Quebec is faring so well in the energy-production sphere that the recent discovery of natural gas in the St. Lawrence Basin was met with tempered enthusiasm. Many Québécois feel they have no need for gas: unlike hydropower, it is a nonrenewable resource that would boost carbon emissions and the chances of widespread water pollution.

THE SOUTHEAST (4)

For more than a century following the U.S. Civil War (1861–1865), the Southeast (Fig. 1-14) remained in economic stagnation and cultural isolation from the rest of the country. During the 1970s, however, things changed so fast that a New South was born almost overnight. The Sunbelt migration stream drove people and enterprises into the long-dormant cities. Core-region companies looking for headquarters or subsidiary offices found Atlanta, Charlotte, Tampa, Miami-Fort Lauderdale, and other urban areas both economical and attractive, swiftly turning them into boomtowns. Racial segregation had been dismantled in the wake of the civil rights movement. A new social order was matched by new facilities ranging from airports (Atlanta's quickly became one of the world's busiest) to theme parks open to all. Soon the nation was watching Atlanta-based CNN on television, vacationing at Orlando's Walt Disney World, and monitoring space flights that originated at Cape Canaveral.

The geography of development in the South, however, is rather uneven. Although many cities and certain agricultural areas have benefited, others have not; the South contains some of the realm's poorest rural areas, and the widening gap between rich and poor here is greater than in any other region in the realm. The 'Silicon Alley' corridor of Northern Virginia's Washington suburbs, central North Carolina's Research Triangle, and metropolitan Atlanta's corporate campuses are locales that typify the New South; but Appalachia and rural Mississippi still represent the Old South, where depressed farming areas and stagnant small industries inhibit change.

Another, relatively small part of this region warrants special attention. South Florida used to be a particularly remote and nearly unpopulated corner of the United States. Before the twentieth century, it was too far from the northeastern Core of the country (and even from closer urban centers such as Atlanta) to warrant development; the heat, humidity, and mosquitoes posed a formidable challenge to prospective settlers; and there were no natural resources to exploit. Indeed, it was not until the 1960s that this subregion of the Southeast became fully integrated into the national economy and urban system—thanks to the widespread introduction of air conditioning and affordable air travel, not to mention the mass immigration of Cubans fleeing Fidel Castro's communist regime. Today, Miami has evolved into a **world-city 16**, a globally linked, southward-oriented conduit between North America and the rest of the hemisphere (Fig. 1-15).

THE SOUTHWEST (5)

As Figure 1-14 shows, by our definition the Southwest begins in East Texas and extends all the way to the eastern edge of Southern California. New Mexico, in the middle of the Southwest region, is the least developed economically and ranks low on many U.S. social indices, but its environmental and cultural attractions benefit Albuquerque and Santa Fe, with Los Alamos a famous name in

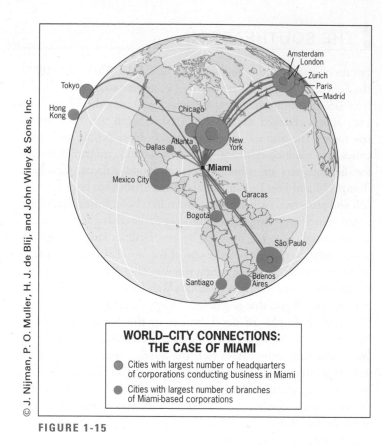

© J. Nijman, P. O. Muller, H. J. de Blij, and John Wiley & Sons, Inc.

**WORLD–CITY CONNECTIONS:
THE CASE OF MIAMI**

● Cities with largest number of headquarters
   of corporations conducting business in Miami

● Cities with largest number of branches
   of Miami-based corporations

FIGURE 1-15

research-and-development circles. To the west, Arizona's technologically transformed desert now harbors two large, coalescing metropolises—Phoenix and Tucson. Growth continues, but rising concerns about climate change add to the uncertainties of maintaining water supplies already stretched to the limit.

Texas leads this region in most respects. Its economy, once all but fully dependent on oil and natural gas, has been restructured

and diversified so that today the Dallas-Fort Worth–Houston–Austin-San Antonio triangle has become one of the world's most productive **technopoles 17**—state-of-the-art, high-technology industrial complexes—especially Austin at the heart of it. This subregion is also a hub of international trade and the northern anchor of a North American Free Trade Agreement (NAFTA)-generated transnational growth corridor that extends into Mexico as far south as Monterrey.

The huge State of Texas, larger than France, is surpassed in areal size only by Alaska and is also the second most populous U.S. State after California (27 and 39 million, respectively). Parts of the State are quite affluent, with much of its wealth still emanating from the energy sector: Texas claims about a quarter of all U.S. oil reserves and nearly a third of its natural gas deposits. The Lone Star State has also become a leader in the generation of wind energy, with wind turbines increasingly dotting the sparsely settled landscape of immense West Texas.

THE PACIFIC HINGE (6)

The Southwest meets the Far West near metropolitan San Diego on California's border with Mexico, and from there the region we call the Pacific Hinge extends all the way north into Canada where Vancouver forms the northern anchor in southwestern British Columbia (Fig. 1-14). We include almost all of California in this region, but (for environmental as well as economic reasons) only the western portions of Oregon and Washington State. This is a particularly important part of North America and increasingly serves as a counterweight to the historic Core in the U.S. Northeast.

The Pacific Hinge is a major economic region not just in this realm but in the global economy too. It includes such leading

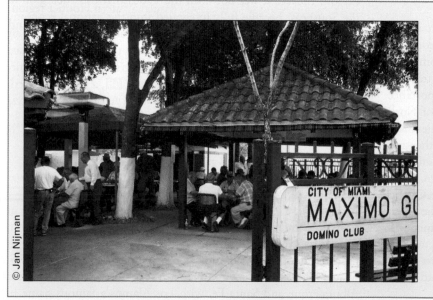

© Jan Nijman

@ FROM THE FIELD NOTES...

"Domino Park lies at the center of Little Havana, perhaps the most iconic neighborhood in Greater Miami. After Fidel Castro came to power in 1959, thousands of Cubans fled the nearby island and resettled here. Today, over one million Cubans reside in the metropolitan area, the biggest concentration of any single ethnic group in the United States. In fact, Miami would easily qualify as the second-largest city in Cuba. Domino Park is situated on Southwest 8th Street, known locally as 'Calle Ocho.' It is only a short walk from the Bay of Pigs Monument that commemorates the fallen soldiers of the failed invasion of Cuba by exiles in 1961. Taking a stroll along Calle Ocho, you can stop at a cafe for a cortadito, watch cigars being made by hand, or get fitted for a guayabera shirt. Domino Park recently opened to women (!), but it is mostly older men who come here to play games, talk politics, and meet with friends—just as they would in Cuba."

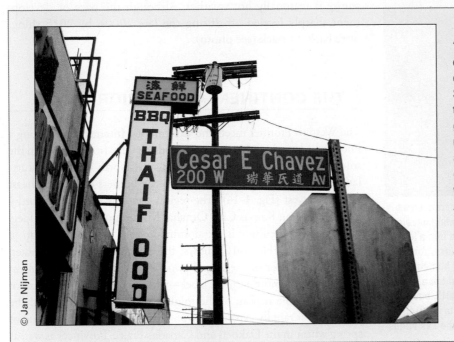

@ FROM THE FIELD NOTES...

"On this street corner in central Los Angeles, four different cultures converge. Los Angeles is one of the most ethnically diverse metropolitan areas in the United States and in the world. More than a third of the population was born abroad, and more than half speak a language other than (or in addition to) English. The largest non-native ethnic group are Hispanics (mainly Mexicans and Central Americans), but there also are many Asians, especially Chinese and Koreans. Cesar E. Chavez, after whom this avenue is named, was a charismatic Mexican-American farm worker turned union organizer and became nationally known as a leading civil rights activist. His birthday, March 31, is a State holiday in both California and Texas—and an optional holiday in Arizona (!). Chavez Avenue is a major east-west thoroughfare just north of downtown Los Angeles, and even a short stroll along it offers a sampling of global cultures."

cities as Los Angeles, San Francisco, San Diego, Portland, and Seattle as well as America's most populous State: in fact, California's economy ranks among the world's ten largest *by country*. This region also encompasses one of the realm's most productive agricultural areas in California's Central Valley, magnificent scenery, and a culturally diverse population drawn by its long-term economic growth and pleasant living conditions.

Our regional definition is mainly based on economic considerations and intensifying trade connections across the Pacific Ocean. This part of North America is deeply involved in the development of countries in eastern Asia. In the postwar era, Japan's success had a salutary impact here, but what has since taken place in China, South Korea, Taiwan, Singapore, and other Asian-Pacific economies has created unprecedented opportunities. The term **Pacific Rim 18** has come into use to describe the discontinuous regions surrounding the great Pacific Ocean that have experienced spectacular economic growth and progress over the past four decades: not only coastal China and various parts of East and Southeast Asia, but also Australia, South America's Chile, and the western shores of Canada and the conterminous United States. That is why in this chapter we speak of the *Pacific Hinge* of North America, which is now the key interface between the realm and the Rim.

The Pacific Rim is therefore a classic example of a ***functional region***, with economic activity in the form of capital flows, raw-material movements, and trading linkages that generate urbanization, industrialization, and labor migration. As part of this process, human landscapes from Sydney to Santiago are being transformed within a 32,000-kilometer (20,000-mi) corridor that rings the globe's largest body of water.

THE WESTERN FRONTIER (7)

Where the forests of the Pacific Hinge yield to the scrub of the rain shadow on the inland slopes of the mountain wall that parallels the west coast, and the terrain turns into intermontane ('between the mountains') basins and plateaus, lies a region stretching eastward from the Sierra Nevada and Cascades to the Rockies, encompassing segments of southern Alberta and British Columbia, eastern Washington State and Oregon, all of Nevada, Utah, and Idaho plus the western components of Montana, Wyoming, and Colorado. As Figure 1-14 shows, Edmonton, Calgary, and Denver are situated along the inland edge of this aptly named Western Frontier region, and Salt Lake City, anchoring the Software Valley technopole and symbolizing the new high-tech era, lies at the heart of it.

Remoteness, dryness, and sparse population typified this region for a very long time. It became the redoubt of the Mormon faith and a place known for boom-and-bust cycles of mining, logging, and, where possible, livestock raising. However, in recent decades advances in communications and transportation technologies, sunny climates, wide-open spaces, lower costs of living, and growing job opportunities have combined to form effective pull factors for myriad outsiders. Every time an earthquake struck in California, eastward migration got a boost. And during the most recent ten-year period, millions of Californians emigrated from the overcrowded and overpriced Pacific coast to the inland high desert.

Development in the Western Frontier centers on its widely dispersed urban areas, which are making this one of the realm's fastest growing regions (albeit from a low base). Here thousands of

Ethan Miller/Getty Images, Inc.

Between 1990 and 2000, this Las Vegas suburb of Henderson ranked as the fastest-growing municipality in the United States. Sprawling development on a massive scale converted this patch of high desert into a low-density urban checkerboard whose expansion seemed endless. Yet by 2009, the growth cycle had ground to a halt as thousands of homeowners were forced into foreclosure, unemployment was rising, and the national economic crisis was clouding the immediate future. Since 2013, however, the city has reemerged (attracting a record 41 million tourists in 2014), suburban growth has been restarted, and the prospects of the Las Vegas Valley have been steadily brightening.

new high-technology manufacturing and specialized service jobs have propelled a two-decade-long influx, slowed but not reversed by the Great Recession. Metropolitan Las Vegas typifies this accelerated growth trajectory. Far more than a gambling and entertainment magnet that draws some 40 million visitors annually, Las Vegas attracts in-migrants because of its history of job creation,

relatively cheap land and low-cost housing, moderate taxes, and sunny if seasonally hot weather. Although the pace of development sharply leveled off during the late 2000s, by 2015 things were back on track (see photo).

THE CONTINENTAL INTERIOR (8)

The Western Frontier meets the Continental Interior region along a line marked by cities that influence both regions, from Denver in the south to Edmonton in the north. This vast North American heartland extends from interior Canada to the borders of the Southeast and Southwest (Fig. 1-14). The region contains many noteworthy cities, including Kansas City, Omaha, Minneapolis, and Winnipeg; nonetheless, agriculture is the dominant story here. As the map of U.S. farmlands (Fig. 1-16) shows, this is the nation's breadbasket. This is the land of the beef- and pork-producing Corn Belt—*Meat Belt* would be a more accurate label—which extends into the western half of the Core region; of the mighty soybean, America's and the world's most rapidly expanding crop of the past half-century; and of spring wheat in the Dakotas and Canada's Prairie Provinces as well as winter wheat in Kansas. Indeed, the cities and towns of this region share histories of food processing, packing, and marketing.

This is also the scene of the struggle for survival of the family farm as the unrelenting incursion of large-scale corporate farming threatens a longstanding way of life. Even a cursory look at demographic statistics makes it clear why many States in the Continental Interior are losing population or gaining far fewer than the national average. The Great Plains, the western component of this region, is especially hard hit: younger people and those better off are leaving, whereas older and less affluent residents stay behind. Villages and small towns continue to die, and the notion of abandoning certain areas to return them to their natural state is being seriously discussed.

But things are not the same everywhere. The Continental Interior may not have as prominent a presence on the national scene as New York or California, and the region is not as wealthy, but with abundant coal and oil deposits (in Wyoming and North Dakota, respectively) as well as South Dakota's booming financial sector, unemployment rates here rank among the lowest in the nation. Buoyed by an economy largely driven by agriculture and mining, and with fossil fuel production a leading growth industry, the effects of the Great Recession were minimal and short-lived in the Continental Interior.

THE NORTHERN FRONTIER (9)

Figure 1-14 leaves no doubt as to the dimensions of this final North American region: it is by far the largest of the realm, covering just about 90 percent of Canada and all of the biggest U.S. State, Alaska. Not only does this region include the northern parts of seven of Canada's provinces, but it also comprises the Yukon and

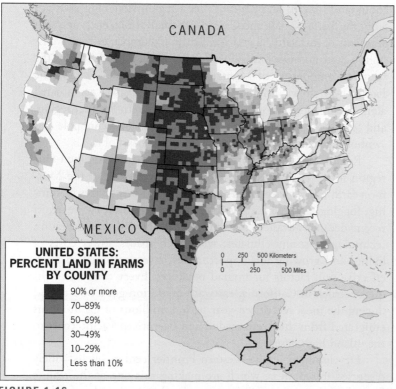

UNITED STATES: PERCENT LAND IN FARMS BY COUNTY

- 90% or more
- 70–89%
- 50–69%
- 30–49%
- 10–29%
- Less than 10%

© J. Nijman, P. O. Muller, and John Wiley & Sons, Inc.

FIGURE 1-16

Northwest Territories as well as recently established Nunavut. The very sparsely populated Northern Frontier remains a land of isolated settlement based on the exploitation of newly discovered resources. The Canadian Shield, underlying most of the eastern half of the region, is a rich storehouse of mineral resources, including metallic ores such as nickel, uranium, copper, gold, silver, lead, and zinc. The Yukon and Northwest Territories have proven especially bountiful, with the mining of gold and especially diamonds (Canada currently ranks third in the world); at the opposite, eastern edge of the Shield, near Voisey's Bay on the central coast of Labrador, the largest body of high-grade nickel ore ever discovered is now open for extraction.

The Province of Alberta has experienced one of the most spectacular economic booms in the realm's history. Recent estimates indicate that Alberta's oil reserves, dominated by the Athabasca **tar sands** 19 (see Fig. 1-17), rank third in the world behind only those of Venezuela and Saudi Arabia. Alberta has become Canada's fastest-growing province, and Calgary its fastest-expanding major city. The boom makes Calgary a magnet for business and highly skilled workers; the city's skyline is in spectacular transformation, comparable to what has recently occurred on the Arabian Peninsula. But much depends on the price of oil, which has fluctuated considerably in the past few years.

Productive locations in the Northern Frontier constitute a far-flung network of mines, oil and gas installations, pulp mills, and hydropower stations that have spawned hundreds of small settlements as well as thousands of kilometers of interconnecting transportation and communications lines. Inevitably, these activities have infringed on the lands of indigenous peoples without the preparation of treaties or agreements, which has led to recent negotiations between the government and leaders of First Nations over resource development in the Northern Frontier. But essentially this gigantic region remains a frontier in the truest sense of that term.

Alaska's regional geography differs from the rest of the Northern Frontier in that the State contains several urban settlements and an incipient core area in the coastal zone around Anchorage. The population of Alaska (750,000) accounts for more than one-third of the region's total. Here also lies the only metropolis of any size, Anchorage (population 300,000), a key node on the aviation network that links the realm to Pacific Asia. The State's internal communications (air and surface) are also better developed.

Climate change is likely to affect the Northern Frontier in as-yet-unforeseeable ways. As the map shows, this region extends all the way north to the Arctic Ocean; it therefore includes waters that, with continued global warming and associated contraction of the Arctic ice cap, will open (possibly year-round) nearshore waterways that would dramatically alter global shipping routes and intercontinental distances—a topic discussed in some detail in Chapter 12. The question of ownership of these waters—and the submerged seafloor beneath them, coming within reach of exploitation—will be a matter of significant international contention in the decades ahead.

In 2010, Canada solidified its reputation as an environmentally conscious nation through the signing of an unprecedented agreement. The Canadian Boreal Forest Agreement involves

BOREAL FOREST OF THE NORTHERN FRONTIER

Boreal forest	—— Suspended timber harvest
Boreal caribou range	—— Athabasca Tar Sands

0   250   500   750   1000 Kilometers
0   125   250   375   500 Miles

80° Longitude West of Greenwich 70°

FIGURE 1-17

© J. Nijman, P. O. Muller, H. J. de Blij, and John Wiley & Sons, Inc.

21 logging firms and 9 environmental organizations as well as the Canadian government. As shown in Figure 1-17, it imposes a moratorium on logging and any other development within a huge zone of 290,000 square kilometers (112,000 sq mi). Greenpeace, which originated in Canada, has called it the biggest agreement so far that preserves nature. The word 'boreal' is derived from the Latin word for north; **boreal forests** 20—identical to the Russian *taiga* (snowforest) discussed in Chapter 5—are dominated by dense stands of coniferous (cone-bearing) needleleaf trees such as spruce, fir, and pine. The boreal forests of the Northern Frontier also constitute the habitat of such endangered wildlife species as the caribou, lynx, American black bears, and wolverines.

POINTS TO PONDER ?

■ Whereas more than 70 percent of Canada's population resides within 200 kilometers (125 mi) of the northern U.S. border, only about 12 percent of the Mexican population lives that close to the southern U.S. border.

■ The United States does not have an official language. Should it?

■ Thanks to the Pacific Rim trade, California's ports are among the busiest in the United States. How will they be affected by the newly enlarged Panama Canal that opens for ships in 2016?

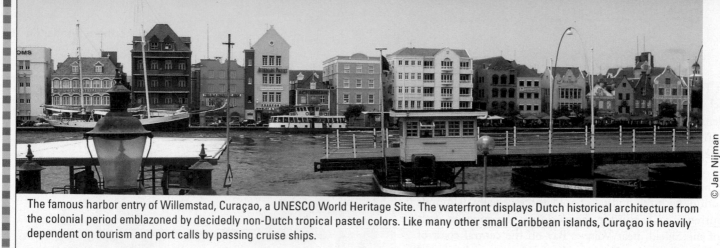

The famous harbor entry of Willemstad, Curaçao, a UNESCO World Heritage Site. The waterfront displays Dutch historical architecture from the colonial period emblazoned by decidedly non-Dutch tropical pastel colors. Like many other small Caribbean islands, Curaçao is heavily dependent on tourism and port calls by passing cruise ships.

© Jan Nijman

IN THIS CHAPTER

- Drug trafficking: Shifting from Mexico to the Caribbean?
- Organized crime and violence in Central America
- The expanded Panama Canal opens for business
- Is small beautiful? The predicament of tiny island-nations

CONCEPTS, IDEAS, AND TERMS

2

THE MIDDLE AMERICAN REALM

FIGURE 2-1 © J. Nijman, P. O. Muller, H. J. de Blij, and John Wiley & Sons, Inc.

Look at a world or hemispheric map, and it is obvious that the Americas comprise two landmasses: North America extending from Alaska to Panama and South America from Colombia to Argentina. But here we are reminded that continents and geographic realms do not necessarily coincide. Between North America and South America lies the small but important geographic realm known as Middle America. Consisting of an impediment-filled mainland corridor and myriad Caribbean islands, Middle America is a highly fragmented realm.

DEFINING THE REALM

From Figure 2-1 it is clear that Middle America is much wider than it is long. The distance from Baja California to Barbados is about 6000 kilometers (3800 mi), but the area from the latitude of Tijuana to Panama City is only half that distance. In terms of total area, Middle America is the second smallest of the world's geographic realms. As the map shows, the dominant state of this realm is Mexico, larger than all its other countries and territories combined.

Middle America may be a small geographic realm by global standards, but comparatively it is densely peopled. Its population passed the 200-million milestone in 2011, more than half of it residing in Mexico alone, and the rate of natural increase (at 1.5 percent) remains above the world average. Figure 2-2 shows Mexico's populous interior core area quite clearly, but note that the population in Guatemala and Nicaragua tends to cluster toward the Pacific rather than the Caribbean coast. Among the islands, this map reveals how crowded Hispaniola (containing Haiti and the Dominican Republic) is, but at this scale we cannot clearly discern the pattern on the smaller islands of the eastern Caribbean and the Bahamas, some of which are also densely populated.

What Middle America lacks in size it makes up in physiographic and cultural diversity. This is a realm of soaring volcanoes and spectacular shorelines, of tropical forests and barren deserts, of windswept plateaus and scenic islands. It holds the architectural and technological legacies of ancient **indigenous** ▮1▮ civilizations. Today it is a mosaic of immigrant cultures from Africa, Europe, and elsewhere, richly reflected in music and the visual arts. Material poverty, however, is endemic: Haiti is often cited as the poorest country in the Americas; Nicaragua, on the mainland, is almost as badly off. As we will discover, a combination of factors has produced a distinctive but challenged realm between North and South America.

Sometimes you will see Middle and South America, together, referred to in combination as 'Latin' America, alluding to their prevailing Spanish-Portuguese heritage. This is an imprecise regional designation, much as Anglo America, a term once commonly used for North America, was also improper. Such culturally based terminologies reflect historic power and dominance, and they tend to make outsiders out of those people they do not represent. Thus you will not find the cultural landscape particularly Latin in the Bahamas, Jamaica, Trinidad, Belize, or lengthy stretches of Guatemala and Mexico.

But is Middle America sufficiently different from either North or South America to merit distinction as a separate realm? Consider this: North America encompasses just a pair of states, and the entire continent of South America has only 12 states (plus France's dependency on the northeast coast). But far smaller Middle America, as we define it, incorporates more than three dozen political entities, including several dependencies (or quasi-dependencies) of the

▮▌ **MAJOR GEOGRAPHIC QUALITIES**

MIDDLE AMERICA

1. Middle America is a relatively small realm consisting of eight mainland countries from Mexico to Panama and all the islands of the Caribbean Basin to the east.

2. Middle America's mainland constitutes a crucial barrier between Atlantic and Pacific waters. In physiographic terms, this is a land bridge connecting the continental landmasses of North and South America.

3. Middle America's cultural geography is complex. Various African and European influences dominate the Caribbean, whereas Spanish and indigenous traditions survive on the mainland.

4. Middle America is a realm of intense cultural and political fragmentation. The presence of many small, insular, and remote countries poses major challenges to economic development.

5. The realm's northern political boundary is in some respects part of a dynamic borderland and transition zone.

6. In terms of area, population, and economic potential, Mexico leads the realm.

7. Many of the countries in this realm find themselves relatively isolated from the rest of the hemisphere.

© J. Nijman, P. O. Muller, H. J. de Blij, and John Wiley & Sons, Inc.

FIGURE 2-2

United Kingdom, the Netherlands, and France as well as a few constituent territories of the United States. Therefore, unlike South America, Middle America is a multilingual patchwork of independent states, territories in political transition, and residual colonial dependencies, with strong continuing ties to the United States and non-Iberian Europe. Middle America is defined in large measure by its vivid multicultural geography.

THE REALM'S NORTHERN BOUNDARY

The 3169-kilometer (1969-mi) land border between North America and Middle America is the longest in the world separating a rich realm from a poor one. The U.S.-Mexican boundary crosses half the continent from the Pacific to the Gulf of Mexico, yet Mexican cultural influences penetrate deeply into the southwestern States and American impacts reach far inside Mexico. To some Mexicans, the border is a constant reminder of territory lost to the United States in historic conflicts; to many Americans, it is a symbol of economic contrasts and illegal immigration. Along the Mexican side, the effects of **NAFTA 2**—the North American Free Trade Agreement between Canada, the United States, and Mexico that went into effect in 1994—have transformed Mexico's economic geography.

Although the two realms are separated by a sharp political (and economic) boundary, this is a good example of a dynamic **borderland 3** marked by considerable interaction, and in some ways the boundary can be more broadly viewed as a transition zone between North and Middle America. The implementation of NAFTA sparked an economic boom as Mexico became part of a free-trade zone and market encompassing some 475 million people. A major beneficiary was the strip of Mexican land that ran along the boundary with the United States. The resulting boom changed urban landscapes all

© David R. Frazier Photolibrary, Inc./Alamy

Automobiles wait at the border to enter the United States from Mexico at the Tijuana-San Ysidro port of entry just south of San Diego—the busiest land border crossing in the world. Here, some 50,000 northbound vehicles and 25,000 pedestrians enter the United States every day. A major expansion of this California facility, located at the southern terminus of Interstate-5, will soon increase its capacity and reduce vehicle waiting times. The number of Mexico-bound inspection lanes doubled from 11 to 22 in 2012; the U.S.-bound lanes will similarly expand from 34 to 63 in 2016.

along the emerging border zone—but it could not, of course, close the massive economic gap between the two sides (see photo).

Under NAFTA, factories based in Mexico can assemble imported, duty-free raw materials and components into finished products, which are then exported back into the U.S. market. Logically, most of these factories, called *maquiladoras* 4, are located as close to the U.S. border as possible. Thus manufacturing employment in the cities and towns along that border, from Tijuana on the Pacific to Matamoros at the mouth of the Rio Grande, expanded rapidly. Mexico now provides roughly 14 percent of all U.S. imports annually, the second-largest source of U.S. imports after China (22 percent). China has been increasing its share more rapidly, even though it faces tariffs that NAFTA-member Mexico does not. This puts pressure on Mexico to improve its competitiveness in order to build an even closer economic relationship with the United States and Canada.

A Troubled Border Zone

For several years, this has also been a troubled border zone in view of illegal border crossings by Mexicans and Central Americans

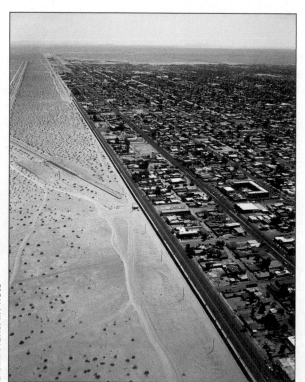

The boundary between Mexico and the United States displays some stunning cross-border contrasts. In this scene, the crowded Mexican town of San Luis Rio Colorado ends at a tall metal fence, beyond which the Arizona desert takes over. This photo, therefore, looks eastward; south is to the right. Like other Mexican towns pressed against the border, San Luis Rio Colorado (locals simply call it San Luis) grew rapidly and chaotically in response to the opening of several *maquiladoras,* assembly plants making products for export to the United States.

© Jim Wark/AirPhoto

looking for a better life in the United States. In 2014, there were an estimated 500,000 illegal crossings, but this number is said to be part of a declining trend since 2010 because of tighter border patrols and the increasing preference of potential migrants to remain in Middle America rather than head north. In Mexico, urbanization continues to rise (from 66 percent of the population in 1980 to 78 percent in 2014), with midsized and smaller cities growing fastest today. Should Mexico succeed in developing and modernizing its economy in a more regionally balanced manner, this population shift away from the countryside may well be sustainable beyond the 2010s.

Another huge concern in the border zone is the violence associated with the cocaine trade and, to a lesser extent, the smuggling of marijuana. Most cocaine is produced in northwestern South America, and the lion's share of what is destined for the U.S. market passes through Mexico (see Fig. 2-12). Much of the marijuana consumed within the United States is grown in western Mexico and similarly finds its way across the border. Contraband narcotics are mainly carried in trucks and cars, often imaginatively disguised as regular cargo (e.g., stuffed in teddy bears, inside hollowed-out furniture, or simply in boxes labeled as anything from candy to engine parts). It is the sheer volume of this illicit cross-border flow that makes it virtually impossible for authorities to conduct sufficient inspections. Unquestionably, the infrastructure of cities and industrial areas on the Mexican side of the border further facilitates the operation of the drug trade.

The city of Ciudad Juárez seems ideally situated to handle this clandestine activity. It is located directly on the international boundary, and together with El Paso, Texas on the U.S. side, forms a single metropolis—even though the two cities are separated by a continuous artificial barrier running across this entire urban area (Fig. 2-3). This map also shows the four ports of entry, which account for about 25 million border crossings every year. Juárez has grown quite rapidly and exhibits numerous manufacturing districts as well as a sprawling mosaic of neighborhoods, ranging from clusters of upper-income **colonias** to the slums that house the poorest of the new migrants from the south. With a plentiful supply of poor people willing to take the risk of driving or carrying drugs across the border, Juarez also has dozens of *maquiladoras* that can be bribed to store, conceal, and transport the contraband north.

In 2006, the U.S. government, responding both to homeland security concerns (potential terrorists may be entering the country via Mexico) and the tidal wave of cocaine smuggling, initiated the construction of a fortified fence along the entire length of the Mexican border. The very idea seems at odds with the ideals of closer economic and political cooperation, and many security experts questioned its efficacy. By 2011, however, with about one-third of the border hardened by fences and walls, the project to build a 'virtual fence' to fill the remaining gaps had become such a boondoggle that it was terminated. Nevertheless, imposing physical barriers are now concentrated around all the main crossing points, and it remains unclear how the longer-term economic health of the border zone they divide will be affected.

© J. Nijman, P. O. Muller, H. J. de Blij, and John Wiley & Sons, Inc.

FIGURE 2-3

PHYSICAL GEOGRAPHY OF MIDDLE AMERICA

A Land Bridge

The funnel-shaped mainland, a 4800-kilometer (3000-mi) connection between North and South America, is wide enough in the north to contain two major mountain chains and a vast interior plateau, but narrows to a slim 65-kilometer (40-mi) ribbon of land in Panama. Here this strip of land—or *isthmus*—bends eastward so that Panama's orientation is east-west. Thus mainland Middle America exemplifies what physical geographers call a **land bridge 5**, an isthmian link between continents.

If you examine a globe, you can see some other present and former land bridges: Egypt's Sinai Peninsula between Asia and Africa, the (now-broken) Bering land bridge between northeast-ernmost Asia and Alaska, and the shallow waters between New Guinea and Australia. Such land bridges, though temporary features in geologic time, have played crucial roles in the dispersal of animals and humans across the planet. But even though mainland Middle America forms a land bridge, its internal fragmentation has always inhibited movement. Mountain ranges, swampy coastlands, and dense rainforests make contact and interaction difficult.

Island Chains

As shown in Figure 2-1, the approximately 7000 islands of the Caribbean Sea stretch in a lengthy arc from Cuba and the Bahamas eastward and then southward to Trinidad, with numerous outliers outside (such as Barbados) and inside (e.g., the Cayman Islands) the main chain. The four large islands in the Caribbean's

east-west segment—Cuba, Hispaniola (containing Haiti and the Dominican Republic), Puerto Rico, and Jamaica—are called the *Greater Antilles*, and all the remaining smaller islands constitute the *Lesser Antilles*. The entire Antillean **archipelago 6** (island chain) consists of the crests and peaks of mountain chains that rise from the floor of the Caribbean, the result of collisions between the Caribbean Plate and its neighbors (Fig. G-4). Some of these crests are relatively stable, but elsewhere they contain active volcanoes, and almost everywhere in this realm earthquakes are an ever-present danger—in the islands as well as on the mainland. Add to this constant hazard the realm's seasonal exposure to Atlantic/Caribbean hurricanes, and it amounts to some of the highest-risk real estate on Earth.

Dangerous Landforms

The danger from below is dramatically illustrated in the Haitian, western half of Hispaniola, which is laced with geologic fault lines associated with the nearby boundary that separates the North American and Caribbean plates (see Fig. 2-16, inset map). On January 12, 2010, Haiti was devastated by a massive earthquake; the epicenter of that 7.0 temblor was located just outside Port-au-Prince and virtually destroyed this teeming, impoverished capital city of 2 million-plus (see photo). At least 300,000 people died, and within a week more than a million had fled to the countryside. It was the worst natural disaster in Haitian history but certainly not the first.

© SWOAN PARKER/Xinhua Press/Corbis

Demonstrators in Port-au-Prince, Haiti's capital, march during a recent protest to demand improvement of their living conditions. Recovery since the catastrophic earthquake of 2010 has been painfully slow. On the fifth anniversary of the disaster in 2015, Haiti continued to be plagued by recurrent outbreaks of cholera and other serious diseases, hundreds of thousands of inadequately sheltered homeless citizens, and the relentlessly crushing effects of a stagnant, still-battered economy. With the rebuilding effort all but stalled, this overwhelmingly vulnerable country was facing its sixth hurricane season since the quake: how much longer could its fortunate string of uneventful storm seasons hold out?

The environmental hazard from above comes in the form of hurricanes, powerful tropical cyclones that annually threaten the Caribbean Basin and its surrounding coastlines. The eastern half of Middle America is one of the most hurricane-prone areas in the world. One of the key conditions for the formation of hurricanes is very warm ocean water, which further heats the hot moist air rising above it to 'fuel' the evolving storm. The prolonged Atlantic/Caribbean hurricane season extends from June 1 to December 1, with the greatest number of tropical cyclones occurring in August and September when seawater reaches its highest temperatures. Most storms travel in a westerly direction from their low-latitude spawning ground off the coast of West Africa, steered across the Atlantic by the trade winds to reach the Caribbean Sea. Once there, many of these cyclones follow similar routes within **Hurricane Alley 7**, whose wide axis lies along all of the Greater Antilles and then broadens to include southern Florida, Mexico's Yucatán Peninsula, and all of the Gulf of Mexico. On average, every season sees the development of four to eight major hurricanes, and rarely does a year go by without a destructive landfall on at least one of these densely populated areas.

Altitudinal Zonation of Environments

Continental Middle America and the western margin of South America are areas of high relief and strong environmental contrasts. Even though settlers have always favored temperate intermontane basins and valleys, people also cluster in hot tropical lowlands as well as high plateaus just below the snow line in South America's Andes Mountains. In each of these zones, distinct local climates, soils, vegetation, crops, domestic animals, and modes of life prevail. Such **altitudinal zones 8** (diagrammed in Fig. 2-4)

are known by specific names as if they were regions with distinguishing properties—as in reality they are.

The lowest of these vertical zones, from sea level to 750 meters (2500 ft), is known as the ***tierra caliente***, the 'hot land' of the coastal plains and low-lying interior basins where tropical agriculture predominates. Above this zone lie the tropical highlands containing Middle and South America's largest population clusters, the ***tierra templada*** of temperate land reaching up to about 1800 meters (6000 ft). Temperatures here are cooler; prominent among the commercial crops is coffee, while corn (maize) and wheat are the staple grains. Still higher, from about 1800 to 3600 meters (6000 to nearly 12,000 ft), is the ***tierra fría***, the cold country of the higher Andes where hardy crops such as potatoes and barley are mainstays. Above the tree line, which marks the upper limit of the *tierra fría*, lies the ***tierra helada***; this fourth altitudinal zone, extending from about 3600 to 4500 meters (12,000 to 15,000 ft), is so cold and barren that it can support only the grazing of sheep and other hardy livestock. The highest zone of all is the uninhabited ***tierra nevada***, a zone of permanent snow and ice associated with the loftiest Andean peaks. As we will see, the varied human geography of mainland Middle and western South America closely reflects these diverse environments.

Tropical Deforestation

Before the Europeans arrived, two-thirds of continental Middle America (at lower altitudes) was covered by tropical rainforests. It is estimated that at present only about 10 percent of this vegetation remains. Between 2000 and 2010 alone, Central America lost almost 12 percent of its woodlands.

The causes of **tropical deforestation 9** are related to the persistent economic and demographic problems of disadvantaged

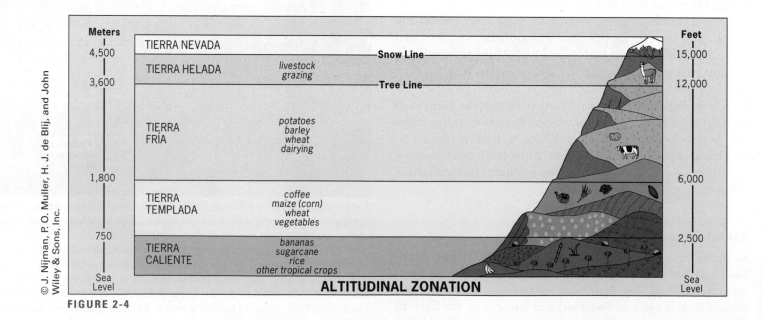

© J. Nijman, P. O. Muller, H. J. de Blij, and John Wiley & Sons, Inc.

FIGURE 2-4

countries. In Central America, the leading cause has been the need to clear rural lands for cattle pasture as many countries, especially Costa Rica, became meat producers and exporters. Because tropical soils are so nutrient-poor, newly deforested areas are able to function as pastures for only a few years at most. These fields are then abandoned for other freshly cut lands and quickly become a ravaged landscape (see photo). Without the protection of binding tree roots, local soil erosion and flooding immediately become problems, affecting still-productive areas nearby. A second cause of deforestation is the rapid logging of tropical woodlands as the timber industry increasingly turns from the exhausted forests of the midlatitudes to harvest the rich tree resources of the equatorial zones, in response to accelerating global demands for housing, paper, and furniture. A third major contributing factor is related to the region's population explosion: as more and more peasants are required to extract a subsistence from inferior lands, they have no choice but to cut down the remaining forest for both firewood and additional crop-raising space, and their intrusion prevents the trees from regenerating.

Despite scattered attempts to reverse this landscape scourge, deforestation continues to afflict Central America. Its worst effects plague the steeper slopes of interior highlands, as here in western Panama. In the wake of recent deforestation, the land near the top of this hill has already begun to erode because in the absence of binding tree roots the copious tropical rains are making short work of the unprotected topsoil.

CULTURAL GEOGRAPHY

Mesoamerican Legacy

Mainland Middle America was the scene of the emergence of a major ancient civilization. Here lay one of the world's true **culture hearths** 🔟, a source area from which new ideas radiated outward and whose population could expand and make significant material as well as intellectual progress. Agricultural specialization, urbanization, trade, and transportation networks developed, and writing, science, art, and other spheres of achievement saw major advances. Anthropologists refer to the Middle American culture hearth as *Mesoamerica*, which extended southeast from the vicinity of present-day Mexico City to central Nicaragua. Its development is particularly remarkable because it occurred in highly different geographic environments, each presenting obstacles that had to be overcome in order to unify and integrate large territories. First, in the low-lying tropical plains of what is now northern Guatemala, Belize, and Mexico's Yucatán Peninsula, and perhaps simultaneously in Guatemala's highlands to the south, the Maya civilization arose more than 3000 years ago. Later, in the fourteenth century AD, far to the northwest on the high plateau in central Mexico, the Aztecs founded a major civilization centered on Tenochtitlán (present-day Mexico City) in the Valley of Mexico. Some of the greatest contributions of Mesoamerica's indigenous peoples came from the agricultural sphere and included the domestication of corn (maize), the sweet potato, cacao beans (the raw material of chocolate), and tobacco.

Spanish Conquest

Spain's defeat of the Aztecs in the early sixteenth century opened the door to Spanish penetration and supremacy. The Spaniards were ruthless colonizers but not more so than other European powers that subjugated other cultures. The Spaniards enslaved the native population and were determined to destroy the strength of their society. Biology accomplished what ruthlessness could not have achieved in so short a time: diseases introduced by the Spaniards and the slaves they imported from Africa killed millions of indigenous people.

Middle America's cultural landscape was drastically modified. Unlike the indigenous peoples, who had utilized stone as their main building material, the Spaniards employed great quantities of wood and used charcoal for heating, cooking, and smelting metal. The onslaught on the forests was immediate, and rings of deforestation swiftly expanded around the colonizers' towns. The Spaniards introduced large numbers of cattle and sheep, and brought over their own crops (notably wheat) and farming equipment. Soon large wheatfields began to encroach upon the small plots of corn that the native people cultivated.

The Spaniards' most far-reaching cultural changes derived from their traditions as town dwellers. The indigenous people were moved off their land into nucleated villages and towns that the Spaniards established and laid out. In these settlements, the Spaniards could exercise the kind of rule and administration to which they were accustomed. The internal focus of each Spanish town was the central *plaza* or market square, around which both the local church and government buildings were located. The surrounding street pattern was deliberately laid out in gridiron form. Each town was located near what was thought to be good agricultural land (which often was not so good), so that the indigenous people could venture out each day and work in the fields. Packed tightly into these towns and villages, they came face to face with Spanish culture. Here they (forcibly) learned the Europeans' Roman Catholic religion and Spanish language, and they paid their taxes and tribute to a new master. Many of Middle America's leading cities still bear this Spanish imprint.

Collision of Cultures

But Middle America is not Spain, and its fragmented cultural mosaic reflects the collision of indigenous, Spanish, and other European influences. Indeed, in more remote areas in southeastern Mexico and interior Guatemala, indigenous societies survived and to this day native languages prevail over Spanish (see Fig. 2-10).

In Middle America outside Mexico, only Panama, with its twin attractions of interoceanic transit and gold deposits, became an early focus of Spanish activity. From there, following the Pacific-fronting side of the isthmus, Spanish influence radiated northwestward through Central America and into Mexico. The major arena of international competition in Middle America, however, lay not on the Pacific side but on the islands and coasts of the Caribbean Sea. Here

the British gained a foothold on the mainland, controlling a narrow coastal strip that extended southeast from Yucatán to what is now Costa Rica. As the colonial-era map (Fig. 2-5) shows, in the Caribbean the Spaniards faced not only the British but also the French and the Dutch, all interested in the lucrative sugar trade, all searching for instant wealth, and all seeking to expand their empires.

Much later, after centuries of European colonial rivalry in the Caribbean Basin, the United States entered the picture and made its influence felt in the coastal areas of the mainland, not through colonial conquest but through the introduction of widespread, large-scale, banana plantation agriculture. The effects of these plantations were as far-reaching as the impact of colonialism on the Caribbean islands. Because the diseases the Europeans had introduced were most rampant in these hot humid lowlands (as well as the Caribbean islands to the east), the indigenous population that survived was too small to provide a sufficient workforce. This labor shortage was quickly remedied through the trans-Atlantic slave trade from Africa that transformed the population composition of the Caribbean Basin.

The cultural variety of the Caribbean Basin is especially striking, and it is hardly an arena of exclusive Hispanic cultural heritage. For example, Cuba's southern neighbor, Jamaica (population 2.7 million, mostly of African ancestry), has a legacy of British involvement,

© J. Nijman, P. O. Muller, H. J. de Blij, and John Wiley & Sons, Inc.

FIGURE 2-5

while to the east in Haiti (10.3 million, overwhelmingly of African ancestry) the strongest imprints have been African and French. The Lesser Antilles also exhibit great cultural diversity. There are the (once Danish) U.S. Virgin Islands; French Guadeloupe and Martinique; a group of British-influenced islands, including Barbados, St. Lucia, and Trinidad and Tobago; and the Dutch St. Maarten (shared with the French) as well as the now-autonomous A-B-C islands of the former Netherlands Antilles—Aruba, Bonaire, and Curaçao— off the northwestern Venezuelan coast.

POLITICAL AND ECONOMIC FRAGMENTATION

Independence

Independence movements stirred Middle America at an early stage. On the mainland, insurrections against Spanish authority (beginning in 1810) achieved independence for Mexico by 1821 and for the Central American republics by the end of the 1820s, resulting in the creation of eight different countries. The United States, concerned over European designs in the realm, proclaimed the Monroe Doctrine in 1823 to deter any European power from reasserting its authority in the newly independent republics or from further expanding its existing domains.

By the end of the nineteenth century, the United States itself had become a major force in Middle America. The Spanish-American War of 1898 made Cuba independent and placed Puerto Rico under the U.S. flag; soon thereafter, the Americans were in Panama constructing the Panama Canal across their newly acquired Canal Zone. Meanwhile, with U.S. corporations propelling a boom based on massive banana plantations, the Central American republics had become colonies of the United States in all but name.

Independence came to the Caribbean Basin in fits and starts. African-Caribbean Jamaica as well as Trinidad and Tobago, where the British had brought in a large South Asian population, attained full sovereignty from the United Kingdom in 1962; other British islands (among them Barbados, St. Vincent, and Dominica) became independent later on. France, however, retains Martinique and Guadeloupe as *Overseas Départements* of the French Republic, and the Dutch islands are at various stages of autonomy. No less than 33 states are found on the political map of the Caribbean Basin today.

Regional Contrasts

There are some important contrasts, socially and economically, between the Middle American mainland on the one hand and the Caribbean coasts and islands on the other. The mainland is dominated by Spanish as well as indigenous influences and also includes **mestizo 11** sectors where the two ancestries mixed. Agriculture on the mainland is often organized around the **hacienda 12**, a Spanish institution through which landowners possessed a domain whose productivity they might never push to its limits; the very possession of such a vast estate brought with it social prestige and a comfortable lifestyle. Native workers lived on the land—which may once have been their land—and had plots where they could grow their own subsistence crops. All this is written as though it is mostly in the past, but the legacy of the hacienda system, with its inefficient use of land and labor, still exists throughout mainland Middle America.

The Caribbean coasts (from Belize southward) and the Caribbean islands possess a very different cultural heritage, based on a fusion of European and African influences. Here, **plantations 13** long dominated the economies. The plantation, in contrast to the hacienda, is all about efficiency and profit. Foreign ownership and investment is the norm, as is production for export. Most plantations grow only a single crop, be it sugar, bananas, or coffee. Much of the labor is seasonal, being needed in large numbers mainly

© Jan Nijman

@ FROM THE FIELD NOTES...

"Driving around the small Caribbean island of Bonaire and coming face to face with this surreal landscape, I first thought I had experienced a mirage. These glistening white, perfectly cone-shaped hills are actually salt piles. This small (formerly Dutch) island off the coast of Venezuela possesses the perfect geographic conditions for salt production: it has a series of salt-water inlets, it is very hot and dry, and lies in the zone of persistent trade winds. Remember that salt, nowadays taken for granted in every household, has long been one of the world's most precious spices, and was widely used for the preservation of meat and fish. The Dutch began large-scale production of salt here in the 1620s; today this local industry is in the hands of the Antilles International Salt Company and continues to be an important source of Bonaire's foreign revenues."

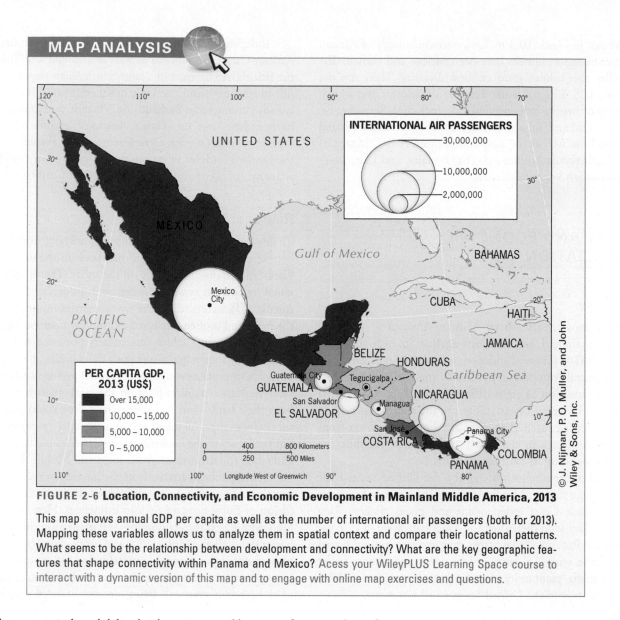

MAP ANALYSIS

FIGURE 2-6 Location, Connectivity, and Economic Development in Mainland Middle America, 2013

This map shows annual GDP per capita as well as the number of international air passengers (both for 2013). Mapping these variables allows us to analyze them in spatial context and compare their locational patterns. What seems to be the relationship between development and connectivity? What are the key geographic features that shape connectivity within Panama and Mexico? Acess your WileyPLUS Learning Space course to interact with a dynamic version of this map and to engage with online map exercises and questions.

during the harvest period; such labor has been imported because of the scarcity of indigenous workers. With its 'factory-in-the-field' operations, the plantation enables far more efficient use of land and labor than the hacienda. Profit and wealth, rather than social prestige, are the dominant motives for the plantation's establishment and operation.

Connections Matter

The unusual layout of the Middle American realm offers an opportunity to consider the spatial dimensions of development. Geographical fragmentation and territorial size play their parts, but so does *accessibility* in the form of **connectivity 14** in both the Caribbean and on the mainland. Figure 2-6 in the Map Analysis box provides a closer look at the relationship between development and connectivity.

Is Small Beautiful?

As noted, the Middle American realm is exceptional in the modest size of its territorial extent and population. But the number of countries is considerable, and they tend to be quite small. Caribbean islands often invoke images of paradise: beautiful scenery, tropical cocktails, and shiny blue waters. But the economic realities are almost invariably harsh. The limited land area and small population of most islands (those of the Lesser Antilles average less than half a million people), together with their insularity, relative accessibility, and low connectivity, pose formidable challenges that are common to **small-island developing economies 15**.

What are the consequences of these geographical properties for economic development? First, natural resources are frequently limited, which requires a heavy reliance on imports made more expensive by added transport costs. Second, the cost of government is relatively high per capita: even the smallest population will require services such as schools, hospitals, and waste disposal. Third, these specialized services must often be brought in from elsewhere. And fourth, local production cannot really benefit from **economies of scale 16** (unit-cost savings resulting from large-scale production). Consequently, local producers can be put out of business by cheaper imports, thereby driving up unemployment.

Given the Caribbean Basin's limited economic options, does the tourist industry offer better opportunities? The resort areas, scenic treasures, and historic locales of Caribbean America attract more than 25 million visitors annually, with about half of them traveling on Florida-based cruise ships. In Jamaica, for example, this industry accounts for about one-third of its gross domestic product and employs more than one-fourth of its labor force.

But Caribbean tourism also has serious drawbacks. The invasion of poor communities by affluent tourists contributes to rising local resentment, which is fueled by the glaring contrasts of shiny new hotels towering over substandard housing, or luxury liners gliding past poverty-stricken villages. At the same time, tourism can debase local culture, which often is adapted to suit the visitors' tastes at hotel-staged 'culture' shows. In addition, the cruise industry tends to monopolize revenues (accommodations, meals, and entertainment), with relatively few dollars flowing into the local economy. Finally, even though tourism does generate income in the Caribbean, the intervention of island governments and multinational corporations can remove opportunities from local entrepreneurs in favor of large operators and major resorts.

The Push for Regional Integration

Another challenge for the countries of the Middle American realm is to foster greater **economic integration** **17** (the benefits of forging new international partnerships). Many of the countries on the mainland as well as in the Caribbean are poorly connected within the realm and are heavily dependent on major outside countries, particularly the United States. For the large majority of these countries, the United States is their primary trading partner. Consider this: of all the trade that engages Middle American countries, less than 10 percent occurs within the realm. And less than 1 percent takes place between the Caribbean Basin and the Middle American mainland.

Over the years, efforts have been made to advance economic integration and convert this realm into more of a *functional region*. There are presently two trade organizations in Middle America (in addition to Mexico's membership in NAFTA). *CARICOM* (Caribbean Community), established in 1989, consists of 15 full

Hundreds of thousands of North American retirees, and many affluent purchasers of second homes, have sought the sun and low-cost residential opportunities of Middle America, converting some areas into virtual exclaves of the North. From Mexico to Panama, waterfront real estate is among the attractions for these permanent and seasonal migrants, as here in Cabo San Lucas at the southern tip of Mexico's Baja California peninsula. This is yet another example of international connections that matter.

© Liane Cary/Age Fotostock America, Inc.

members, including nearby Guyana and Suriname in South America. To a certain degree, CARICOM emulates the European Union (EU), and in 2009 it even introduced a common passport; but unlike the EU, this organization has had only a superficial economic impact. The more influential *CAFTA-DR* (Central American Free Trade Agreement) includes the U.S., five Central American countries, and the Caribbean's Dominican Republic (hence the "DR" suffix). But this organization seems to be a mixed blessing: it may increase access to U.S. markets and lead to cheaper imports, yet it also reinforces the United States' dominant position at the expense of greater intra-realm integration.

REGIONS OF MIDDLE AMERICA

The Middle American realm can be subdivided into four distinct geographic regions (the first two on the mainland; the other two in the Caribbean Basin): (1) *Mexico*, the giant of the realm in every respect; (2) *Central America*, the string of seven small republics occupying the land bridge from Mexico to South America; (3) the four islands of the *Greater Antilles*—Cuba, Jamaica, Hispaniola (containing Haiti and the Dominican Republic), and Puerto Rico; and (4) the numerous small islands of the *Lesser Antilles* (Fig. 2-7).

▌MEXICO

Physiography

The physiography of Mexico is reminiscent of that of the conterminous western United States, although its environments are more tropical. Figure 2-8 shows several prominent features: the elongated peninsula of Baja (Lower) California in the northwest, the far eastern Yucatán Peninsula, and the Isthmus of Tehuantepec in

REGIONS OF MIDDLE AMERICA

Mexico
Central America
Greater Antilles } Caribbean
Lesser Antilles

POPULATION
· Under 50,000
● 50,000–250,000
● 250,000–1,000,000
● 1,000,000–5,000,000
● Over 5,000,000

National capitals are underlined

0 100 200 300 Miles
0 200 400 600 Kilometers

FIGURE 2-7

© J. Nijman, P. O. Muller, H. J. de Blij, and John Wiley & Sons, Inc.

the southeast where the Mexican landmass tapers to its narrowest extent. Here in the southeast, Mexico's physiography most resembles that of Central America: a mountainous backbone forms the isthmus, curves southeastward into Guatemala, and extends northwestward toward Mexico City. Shortly before reaching the capital, this mountain range divides into two chains, the Sierra Madre Occidental to the west and the Sierra Madre Oriental to the east. These diverging ranges frame the funnel-shaped Mexican heartland, the center of which consists of the rugged, extensive Plateau of Mexico (the important Valley of Mexico lies near its southeastern end). Mexico's climates are characterized by dryness, particularly in the broad, mountain-flanked north (also see Fig. G-6). Most of the better-watered areas lie in the southern half of the

country where a number of major population concentrations have developed (see Fig. 2-2).

Regional Diversity

Physiographic, demographic, economic, historical, and cultural criteria combine to reveal a regionally diverse Mexico extending from the lengthy ridge of Baja California to the tropical lowlands of the Yucatán Peninsula, and from the economic dynamism of the U.S. borderland to the indigenous traditionalism of far southeastern Chiapas. The country's core area, anchored by Mexico City and extending westward to Guadalajara, lies between the more Hispanic-mestizo north and the dominantly indigenous-mestizo

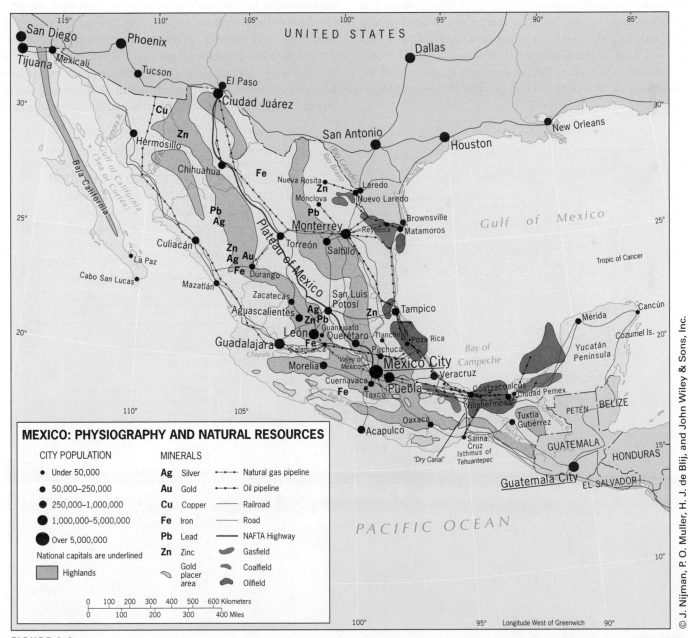

MEXICO: PHYSIOGRAPHY AND NATURAL RESOURCES

FIGURE 2-8

south. East of the core area is the Gulf Coast, once dominated by major irrigation projects and sprawling livestock-raising schemes but now the mainland center of Mexico's petroleum industry. The region south of the core is dominated by the rugged, Pacific-fronting Southern Highlands, where coastal Acapulco's luxurious resorts stand in stark contrast to the indigenous villages and communal-farm settlements of the interior. The dry, far-flung north stands in particularly sharp contrast to these southern regions. To go from comparatively well-off Monterrey in the far north to poverty-mired Chiapas adjacent to southern Guatemala is to observe the entire range of Mexico's regional geography.

Population and Urbanization Patterns

Mexico's population expanded rapidly throughout the closing decades of the twentieth century, doubling in just 28 years; but demographers have recently noted a sharp drop in fertility, and they are predicting that Mexico's population (currently at 122.3 million) will cease growing altogether by about 2050. That will have important implications for the country's economy and for the United States as well since it is likely to affect long-term migration patterns.

The distribution of Mexico's population relative to the country's 31 internal States is shown in Figures 2-2 and 2-9. The largest concentration, containing the core area and more than half of the Mexican people, extends across the densely populated 'waist' of the country from Veracruz State on the eastern Gulf Coast to Jalisco State on the Pacific. The center of this corridor is dominated by the most populous State, Mexico (**3** on the map), at whose heart lies the Federal District of Mexico City (**9**). In the dry and rugged terrain to the north of this central corridor lie Mexico's least-populated States. Southern Mexico also exhibits a sparsely peopled periphery in the hot and humid lowlands of the Yucatán, but to the southwest most of the highlands of the continental spine contain sizeable populations.

Another major feature of Mexico's population map is urbanization, driven by the pull of the cities (with their perceived opportunities for upward mobility) in tandem with the push of the economically stagnant countryside. Today, 79 percent of the Mexican people reside in towns and cities, a surprisingly high proportion for a less-developed country. Undoubtedly, these numbers are affected by the explosive recent growth of the region around Mexico City, which has surpassed 30 million (making it one of the largest urban concentrations on Earth) and is home to a quarter of

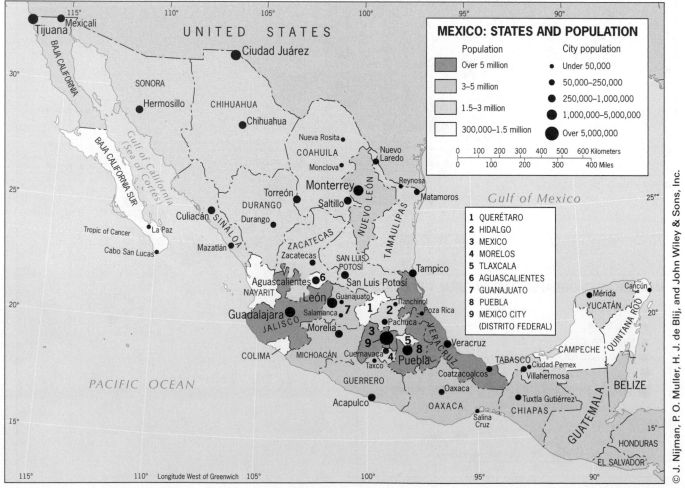

© J. Nijman, P. O. Muller, H. J. de Blij, and John Wiley & Sons, Inc.

FIGURE 2-9

the national population. Urbanization rates are at their lowest in the peripheral southern uplands where indigenous society has been least affected by modernization.

A Mix of Cultures

Nationally, the indigenous imprint on Mexican culture remains strong. Today, 62 percent of all Mexicans are mestizos, 21 percent are predominantly indigenous, and another 7 percent are full-blooded indigenous; almost all of the remaining 10 percent are Europeans. The Spanish influence in Mexico has been profound, but it has been met with an equally powerful thrust of indigenous culture. It has therefore not been a case of one-way European-dominated **acculturation 18** but rather **transculturation 19**—the two-way exchange of culture traits between societies in close contact. In the southeastern periphery (Fig. 2-10), several hundred thousand Mexicans still speak only an indigenous language, and millions more still utilize these languages daily even though they also speak Mexican Spanish. The latter has been strongly shaped by indigenous influences, as have Mexican modes of dress, foods and cuisine, artistic and architectural styles, and folkways. This fusion of heritages, which makes Mexico unique, is the end product of an upheaval that began to reshape the country one century ago.

Agriculture: Fragmented Modernization

Following the Mexican revolution of the 1910s and the ensuing land redistribution schemes, around half of the land was organized into so-called *ejidos* **20**, communally owned farms of 20 or more families. It meant an abrupt end for many large, single-family-owned large haciendas. This system of land management is an indigenous legacy, and not surprisingly most *ejidos* lie in central and southern Mexico where nativist social and agricultural traditions are strongest. About half of Mexico's land continues to be held in such 'social landholdings.' But some of the land is excessively fragmented and/or managed inefficiently with scarce capital, resulting in both inferior crop yields and widespread rural poverty. In the 1990s, the Mexican government attempted to privatize *ejidos*, hoping to promote consolidation and increase productivity, but that effort did not succeed.

At the same time, larger-scale commercial agriculture has diversified during the past three decades and made major gains in both domestic and export markets. The country's arid northern tier has led the way as major irrigation projects have been built on streams flowing down from the interior highlands. Along the booming northwestern coast of the mainland, which lies within a day's drive of Southern California, large-scale production of fruits, vegetables, cut flowers, and cotton has become the cornerstone of an increasingly profitable export trade to the United States and elsewhere.

Evolving Economic Geographies

The *maquiladoras* in the country's northern borderland—foreign-owned factories that assemble imported raw materials and components and then export finished manufactures back to the United States—account for over one-fifth of Mexico's industrial jobs and a major share of its total exports. *Maquiladoras* presently employ more than two million Mexican workers. Many companies are headquartered in cities just north of the border, such as San Diego and El Paso, and this is where design, marketing, and other strategic work takes place. But most of the workforce is located south of the border at the assembly plants (which is why the cities on the Mexican side have grown so much faster than their U.S. counterparts). Think electronics, machinery, garments, construction materials, automobile parts, and much more. Although the impact of NAFTA has been widespread, it should be noted that Mexican employees work long hours for low wages with meager fringe benefits. Moreover, most reside in the over-crowded rudimentary settlements that encircle the burgeoning urban centers of the border zone. In 2015, the minimum daily wage in Mexico was still below six U.S. dollars. It is estimated that 15 percent actually earn less, whereas more highly skilled workers in the north might make three times the minimum wage. The pay-scale disparity with the adjacent United States, however, remains enormous.

Mexico's government seeks to encourage the growth of higher-paying jobs in more advanced economic sectors—particularly electronic goods—that are also less likely to be lost to other parts of the world. Expansion of education and training in high-technology and management fields is a critical first step. The northeastern city of Monterrey in the relatively high-income State of Nuevo Léon has become a successful model in this effort, nurturing both an international business community and an ultra-modern industrial complex that has attracted leading multinational corporations. Here indeed lies hope for Mexico's future, and one frequently hears reports these days that rising labor costs in China may induce some manufacturers to return to northern Mexico, allowing them to more directly control flexible (so-called on-time) production at a close distance.

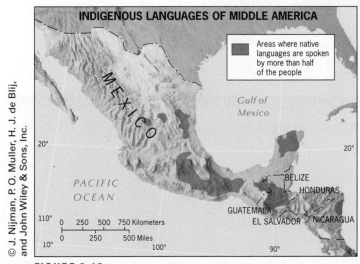

INDIGENOUS LANGUAGES OF MIDDLE AMERICA

Areas where native languages are spoken by more than half of the people

MEXICO

Gulf of Mexico

PACIFIC OCEAN

BELIZE
HONDURAS
GUATEMALA
EL SALVADOR
NICARAGUA

0 250 500 750 Kilometers
0 250 500 Miles

© J. Nijman, P. O. Muller, H. J. de Blij, and John Wiley & Sons, Inc.

FIGURE 2-10

MEXICO: ECONOMIC AND POLITICAL DISPARITIES

GDP PER CAPITA IN US$ (2010)
National average=10,000

- Above 12,000
- 10,000–12,000
- 8,000–10,000
- 6,000–8,000
- 4,000–6,000
- Below 4,000

BORDER-ZONE MAQUILADORA EMPLOYMENT

- 100,000–300,000
- 50,000–100,000
- 15,000–50,000
- 5,000–15,000

— Railroad
— Road
— NAFTA Highway

State figures exclude oil industry
Source: INEGI

0   100   200   300   400   500   600 Kilometers
0   100   200   300   400 Miles

1 QUERÉTARO
2 HIDALGO
3 MEXICO
4 MORELOS
5 TLAXCALA
6 AGUASCALIENTES
7 GUANAJUATO
8 MEXICO CITY (DISTRITO FEDERAL)

2012 PRESIDENTIAL VOTE BY STATE
WINNER OF EACH STATE

- **PAN** National Action Party
- **PRD** Party of the Democratic Revolution
- **PRI** Institutional Revolutionary Party

FIGURE 2-11

States of Contrast

Countries with strong internal regional disparities face serious challenges that can be difficult to overcome and may deteriorate over time. Mexico's southernmost States—Chiapas, Oaxaca, and Guerrero, all bordering the Pacific Ocean—are by far the poorest. The States bordering the United States in the north, including Nuevo Léon, Chihuahua, and Baja California, have the highest incomes. Using rural poverty as a measure, we find that only about 10 percent of people in the countryside in the north are in the poorest category, but nearly 50 percent in the south.

Mexico's north-south divide is especially noticeable in the economic data mapped in Figure 2-11. In general terms, annual per capita income in the northern States exceeds U.S. $10,000; but in the southern States it falls below $5000. Moreover, Mexico's infrastructure, already substandard, serves the south far less well than the north. Whatever the index—literacy, electricity use, water availability—the south lags by a wide margin. The population of the south not only trails far behind the rest of Mexico in overall development, but it is also the least well educated, the least productive agriculturally, and the most isolated in the country. Since the early 1990s, a radical group of Maya peasant farmers in Chiapas State calling themselves the Zapatista National Liberation Army (ZNLA) has engaged in guerrilla warfare, demanding better treatment for Mexico's 33 million indigenous citizens. Despite substantial public support, their quarter-century struggle has not yielded noticeable results.

These regional contrasts were thrown into sharp relief in 2012, when Mexico's presidential elections were contested by three candidates broadly representing the conservative elite, the working classes, and the middle classes. When the ballots were counted, there was a clear spatial divide in the voting patterns, which reflected the varying economic fortunes across the country. Generally, most of the votes in the north went to the conservative and centrist candidates, whereas the poorer southern States supported the leftist candidate (see inset map in Fig. 2-11). It was the centrist candidate, Enrique Peña Nieto, who narrowly won the presidency in 2012—providing the latest evidence of Mexico's sharp politico-geographical cleavages.

The Drug Wars

Since the mid-2000s, Mexico has been plagued by yet another obstacle in its struggle to achieve sustained economic progress. This problem has received worldwide attention as the cocaine-producing drug cartels centered in Colombia established new bases in U.S.-border cities in northern Mexico and launched a vicious war for supremacy. They responded in part to the success of the antidrug campaign in Colombia, but the cartels also saw

At the height of the drug war, groups of students in Mexico City staged this 2010 protest against cocaine trafficking violence. An estimated 2738 people were murdered that year in Ciudad Juárez alone. Since then, Mexican authorities have clamped down on the narcotics trade, which has in part shifted from Mexico to the Caribbean islands. The number of drug-related deaths has seen a steady decline, but the violence remains a serious threat to Mexican society.

© ALEX CRUZ/epa/Corbis

new opportunities beckoning in Mexican territory adjacent to their main market in the United States.

Figure 2-12 provides an overview of the geography of this drug war. Cocaine is produced in Colombia, Bolivia, and Peru, and a large share of it enters the United States through Mexico. Bolivian and Peruvian cocaine, constituting approximately half of Mexico's 'imports,' is shipped by sea and illegally enters the country in the Pacific Coast States of Guerrero, Michoacán, Colima, and Sinaloa. The other half originates in Colombia and enters either overland via the Middle American land bridge through Guatemala, or by boat across the Caribbean and Gulf of Mexico to be smuggled into the eastern States of Yucatán, Quintana Róo, and Veracruz. The illicit powder is then repackaged (and/or crystallized to form crack) in smaller quantities for 'retailing' purposes prior to being smuggled into the United States, mainly through the border cities of Reynosa, Nuevo Laredo, Ciudad Juárez, and Tijuana (Fig. 2-12).

Interestingly, when Mexican criminal organizations were co-opted by the Colombian cartels, they demanded to be paid off in cocaine. This allowed the Mexicans to start up their own distribution networks—it also meant that a good amount of the product 'stuck around' and ended up on the streets of Mexican cities, resulting in heightened drug addiction and rising crime. Several major Mexican cartels sprang up and got involved in this drug trafficking operation, and competition continues to literally be murderous. Territorial control is key in this 'business': each cartel dominates its own turf, with various components being heavily contested from time to time. The cartels themselves are not particularly stable, with countless mergers and splits occurring over time. The two leading rivals are Los Zetas and the Sinaloa Cartel, which control huge swaths of eastern and western Mexico, respectively.

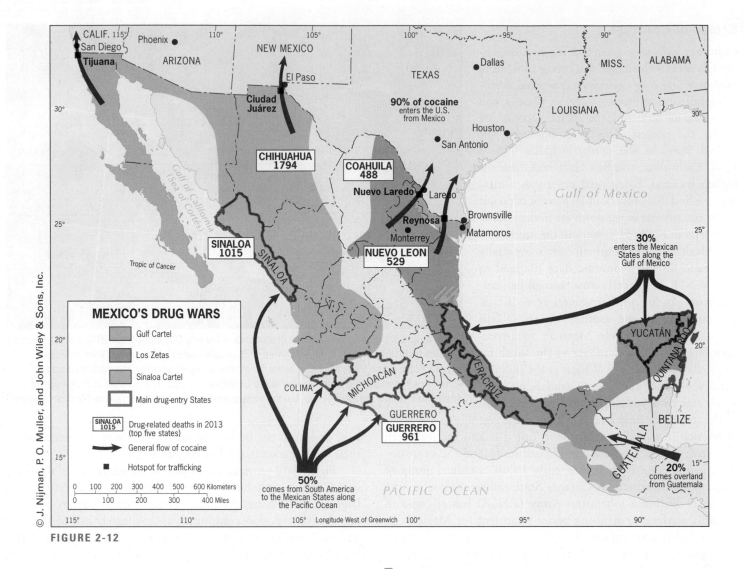

FIGURE 2-12

Much of the worst violence occurs where control is contested, especially around the most crucial U.S. ports of entry and along primary transport routes. About half of Mexico's drug-related killings occur in the States of Sinaloa (cocaine inflow; marijuana production) and Chihuahua (which contains Ciudad Juárez, the leading U.S. port of entry). By mid-2015, the murder total since the war erupted was estimated to be more than 140,000.

After peaking around 2010, the level of violence has steadily declined in this decade, which many attribute to the apparent shift of cocaine trafficking from Mexico to the Caribbean Basin. Relentless crackdowns by the Mexican government may finally have induced the drug suppliers and the cartels to forge easier new routes via the Caribbean's small, impoverished, and far more vulnerable island-states. This ongoing realignment might just provide the breathing space that Mexico needs to strengthen its economic base, address the challenges of uneven regional development, and capitalize on its deepening trade relationship with the United States (and Canada). Mexico's recent upheavals, instabilities, and setbacks notwithstanding, the country's developmental potential remains promising.

THE CENTRAL AMERICAN REPUBLICS

A Land Bridge

Crowded onto the narrow segment of the Middle American land bridge between Mexico and the South American continent are the seven countries of Central America (Fig. 2-13). Territorially, they are all quite small; their population sizes range from Guatemala's 15.5 million down to Belize's 332,000. The land bridge here consists of a highland belt flanked by coastal lowlands on both the Caribbean and Pacific sides (Fig. 2-14). These uplands are studded with volcanoes, and local areas of fertile volcanic soils are scattered throughout them.

The land bridge has a fascinating geologic and evolutionary history, and one famous study referred to it as the Monkey Bridge. For some 50 million years, North and South America were separated; the land bridge was formed only 3 million years ago, becoming a biologic highway of sorts for evolutionary exchange. The region contains only 1 percent of the Earth's land but 7 percent of all the world's natural species. The southern part (Costa Rica and Panama) is known as a global **biodiversity hot spot 21**, even

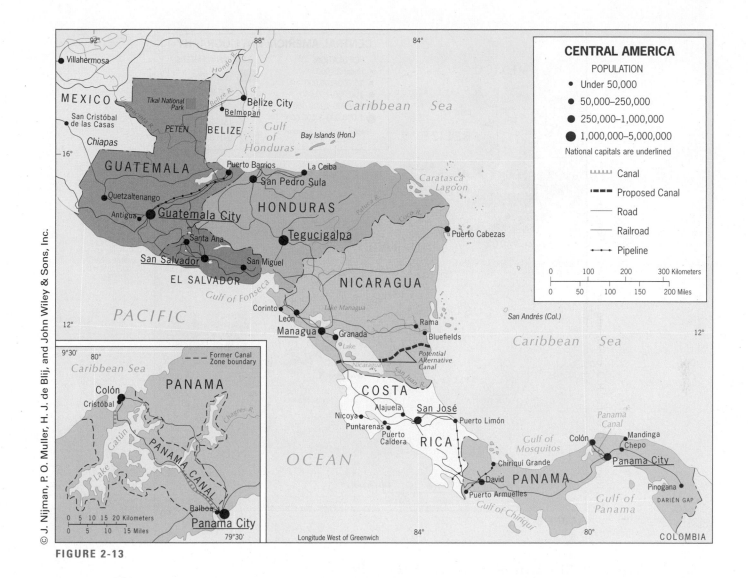

FIGURE 2-13

though deforestation has been a major problem. Human inhabitants have always been concentrated in the upland zone, where tropical temperatures are moderated by elevation and rainfall is sufficient to support a variety of crops.

Trapped in the Narcotics Trade

Unfortunately for Central America, over the past decade the region has increasingly been forced to accommodate the cocaine trade. Its intermediate location between northern South America and Mexico made the Central American land bridge the only possible route when narco-traffickers decided to expand their overland transit operations. Consequently, as much as three-quarters of all the cocaine that now reaches the United States through Mexico has passed through at least one Central American country. Given the region's endemic poverty and severely limited economic opportunities, large numbers of locals are believed to have been drawn into the potentially lucrative drug trade, sparking a steep rise in organized criminal activity and an explosion of violence, especially in the countries closest to Mexico—Guatemala, Honduras,

El Salvador, and Belize. Since 2011, Honduras has recorded the highest murder rate in the world of around 90 per 100,000 (in the U.S., not one of the safest countries in the world, it is under 5 per 100,000); at the same time, Belize, El Salvador, and Guatemala were joining Honduras to all rank in the world's top five on this unenviable list. Central American governments, with few resources to call upon, were virtually powerless to respond to the criminal cartels and affiliated vicious youth gangs, while most businesspeople and civilians had learned long ago never to depend on the police or military. Not surprisingly, in the Honduran capital of Tegucigalpa, private security forces (for those able to afford them) today outnumber police personnel four to one.

Guatemala

The westernmost of Central America's republics, Guatemala has more land neighbors than any other. Straight-line boundaries lying across the tropical forest mark much of the border with Mexico, creating the box-like Petén region between Chiapas State on the west and Belize on the east; also to the east lie Honduras

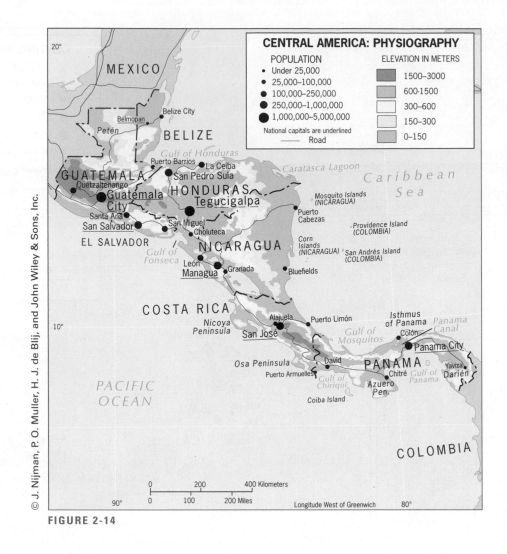

CENTRAL AMERICA: PHYSIOGRAPHY

POPULATION
- Under 25,000
- 25,000–100,000
- 100,000–250,000
- 250,000–1,000,000
- 1,000,000–5,000,000

National capitals are underlined
— Road

ELEVATION IN METERS
- 1500–3000
- 600-1500
- 300–600
- 150–300
- 0–150

© J. Nijman, P. O. Muller, H. J. de Blij, and John Wiley & Sons, Inc.

FIGURE 2-14

and El Salvador (Fig. 2-13). This heart of the ancient Maya Empire, which remains strongly permeated by indigenous culture and traditions, has only a small window on the Caribbean but a longer Pacific coastline. Most populous of the seven republics with 15.5 million inhabitants (mestizos are in the majority with 57 percent, indigenous 43 percent), Guatemala has seen a great deal of conflict, and military regimes have dominated political life. There is a deepening split between the wretchedly poor indigenous populations and the better-off mestizos who continue to control the government, army, and land-tenure system.

Guatemala's economic geography demonstrates considerable potential but has long been shackled by unrelenting internal conflicts and the widening of one of the hemisphere's biggest gaps between rich and poor. In tandem, they keep the income of at least 60 percent of the population below the poverty line (some estimates range as high as 75 percent). The country's mineral wealth includes nickel in the highlands and oil in the lower-lying north. Agriculturally, soils are fertile and moisture is ample over highland areas large enough to produce a wide range of crops, including excellent coffee. The 929-kilometer (577-mi) border with Mexico offers opportunities as well, but progress is presently paralyzed by out-of-control drug trafficking and violent crime.

Belize

Strictly speaking, Belize is not a Central American republic in the same tradition as the other six. Until 1981, this country, a wedge of land between northern Guatemala, Mexico's Yucatán Peninsula, and the Caribbean Sea (Fig. 2-13), was a dependency of the United Kingdom known as British Honduras. Slightly larger than Massachusetts and containing a minuscule population of only 332,000 (many of African descent), Belize is much more reminiscent of a Caribbean island than of a continental Middle American state. Today, all that is changing as the demographic complexion of Belize is being reshaped. Thousands of residents of African descent have emigrated (many went to the United States) and were replaced by tens of thousands of Spanish-speaking immigrants. Most of the latter have been land seekers and refugees from strife in nearby Guatemala, El Salvador, and Honduras, and their proportion of the Belizean population has risen from 33 in 1980 to well over 50 percent today. With the newcomers now in the majority, Belize's cultural geography is becoming increasingly Hispanicized; in fact, since 2012 Spanish speakers outnumber those who use English or creole as a first language.

The Belizean transformation extends to the economic sphere as well. No longer just an exporter of sugar and bananas, Belize today is producing new commercial crops, and its seafood-processing and clothing industries have become major revenue earners. Also important is tourism, which annually lures more than 200,000 vacationers to the country's Mayan ruins, resorts, and newly legalized casinos; a booming specialty is *ecotourism*, based on the natural attractions of the country's near-pristine environment, including its magnificent offshore coral reefs. Belize is also known as a center for **offshore banking** 22, a financial haven for foreign companies and individuals who want to avoid paying taxes in their home countries.

Honduras

Honduras, in contrast with Guatemala, has a long Caribbean coastline and a relatively small window on the Pacific (Fig. 2-13). The country also occupies a critical place in the political geography of Central America, flanked as it is by Nicaragua, El Salvador, and Guatemala—all continuing to grapple with the aftermath of years of internal conflict, recent natural disasters, and the ongoing ravages of drug trafficking. Comparable in natural beauty and biodiversity to Costa Rica, Honduras hopes to exploit its potential for ecotourism but is constrained by a defective infrastructure and the lack of funding for new facilities (see Field Note).

Agriculture, livestock, forestry, and limited mining form the mainstays of the economy, with the now-familiar Central American products—apparel, bananas, coffee, and shellfish—earning most of the external income. With 8.1 million inhabitants, about 90 percent mestizo, Honduras faces huge development challenges that are complicated by escalating crime and social dislocation.

In an effort to move forward, Honduras has unveiled a development strategy aimed at the creation of quasi-independent 'special

development regions.' The idea is to attract foreign investment into these protected zones, which are to be independent in their economic, fiscal, and budgetary policies. This initiative follows the charter-cities model that has proven successful in Hong Kong, but critics fear that it may only clear the way for organized crime to take full control of these new economic enclaves—particularly in and around the besieged city of San Pedro Sula (since 2011 the world's 'murder capital').

El Salvador

El Salvador is Central America's smallest country territorially, smaller even than Belize, but with a population 20 times as large (6.3 million) it is the region's most densely peopled. El Salvador adjoins the Pacific in a narrow coastal plain backed by a chain of volcanic mountains, behind which lies the country's heartland (Fig. 2-13). Unlike neighboring Guatemala, El Salvador has a far more homogeneous population (86 percent mestizo and just 1 percent indigenous). Yet ethnic uniformity has not translated into social or economic equality or even opportunity. Whereas other Central American countries were called banana republics, El Salvador was a coffee republic, and the coffee was produced on the huge landholdings of a few landowners and on the backs of a subjugated peasant labor force.

More affluent El Salvadorans who left the country in past decades, and who have succeeded in the United States and elsewhere, send substantial funds back home; these **remittances** 23 now provide the largest single source of foreign revenues. This has helped stimulate such industries as apparel and footwear manufacturing as well as food processing. But a major stumbling block to revitalization of the agricultural sector has again been land reform, and in this regard El Salvador's future still hangs in the balance.

© Jan Nijman

@ **FROM THE FIELD NOTES...**

"Invited by the Honduran government to survey the country's ecotourism potential, I toured the beautiful countryside with a National Geographic Society delegation. There is no doubt that Honduras is exceptionally endowed by nature and that there is considerable potential for a vibrant tourist industry. But this is a poor country, and the government has great difficulty making the required investments—in infrastructure, for example. We flew in these small propeller planes to get from the capital, Tegucigalpa, to the Mayan ruins of Copán. There was no airport nearby, and overland travel is arduous and time consuming. Ecotourism by its very nature is small scale and the revenues tend to be relatively modest, so it is a major challenge to raise the substantial funds needed for costly infrastructural improvements."

Nicaragua

This country is best approached by reexamining the map (Fig. 2-13), which underscores Nicaragua's position tucked away in the heart of Central America. The Pacific coast follows a southeasterly direction here, but the Caribbean coast is oriented north-south so that Nicaragua forms a triangle of land, with its lakeside capital, Managua, situated in a valley on the mountainous, earthquake-prone, Pacific side (the country's core area has always been located here). The Caribbean side, where the uplands yield to a wide coastal plain of tropical rainforest, savanna, and swampland, has for centuries been home to indigenous peoples such as the Miskito, who have lived remote from the focus of national life. Options for this country of 6.1 million are limited: agriculture dominates the economy and manufacturing is weak. Similar to El Salvador, there is a growing reliance on remittances since 2010 from those who have emigrated.

For decades, there has been talk of an interoceanic canal to rival the Pacific-Atlantic waterway across Panama, even though the land bridge here is three times wider. What long seemed like daydreaming took a turn toward reality in 2013 when the Nicaraguan government commissioned a Chinese company to plan and build a canal capable of accommodating ships twice the new maximum size of Panama's. In December 2014, the government announced that ground had been broken for a huge construction project (Fig. 2-13 shows the route of the planned canal). For China, this enormous investment and undertaking is strategically important because the future canal could not only handle its largest freighters, but also significantly enhance Chinese economic and geopolitical influence throughout Middle America's mainland.

Costa Rica

If there is one country that underscores Middle America's variety and diversity it is Costa Rica—because it differs significantly from its neighbors and from the norms of Central America as well. Bordered by two volatile countries (Nicaragua to the north and Panama to the east), Costa Rica is a nation with an old democratic tradition and, in this boiling cauldron, no standing army since 1949! Although the country's Hispanic imprint is similar to that found elsewhere on the mainland, its early independence, its good fortune to lie remote from regional strife, and its leisurely pace of settlement allowed Costa Rica the luxury of concentrating on its economic development. Perhaps most important, internal political stability has predominated over much of the nearly 200 years since its independence from Spain.

Like its neighbors, Costa Rica (2016 population: 4.9 million) is divided into environmental zones that parallel its coastlines. The most densely settled is the central highland zone, lying in the cooler *tierra templada*, whose heartland is the Valle Central (Central Valley), a fertile basin that contains Costa Rica's main coffee-growing area and the leading population cluster focused on the cosmopolitan capital of San José (Fig. 2-13).

The long-term development of Costa Rica's economy has given this country the region's highest standard of living, literacy rate, and life expectancy (though even here, one-fifth of the population is trapped in poverty). Agriculture continues to dominate (with bananas, coffee, tropical fruits, and seafood among the leading exports), and tourism—especially ecotourism—has expanded steadily. Despite the deforestation of more than 80 percent of its original woodland cover, Costa Rica is widely known for its superb scenery and for its (belated) efforts to protect what is left of its diverse tropical flora and fauna.

Panama

Panama reflects some of the usual geographic features of the Central American republics. Its population of 3.9 million is about two-thirds mestizo but also contains substantial indigenous, European, and black minorities. Spanish is the official language, but English is also widely used. Ribbon-like and oriented east-west, Panama's topography is mountainous and hilly. Eastern Panama, especially Darien Province adjoining Colombia, is densely covered by tropical rainforest, and contains the only remaining gap in the intercontinental Pan American Highway (Fig. 2-13). Most of the rural population lives in the uplands west of the canal; there, Panama produces bananas, coffee, sugarcane, rice, and, along its narrow coastal lowlands, shrimps and other seafood. Much of the urban population is concentrated in the vicinity of the famous Canal, anchored by the cities at each end of this waterway.

Panama owes its existence to the idea of an artificial waterway connecting the Atlantic and Pacific oceans to avoid the lengthy circumnavigation of South America. The Panama Canal (see the inset map in Fig. 2-13) was opened in 1914, a symbol of U.S. power and influence in Middle America. It was both a massive feat of engineering and an especially dangerous undertaking in a hostile tropical environment. It is estimated that 25,000 workers died during the 10 years of excavation and construction. The canal itself was contained within the cross-isthmian *Canal Zone*, an elongated strip of land 16 kilometers (10 mi) wide, with the waterway forming its centerline (Fig. 2-13 inset). This corridor was ceded to the United States in 1904 by newly independent Panama under a treaty granting the United States 'all the rights, powers, and authority' in the Zone 'as if it were the sovereign of the territory.' In the 1970s, Panama sought to terminate U.S. control in the Canal Zone, and delicate negotiations were opened. In 1977, an agreement was reached on a staged withdrawal by the United States from the territory, first from the Canal Zone and then from the Panama Canal itself (a process completed on December 31, 1999).

Near the northern, Caribbean end of the Panama Canal lies the city of Colón, site of the Colón Free Zone, a huge trading hub designed to transfer and distribute goods on their way to South America. Near the southern, Pacific end lies the capital, Panama City, often likened to Miami because of its waterfront location and high-rise-dominated skyline. The capital is the financial center that handles the revenues generated by the canal, but its skyscrapered profile also reflects the proximity of Colombia's illicit drug industry and associated money-laundering and corruption.

Its 5000 containers piled high, a maximally-sized cargo carrier squeezes through a lock in one of the Panama Canal's original lanes in 2014 (left photo). The right photo, also taken in 2014, shows the much wider and deeper new third lane, about a year away from completion, near the city of Colón at the Caribbean outlet of the Canal. The long-awaited grand opening of the expanded waterway is slated to occur during the first quarter of 2016.

NEW FEATURES

1 Retaining basins fill lock chambers and retain 60 percent of the water for re-use.

2 Rolling lock gates slide in recesses, saving space and making maintenance work much easier.

3 Tugboats maneuver vessels into position, replacing costly towing locomotives.

Locomotives

Current locks ▼

Proposed new lane ▶

Locks transfer vessels from one elevation to another by filling closed chambers with water, or emptying them. When the level is even with that of the next chamber, the gates open.

3 *Tugboat*

POST-PANAMAX SHIP

PANAMAX SHIP Current maximum size

60 ft ← 160 ft → ← 180 ft →

← 106 ft → 42 ft ← 110 ft →

Adapting to larger ships

The most recent container vessels and tankers are known as post-Panamax. They can hold up to 12,000 containers, 2.5 times the cargo capacity of ships able to fit in the existing locks.

Mika Grondahl/NG Image Collection

Diagram showing the expansion and upgrading of the Panama Canal's lock system in order to accommodate ships that currently exceed the waterway's ('Panamax') capacity. The delayed 2016 target date for completing the new third lane was originally set for August 15, 2014, the Canal's one hundredth anniversary.

Between 13,000 and 14,000 ships sail through the Panama Canal each year, but all must be less than 32 meters (106 ft) in width. The Canal's much wider third lane, scheduled to open in early 2016, will accommodate vessels up to 55 meters (180 ft) in width that can carry two-and-a-half times as many shipping containers (up to 13,000)(see diagram and photos). This will boost interoceanic traffic and increase business opportunities throughout central Panama. U.S. ports on the Gulf of Mexico and the Atlantic seaboard will also get busier as they handle this heightened volume of cargoes. The ports of Miami and the New York–New Jersey metropolitan region have already begun to expand their capacities, and leading retailers such as Walmart (a prodigious importer of Chinese goods) are building additional storage facilities at the Port of Houston and other key coastal distribution points.

Even if its Nicaraguan canal never comes to fruition, China is likely to be the leading beneficiary of the upgraded Panama Canal because its rapid recent development has made it the world's number-one maritime exporter. That also underscores why this is such an outstanding example of just how tightly interconnected the world economy is becoming—and how the relative location of a tiny country in continental Middle America is of the utmost global significance.

THE CARIBBEAN BASIN

Fragmentation, Insularity, and Vulnerability

As Figure 2-7 reveals, the Caribbean Basin, Middle America's island arena, consists of a broad arc of numerous islands extending from the western tip of Cuba to the southern coast of Trinidad. The four larger islands or **Greater Antilles** (Cuba, Hispaniola, Jamaica, and Puerto Rico) are clustered in the east-west segment of this arc. The smaller islands or **Lesser Antilles** of the eastern segment extend southward within a crescent-shaped chain from the Virgin Islands to Trinidad and Tobago. (Breaking this tectonic-plate-related regularity are the Bahamas as well as the Turks and Caicos, located north of the Greater Antilles, and many other islands too small to appear on a map at the scale of Fig. 2-7.)

On these myriad islands, whose combined land area accounts for only 9 percent of the Middle American realm, lie 33 states and numerous other political entities. (Europe's colonial flags have not totally disappeared from this archipelago, and the U.S. flag flies over both Puerto Rico and the Virgin Islands.) The inhabitants of these states and territories, however, constitute 21 percent of Middle America's total population, making this the most densely peopled part of the Western Hemisphere.

The island-nations of the Caribbean Basin are generally very small, and their territories are separated, often at considerable distances from other islands. As noted earlier in this chapter, these geographic conditions create a set of circumstances that have proven to be quite challenging. Economic opportunities are minimal, most goods and services are relatively expensive, and due to limited interaction with the outside world, the island societies tend to be static.

It is this vulnerability of the small Caribbean islands, perhaps even more so than is the case for the Central American countries, that has triggered the recent sharp intensification of cocaine trafficking across this far-flung island chain. From the main cocaine distribution source in Colombia, a growing amount of the product is now being routed eastward across the South American continent—via Venezuela, northernmost Brazil, and the three coastal Guianas—in order to enter the Caribbean archipelago from the southeast. From there it proceeds to the United States, either directly or, more likely, indirectly through northern Central America and Mexico. The stressful effects of the growing drug trade on the fragmented, poverty-mired societies of the Lesser and Greater Antilles are identical to those that plague the central segment of the continental land bridge: expanding cocaine use and addiction among local populations; surges of organized criminal activity; and an alarming rise in the overall level of violence.

Social Cleavages

Social stratification **24** in most Caribbean islands is rigid, and upward mobility is decidedly limited. Class structures tend to be closely associated with ethnicity, and island societies still carry imprints of colonial times. The historical geography of Cuba, the Dominican Republic, and Puerto Rico is suffused with Hispanic culture; Haiti and Jamaica carry stronger African legacies. But the reality of this ethnic diversity is that European lineages still hold the advantage. Hispanics tend to be in the best positions in the Greater Antilles; people who have mixed European-African ancestries, and who are often described as **mulatto** **25**, rank next. The great bulk of this social pyramid is constituted by the most underprivileged: the Afro-Caribbean majority. In virtually all Caribbean societies, the advantaged minorities hold disproportionate power and exert overriding influence. In Haiti, the mulatto minority accounts for less than 5 percent of the population but has long held most of the power. In the neighboring Dominican Republic, the pyramid of power puts Hispanics (16 percent) at the top, the mixed sector (73 percent) in the middle, and the Afro-Caribbean minority (11 percent) at the bottom. In the Caribbean social mosaic, historical advantage has a way of perpetuating itself.

The composition of the population of the islands is further complicated by the presence of Asians from both China and India. During the nineteenth century, the emancipation of slaves and subsequent local labor shortages brought some far-reaching solutions. More than 100,000 Chinese emigrated to Cuba as indentured laborers; and Jamaica, Guadeloupe, and especially Trinidad saw nearly 250,000 South Asians arrive for similar purposes. To the African-modified forms of English and French heard in the Caribbean Basin, therefore, can be added several Asian languages. The ethnic and cultural diversity of the societies of Caribbean Middle America seems endless.

THE GREATER ANTILLES

The four islands of the Greater Antilles (whose populations constitute 90 percent of the Caribbean Basin's total) contain five political entities: Cuba, Haiti, the Dominican Republic, Jamaica, and Puerto Rico (Fig. 2-7). Haiti and the Dominican Republic share the island of Hispaniola.

Cuba

The largest Caribbean island-state in terms of both territory (111,000 square kilometers/43,000 sq mi) and population (11.3 million), Cuba lies only 145 kilometers (90 mi) from the southernmost island of the Florida Keys (Fig. 2-15). Havana, the now-dilapidated capital, lies almost directly south of outermost Key West on the northwestern coast of the elongated island.

Cuba was a Spanish possession until 1898 when, with U.S. help in the ten-week-long Spanish-American War, it achieved independence. Fifty years later, an American-backed dictator was fully in control, and by the 1950s Havana had become an American playground. The island was ripe for revolution, and in 1959 Fidel Castro's insurgents gained control, thereby converting Cuba into a communist dictatorship and a client-state of the Sovier Union. Nearly a million Cubans fled the island for the United States, and Miami swiftly became the second-largest Cuban city after Havana.

In the fall of 1962, the world was on the brink of nuclear war when the United States called on the Soviet Union to remove its newly installed nuclear missiles from Cuba (aimed at the United States) or face reprisals. In the end, the Soviets conceded and in return Washington promised not to attack Cuba. The United States

also continued to retain the Guantánamo Bay Naval Base located near the southeastern tip of the island (Fig. 2-15), which was perpetually leased to the United States by the Cuban government in 1903. Since 2002, the Guantánamo base has acquired notoriety as a detention camp for 9/11-related terrorists as well as prisoners of war from both the Iraq and Afghanistan conflicts.

Castro's rule survived the 1991 demise of the Soviet Empire despite the loss of its benefactor's subsidies and sugar markets on which the country had long relied. As the map indicates, sugar was Cuba's economic mainstay for many years; the plantations, once the property of rich landowners, extend all across the island. But sugarcane is losing its position as the top-ranked Cuban foreign exchange earner. Mills have been closing down, and cane fields are being cleared for other crops as well as for pastures. However, Cuba has additional economic opportunities, especially in its highlands.

Cuba has three mountainous zones, of which the southeastern Sierra Maestra is the highest and most extensive. These highlands create considerable environmental diversity as reflected by sizeable timber-producing tropical forests and varied soils on which crops ranging from tobacco to coffee to rice to subtropical and tropical fruits are grown. Rice and beans are the staples, but Cuba is unable to meet its dietary needs and so must import food. The central and western savannas support livestock-raising. Even though Cuba has only limited mineral reserves, its nickel deposits are extensive and have been mined for more than a century.

In 2006, Fidel Castro became too ill to lead the country and turned over the reins to his brother Raúl. Since then, a series of (modest) liberalization measures have slowly taken effect. These changes were undoubtedly necessitated by the persistently deplorable condition of the Cuban economy. In late 2014, encouraged

FIGURE 2-15

© Jerry Ginsberg/Danita Delimont

Tourists strolling through the plaza in front of the Cuban capital's Cathedral of San Cristóbal de la Habana, a Spanish Baroque architectural gem completed in 1777. With the recent lifting of the U.S. embargo, the ranks of these tourists are sure to swell as a new wave of American visitors is expected to descend on Cuba in the second half of this decade.

by signs of reform in Cuba and a growing consensus that the American embargo of the island was now hurting the people more than the regime by depriving the country of the means to develop, the U.S. government announced the lifting of its embargo in a series of stages. Normal diplomatic relations, marked by fully functioning embassies, were on course to be restored during 2015-2016.

Jamaica

Across the deep Cayman Trench from southern Cuba lies Jamaica, and a cultural gulf separates these two countries as well. A former British dependency, Jamaica has an almost entirely Afro-Caribbean population. As a member of the British Commonwealth, it still recognizes the British monarch as the chief of state, represented by a governor-general. The effective head of government in this democratic country, however, is the prime minister. English remains the official language here, and British traditions still linger.

Smaller than Connecticut and with 2.7 million people, Jamaica has experienced a steadily declining national income over the past few decades despite its relatively slow population growth. Tourism has become the largest source of income, but the markets for bauxite (aluminum ore), of which this island was once a leading exporter, have dwindled. And like other Caribbean countries, Jamaica has difficulty making money from its sugar exports. Jamaican farmers also produce crops ranging from bananas to tobacco, but the country faces the disadvantages on world markets common to those in the global periphery. Meanwhile, Jamaica must import all of its oil and much of its food because the densely populated coastal flatlands suffer from overuse and shrinking harvests.

The capital, Kingston, lies on the southeastern coast and reflects Jamaica's economic struggle. Almost none of the hundreds of thousands of tourists who visit the country's beaches, explore its

Cockpit Country of awesome limestone towers and caverns, or populate the many cruise ships calling at Montego Bay or other points along the north coast get even a glimpse of what life is like for the ordinary Jamaican.

Haiti

Already the long-time poorest state in the Western Hemisphere, Haiti in the fall of 2008 was staggered by no less than four tropical cyclones that struck within a few weeks of each other. The storms and the floods they unleashed killed more than 800 people and uprooted some 800,000. Then, barely 15 months later came the monstrous earthquake of January 12, 2010, which produced devastation on a truly unimaginable scale. The capital city, Port-au-Prince, lay in ruins, most of the country's infrastructure collapsed, and the Haitian government became effectively invisible. Schools, hospitals, and railroads as well as the seaport and the airport all came to a standstill. Millions became unemployed overnight. But all of that paled in comparison to the scale of human suffering. In the months following the quake, the estimated death toll reached and then surpassed 300,000; at least another 300,000 victims were injured, many permanently disabled. About 1.5 million homeless people initially fled the capital for the countryside, and, as noted earlier in this chapter, five years after the disaster these homeless still numbered in the hundreds of thousands (see earlier photo of Port-au-Prince).

It was a tragedy of colossal proportions, its horrendous images broadcast around the world on television and the Internet. Large-scale international emergency aid arrived, but it became clear that, in the long run, much of the affected area would have to be totally rebuilt. In the catastrophe's aftermath, millions of Haitians continue to subsist in the most wretched circumstances imaginable.

The challenge of recovery remains a gargantuan undertaking because nature has not provided this country of 10.3 million with any breaks. Haiti lies along a particularly dangerous tectonic fault zone where the North American and Caribbean plates meet (Fig. 2-16, inset map). The quake of 2010, geologists say, had been in the making for centuries, and others could easily follow in the future. This country also lies astride the axis of Hurricane Alley, and even a single year without a major storm is something to be grateful for (fortunately including the five hurricane seasons following the earthquake).

Few countries in the world have a more checkered history than Haiti, where political instability, repression, and material deprivation have been constants since the nineteenth century. Haiti's GNI per capita is less than one-sixth of Jamaica's, a level below even that of many impoverished African countries. Compared to other countries in Middle and South America, its infant mortality rate is nearly three times as high. A shocking one-third of elementary-school-age girls have never even attended school. And because it has so few natural resources and almost nothing to offer in the international trading arena, most of Haiti's limited public expenditures depend on direct foreign aid. As for the limited private consumption that takes place, it relies heavily on remittances from the Haitian Diaspora in such cities as Miami, Paris, New York, and Montreal.

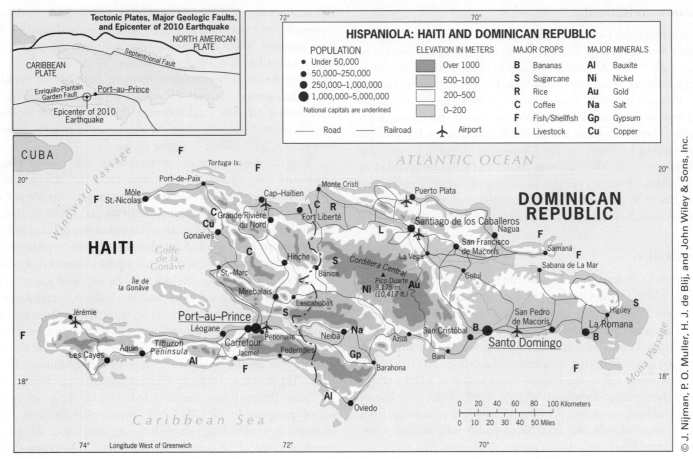

Tectonic Plates, Major Geologic Faults, and Epicenter of 2010 Earthquake

NORTH AMERICAN PLATE

Septentrional Fault

CARIBBEAN PLATE

Enriquillo-Plantain Garden Fault — Port-au-Prince

Epicenter of 2010 Earthquake

HISPANIOLA: HAITI AND DOMINICAN REPUBLIC

POPULATION
- Under 50,000
- 50,000–250,000
- 250,000–1,000,000
- 1,000,000–5,000,000

National capitals are underlined

ELEVATION IN METERS
- Over 1000
- 500–1000
- 200–500
- 0–200

MAJOR CROPS
- B Bananas
- S Sugarcane
- R Rice
- C Coffee
- F Fish/Shellfish
- L Livestock

MAJOR MINERALS
- Al Bauxite
- Ni Nickel
- Au Gold
- Na Salt
- Gp Gypsum
- Cu Copper

— Road — Railroad ✈ Airport

© J. Nijman, P. O. Muller, H. J. de Blij, and John Wiley & Sons, Inc.

FIGURE 2-16

© H.J. de Blij

@ FROM THE FIELD NOTES...

"Fly along the political boundary between Haiti and the Dominican Republic, and you see long stretches of the border marked by the starkest possible contrast in vegetation: denudation prevails to the west in Haiti while the forest prospers on the Dominican (eastern) side. On the Haitian side, overpopulation, widespread grinding poverty, and chronic government mismanagement—all exacerbated by the snail's pace of earthquake recovery—combine to create one of the Caribbean's most jarring cultural-landscape divides."

Dominican Republic

The mountainous Dominican Republic has a larger share of Hispaniola than Haiti (Fig. 2-16) in terms of territory (64 percent of the island), but at 10.4 million is nearly equal in size to the Haitian population. 'The Dominican,' too, has a wide range of natural environments but also a much stronger resource base than its neighbor to the west. Nickel, gold, and silver have long been exported along with sugar, tobacco, coffee, and cocoa, but tourism is the leading industry. Yet the Dominican Republic's economic performance since 2000 has been erratic and today remains below its potential. However, compared to near-destitute, low-income Haiti, the D.R. still ranks two developmental categories higher as an upper-middle-income country (see Fig. G-10).

Puerto Rico

The largest U.S. domain in Middle America, this easternmost and smallest island of the Greater Antilles region covers 9000 square kilometers (3500 sq mi) and has a population of 3.6 million. Puerto Rico is larger than Delaware and more populous than Iowa. Most Puerto Ricans are concentrated in the urbanized northeastern sector of this rectangular island (Fig. 2-17).

Puerto Rico fell to the United States well over a century ago during the Spanish-American War of 1898. Since the Puerto Ricans had been struggling for some time to free themselves from Spanish imperial rule, this transfer of power was, in their view, only a change from one colonial master to another. As a result, the first half-century of U.S. administration was difficult, and it was not until 1948 that Puerto Ricans were permitted to elect their own governor. When the island's voters approved the creation of a Commonwealth in a 1952 referendum, Washington, D.C. and San Juan, the seats of government involved, entered into a complicated arrangement. Puerto Ricans are U.S. citizens but pay no federal taxes on their local incomes. Puerto Rico also receives a sizeable annual subsidy from Washington, averaging approximately U.S. $4.2 billion in recent years.

Despite these apparent advantages in the poverty-plagued Caribbean, Puerto Rico has not thrived under U.S. administration. Long dependent on a single-crop economy (sugar), the island based its industrial development during the 1950s and 1960s on its comparatively cheap labor, tax breaks for corporations, political stability, and special access to the U.S. market. Consequently, pharmaceuticals, electronic equipment, and apparel top the current list of exports, not sugar or bananas. But this industrialization failed to stem a tide of emigration that carried more than one million Puerto Ricans to the New York City area alone. The same wage advantages that favored corporations kept many Puerto Ricans poor or unemployed. The level of unemployment on the island in 2015 stood at about 12 percent, just around twice the average in the continental United States. The private sector today remains severely underdeveloped.

Most Puerto Ricans are now fed up with their lack of economic progress. In the 2012 U.S. presidential election, which included a nonbinding referendum on Statehood, 61 percent voted in favor of becoming the 51st State. Discussions about

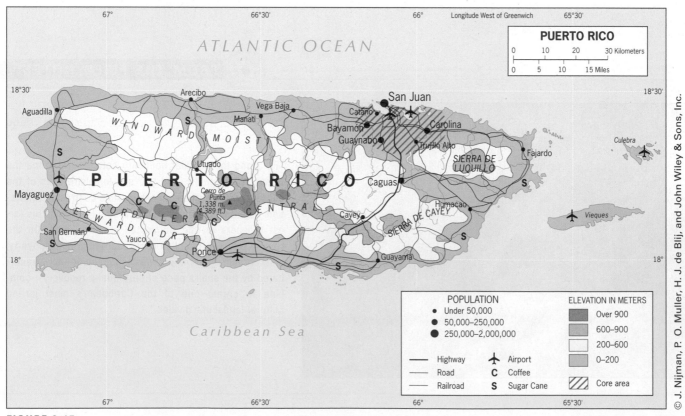

FIGURE 2-17

implementation (which has political implications for the deeply divided U.S. Congress) as well as the process and conditions of accession are likely to continue for at least the rest of this decade.

THE LESSER ANTILLES

As Figure 2-7 shows, the Greater Antilles are flanked by two clusters of islands: the extensive Bahamas/Turks and Caicos archipelago to the north and the Lesser Antilles to the southeast. **The Bahamas**, a former British colony that is now the closest Caribbean neighbor to the United States, alone consists of nearly 3000 coral islands—most of them rocky, barren, and uninhabited—but approximately 700 carry vegetation, of which roughly 30 are inhabited. Centrally positioned New Providence Island houses most of the country's 380,000 inhabitants and also contains the capital, Nassau, a leading tourist attraction.

The Lesser Antilles are grouped geographically into the Leeward Islands and the Windward Islands, a (climatologically incorrect) reference to the prevailing airflows in this tropical zone. The Leeward Islands extend southward from the U.S. Virgin Islands to the French dependencies of Guadeloupe and Martinique, and the Windward Islands from St. Lucia south and then west to the Dutch-affiliated islands of Aruba and Curaçao off the northwestern Venezuelan coast (Fig. 2-7).

It would be impractical to detail the individual geographic characteristics of each island of the Lesser Antilles, but we should note that nearly all share the insularity and environmental risks of the Caribbean Basin—major earthquakes, volcanic eruptions, and hurricanes; that they confront, to varying degrees, similar social and economic problems in the form of limited domestic resources, overpopulation, soil exhaustion, land fragmentation, and rising levels of crime and violence; and that tourism has become the leading industry for many.

Under these kinds of circumstances one looks for any hopeful sign, and such an indication comes from **Trinidad and Tobago**, the two-island republic at the southern end of the Lesser Antilles (Fig. 2-7). This country (population: 1.4 million) has embarked on a natural-gas-driven industrialization boom that could turn it into a future economic tiger. Trinidad has long been an oil producer and has also become the largest supplier of liquefied natural gas for the United States. Many of the new industrial facilities have agglomerated at the state-of-the-art Point Lisas Industrial Estate outside the capital, Port of Spain, and they have propelled Trinidad to become the world's leading exporter of ammonia and methanol. Natural gas also is an efficient fuel for the manufacturing of metals, and steelmakers as well as aluminum refiners have been attracted to locate here. With Trinidad lying just off the Venezuelan coast of South America, it is also a sea-lane crossroads that is strongly connected to the vast, near-coastal supplies of iron ore and bauxite that are mined in nearby countries, most importantly the Brazilian Amazon.

But Trinidad is an exception. It is a reflection of the predicament of small-island developing economies that, after the initial wave of decolonization in the 1950s and 1960s, several territories

Boutin/Sipa Press

The capital city of Trinidad and Tobago may have an historic colonial name (Port of Spain), but after nearly three centuries of Spanish rule, the British took control here. English became the *lingua franca* (everyone's second language), democratic government followed independence in 1962, and the development of natural-gas reserves has recently come to propel a thriving economy. As can be seen here, the harbor of Port of Spain is bursting at the seams as a car carrier delivers automobiles from Japan, and containers are stacked high on the docks.

decided that they were probably better off affiliating with the European country that had ruled them. Thus, for example, Guadeloupe remained with France and the Cayman Islands are still British. The former Netherlands Antilles passed up full independence to continue their affiliation with the Dutch but did so individually and with considerable autonomy: **Curaçao**, **Aruba**, and **St. Maarten** are now 'countries' within the Kingdom of the Netherlands, whereas the other Caribbean islands of the former Dutch Empire have acquired the status of overseas municipalities. Although economic logic seems to dictate continued association with the former colonial powers, local politics and pride demand some minimal degree of autonomy and avoidance of paternalistic European interference. The combined result is often a complicated legal framework that shapes the contemporary political status of the islands.

POINTS TO PONDER ?

- The United States is the single most important economic partner of just about every country and territory in this realm—is that a positive or a negative for Middle America?

- The 3169-kilometer (1969-mi) land border between North America and Middle America is the longest in the world that separates a rich realm from a poor one.

- The relative location of a small Middle American country, Panama, significantly impacts the global economy and especially its two biggest players, the United States and China.

- Most of the small Caribbean island-nations contain less than half a million people—is that enough to sustain a viable economy?

Led by its many European-heritage communities, Columbus Day (October 12) has long been celebrated throughout South America. But the tide is now steadily turning as the realm's indigenous peoples gain greater social traction, increasingly asserting their identities and traditions. In Chile, rising numbers now refer to this holiday as *The Day of the Encounter Between Two Worlds*. In the capital city, Santiago, on October 12, 2013, major clashes broke out between police and hundreds of protesters marching for indigenous rights. One year later, Santiago's official Columbus Day celebration was joined by a much larger but more peaceful march to raise the nation's consciousness about continuing discrimination against native minorities. Here on the capital's Alameda Avenue, this 2014 march was led by Chile's biggest indigenous minority—the Mapuche Amerindians. The child is holding the Mapuche flag, symbol of resistance as well as demands for autonomy arising from the government's persistently heavy-handed approach to handling land-rights disputes in south-central Chile's Mapuche homeland.

3

THE SOUTH AMERICAN REALM

IN THIS CHAPTER

- Challenges of regional integration
- South America after the commodity boom
- The strengthening indigenous peoples' movement
- China's economic presence in South America
- Medellín: From cocaine capital to model city
- Brazil sets an example in controlling deforestation

CONCEPTS, IDEAS, AND TERMS

© Mario Tellez/Demotix / Corbis Images

FIGURE 3-1

O f all the continents, South America has the most familiar shape—a gigantic triangle connected by mainland Middle America's tenuous land bridge to its neighbor in the north. South America also lies not only south but mostly east of its northern counterpart. Lima, the capital of Peru—one of the continent's westernmost cities—lies farther east than Miami, Florida. Thus South America juts out much more prominently into the Atlantic Ocean toward southern Europe and Africa than does North America. But lying so far eastward means that South America's western flank faces a much wider Pacific Ocean, with the distance from Peru to Australia nearly twice that from California to Japan.

DEFINING THE REALM

As if to reaffirm South America's northward and eastward orientation, the western margins of the continent are rimmed by one of the world's longest and highest mountain ranges, the Andes, a giant wall that extends unbroken from Tierra del Fuego near the continent's southern tip in Chile to northeastern Venezuela in the far north (Fig. 3-1). The other major physiographic feature of South America dominates its central north—the Amazon Basin; this vast humid-tropical amphitheater is drained by the mighty Amazon River, which is fed by several major tributaries. Much of the remainder of the continent can be classified as plateau, with the most important components being the Brazilian Highlands that cover most of Brazil southeast of the Amazon Basin, the Guiana Highlands located north of the lower Amazon Basin, and the cold Patagonian Plateau that blankets the southern third of Argentina. Figure 3-1 also reveals two other noteworthy river basins beyond Amazonia: the Paraná-Paraguay Basin of south-central South America, and the Orinoco Basin in the far north that drains interior Colombia and Venezuela.

PHYSIOGRAPHY

Explorers' Continent

It was here in northern South America that the great German explorer and scientist Alexander von Humboldt, one of the founders of the modern discipline of geography, embarked on his legendary expeditions in the early nineteenth century. After landing on the coast of Venezuela and trekking across the continent's northern interior, the 30-year-old Humboldt was struck by the area's biodiversity, its majestic natural beauty, and the adaptive abilities of the human populations. He discovered and named many species of flora and fauna, traversed tropical grasslands and jungles, met with indigenous peoples, crossed dangerous rivers, and reached the summit of the highest mountain in the Americas climbed by Europeans at the time (Ecuador's Chimborazo). He compiled large numbers of maps and was one of the first scientists to note how the coastlines of eastern South America and western Africa fitted together like pieces of a jigsaw puzzle, speculating about the

---

MAJOR GEOGRAPHIC QUALITIES

SOUTH AMERICA

1. South America's physiography is dominated by the Andes Mountains in the west and the Amazon Basin in the central north. Much of the remainder is plateau country.

2. Almost half of the realm's area and just under half of its total population are concentrated in one country—Brazil.

3. South America's population remains concentrated along the continent's margins. Most of the interior is sparsely peopled, but sections of it are now undergoing significant development.

4. Interconnections among the states of the realm are improving. Economic integration has become a growing force but is still at an early stage.

5. Regional economic contrasts and disparities remain strong, both in the realm as a whole and within individual countries.

6. Cultural pluralism prevails in almost all of the realm's countries and is often expressed regionally.

7. Rapid urban growth continues to mark much of the South American realm, and urbanization overall is today on a par with the levels of the United States and western Europe.

8. This realm contains abundant natural riches that, although important for exports, tend to create considerable dependency on economic conditions in world markets.

continents' geologic movements. Humboldt was a key figure in the rise of the modern discipline of geography because of his many discoveries as well as his views on the **unity of place** **1**—that in a particular locale or region intricate connections exist between climate, geology, biology, and human cultures. As such, he laid the foundation for modern geography as an *integrative discipline* with a spatial perspective.

Almost three centuries earlier, the opposite end of South America was the scene of a crucial stage in the first circumnavigation of the globe and expedition led by Ferdinand Magellan. Once it became clear that Columbus had stumbled upon America, not India, Spanish and Portuguese explorers continued their efforts to discover a westward passage from the Atlantic to the Pacific. This took them to the far south along what is now the Argentinean coast. Magellan's five ships set sail for the treacherous waters of what he named the *Estrecho de Todos los Santos*—now called the Strait of Magellan (Fig. 3-1). Of the (remaining) four ships in Magellan's fleet, only three survived the daring, 600-kilometer (375-mi)-long passage; the fourth crashed on the rocks in the icy waters of the Southern Ocean.

Myriad Climates and Habitats

On the world map, if the Russia/Central Asia realm is the widest in east-west extent, South America is the longest measured from north to south. (Within South America, no other country is more emblematic of this elongated geography than Chile, averaging only about 150 kilometers [90 mi] in width but 4000 kilometers [2500 mi] in length.) As a consequence of this huge latitudinal span, from about 12°N to 54°S, this realm contains an enormous variety of climates and vegetation. Combine this with substantial variation in elevation from west to east, and it is clear why South America has such an impressive range of natural habitats.

Take another look at the map showing the global distribution of climates (Fig. G-6) and note the variation in climatic types, particularly in the realm's northwest and southern half. Travel northeast from Lima, Peru, for about 600 kilometers (375 mi) and you encounter no less than four different climate zones: arid, highland, and two varieties of humid tropical. A transect of similar length from Santiago, Chile, eastward across the continent to Buenos Aires, Argentina, will take you through five climate zones: interior highland, arid, and semiarid environments bracketed by a different humid-temperate climate along each coast. Vegetation in South America varies accordingly, from lush tropical rainforests to rocky and barren snow-covered mountaintops to grasslands, fertile as well as parched. This natural diversity also makes for considerable cultural differences, as we shall soon see.

STATES ANCIENT AND MODERN

Thousands of years before the first European invaders appeared on the shores of South America, peoples now referred to as indigenous **2** had migrated into the continent via North and Middle America and founded societies in coastal valleys, in river basins, on plateaus, and in mountainous locales. These societies achieved different and remarkable adaptations to their diverse natural environments, and by about 1000 years ago, a number of regional cultures thrived in the elongated valleys between mountain ranges of the Andes from present-day Colombia southward to Bolivia and Chile. These high-altitude valleys, called **altiplanos** **3**, provided fertile soils, reliable water supplies, building materials, and natural protection to their inhabitants.

The Inca State

One of these *altiplanos*, surrounding the city of Cuzco in what is now southeastern Peru, became the core area of South America's greatest indigenous empire, that of the **Inca**. The Inca were expert builders whose stone structures (among which Machu Picchu near Cuzco is the most famous), roads, and bridges helped to unify their vast empire; they also proved themselves to be efficient administrators, successful farmers and herders, and skilled manufacturers; their scholars studied the heavens, and their physicians even experimented with brain surgery. Great military strategists, the Inca integrated the peoples they vanquished into a stable and well-functioning state, an amazing accomplishment given their need to contend with treacherous, high-relief terrain.

As a minority ruling elite in their far-flung empire, the Inca were at the pinnacle in their rigidly class-structured, highly centralized society. So centralized and authoritarian was their state that a takeover at the top was enough to gain immediate power over all of it—as a small army of Spanish invaders discovered in the 1530s. The European invasion brought a quick end to thousands of years of indigenous cultural development and changed the social order forever.

The Iberian Invaders

The modern map of South America started to take shape when the Iberian colonizers began to understand the location and economies of the indigenous societies. The Inca, like Mexico's Maya and Aztec peoples, had accumulated gold and silver at their headquarters, possessed productive farmlands, and constituted a ready labor force. Not long after the defeat of the Aztecs in 1521, Francisco Pizarro sailed southward along the continent's northwestern coast, learned of the existence of the Inca Empire, and withdrew to Spain to organize its overthrow. He returned to the Peruvian coast in 1531 with 183 men and two dozen horses, and the events that followed are well known. Only two years later, his party rode victorious into Cuzco.

The new order that emerged in western South America placed the indigenous peoples in serfdom to the Spaniards. Great haciendas were formed by **land alienation** **4** (the takeover of indigenously held land by foreigners), taxes were instituted, and a forced-labor system was introduced to maximize the profits of exploitation. Lima, the west-coast headquarters of the Spanish conquerors, soon became one of the richest cities in the world, its wealth based on the exploitation of vast Andean silver deposits. Lima also became the seat of the Viceroyalty of Peru, the political administrative district (later subdivided into a triad of viceroyalties) whereby its South American territories were incorporated into the Spanish Empire.

Meanwhile, another vanguard of the Iberian invasion was penetrating the east-central part of the continent, the coastlands of present-day Brazil. This area had become a Portuguese sphere of influence, and Brazil's boundaries were bent far inland to include almost the entire Amazon Basin. Eventually, the country came to be only slightly smaller in territorial size than all the other South American countries combined. This westward thrust was the result of Portuguese and Brazilian penetration, particularly by the *Paulistas*, the settlers of São Paulo who needed indigenous slave labor to run their plantations.

Independence and Isolation

Despite their adjacent location on the same continent, their common language and cultural heritage, and their shared national problems, the countries that arose out of South America's Spanish viceroyalties (together with Brazil) until quite recently existed in a considerable degree of isolation from one another. Distance as well as physiographic barriers reinforced this separation, and the realm's major population agglomerations still adjoin the coast, mainly the eastern and northern coasts (Fig. 3-2). The viceroyalties existed primarily to extract riches and fill Spanish coffers. In Iberia, there was little interest in developing the American lands for their own sake. Only after those who had a stake in Spanish and Portuguese America, having made it their new home and had therefore rebelled against Iberian authority, did things begin to change, and then very slowly. Thus South America was saddled with the values, economic outlook, and social attitudes of eighteenth-century Iberia—not the best tradition from which to begin the task of forging modern countries.

Certain isolating factors had their effect even during the wars for independence. Spanish military strength was always concentrated at Lima, and those territories that lay farthest from their center of power—Argentina as well as Chile—were the first to gain independence from Spain (in 1816 and 1818, respectively). In the north, Simón Bolívar led the burgeoning independence movement, and in 1824 two decisive military defeats there spelled the end of Spanish power in South America.

This joint struggle, however, did not produce unity because no fewer than nine countries emerged from the three former viceroyalties. It is not difficult to understand why this fragmentation took place. With the Andes intervening between Argentina and Chile and the Atacama Desert between Chile and Peru, overland distances seemed even greater than they really were, and these obstacles to contact proved quite effective. Hence, from their outset the new countries of South America began to grow apart amid friction that occasionally escalated into local wars. In fact, only within the past few decades have the countries of this realm finally begun to recognize the mutual advantages of increasing cooperation and to make lasting efforts to steer their relationships in this direction.

The Population Map—Then and Now

If we were able to reconstruct a (pre-Columbian) map of South America's population before the arrival of the Europeans, it would vary sharply from the current one shown in Figure 3-2. Indigenous societies inhabited not only the Andes and adjacent lowlands but also riverbanks in the Amazon Basin, where settlements numbering in the thousands subsisted on fishing and farming. They did not shy away from harsh environments such as those of the island of Tierra del Fuego in the far south, where the fires they kept going against the bitter cold led the Europeans to name the place 'land of fire.'

Today, the realm's population distribution looks quite different. Many of the indigenous societies succumbed to the European invaders, not just through warfare but also because of the diseases

SOUTH AMERICA: POPULATION DISTRIBUTION, 2016
One dot represents 50,000 persons

0   600   1200 Kilometers

0   300   600 Miles

FIGURE 3-2

© Jan Nijman

@ **FROM THE FIELD NOTES...**

"Cuzco is the historic capital of the Inca Empire, located high in the Peruvian Andes at an elevation of 3400 meters (11,200 ft). The city center is a UNESCO-designated World Heritage Site and attracts well over a million tourists each year. I drove out of the valley up the hillside to get a good view of the layout of the center, so typical of Spanish colonial cities. The focal Plaza de Armas with its parks and benches is surrounded by the Cathedral of Santo Domingo and other religious and municipal buildings, and adjacent blocks around them are used for commerce and retailing. Located farther away from the Plaza are newer residential areas. This historic Inca capital was sacked in the early sixteenth century by Pizarro's conquistadores; few Inca ruins remain in the city center, but just outside Cuzco lies the impressive Sacsayhuaman fortress, with most of its walls still intact. This is not a very large city—it is too remote for that—but Cuzco's population has tripled over the past quarter-century to nearly 500,000."

the Iberian conquerors brought with them. It is estimated that 90 percent of native Amazonians died within a few years of contact, and even the peoples of isolated Tierra del Fuego are no longer there to build their fires. From one end of South America to the other, the European intrusion spelled disaster.

Spanish and Portuguese colonists penetrated the interior of South America, but the great majority of the settlers stayed on or near the coast, a pattern still evident today. Almost all of the realm's major cities have coastal or near-coastal locations, and the current population distribution map gives you the impression of a continent still awaiting penetration and settlement. But another look at Figure 3-2 will reveal a swath of population located well inland from the settlements along the northwestern coast, most clearly in Peru but also extending northward into Ecuador and southward into Bolivia. That is the legacy of the Inca Empire and its incorporated peoples, surviving in their highland redoubt and still numbering in the tens of millions.

THE CULTURAL MOSAIC

Indigenous Reawakening

In the early twenty-first century, South America's long-downtrodden indigenous peoples are staging a social, political, and economic reawakening (see chapter-opening photo). In several of the realm's countries, where their numbers are large enough to translate into political strength, they have begun to realize their potential. This is

especially the case in the Andean countries: indigenous peoples in Peru constitute no less than 45 percent of the national population, and in Bolivia they are in the majority at 55 percent.

Newly empowered indigenous political leaders have emerged and are now bringing the plight of the realm's aboriginal peoples to both local and international attention. South America's native peoples were conquered, decimated by foreign diseases, robbed of their best lands, subjected to involuntary labor, denied the right to grow their traditional crops, socially discriminated against, and swindled out of their fair share of the revenues from resources in their traditional domains. They may still rank among the poorest of the realm's poor, but they are increasingly asserting themselves. For a number of South American states, the consequences of this movement will be far-reaching.

The indigenous reawakening is in part related to changing religious practices in South America. Officially, just over 80 percent of the population is Roman Catholic, and traditionally South Americans tend to be viewed as devout followers of the Vatican. But recent surveys show that many do not attend church regularly and that more than half of all Catholics describe themselves as believers of the doctrine, rather than adherents of the Church. Since the late nineteenth century, the Catholic Church has often been criticized for its conservative position on social issues and for always siding with the establishment. During the 1950s, a powerful movement known as **liberation theology** 5 emerged in South America and subsequently gained followers around the world. This movement was a blend of Christian religion and humanist philosophy that interpreted the teachings of Christ as a quest to liberate the impoverished masses from oppression. Although the Church has lately been trying to make amends, it could not avoid losing popular support, especially among indigenous peoples.

African Descendants

When the Portuguese began to develop their territory on the coast of what is now Brazil, they turned to the same lucrative activity that their Spanish rivals had pursued in the Caribbean—the plantation cultivation of sugar for the European market. And they, too, found their workforce in the same source region, as millions of Africans (nearly half of all who came to the Americas) were brought in bondage to the tropical Brazilian coast north of Rio de Janeiro. Not surprisingly, Brazil now has South America's largest black population, which is still heavily concentrated in the country's poverty-mired northeastern States. With Brazilians of direct or mixed African ancestry today accounting for just over half of the population of 201 million, the Africans decidedly constitute a major component in the immigration of foreign peoples into South America.

Ethnic Landscapes

The cultural landscape of South America, similar to that of Middle America, is a layered one. Indigenous inhabitants cultivated and crafted diverse landscapes throughout the continent, some producing greater impacts than others. When the Europeans arrived, the cultural transformation that resulted from depopulation severely impacted the environment. Native peoples now became minorities in their own lands, and Europeans introduced crops, animals, and ideas about land ownership and land use that irreversibly changed South America. They also brought in African slaves from various parts of Subsaharan Africa. Europeans from non-Iberian Europe also began emigrating to South America, particularly during the first half of the twentieth century. Japanese settlers arrived in Brazil and Peru during the same era. All of these elements have contributed to the present-day ethnic complexion of this realm. Figure 3-3 shows the distinct concentrations of indigenous and African cultural dominance, as well as areas where these groups are largely absent and people of European ancestry predominate.

Of course, ethnic origins are not always so straightforward, and patterns can change as a result of internal migrations and ethnic mixing. In recent years, research on individual DNA and genetics has taught us a lot about the regional and group origins of people. On the map, Argentina is indicated as having a predominantly European ancestry. More specifically, recent research shows that, for the average Argentine, nearly 80 percent of his or her genetic structure is European, 18 percent aboriginal, and 2 percent African. If nobody had any mixed ancestors, these percentages would translate perfectly into European, indigenous, and African shares of the population—but of course that is not the case. Many people do not have perfect knowledge about their ethnic ancestry, and although some may be descendants of people with a single origin, many others exhibit at least some degree of mixed ancestry. In the aggregate, however, there is no doubt that the population of cone-shaped southern South America is

FIGURE 3-3

© J. Nijman, P. O. Muller, H. J. de Blij, and John Wiley & Sons, Inc.

predominantly of European origin. South America at large is a realm marked by **cultural pluralism** **6**, where peoples of various ethnicities and cultures cluster in adjacent areas but generally do not mix. The result is a cultural mosaic of almost endless variety.

ECONOMIC GEOGRAPHY

Agricultural Land Use and Deforestation

The internal divisions of this cultural kaleidoscope are further reflected in the realm's economic landscape. In South America, larger-scale **commercial** **7** or market (for-profit) and smaller-scale **subsistence agriculture** **8** (primarily for household use) exist side by side to a greater degree than anywhere else in the world (Fig. 3-4). The geography of plantations and other commercial agricultural systems was initially tied to the distribution of landholders of European background, whereas subsistence farming (such as highland mixed subsistence-market, agroforestry, and shifting cultivation) is historically associated with the spatial patterns of indigenous peoples as well as populations of African and Asian descent.

At a broader level of generalization, agricultural systems and land use in South America also vary in close relationship to physiography, as one would expect. Forestry and agroforestry prevail in the Amazon Basin; ranching dominates in the grasslands of the Southern Cone; and different forms of agriculture, with or without irrigation, are found in an extensive zone from northeastern Brazil to northern Argentina, as well as scattered among pockets of moderate elevation in the Andean highlands. Land uses throughout South America are changing rapidly today, owing mainly to ongoing deforestation and the introduction or expansion of new crops. The fastest-growing crop is soybeans, the cultivation of which now dominates much of east-central Brazil and spills over into adjacent areas of Paraguay, Uruguay, and Argentina (Fig. 3-4).

Deforestation is a global problem, but it has been a particularly acute problem in northern Brazil. In the past, deforestation was attributed mainly to small-scale landholders, colonists who had made their way into the Amazon rainforest to eke out a new living. The deforestation crisis intensified as *agro-industrial* operations engaged in large-scale production for the export market increasingly penetrated Amazonia and transformed its land cover. The exact loss of rainforest is difficult to measure, but it has been estimated that Brazil lost an area of rainforest the size of Ohio every year between 1980 and 2000. Since then, the government has succeeded in reducing the rate of deforestation while increasing the area of protected forests—but in absolute terms, deforestation continues.

As South American economies modernize, agriculture remains highly important. Almost one-fifth of this realm's workforce is still employed in the primary sector that includes farming, cattle-raising, and fishing—a much greater share than in North America. The South American contribution to global trade in grains, soybeans, coffee, orange juice, sugar, and many other crops is significant and still increasing. And in addition to all these successful agricultural endeavors, illegal farming thrives in the form of narcotics production, particularly cocaine.

The Geography of Cocaine

All of the cocaine that enters the United States comes from South America, mainly from Colombia, Peru, and Bolivia. Within these three countries, cocaine annually brings in billions of (U.S.) dollars and 'employs' tens of thousands of workers, constituting a powerful industry. The first of the three stages of cocaine production involves the extraction of coca paste from the coca plant, a raw-material-oriented activity that is located near the areas where the plant is grown. The main zone of coca plant cultivation is along the eastern slopes of the Andes and in adjacent tropical lowlands in Bolivia, Peru, and Colombia (Fig. 3-5).

Coca leaves harvested in the source areas of the Andes and nearby interior lowlands make their way to local collection centers, located at the convergence of rivers and trails, where coca paste is extracted and prepared. The second stage of production involves the refining of that coca paste (about 40 percent pure cocaine) into cocaine hydrochloride (more than 90 percent pure), a lethal concentrate that is diluted with substances such as sugar or flour before being sold on the streets to consumers. Cocaine refining requires sophisticated chemicals, carefully controlled processes, and a labor force skilled in their supervision. Most of this processing stage is

SOUTH AMERICA: AGRICULTURAL SYSTEMS

- Oases
- Scattered Pastoralism
- Non-Soybean Grain Crops
- Dairy Products and Flowers for Export
- Mixed Dry Farming
- Soybeans
- Cattle
- Plantation
- Agroforestry & Shifting Cultivation
- Highland Mixed Subsistence-Market
- Mixed
- Nonagricultural areas

0   500   1000   1500 Kilometers
0   500   1000 Miles

© J. Nijman, P. O. Muller, H. J. de Blij, and John Wiley & Sons, Inc.

FIGURE 3-4

dominated by Colombia and takes place in the lowland central-south and east, beyond the reach of the Bogotá government.

The final stage of production entails the distribution of cocaine to the marketplace, which depends on an efficient, clandestine transportation network that leads into the United States. Private planes operating out of remote airstrips were the preferred 'exporting' method until about 1990, and in the 1970s and 1980s Miami was a major center of cocaine distribution in the United States. But aggressive U.S. measures along the coasts of Florida caused a shift toward Mexico (and also, since the mid-2000s, via countries of northern Central America). Today, most cocaine still reaches the United States through smuggling across the Mexican border (a topic elaborated in Chapter 2). In this decade, another route has opened: small planes fly the product eastward from

**COCA PLANT PRODUCTION
IN NORTHWESTERN SOUTH AMERICA**

Coca plant growng area
(Colombian territories are not named)

FIGURE 3-5

Colombia to Venezuela and the Guianas; from there it is transshipped northward by boat or plane to various Caribbean islands; subsequently, it is routed westward to Central America or Mexico, sometimes even directly into the United States.

Industrial Development

The geography of industrial production varies markedly in this realm. Brazil and Chile have shown rapid growth; Argentina has been in long-term decline; and countries such as Peru, Ecuador, and Bolivia have long struggled to modernize their economies and improve standards of living. Within countries, there is substantial **uneven development 9** as well. Most manufacturing is concentrated in and around major urban centers, leaving vast empty spaces in the interior of the realm.

Brazil, in particular, increasingly drew attention over the past decade for its momentous growth and the rising sophistication of a number of economic sectors. In the past, typical Brazilian exports were foods and footwear. Today, the leading exports include oil (thanks to the fortuitous discovery of huge offshore reserves), steel, and state-of-the-art Embraer aircraft.

Despite recent impressive economic growth across much of the realm, it is clear that the longstanding isolation of its individual countries remains a serious impediment. Most governments agree that tighter integration, political as well as in terms of transportation infrastructure, would greatly enhance economic opportunities.

Economic Integration

The realm continues to be economically fragmented, and only about 25 percent of its exports stay within South America (a very low share compared to Europe, where intra-realm exports account for nearly 70 percent). Nonetheless, things are moving in the right direction. With mutually advantageous trade the catalyst, a new continent-wide spirit of cooperation is blossoming at every level. Cross-border rail, road, and pipeline projects, stalled for years, are multiplying steadily. And importantly, investments now flow more freely than ever from one country to another, most notably in the agricultural sector.

Recognizing that free trade may well be able to solve many of the realm's economic-geographic problems, governments are now pursuing multiple avenues of heightened international cooperation. This is a good example of the concept of **supranationalism 10**, which geographers define as a voluntary association in economic, political, or cultural spheres of three or more independent countries willing to yield some measure of sovereignty for their mutual benefit (a concept elaborated in Chapter 4).

In 2015, South America's republics were affiliating with a number of trading blocs, of which the following two are the most important:

- *Mercosur/l* (*Mercosur* in Spanish; *Mercosul* in Portuguese): Launched in 1995, this Common Market established a free-trade zone and customs union that now links Brazil, Argentina, Uruguay, Paraguay, and Venezuela (five additional western South American countries are associate members). The organization aspires to be the dominant free trade association for all of South America, but political agendas tend to prevail over economic needs. As a result, the path toward full economic integration is proving to be both slow-paced and obstacle-ridden.

- *Pacific Alliance (PA):* Inaugurated in 2012, this new bloc was founded by Mexico, Colombia, Peru, and Chile. They swiftly formed a free-trade area along the western edge of the continent and announced their commitment to further economic integration 'with a clear orientation to Asia.' Whereas Mercosur/l countries are often entangled in political debates, the PA has advanced steadily with a collective economic growth rate almost twice as high. Candidates for new membership, led by Costa Rica, Panama, and Guatemala, are lining up; other countries, including Canada and the United States, have acquired observer status. The rise of the Pacific Alliance in this decade, and its articulation of a strong liberal economic stance in comparison with Mercosur/l have spawned notions of an east-west 'continental divide' regarding the realm's evolving economic geography.

The Commodities Boom and Its Aftermath

During the first decade of this century, South America's economy in the aggregate grew robustly by about 5 percent annually. Much of this growth occurred in Brazil, Chile, and Peru, but other parts of the continent also fared well. A leading driver was the steadily rising demand for the realm's abundant raw materials in the global marketplace, in no small part resulting from the explosive growth (and accompanying raw-material appetite) of China. The Asian giant gobbled up Brazilian soybeans, Chilean copper, Peruvian silver, Venezuelan oil, and so much more. Commodity prices skyrocketed and South America reaped the benefits.

The economic picture in this decade, however, is quite different. Commodity-driven growth has leveled off markedly since 2011, much of it in response to slowing Asian demand and lowering prices on the global market. Although exports of raw materials are still substantial, the heyday hype of the commodity boom has faded. In its place has been a resurfacing of the pre-boom realization that heavy reliance on the production of raw materials (slightly more than 50 percent of the realm's exports today) is a risky longer-term proposition—for a variety of reasons.

First, the increasing demand for commodities tends to drive up the value of a producing country's currency, which in turn limits sales of other exports as they become more expensive for foreign trading partners. Second, higher revenues from the export of raw materials can undermine the development of local manufacturing and other processing industries. Some economists now employ the term *premature deindustrialization*, wherein domestic industries erode as government policies overly emphasize the exporting of commodities at a time of rising value in the local currency. Third, resource management, because of complex ownership issues and exploitation rights, invites both corruption and complacency (note, for instance, that major discoveries of oil reserves very rarely result in a society's advancement). Finally, and most significantly, the great majority of resources (especially key minerals and energy deposits) are nonrenewable, making it vital that producing countries plan for an economic future without them. Thus, in the aftermath of South America's commodity boom in the 2000s, its longer-term benefits may well depend on the ability of national governments to successfully refocus their energies on nurturing their home-grown industries.

URBANIZATION AND ITS REGIONAL EXPRESSIONS

Rural-Urban Migration

As in most other realms, South Americans are leaving the land and migrating to towns and cities. This realm's modern urbanization process got an early start and has been intensifying since 1950 as the growth of urban settlements has averaged 5 percent a year. Meanwhile, rural areas have grown by less than 2 percent annually over those same six-plus decades. As a result, the realm's urban population has climbed to its current level of 82 percent, ranking it with those of western Europe and the United States. These

An aerial image of one of South America's most spectacular raw-material sites: northern Brazil's Carajás Iron Ore Mine, located where the southeastern rim of the Amazon Basin meets the Brazilian Highlands in the center of Pará State. This may be the world's richest as well as largest proven deposit of iron ore, whose open-pit mines steadily deepen and sprawl outward as minerals are extracted from the surface—one stripped-away, environmentally degrading layer at a time. In 2012, 320 million metric tons were hauled away via a dedicated, 850-kilometer (535-mi) railroad to the Atlantic port of São Luis (see Fig. 3-16).

NASA image created by Jesse Allen, using EO-1 ALI data provided courtesy of the NASA EO-1 team

numbers underscore not only the dimensions but also the durability of the **rural-to-urban migration 11** from the countryside to the cities.

In South America, as in Middle America, Africa, and Asia, people, driven by the poverty of the rural areas, are attracted to the cities. Both ***push*** and ***pull*** factors are at work. Rural land reform has been very slow in coming, and for this and other reasons every year tens of thousands of farmers simply give up and leave, seeing little or no possibility of economic advancement. The urban centers lure them because they are perceived to provide opportunity—the chance to earn a regular wage, visions of education for their children, better medical care, upward social mobility, and the excitement of life in a big city.

But the move itself can be traumatic. Cities in developing countries are surrounded and often invaded by squalid slums, and this is where the urban immigrant most often finds a first—and sometimes permanent—abode in a makeshift shack without even the most basic amenities and sanitary facilities. But still the people come, hopeful for a better life. Many newcomers earn their first cash income by becoming part of the **informal sector 12**, in which workers are undocumented and money transactions are beyond the control of government. The entry-level settlements consist mostly of self-help housing, vast shantytowns known as **barrios 13** in Spanish-speaking South America and *favelas* **14** in Brazil. Some of their entrepreneurial inhabitants succeed more than others, transforming parts of these shantytowns into beehives of activity that can propel resourceful workers toward a middle-class existence.

Regional Patterns

The generalized spatial pattern of South America's urban transformation is displayed in Figure 3-6, a *cartogram* of the continent's population. Here we see not only the realm's countries in population space relative to one another, but also the proportionate sizes of individual large cities within their total national populations.

Regionally, the Southern Cone (countries colored green) is the most highly urbanized. Today in Argentina, Chile, and Uruguay, almost all of the population resides in cities. Ranking below them in urbanization is Brazil (tan colored). The next highest group of countries (beige) borders the Caribbean Sea in the north. Not surprisingly, the Andean countries (brown) constitute the realm's least urbanized zone. Figure 3-6 tells us a great deal about the relative positions of the largest metropolises within their countries. Three of them—Brazil's São Paulo and Rio de Janeiro, and Argentina's Buenos Aires—rank among the world's **megacities** **15**, in which

FIGURE 3-6

© J. Nijman, P. O. Muller, H. J. de Blij, and John Wiley & Sons, Inc.

Bambu Productions / Getty Images, Inc.

Brazil's São Paulo is the third-largest metropolis in the world and, with more than 26 million people, by far the biggest in this realm. Most of its growth over the past half-century is the result of massive rural in-migration. Severely overcrowded São Paulo has grown so rapidly that certain neighborhoods exhibit some of the highest residential densities on Earth. Moreover, this sprawling megacity has become notorious for its gridlocked automobile traffic and choked transit systems, which produce commuting times that are just about triple those in major U.S. cities. This photo was taken in late 2014 from a taxi helicopter, an elitist new mode of transportation used routinely by high-level business executives to beat the congestion below.

the metropolitan population exceeds 10 million. The size and density of these megacities pose enormous challenges in terms of housing, mobility, pollution, water provision, and overall environmental sustainability (see photo).

Problems of Inequality

An important indicator of South America's economic growth in this century has been the steady expansion of its middle class. The World Bank recently reported that this socioeconomic stratum had increased its share of the realm's income-earning population from 20 to 29 percent between 2000 and 2010. Nevertheless, longstanding deep divisions not only persist but continue to widen, requiring the government of every country to address these inequalities far more aggressively. By most measures, the astonishing disparity between rich and poor is still wider in this realm than in any other, and wealth remains more disproportionately concentrated than ever in the hands of a tiny minority.

EXTERNAL RELATIONS

The Shadow of the United States

The United States has long played a key role in this realm, beginning in 1823 with the Monroe Doctrine's assertion that European

MAP ANALYSIS

© J. Nijman, P. O. Muller, and John Wiley & Sons, Inc.

INCOMES AND THE PRICE OF iPHONE5 IN SOUTH AMERICA
Monthly Wage in US$

- Over 1,100
- 800 – 1,100
- 500 – 800

Price of iPhone5, US$
Unlocked, prices at official exchange rates
(Venezuela at black market rate)

FIGURE 3-7

There is considerable spatial variation in levels of development among and within South American countries. Figure 3-7 provides information on gross domestic incomes per month and the price of the iPhone5 in 2014. As a point of reference, note that in the United States the iPhone5 in 2014 cost around $700 and that monthly income per capita was $4,550. Now observe the spatial variation of incomes and the price of the iPhone5 in South America.

What is the relationship between these two variables? Is it what you would have expected? Access your WileyPLUS Learning Space course to interact with a dynamic version of this map and to engage with online map exercises and questions.

colonial powers no longer had rights to South America. During the Cold War (1945–1990), the United States became politically involved in a number of countries, mainly to keep Soviet influence out of the Western Hemisphere (above all in Chile during the early 1970s). This effort did not always enhance the local standing of the United States. In fact, anti-Americanism never seems far from the surface in South America, based heavily on the U.S. record of supporting right-wing dictatorships as well as continuing suspicions that the Americans still harbor imperialistic intentions.

Relations with the United States have never been smooth and were consistently asymmetrical. For instance, South America today attracts only about 4 percent of all U.S. foreign trade. But the United States remains the biggest trading partner for the realm as a whole, accounting for almost one-fifth of the exports and imports of all South American countries. It was hardly a coincidence that *dependencia* theory 16 originated in South America during the 1960s. It was a new way of thinking about economic development that interpreted the persistent poverty of certain countries as the outcome of their unequal relations with the world's more affluent countries. Whatever the current validity of *dependencia* theory, the asymmetry remains and U.S. foreign policy, as viewed from South America, is not always credible. Ironically, America's special interests in this realm are now under increased pressure from another major power that would have been furthest from President Monroe's mind back in 1823: China.

China's Growing Economic Clout

China has displaced the United States as the leading trading partner of Brazil, Chile, and Peru. Other South American countries may soon be next. This remarkable economic rebalancing, and its possible political ramifications, has yet to fully register in the United States. Nearly invisible just ten years ago, China has made its presence felt all across South America by establishing new embassies and consulates, buying up companies, partnering joint ventures, financing infrastructure projects and development assistance, and sending as well as inviting highest-level trade delegations. Their motives are clear: the Chinese still need large supplies of key raw materials such as oil, copper, and a plethora of minerals to fuel the ongoing economic development of their country. Just as importantly, they are also seeking to expand their markets for Chinese exports, and the steadily expanding middle-income consumer classes of Brazil and other countries make enticing targets.

It is worth noting as well that for several South American countries, China in some ways poses the same challenges as the United States did in the past—an asymmetry of relations that can lead to dependency and undercut the development of domestic economic sectors. Despite that concern, much needed growth of manufacturing and other value-added production has taken off in certain parts of South America but severely lags in others. We are now ready to turn to a geographic analysis of the major regions of this dynamic realm.

REGIONS OF SOUTH AMERICA

On the basis of physiographic, cultural, and political criteria, South America can be divided into four rather clearly defined regions (Fig. 3-8):

1. *The Caribbean North*, located almost entirely north of the equator, consists of five entities that display a combination of Caribbean and South American features: Colombia, Venezuela, and those that represent three historic colonial footholds by Britain (Guyana), the Netherlands (Suriname), and France (French Guiana).

2. *The Andean West* is formed by four republics that share a strong indigenous cultural heritage as well as powerful influences resulting from their Andean physiography: Ecuador, Peru, Bolivia, and, transitionally, Paraguay.

3. *The Southern Cone*, for the most part located south of the Tropic of Capricorn (latitude 23½°S), includes three countries: Argentina, Chile, and Uruguay (all with strong European imprints and little remaining indigenous influence) plus aspects of Paraguay.

4. *Brazil* occupies an enormous part of interior and eastern South America. In the Amazon Basin of the north, its own interior overlaps with the continent's 'green heart'—often referred to as the world's 'lungs'—the largest tropical rainforest on Earth. Brazil is South America's giant, accounting for just about half the realm's territory as well as population. In Brazil the dominant Iberian influence is Portuguese, not Spanish, and here Africans, not indigenous peoples, form a significant component of demography and culture.

Each of these four regions shares some important commonalities in terms of physical environment, ethnic origins, cultural milieu, and international outlook. But there also are important differences, especially involving economic performance, democratic functioning, and social stability.

THE CARIBBEAN NORTH

The countries of South America's northern tier have something in common besides their coastal location: each has a coastal tropical-plantation zone based on the Caribbean colonial model. Especially in the three Guianas, early European plantation development

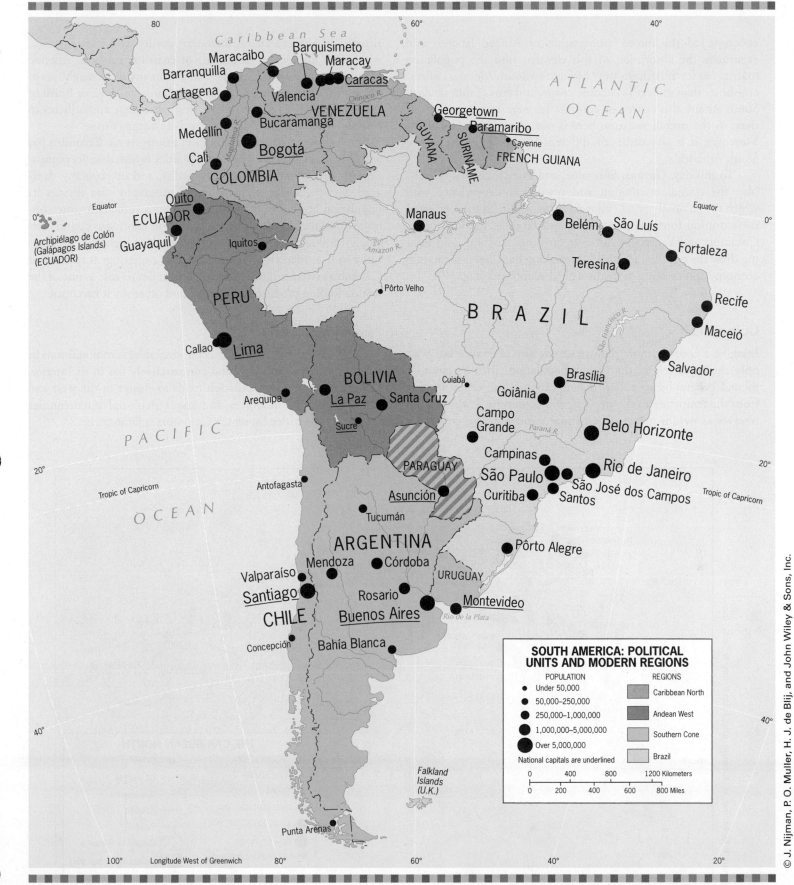

FIGURE 3-8

SOUTH AMERICA: POLITICAL UNITS AND MODERN REGIONS

POPULATION
- Under 50,000
- 50,000–250,000
- 250,000–1,000,000
- 1,000,000–5,000,000
- Over 5,000,000

National capitals are underlined

REGIONS
- Caribbean North
- Andean West
- Southern Cone
- Brazil

0   400   800   1200 Kilometers

0   200   400   600   800 Miles

© J. Nijman, P. O. Muller, H. J. de Blij, and John Wiley & Sons, Inc.

encompassed the forced immigration of African laborers and eventually the absorption of this element into the population matrix. Far fewer Africans were brought to South America's northern shores than to Brazil's Atlantic coasts, and tens of thousands of South Asians also arrived as contract laborers and stayed as settlers, so the overall situation here is not comparable to Brazil's. Moreover, it is also distinctly different from that of the rest of South America.

To this day, Guyana, Suriname, and French Guiana still display the coastal orientation and plantation dependency with which the colonial period endowed them, although the logging of their tropical forests is penetrating and ravaging the interior. In Colombia and Venezuela, however, farming, ranching, and mining drew the population inland, overtaking the coastal-plantation economy and creating more diversified economies (Fig. 3-9).

Colombia

Imagine a country more than twice the size of France but with only three-quarters of the French population, with an environmental geography so varied that it can produce crops ranging from the temperate to the tropical, and possessing world-class oil reserves as well as many other natural resources. This country is situated in the crucial northwestern corner of South America, with 3200 kilometers (2000 mi) of coastline on both Atlantic (Caribbean) and Pacific waters, closer than any of its neighbors to the markets of the north and sharing a border with giant Brazil to the southeast. This nation uses a single language and adheres to one dominant religion. Wouldn't such a country thrive?

The answer, for most of its recent history, is no. Colombia has a history of strife and violence; its politics is unstable, its population divided through extreme inequality, and its economy damaged. Its spectacular, scenic physical geography also divides its population of 48.3 million into clusters not sufficiently interconnected to foster integration; even today, this enormous country has barely 1600 kilometers (1000 mi) of four-lane highways. Its proximity to U.S. markets is a curse as well as a blessing: at the root of Colombia's latest surge of internal conflict lies its role as one of the world's leading producers of cocaine and other illicit narcotics.

People and Resources

As Figure 3-9 reveals, Colombia's physiography is mountainous in its Andean west and north and comparatively flat in its interior. Colombia's scattered population tends to cluster in the west and north, where the resources and the agricultural opportunities (including the coffee for which Colombia is famous) lie.

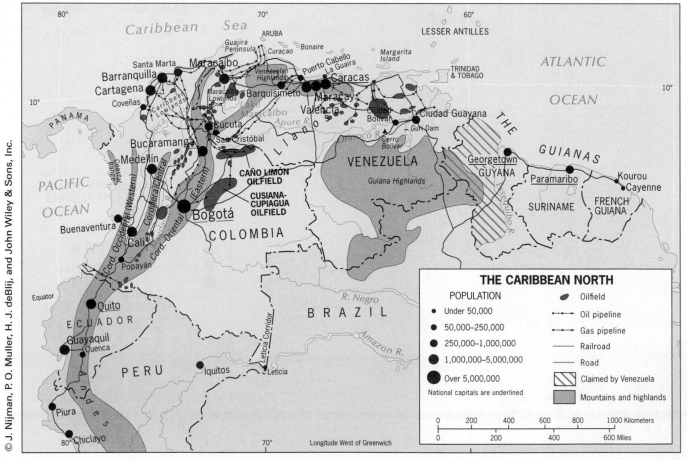

FIGURE 3-9

Colombia's population clusters continue to be poorly interconnected. Several lie on the Caribbean coast, centered on Barranquilla, Cartagena, and Santa Marta, old colonial entry points. Others are anchored by major cities such as Medellín and Cali. What is especially interesting about Figure 3-9 in this context is how little development has taken place along the country's lengthy Pacific coast, where the port city of Buenaventura, across the mountains from Cali, is the only settlement of any size.

The map also shows that Colombia and neighboring Venezuela share the oil and gas reserves in and around the 'Lake' Maracaibo area ('Lake' because this is really a gulf with a narrow opening to the sea), but Venezuela has the bigger portion. Recent discoveries, however, have boosted Colombia's production along the base of the easternmost Andean cordillera, and our map shows a growing system of pipelines from the interior to the coast. Meanwhile, the vast and remote southeastern interior proved fertile ground for that other big Colombian money-maker—illicit drugs.

Cocaine's Curse

The rise of the narcotics industry, fueled by outside (especially U.S. but also Brazilian and European) demand and coupled with its legacy of violence, crippled the state for decades and threatened its very survival. Drug cartels based in the cities controlled vast networks of producers and exporters; they infiltrated the political system, corrupted the military and police, and waged wars with each other that cost tens of thousands of lives and destroyed much of Colombia's social order. The drug cartels also organized their own armed forces to combat initiatives by the Colombian government to control the illegal narcotics economy. Meanwhile, the owners of large haciendas in the countryside hired private security guards to protect their properties, banding together to expand these units into what quickly became, in effect, private armies. Colombia was in chaos as narcoterrorists committed appalling acts of violence in the cities, rebel forces and drug-financed armies of the political 'left' fought against paramilitaries of the political 'right' in the countryside, and Colombia's legitimate economy, from coffee-growing to tourism, suffered grievously. To further weaken the national government, the rebels even took to blowing up oil pipelines.

Threats of Insurgency

Colombia's problems were compounded by the presence of various rebel groups (most notably, the Revolutionary Armed Forces of Colombia, or FARC) that contested government control of several provinces and that established mutually convenient partnerships with the cocaine cartels.

FARC initially pursued a purely political or ideological agenda, but their widespread guerrilla activities soon brought them into contact with the cocaine cartels, which had their own reasons for wanting to exercise territorial control. In essence, this led to close, if murky, ties between the cartels and the rebel armies: the insurgents had found themselves a source of funding, while the cartels linked up with a convenient source of military support against government forces in times of need. In the course of the 1990s, FARC assumed the role of what political geographers call an **insurgent state 17**, whereby rebels have total control of part of the state's territory. The area controlled by FARC was named *Farclandia*, the red-striped zone mapped in Figure 3-10. It appeared to be just a matter of time before the national government in Bogotá would lose control, the status quo would slide into disintegration, and Colombia would devolve into a **failed state 18** (a country whose institutions have collapsed and in which anarchy prevails).

Colombia's Revival

Since 2002, however, Colombia has mounted successful campaigns to defeat the rebels militarily and to persuade them through legal means to give up their arms. In 2015, long-running peace talks were continuing, yet the influence of the rebels was dwindling as the government had restored control over the bulk of its

FIGURE 3-10

© J. Nijman, P. O. Muller, H. J. de Blij, and John Wiley & Sons, Inc.

territory—effectively transforming Colombia from an incipient failed state into a more stable, secure, and potentially prosperous country.

The government matched its domestic counteroffensive against the rebels with an international campaign to help revive the economy, promoting market-oriented, business-friendly policies. The country's impressive economic performance in this decade is based on revenues from fossil fuels, metallic ores, tourism, and farm products, particularly exports of coffee and cut flowers (Colombia is the leading U.S. importer). Bogotá, the capital, centers the country's leading economic region and has become one of the realm's primary business complexes. The city of Medellín—former base of the legendary drug kingpin, Pablo Escobar, and designated the world's murder capital in the late 1980s—has made an comeback in recent years and is now internationally regarded as a model for innovative urban design (see photo).

Venezuela

A long boundary separates Colombia from Venezuela, its neighbor to the northeast. Much of what is important in Venezuela is concentrated in the northern and western parts of the country, where the Venezuelan Highlands form the eastern spur of the northern end of the Andes system. Most of Venezuela's 30.4 million people are concentrated in these uplands, which include the capital of Caracas, its early rival Valencia, and the commercial/industrial centers of Barquisimeto and Maracay.

The Venezuelan Highlands are flanked by the Maracaibo Lowlands and 'Lake' Maracaibo to the northwest and by an immense plainland of savanna country, known as the Llanos, in the Orinoco Basin to the south and east (Fig. 3-9). The Maracaibo Lowlands, at

In 2013, Medellín, Colombia was voted the world's most innovative city by the U.S. Urban Land Institute. Striking innovations include cable cars and escalators on the city's steep hillsides that have drastically reduced travel time. Through thriving public-private partnerships, the city has invested in new parks, schools, and libraries, especially in poor neighborhoods. This erstwhile capital of Colombia's cocaine industry may have been plagued by violence in the 1980s and 1990s, but 25 years later, Medellín's turnaround has attracted attention and acclaim from urban planners around the world.

one time a disease-infested, sparsely peopled coastland, today constitute one of the world's leading oil-producing areas; much of this fuel is drawn from reserves that lie beneath the shallow waters of the 'lake' itself. The country's third-largest city, Maracaibo, is the focus of the petroleum industry that transformed the Venezuelan economy in the 1970s; however, as we shall see, since then oil has been more of a curse than a blessing.

The Llanos on the southern side of the Venezuelan Highlands and the Guiana Highlands in the country's far southeast, like much of Brazil's interior, are in a nascent stage of development. The 300- to 650-kilometer- (180- to 400-mi) long Llanos slope gently from the base of the northernmost Andean spur to the Orinoco's floodplain. Their mixture of savanna grasses and scrub woodland facilitates cattle grazing on higher ground, but widespread wet-season flooding of the more fertile lower-lying areas has thus far inhibited the plainland's commercial farming potential (much of the development of the Llanos to date has been limited to the exploitation of its massive oil reserves). Crop-raising conditions are more favorable in the *tierra templada* areas at moderate altitudes in the Guiana Highlands. Economic integration of this even more remote interior zone with the heartland of Venezuela was spearheaded by the discovery of rich iron ores during the 1950s on the northern side of the Guiana Highlands southwest of Ciudad Guayana near the mouth of the Orinoco River. Local railroads were built to connect the mines with the Orinoco, and from there ores have long been shipped directly by sea to foreign markets.

A Mounting Crisis

Venezuela is one of those countries in which a rich natural resource base can have destructive consequences for reasons elaborated earlier in this chapter. The country has long been in political turmoil because its governments had the bad habit of living off oil profits while neglecting long-term productive investments in the economy to better the lives of most Venezuelans. This overreliance on oil exports was felt most painfully when prices on the world market fell.

By the mid-1990s, the government was required to sharply devalue the currency, and a political crisis ensued that resulted in a severe recession and widespread social unrest. With more and more Venezuelans enraged at the way their oil-rich country was approaching bankruptcy without making progress toward a more equitable distribution of the national wealth, the way was cleared for the rise to power of the socialist radical, Hugo Chávez, in 1999 (with an apparent mandate from the urban poor and the increasingly penurious middle class).

Chávez's rule, however, quickly turned both autocratic and economically destructive. During his long presidency that ended with his death in 2013, relatively high oil prices brought in consistently high revenues totaling well over (U.S.) $800 billion. But there was very little to show for it: infrastructural improvements were hardly to be seen, bloated public-sector companies chronically underperformed, and the entrepreneurial sector of the population had fled the country soon after Chávez came to power.

As for the poor, they intermittently benefited from new social programs, but life did not improve for them. Today, all Venezuelans are worse off, bedeviled by constant shortages of basic foods and vital retail products under the chaotic maladministration of the regime of Nicolas Maduro, Chávez's successor. On the larger South American stage, twenty-first-century Venezuela isolated itself with its 'Bolivarian Revolution' at a time when most other South American countries embraced **neoliberal policies** 19 aimed at supporting private entrepreneurship and free markets.

The 'Three Guianas'

Three small political entities form the eastern flank of the realm's northern region: Guyana, Suriname, and French Guiana. They are good reminders of why the name 'Latin' America is inappropriate: the first is a legacy of British colonialism and employs English as its official language; the second is a remnant of Dutch influence where Dutch is still official among its polyglot of tongues; and the third is still a dependency—of France. None has a population exceeding 1 million, and all three exhibit social characteristics and cultural landscapes far more representative of the Caribbean Basin than South America. Here British, Dutch, and French colonial powers acquired possessions and established plantations, brought in African and Asian workers, and created economies similar to those that mark a number of the nearby Caribbean islands.

Guyana, with 800,000 people, has about as many inhabitants as Suriname and French Guiana combined. When Guyana became independent in 1966, its British rulers left behind an ethnically and culturally divided population in which people of South Asian (Indian) ancestry now make up about 44 percent and those of African background (including Afro-European ancestry) 30 percent. This makes for contentious politics, given the religious mix that is approximately 57 percent Christian, 28 percent Hindu, and 7 percent Muslim. Guyana remains dominantly rural, and plantation crops continue to figure prominently among exports (gold from the interior is the most valuable single product).

Oil may soon become a factor in the economy, however, because a recently discovered reserve that lies offshore from Suriname extends westward beneath Guyana's waters. Still, Guyana is among the realm's poorest and least urbanized countries, and it is ever more strongly affected by the expanding drug industry of South America's north. Its thinly populated interior, beyond the reach of antidrug campaigns, has become a staging area for cocaine distribution to the United States via Middle America (and increasingly to Brazil and Europe). A recent study suggests that Guyana's take from the drug trade already amounts to at least one-fifth of the country's total economy.

Suriname actually progressed more rapidly than Guyana did after it was granted independence in 1975, but unending political instability swiftly ensued. The Dutch colonists brought South Asians, Indonesians, Africans, and even some Chinese to their colony, making for a notably fractious state. More than 100,000 residents—about one-quarter of the entire population—emigrated to the Netherlands, and were it not for support from its former colonial ruler, Suriname's economy would have collapsed. Agriculturally, its rice farms provide Suriname with self-sufficiency, even allowing for some exporting (which is dominated by plantation crops). Suriname's leading income producer continues to be its bauxite (aluminum ore) that is mined in a zone across the middle of the country. To this short list can also be added the promise of those offshore oil finds noted above, which may soon be earning important revenues.

Suriname's cultural geography is spiced by the many languages spoken by its 540,000 citizens. Dutch is the official tongue, but other than by officials and in schools it is not heard much. A mixture of Dutch and English, *Sranan Tongo*, serves as a common language, but one also hears indigenous, Hindi, Chinese, Indonesian, and even some French Creole on the streets.

French Guiana the easternmost outpost of the Caribbean North region, is the last remaining dependency on the mainland of South America. This territory is an anomaly in other ways as well. Consider this fact: it is almost as large as South Korea (population: just above 50 million) with a population totaling only 280,000. Its status is that of an Overseas *Département* of France, and its official language is French. Nearly half the population resides in the immediate vicinity of the coastal capital, Cayenne.

In 2015, there still was no prospect of independence for this decidedly underdeveloped relic of the former French Empire. Gold remains the most valuable export, and a small fishing industry sends some exports to France. But what really matters here in French Guiana is the European Space Agency's launch complex at Kourou on the Atlantic coast, which accounts for more than half of the territory's entire economic activity—from plantation farming to spaceport . . . but with minimal involvement, or benefits, for most ordinary people.

THE ANDEAN WEST

The second regional grouping of South American states—the Andean West (Fig. 3-11)—is dominated physiographically by the great Andes mountain chain and historically by indigenous peoples. This region encompasses Peru, Ecuador, Bolivia, and transitional Paraguay, the last with one foot in the West and the other in the South (Fig. 3-8). Bolivia and Paraguay also constitute South America's only two landlocked countries.

Spanish conquerors overpowered the indigenous nations, but they did not reduce them to small minorities as happened to so many indigenous peoples in other parts of the world. Today, 45 percent of the residents of Peru, the region's most populous country, are indigenous; in Bolivia, they form a majority of 55 percent (although there is considerable mixing of populations). Approximately 25 percent of Ecuador's population can be classified as indigenous, and in Paraguay the ethnic mix, not regionally clustered as in the other three countries, is dominated by those of indigenous ancestry.

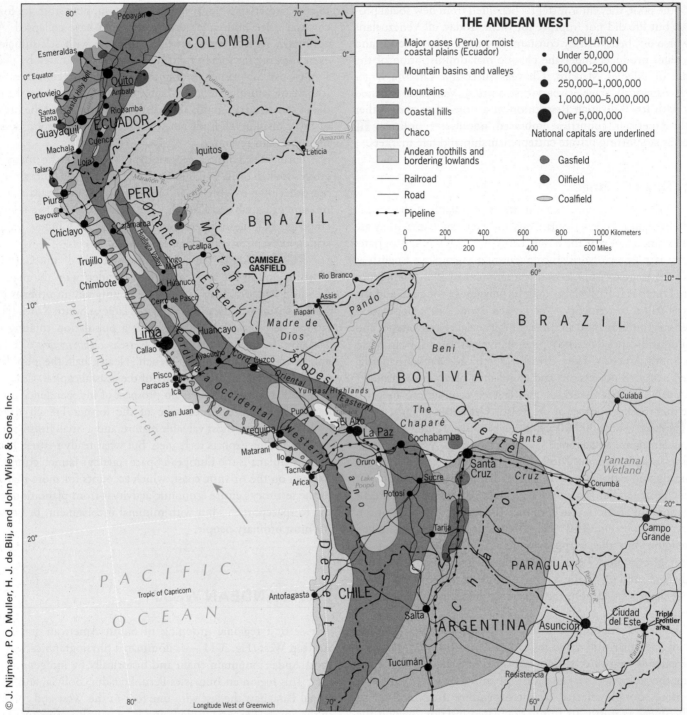

FIGURE 3-11

This region has long been South America's poorest economically, with lower incomes, higher numbers of subsistence farmers, and fewer opportunities for job-seekers than elsewhere in the realm. For a very long time, the urbane lives of the land-owning elite have been worlds apart from the hard-scrabble existence of the landless peonage (the word **peón** is an old Spanish term for an indebted day laborer). But in recent years, the Andean West, like the realm as a whole, has been stirring, and oil and natural gas are leading the way.

Peru

Peru straddles the Andean spine for more than 1600 kilometers (1000 mi) and is the largest of the region's four republics in both territory and population (30.4 million). Physiographically and culturally, Peru divides into three parts (Fig. 3-11). Lima and its port, Callao, lie at the center of the **desert coast** subregion, and it is symptomatic of the cultural division prevailing in Peru that for nearly 500 years the capital city has been positioned on the western periphery, not in a central location in a basin of the Andes (and Peru has never elected an indigenous president despite the fact that aboriginal people constitute nearly half of the population). From an economic point of view, however, the Spaniards' choice of a headquarters on the Pacific coast proved to be sound, for the coastal strip has become commercially the most productive part of the country. A thriving fishing industry contributes significantly to the export trade; so do the products of irrigated agriculture in some 40 oases dispersed across the arid coastal plain, which include fruits such as citrus and olives, and vegetables such as asparagus (a big money-maker) and lettuce.

The Andean or **Sierra** subregion occupies just about one-third of the country and is the ancestral home of the largest component in the total population, the speakers of Quechua, the *lingua franca* that emerged during the Inca Empire. Their physical survival during the harsh Spanish colonial regime was made possible by their adaptation to the high-altitude environments they inhabited. The Andean subregion is not nearly as important economically as the coast, but it does yield copper, zinc, and lead from mining centers, the largest of which is Cerro de Pasco. Peru is also the world's sixth-largest producer of gold, the price of which has been soaring in recent years, contributing to this country's steadily growing economy; most of the mines are located in the 'gold belt' of the northernmost Peruvian Andes, but illegal strip mining is widespread in the Amazon rainforest of the southeastern interior. In the high valleys and intermontane basins, the indigenous population is concentrated either in isolated villages, around which many people engage in subsistence agriculture, or in the more favorably located and fertile areas where they are tenants, peons on white- or mestizo-owned haciendas.

Of Peru's three subregions, the East or **Oriente**—the inland-facing slopes of the Andean ranges that lead down to the Amazon-drained, rainforest-covered *montaña*—is the most isolated. The focus of the eastern subregion, in fact, is Iquitos in the far north, a city that looks east rather than west because it can be reached by oceangoing vessels sailing 3700 kilometers (2300 mi) up the Amazon River across northern Brazil. Iquitos grew rapidly during the Amazon's wild-rubber boom of the early twentieth century but swiftly declined thereafter; now it is finally growing again, reflecting Peruvian plans to open up the eastern interior. Today, this subregion, and perhaps Peru as a whole, appears to be on the threshold of a new era owing to major new discoveries of oil and gas reserves in the *Oriente*. Indeed, while still one of the poorer countries in this realm, since the mid-2000s Peru has experienced a higher economic growth rate than any other. A trans-Andean pipeline already transports natural gas from the Camisea reserve (north of Cuzco) to a conversion plant on the coast south of Lima. From there, an underwater connector pipeline extends to an off-shore loading platform for tankers taking it to the U.S. market (Fig. 3-11). Additional interior gas reserves are now being tapped, and they only reinforce Peruvian aspirations to become a major energy producer.

Peru's economic development is not proceeding without adversity, however, as the government is being challenged on at least two fronts. Environmentalists and supporters of indigenous

@ FROM THE FIELD NOTES...

"It is late January of 2015 and this *altiplano* of south-central Peru offers some panoramic views. Down below, just right of center, is the old Andean town of Chinchero, lying at an elevation of 3600 meters (12,000 ft) and surrounded by mountains up to 5800 meters (19,000 ft) high. The original name of the town has been lost, but it belongs to what the Incas called the 'Sacred Valley' or the 'Land of the Rainbow.' It is easy to get a sense of how the Incas capitalized on the advantages of this highland environment during their pre-Columbian rule. Soils here are fertile, temperatures sufficiently moderate, and the snowpack and glaciers of the enveloping Andean ranges keep local streams and rivers flowing year-round. True to their heritage, the lower slopes all around Chinchero are still being farmed as they have been for centuries, growing potatoes, wheat, and other hardy *tierra fría* crops."

rights are raising issues throughout the *Oriente* (even as political activists argue against the terms of trade that Peru has accepted with the big oil companies). They complain that the oil- and gas-led exploitation of the unprotected east will harm both the environment and local communities, and that the new energy revenues will only further benefit the already-favored coastal residents of Peru, leaving the ever more disadvantaged inhabitants of the interior even farther behind.

Ecuador

On the map, Ecuador, smallest of the three Andean West republics, appears to be just a northern corner of Peru. But that would be a misrepresentation because Ecuador also possesses a complete range of regional variations (Fig. 3-11). It has a coastal belt; an Andean zone that may be narrow (less than 250 kilometers [150 mi]) but by no means of lower elevation than elsewhere in the region; and an *Oriente*—an eastern subregion that is as sparsely settled and as economically marginalized as that of Peru. As with Peru, just about half of Ecuador's population (which totals 15.7 million) is concentrated in the Andean intermontane basins and valleys, and the most productive subregion is the coastal strip. There, however, the similarities end.

Ecuador's Pacific coastal zone consists of a belt of hills interrupted by lowlands, of which the most important lies in the south between the hills and the Andes, drained by the Guayas River. Guayaquil—the country's largest city, main port, and leading commercial center—forms the focus of this subregion. Unlike Peru's coastal strip, Ecuador's is not a desert: it consists of fertile tropical plains not afflicted by excessive rainfall. Seafood (especially shrimp) is a leading product, and these lowlands support a thriving commercial agricultural economy built around bananas, cacao, cattle raising, and coffee on the hillsides. Moreover, Ecuador's western subregion is also far less Europeanized because this white component of the national population is only about one-third the size of Peru's.

A greater proportion of whites are engaged in administration and hacienda ownership in the central Andean zone, where most of the Ecuadorians who are indigenous also reside—and, not surprisingly, where land-tenure reform is an explosive issue. The differing interests of the Guayaquil-dominated coastal lowland and the Andean-highland subregion focused on the capital (Quito) have long fostered a deep regional cleavage between the two. This schism has not lessened in recent years, and autonomy and other devolutionary remedies often are openly discussed in the coastlands.

In the rainforests of the *Oriente* subregion, oil production is expanding as a result of the discovery of additional reserves. Some analysts predict that eastern Ecuador as well as Peru will prove to contain energy resources comparable to those buried along the inland side of the Andes in Colombia and Bolivia. These developments could well portend the dawning of a new 'oil era' accompanied by the economic transformation of the Ecuadorian *Oriente*. Indeed, oil already tops the country's list of exports, and to prepare for the next stage, the energy industry's infrastructure is being modernized and upgraded.

Bolivia

Nowhere are the problems typical of the Andean West more acute than in landlocked, volatile Bolivia. Before reading further, take a careful look at Bolivia's regional geography in both Figures 3–11 and 3–12. Bolivia is bounded by remote peripheries of both Brazil and Argentina, mountainous Andean highlands as well as intermontane **altiplanos** (high-elevation basins and valleys) of Peru, and coveted coastal zones of northern Chile. As the maps show, the Andes in this zone broaden into a vast mountainous complex some 700 kilometers (450 mi) wide.

On the boundary between Peru and Bolivia, freshwater Lake Titicaca lies at 3700 meters (12,500 ft) above sea level and helps make the adjacent *Altiplano* (see map) livable by ameliorating the coldness in its vicinity, where the snow line lies just above the plateau surface. On the lake's surrounding cultivable land, potatoes and *tierra fría* grains have been raised for centuries dating back to pre-Inca times, and the Titicaca Basin still supports a major cluster of

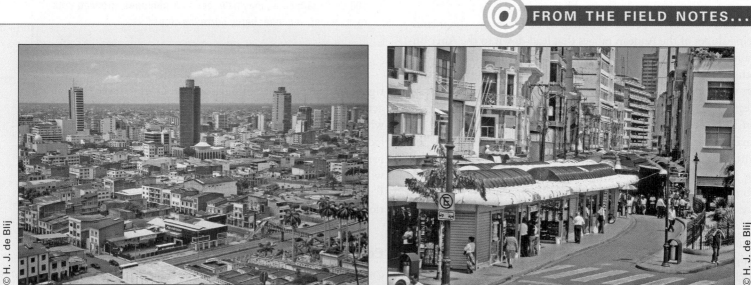

FROM THE FIELD NOTES...

© H. J. de Blij

"I can't remember being hotter anyplace on Earth, not in Kinshasa, not in Singapore . . . not only are you near the equator here in steamy Guayaquil, but the city lies in a swampy, riverine lowland too far from the Pacific to benefit from any cooling breezes and too far from the Andean foothills to enjoy the benefits of suburban elevation. Guayaquil is not the disease-ridden backwater it used to be (left photo): its port is modern, its city-center waterfront on the Guayas River has been renovated, its international airport is the hub of a commercial complex. This has grown into a metropolis of 2.8 million, made possible in part by oil revenues. Beating the heat and glare is an everyday priority: entire streets have been covered by makeshift tarpaulins and more permanent awnings to protect shoppers (right). Talk to the locals, though, and you find that there is another daily concern: the people 'up there in the mountains who rule this country always put us in second place.' Take the 45-minute flight from Guayaquil to cool and comfortable Quito, the capital, and you're in another world, and you quickly forget Guayaquil's problems. That's just what the locals here say the politicians do."

indigenous (Aymara) subsistence farmers. This portion of the *Altiplano* is the heart of modern Bolivia and also contains the capital.

The Bolivian state is the product of the Hispanic impact, and the country's indigenous peoples (who now comprise over half of the national population of 10.7 million) no more escaped the loss of their land than did their Peruvian or Ecuadorian counterparts. What made the richest Europeans in Bolivia wealthy, however, was not land but minerals. The town of Potosí in the eastern cordillera became a legend for the immense deposits of silver in its vicinity; tin, zinc, copper, lead, and several ferroalloys were also discovered there. Indigenous workers were coerced into toiling in the mines under the most appalling conditions. Today, natural gas, oil, and zinc are leading sources of foreign revenues.

Landlocked, physiographically bisected, culturally bifurcated, and economically divided, Bolivia is a severely challenged state. Moreover, the country's prospects are worsened by its political geography: look again at Figure 3-12 and you can see that Bolivia's nine provinces are regionally divided between indigenous-majority *departments* (as subnational units are called here) in the upland west and departments with mestizo majorities in the much-lower-lying east. The capital city, La Paz, is located in a basin on the indigenous-majority Andean *Altiplano*, but many mestizo Bolivians do not recognize it as such.

The Santa Cruz Department, like the others in the *Oriente*, stands in sharp contrast to those of the Andes in the west. Here the hacienda system (discussed in Chapter 2) persists almost unchanged from colonial times, its profitable agriculture supporting a wealthy aristocracy. Recently, this eastern sector of Bolivia has proved to contain abundant energy resources as well, adding to its economic advantages. So there is talk—abetted by frequent public demonstrations—in support of autonomy, even secession here (see photo pair). But it must also be kept in mind that neither the haciendas nor the energy industry could operate without the indigenous labor force.

Paraguay

Paraguay, Bolivia's landlocked neighbor to the southeast, is one of those transitional countries lying in between regions, exhibiting properties of each (see Fig. 3-8). Certainly, Paraguay (population: 6.8 million) is not an 'Andean' country. It has no highlands of consequence. Its well-watered eastern plains give way to the dry scrub of the Gran Chaco in the northwest. Nor does it have clear, spatially entrenched ethnic divisions between indigenous, mestizo, and other peoples as do Bolivia and Peru. But aboriginal ancestry dominates the ethnic complexion of Paraguay, and continuing protests by landless peasants are similar to those taking place elsewhere in the Andean West. Moreover, the native language, Guaraní, is so widely spoken (by over 90 percent of the people) that this is one of the world's most thoroughly bilingual societies.

FIGURE 3-12

Today, Paraguay's longstanding isolation is eroding, most evident in the hot, semiarid Chaco of the northwest. Here, rapid deforestation has opened up a burgeoning new cattle-ranching frontier. For aboriginal peoples like the Ayoreo, however, the impact is disastrous because their traditional livelihoods are threatened with imminent destruction (the word for bulldozer in their language means 'attacker of the world'). The contraction of the Chaco has also sounded alarms among biologists and conservationists because so much of it remains unexplored and is certain to contain myriad undiscovered plant and animal species.

Looking southward, there is little in Paraguay's economic geography to compare to Argentina, Uruguay, or Chile. Paraguay's low GNI resembles that of countries of the Andean West, not the far more advantaged Southern Cone region. This is also one of South America's least urbanized states, and poverty dominates the countryside as well as the slums encircling the capital, Asunción, and other towns (Fig. 3-11). In 2013, about 24 percent of the population lived at or below the official poverty level. Moreover, research suggests that 1 percent of the population owns about 77 percent of the land, which may be a record for inequality in the South American realm.

In the east, more than 300,000 Brazilians have crossed the border to settle in Paraguay, where they have created a thriving commercial agricultural economy that produces soybeans, livestock, and other farm products exported to or through Brazil. This

Aizar Raldes/AFP/Getty Images, Inc.

Rodrigo Buendia/AFP/Getty Images, Inc.

Bolivia's political geography is as divided as its physical geography, the power base of the indigenous population lying in the altiplanos and mountains, and that of the mestizo minority centered in the interior lowlands. These photos, taken a day apart in Santa Cruz where mestizo strength is concentrated, show the crucial difference of opinion: a majority of mestizos want autonomy for their eastern provinces (left); virtually all indigenous people protest the prospect of such autonomy (right), seen as a prelude to secession and independence that would fatally fracture their country.

puts Paraguay in a geographically difficult position, and it often complains that Brazil is not interested in living up to its regional-trade obligations, thereby creating unacceptable difficulties for Paraguayan exporters. Meanwhile, politicians raise fears that Brazilian immigration is creating a steadily expanding foreign enclave within Paraguay, where people speak Brazilian Portuguese (including in the local schools), Brazil's rather than Paraguay's flag flies over public buildings, and a Brazilian cultural landscape is steadily emerging in the eastern border zone.

Another quintessentially geographical problem stemming from Paraguay's long-term weakness lies in the southeast, where the borders of Brazil, Argentina, and Paraguay converge in a chaotic scene of smuggling, money-laundering, political intrigue, and even terrorist activity. This hub of lawlessness—centered on the town of Ciudad del Este, but with local spillovers into the territory of all three countries—is known as the *Triple Frontera* or **Triple Frontier** 20 (Fig. 3-11). It should also be noted that the Triple Frontier contains a sizeable, tristate Middle Eastern community, identified by the U.S. government as a source of funding for Islamic militant groups—an allegation repeatedly rejected by the governments of Argentina, Brazil, and Paraguay.

THE SOUTHERN CONE

Argentina

The largest Southern Cone country by far is Argentina, whose territorial size ranks second only to Brazil in this geographic realm; its population of 41.5 million ranks third overall after Brazil and Colombia. Although Argentina exhibits a great deal of physical-environmental variety within its boundaries, the overwhelming majority of the Argentines are concentrated in

the physiographic subregion known as the *Pampa* (a Spanish word meaning 'plain'). Figure 3-13 underscores the degree of clustering of Argentina's inhabitants on the land and in the cities of the Pampa. It also shows the relative emptiness of the other six subregions (shown in Fig. 3-13): the scrub-forested *Chaco* in the northwest; the mountainous *Andes* in the west, along whose crestline lies the boundary with Chile; the arid plateaus of *Patagonia* south of the Rio Colorado; and the undulating transitional terrain of intermediate *Cuyo, Entre Rios* (also known as Mesopotamia because it lies between the Paraná and Uruguay rivers), and the *North*.

The Argentine Pampa is the product of the past 150 years. During the second half of the nineteenth century, when the great grasslands of the world were being opened up (including those of interior North America, Russia, and Australia), the economy of the long-dormant Pampa began to emerge. The food needs of industrializing Europe grew by leaps and bounds, and the advances of the Industrial Revolution—railroads, more efficient ocean transport, refrigerated ships, and agricultural machinery—helped to make large-scale commercial meat and grain production in the Pampa not only feasible but also highly profitable. Large haciendas were laid out and farmed by tenant workers; railroads were built and radiated ever farther outward from the booming capital of Buenos Aires, soon bringing the entire Pampa into production.

Argentina once was one of the richest countries in the world. Its historic affluence is still reflected in its architecturally splendid cities whose plazas and avenues are flanked by ornate public buildings and private mansions. This is true not only of the capital, Buenos Aires, at the head of the Rio de la Plata estuary—it also applies to interior cities such as Mendoza and Córdoba. The cultural imprint is dominantly Spanish, but the cultural landscape was diversified by a massive influx of Italians and smaller but influential numbers of British, French, German, and Lebanese immigrants. This has also long been one of the realm's most urbanized

FIGURE 3-13

countries: 92 percent of its population is concentrated in cities and towns, a higher percentage even than western Europe or the United States. Fully one-third of all Argentinians live in metropolitan Buenos Aires, by far the leading industrial complex.

But political infighting and economic mismanagement have combined over the long term to shackle a vibrant and varied economy. Part of the problem, it appears, lies in the country's lopsided geography with an enormous gap between a few highly populated and urbanized provinces that hold almost all the political power

and a large number of provinces with very small populations (Fig. 3-14). Buenos Aires Province has 16.8 million people (out of a total 41.5 million), whereas Tierra del Fuego has barely more than 150,000. A dozen additional provinces contain less than 750,000 inhabitants, so that the larger ones besides dominant Buenos Aires are also disproportionately influential in domestic politics, especially Córdoba and Santa Fe. Corruption and mismanagement reached so high a level in 2012 that the authoritative international weekly, *The Economist*, decided it would no longer

"The name Mendoza is synonymous with Argentinean wine," wrote our coauthor Harm de Blij, an expert on the geography of viticulture, in his classic book entitled *Wine Regions of the Southern Hemisphere*. As our maps show, Mendoza is a province as well as a city in the Cuyo subregion of west-central Argentina, and both are massively involved in wine production at some of South America's biggest winemaking facilities. This has always been red-wine country, and its best-known varietals are Malbec and Cabernet Sauvignon. Even though Mendoza Province produces more than 80 percent of Argentina's wine, this still-expanding industry must contend with the area's semiarid environment and increasingly rely on irrigation networks to meet its growing water needs. This view is toward the west, whose horizon is filled entirely by the mighty wall of the Andes; beyond the mountain range lies central Chile, another South American producer of world-class wines.

include statistics provided by the Argentine government; pointedly, the newsmagazine also announced that it did not want to be part of the government's 'deliberate attempt to deceive voters and swindle investors—a reflection of the country's sorry condition in this decade.

If one compares Argentina to Paraguay, it is striking how much better Argentina's position is in terms of natural endowments and geographic opportunities. But as things now stand, the Argentinean state is more reminiscent of Venezuela: things could be so much better than they presently are.

FIGURE 3-14

Chile

For over 4000 kilometers (2500 mi) between the crestline of the Andes and the coastline of the Pacific lies the narrow strip of land that is the Republic of Chile (Fig. 3-15). On average, just 150 kilometers (90 mi) wide (and only rarely over 250 kilometers [150 mi] wide), Chile is the world's quintessential example of what **elongation** 21 means to the functioning of a state. Accentuated by its north-south orientation, this severe territorial attenuation not only results in Chile extending across numerous environmental zones; it has also contributed to the country's external political, internal administrative, and general economic challenges. Nevertheless, throughout most of their modern history, the Chileans have made the best of this potentially disruptive centrifugal force: from the beginning, the sea has constituted an avenue of

longitudinal communication; the Andes Mountains continue to form a barrier to prevent encroachment from the east; and when confrontations loomed at the far ends of the country, Chile proved to be quite capable of coping with aggressive neighbors.

As Figures 3-13 and 3-15 indicate, Chile is a three-subregion country. About 90 percent of its 17.6 million people are concentrated in what is called Middle Chile, where Santiago, the capital and largest city, and Valparaíso, the chief port, are located. North of Middle Chile lies the Atacama Desert, wider, drier, and colder than the coastal desert of Peru. To the south of Middle Chile, the coast is punctuated by a plethora of fjords and islands, the topography is mountainous, and the climate—wet and cool near the Pacific—soon turns drier and much colder against the Andean interior.

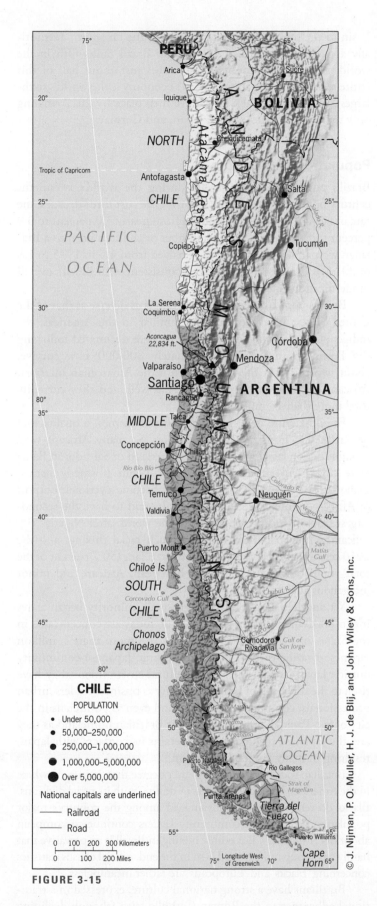

FIGURE 3-15

Within the map:
PERU
BOLIVIA
Arica
Sucre
Iquique
NORTH
Atacama Desert
Tropic of Capricorn
Antofagasta
Salta
CHILE
Chuquicamata
PACIFIC
OCEAN
Copiapó
Tucumán
La Serena
Coquimbo
Córdoba
Aconcagua 22,834 ft.
Mendoza
Valparaíso
Santiago
ARGENTINA
Rancagua
MIDDLE
Talca
Concepción
Chillán
Río Bío Bío
CHILE
Colorado R.
Temuco
Neuquén
Valdivia
Negro R.
Puerto Montt
San Matías Gulf
Chiloé Is.
SOUTH
Chubut R.
Corcovado Gulf
CHILE
Chonos Archipelago
Comodoro Rivadavia
Gulf of San Jorge
Deseado R.
ATLANTIC OCEAN
Lake San Martín
Lake Viedma
Lake Argentino
Puerto Natales
Rio Gallegos
Strait of Magellan
Punta Arenas
Tierra del Fuego
Puerto Williams
Cape Horn
Longitude West of Greenwich

CHILE
POPULATION
· Under 50,000
· 50,000–250,000
● 250,000–1,000,000
● 1,000,000–5,000,000
● Over 5,000,000
National capitals are underlined
— Railroad
— Road
0 100 200 300 Kilometers
0 100 200 Miles

© J. Nijman, P. O. Muller, H. J. de Blij, and John Wiley & Sons, Inc.

Not surprisingly, the land in Middle Chile is the most fertile and valuable, with hardly any agriculture to be found in either the North or South (Fig. 3-4). The three subregions are also apparent on the realm's cultural map, with Europeans dominating the Middle, mestizos prevailing in the North, and indigenous groups forming the majority in the South (Fig. 3-3).

Still, prior to the 1990s, the arid Atacama region in the North accounted for more than half of Chile's foreign revenues. It contains the world's largest exploitable deposits of nitrates, which was the country's economic mainstay before the discovery of methods of synthetic nitrate production in the twentieth century. Subsequently, copper became the dominant export (Chile again possesses the world's largest reserves, which now accounts for more than half of all export revenues). It is mined in several places, but the main concentration lies on the eastern margin of the Atacama near Chuquicamata (see photo), not far from the port of Antofagasta.

Chile has in recent decades embarked on a highly successful program of free-market economic reform that brought stable growth, lowered inflation as well as unemployment, reduced poverty, and attracted foreign investment. The last is of particular significance because these new international connections enabled the export-led Chilean economy to diversify and develop in some badly needed new directions. Copper remains at the top of the export list, but many other mining ventures have been launched. In the agricultural sphere, fruit and vegetable production for export has soared. Industrial expansion is taking place as well, though at a more leisurely pace, and new manufactures have included an array of goods that range from basic chemicals to computer software.

Chile's increasingly globalized economy has propelled the country into a prominent role on the international scene. Its regional commerce has also been growing steadily in conjunction with Chile's new role as a full member of the recently formed Pacific Alliance, which today seems to be outpacing Mercosur/l.

Uruguay

Uruguay, in comparison with Argentina or Chile, is compact, small, and rather densely populated. It is about the size of Florida but with one-sixth the population. This **buffer state 22** of the early independence era, lying between (then) potentially hostile Argentina and Brazil, has evolved into a fairly prosperous agricultural country—in effect a smaller-scale Pampa, though possessing less favorable soils and topography. Montevideo, the coastal capital, contains half of the country's population of 3.4 million; from here, railroads and roads radiate outward into the productive agricultural hinterland (Fig. 3-13). In the immediate vicinity of Montevideo lies Uruguay's major farming zone, producing vegetables and fruits for the metropolis as well as wheat and fodder crops; most of the rest of the country is used for grazing cattle and sheep, with beef products, wool and textile manufactures, and hides dominating the export trade (Fig. 3-13, inset map). Tourism is another major economic activity as Argentines, Brazilians, and other visitors flock to the Atlantic beaches at Punta del Este and other

Peter McBride/Aurora Photos, Inc.

A view into the gaping maw of northern Chile's Chuquicamata copper mine, the world's largest and second-deepest open-pit mine. For most of the century since mining began here in the 1910s, 'Chuqui' was the single leading source of copper on Earth. But the increasingly negative environmental impact of its moonscape of slag heaps and toxic materials has now forced the relocation of the town of Chuquicamata and the 2013 closure of the surface mine. Nonetheless, huge quantities of copper will continue to be extracted from this still-rich deposit after a major new underground mine is built below this site: it will begin operating in 2018—with plans to keep enlarging it through at least 2060.

thriving seaside resorts. In terms of per-capita incomes, Uruguay has long been one the realm's better performing countries.

▌ BRAZIL: GIANT OF SOUTH AMERICA

The emergence of Brazil as the regional superpower of South America and an economic superpower in the world at large has been one of the main stories of the early twenty-first century. In the past, the country was known for producing coffee, citrus, and leather; today, it produces, in addition, machinery, chemicals, high-tech goods, and even airplanes. In fact, since 2000 Brazil's economic performance has consistently ranked it among the most productive emerging-market countries on the international scene. And most recently, in 2013, Brazil surpassed the United Kingdom to become the world's sixth-largest economy.

Why is Brazil so upward-bound today? First, after a long period of dictatorial rule by a minority elite that used the military to stay in power, Brazil embraced democratic government in 1989 and has not looked back since. The era of military coups and repeated crushing of civil liberties is over. Second, with its vast storehouse of natural resources, Brazil has benefited enormously from increased demands for commodities in the world market, mainly from China and India. In some ways, Brazil's political and economic turnaround runs parallel to that of Chile, but because of its sheer size Brazil matters far more to the rest of the realm and, indeed, the world.

By every measure, Brazil is South America's giant. It is so large that it has common boundaries with all the realm's other countries except Ecuador and Chile (Fig. 3-8). Its tropical and subtropical environments range from the equatorial rainforest of the Amazon Basin to the humid-temperate climate of the far south. Territorially as well as in terms of population, Brazil ranks fifth in the world, and on both counts it represents just about half of this entire realm. As noted above, Brazil's economy is the world's sixth-largest, and it is on track to achieve fifth place by 2017, trailing only the United States, China, Japan, and Germany.

Population and Culture

Brazil's population grew rapidly during the world's twentieth-century population explosion. But over the past three decades, the rate of natural increase has plunged from nearly 3.0 percent to 0.9 percent today, and the average number of children born to a Brazilian woman has been more than halved from 4.5 in 1975 to 1.8 in 2013. It is a demographic trend consistent with Brazil's overall modernization since 1980.

Brazil's population of 200.6 million is as diverse as that of the United States. In a pattern quite familiar in the Americas, the indigenous inhabitants of the country were decimated following the European invasion (approximately 800,000 now survive, about two-thirds of them deep within the Amazonian interior). Africans came in very large numbers as well, and they currently total more than 15 million.

Brazil's culture is infused with African themes, a quality that has marked it from the very beginning. So many Africans were brought in bondage to the city and hinterland of Salvador in Bahia State (Fig. 3-16) from what is today Benin (formerly named Dahomey) in West Africa that Bahia has become a veritable outpost of African culture. Indeed, Brazil can be said to have the second-largest black (African) population in the world, after Nigeria. Significantly, however, there was also much racial mixing, and the 2010 census reported that 97 million Brazilians (50.7 percent of the total population) have combined European, African, and minor indigenous ancestries.

Yet another significant, though small, minority began arriving in Brazil in 1908: the Japanese, who today are concentrated in the States of São Paulo and Paraná. The more than 1 million Japanese-Brazilians form the largest ethnic Japanese community outside Japan, and in their multicultural environment they have risen to the top ranks of Brazilian society as business leaders, urban professionals, commercial farmers—and even as politicians in the city of São Paulo. Committed to their Brazilian homeland as they are, the Japanese community also retains its contacts with Japan, resulting in many a trade connection.

Brazilian society, to a much greater degree than is true elsewhere in the Americas, has made progress in dealing with its racial divisions. Blacks do remain the least advantaged among the country's major population groups, and community leaders continue to complain about discrimination. But ethnic mixing in Brazil is so pervasive that hardly any group is unaffected—so in the end, official census statistics concerning 'blacks' and 'Europeans' are rather meaningless.

Brazilians have a strong national culture, expressed in a traditional adherence to the Roman Catholic faith (though declining under Protestant-evangelical and secular pressures), universal use

FIGURE 3-16

of a modified form of Portuguese as the common language ('Brazilian'); and a set of lifestyles in which soccer, 'beach culture,' distinctive music and dance, and an intensifying national consciousness and pride are fundamental ingredients.

Inequality and Poverty

For all its accomplishments in multiculturalism, Brazil remains a country of stark social inequalities (Fig. 3-17). Although such inequality is hard to measure precisely, South America is frequently cited as the geographic realm exhibiting the world's widest gap between affluence and poverty. And within South America, Brazil is reputed to have the most extensive gap of all.

Since around the turn of this century, the Brazilian government has been implementing a set of policies aimed at bringing relief to the poor while maintaining robust economic growth. It has enacted land reform and increased access to education for Brazil's masses. It has also launched a subsidy program that has

BRAZIL: THE GEOGRAPHY OF WEALTH, 2012
(GDP PER CAPITA)

- Over 20,000
- 15,000–20,000
- 10,000–15,000
- Under 10,000

Source: Brazil Institute of Geography and Statistics. Regional Account Reports. Figures used are Brazilian reais. 1.95 R$= 1 US$ approx. (2012)

FIGURE 3-17

© J. Nijman, P. O. Muller, H. J. de Blij, and John Wiley & Sons, Inc.

had significant results in the poorest States. This *Bolsa Familia* (Family Fund) plan, instituted under former President Fernando Cardoso in the 1990s, provides families with small payments of cash to keep their children in school and ensure their vaccinations against diseases that especially afflict the poor. In just a few years this program became so successful that it now serves as a model for antipoverty campaigns the world over.

Regional inequalities are being targeted as well by government policies. Major new legislation was passed in 2012 that would allow revenues from oil to be shared by all of Brazil's States (previously this income was mainly channeled to the three oil-producing States of Rio de Janeiro, Espirito Santo, and São Paulo). In addition, overall federal spending is increasingly being steered to poorer States—particularly the Northeast—in order to accelerate economic growth.

Development Prospects

Brazil is richly endowed with mineral resources, including enormous iron and aluminum ore reserves, extensive tin and manganese deposits, and highly promising oil- and gasfields (Fig. 3-16). Other significant energy developments involve massive state-of-the-art hydroelectric facilities and the successful integration of sugarcane-based anhydrous ethanol with gasoline, allowing Brazil's motor-vehicle owners to use this 'gasohol' instead of costlier imported petroleum (all Brazilian cars today are configured to use this biofuel, whose ethanol content must be at least 18 percent).

Besides these natural endowments, Brazil has more arable land than any other country in the world. Brazilian soils sustain a bountiful agricultural output that makes the country the world's leading exporter of coffee, orange juice, sugar, tobacco, ethanol, beef, and chicken. Commercial agriculture is now the fastest growing economic sector, driven by mechanization and the opening of a booming new farming frontier in the fertile grasslands of southwestern Brazil (discussed below in conjunction with the Interior subregion). Industrialization has driven Brazil's ascendancy as a global economic power. As a result, annual revenues from industrial exports surpassed those from agriculture more than a decade ago.

On the global stage, Brazil has become a formidable presence in other ways. The country's enormous and easily accessible iron ore deposits, the relatively low wages of its workers, and the mechanized efficiency of its steelmakers enable Brazil to produce that commodity at half the cost of steel made in the United States (which caused American steel producers to demand protectionist measures). The most ravenous consumer of Brazil's iron, steel, and most other raw materials is—you guessed it—China.

Brazil already stands out for its huge storehouse of natural resources, and to these may now be added the biggest bounty of all—oil, and lots of it. Through the recent past, a leading concern for Brazil was energy and its rising costs. Recessions were deepened by high oil and gas prices, and other than modest domestic reserves and those in neighboring Bolivia, there were few nearby sources to acquire what was needed. But then in 2009, the state-owned (but publicly traded) petroleum company, Petrobras, confirmed the discovery of an enormous offshore oil reserve, quite possibly among the world's largest, capable of yielding an estimated 2 billion barrels. It was proof, said then-President Luiz Inácio Lula with a broad smile on his face, that 'God is Brazilian.'

Brazil's Subregions

Brazil is a federal republic consisting of 26 States and the federal district of the capital, Brasília (Fig. 3-16). As in the United States, the smallest States lie in the northeast and the larger ones farther west; their populations range from about 500,000 in the northernmost, peripheral Amazon State of Roraima to more than 45 million in burgeoning São Paulo State. Although Brazil is about as large as the 48 contiguous United States, it does not exhibit a clear physiographic regionalism. Even the Amazon Basin, which covers just about 60 percent of the country, is not entirely a plain: between the tributaries of the great river lie low but extensive tablelands. Given this physiographic ambiguity, the six Brazilian subregions that will now be discussed exhibit no absolute or even generally accepted boundaries. In Figure 3-16, those boundaries have been drawn to coincide with the borders of States, making identifications easier.

The **Northeast** was Brazil's source area, its culture hearth. The plantation economy took root here at an early date, attracting

Portuguese planters, who soon began to import the country's largest contingent of African slaves to work in the sugarcane fields. But the ample rainfall occurring along the coast soon gives way to drier and more variable patterns in the interior, which is home to about half of the region's 50-plus million people. This semiarid inland back-country—called the **sertão**—is not only seriously overpopulated but also contains some of the worst poverty to be found anywhere in the Americas. The Northeast produces less than one-sixth of Brazil's gross domestic product, but its inhabitants constitute more than one-fourth of the national population. Moreover, this subregion contains half of the country's poor and exhibits a literacy rate 20 percent below Brazil's mean as well as an infant mortality rate twice the national average.

Much of the Northeast's misery is rooted in its unequal system of land tenure. Farms must be at least 100 hectares (250 acres) to be profitable in the hard-scrabble *sertão*, a size that only large landowners are able to afford. Moreover, the Northeast is tormented by a monumental environmental problem: the cyclical recurrence of devastating droughts at least partly attributable to *El Niño* 23 (periodic events of sea-surface warming off the continent's northwestern coast that skew regional weather patterns).

The Northeast is finally receiving greater attention from the central government, largely in the form of federally funded infrastructure projects and incentive-driven investment (think oil money). In cities such as Recife and Salvador, hordes of peasants driven from the land constantly arrive to expand the surrounding shantytowns. As yet, few of the generalizations about emerging Brazil apply here, but there are some bright spots. A petrochemical complex has been built near Salvador, creating thousands of jobs and luring foreign investors. Irrigation projects have nurtured a number of productive new commercial agricultural ventures. Tourism is booming along the entire Northeast coast, whose thriving beachside resorts attract tens of thousands of vacationing Europeans (flight times are under 8 hours). Recife has now spawned a budding software industry and a major medical complex. And Fortaleza is the center of new clothing and shoe industries that have already put the city on the global economic map.

The **Southeast** has been modern Brazil's *core area*, with its major cities and leading population clusters. Gold first drew many thousands of settlers, and other mineral finds also contributed to the influx—with Rio de Janeiro serving as the terminus of the 'Gold Trail' and then as the longtime capital of Brazil until 1960. 'Rio' became the cultural capital as well, the country's most international center, *entrepôt*, and tourist hub. The third quarter of the twentieth century brought another mineral age to the Southeast, based on the iron ores around Lafaiete carried to the nearby steelmaking complex at Volta Redonda (Fig. 3-16). The surrounding State of Minas Gerais (the name means 'General Mines') formed the base from which industrial diversification in the Southeast has steadily mushroomed. The burgeoning metallurgical center of Belo Horizonte paved the way and is now the endpoint of a rapidly developing, ultramodern manufacturing corridor that stretches 500 kilometers (300 mi) southwest to metropolitan São Paulo (Fig. 3-16, striped zone).

São Paulo State is both the leading industrial producer and the primary focus of ongoing Brazilian development. This economic-geographic powerhouse accounts for nearly half of the country's GDP, with a booming economy that today matches Argentina's in overall size. Not surprisingly, this subregion is growing phenomenally (it contains more than 20 percent of Brazil's population) as a magnet for upwardly mobile migrants, especially from the Northeast.

The wealth of São Paulo State was built on its coffee plantations (known as **fazendas**), and Brazil is still the world's leading producer. But coffee today has been eclipsed by other farm commodities. One of them is orange juice concentrate (here, too, Brazil leads the world). São Paulo State now produces more than double the annual output of Florida, thanks to a climate all but devoid of winter freezes, ultramodern processing plants, and a fleet of specially equipped tankers that ship the concentrate to markets around the globe. Another major crop is soybeans, contributing to Brazil's ranking today as the world's number-two producer.

Matching this agricultural prowess is the State's industrial strength. The revenues derived from the coffee plantations provided the necessary investment capital, ores from Minas Gerais supplied the vital raw materials, São Paulo City's outport of Santos facilitated access to the ocean, and immigration flows from Europe, Japan, and other parts of Brazil contributed the increasingly skilled labor force. As the capacity of the expanding domestic market grew, the advantages of central location and large-scale agglomeration nailed down São Paulo's primacy. This also resulted in metropolitan São Paulo becoming the country's—and South America's—leading industrial complex and megacity. But this rapid progress was accompanied by massive problems of overcrowding, pollution, and congestion, which now severely challenge the sustainability of this gigantic urban dynamo.

The **South** consists of three States, whose combined population now exceeds 30 million: Paraná, Santa Catarina, and Rio Grande do Sul (Fig. 3-16). Southernmost Brazil's exceptional agricultural potential has long attracted a steady flow of European immigrants. They introduced their advanced farming methods to several areas in this part of the country. Portuguese rice farmers clustered in the valleys of Rio Grande do Sul, where tobacco production has now propelled Brazil to become the world's number-one exporter. The Germans, specialists in raising grain and cattle, occupied the somewhat higher areas to the north and Santa Catarina. The Italians selected the highest slopes, where they established thriving vineyards. All of these fertile lands proved highly productive, and with growing markets in the mushrooming urban areas to the north, this tri-State subregion has become Brazil's most affluent corner.

Economic development within the South is not limited to the agricultural sector. Coal from Santa Catarina and Rio Grande do Sul is shipped north to the steel plants of Minas Gerais. Local manufacturing is growing as well, especially in Pôrto Alegre and Tubarão. During the 1990s, a major center of the computer software industry was established in Florianópolis, the island-city as well as State capital just off Santa Catarina's coast. Known as *Tecnópolis*, this budding technopole continues to grow by capitalizing on its seaside amenities, skilled labor force, superior air-travel and global communications linkages, and government and private-sector initiatives to support new companies.

The **Interior** subregion—constituted by the States of Goiás, Mato Grosso, and Mato Grosso do Sul—is also known as the **Central-West**. This is the subregion that Brazil's developers long sought to make a part of the country's productive heartland, and in 1960 the new capital of Brasília was deliberately situated on its margins (Fig. 3-16).

By locating the new capital city in the wilderness 650 kilometers (400 mi) inland from its predecessor, Rio de Janeiro, the nation's leaders dramatically signaled the opening of Brazil's development thrust toward the west. Brasília is noteworthy in another regard because it represents what political geographers call a **forward capital 24**. A state will sometimes relocate its capital to a sensitive area, perhaps near a peripheral zone under dispute with an unfriendly neighbor, in part to confirm its determination to sustain its position in that contested zone. Brasília does not lie close to a contested area, but Brazil's interior was an internal frontier to be conquered by a growing nation. Spearheading that drive, the newly built capital occupied a decidedly forward position.

Despite the subsequent growth of Brasília to 4.2 million inhabitants today (which includes a sizeable ring of peripheral squatter settlements), it was not until the 1990s that the Interior subregion began its economic integration with the rest of Brazil. The catalyst was the exploitation of the vast *cerrado* **25**—the fertile savannas that blanket the Central-West and make it one of the world's most promising agricultural frontiers (at least two-thirds of its arable land still awaits development). As with the U.S. Great Plains, the flat terrain of the *cerrado* is one of its main advantages because it facilitates the large-scale mechanization of farming with a minimal labor force. Another benefit is rainfall, more prevalent here than in the Great Plains or Argentine Pampa. The increasingly dominant local crop is soybeans (see Fig. 3-4), and today only the United States produces more than Brazil.

The **North** is Brazil's territorially largest and most rapidly developing subregion, which consists of the seven States of the Amazon Basin (Fig. 3-16). This was the scene of the great rubber boom of a century ago, when the wild rubber trees in the *selvas* (tropical rainforests) produced huge profits and the central Amazon city of Manaus enjoyed a brief period of wealth and splendor. But the rubber boom had ended by 1920 and for most of the six decades that followed, Amazonia was a stagnant hinterland lying remote from the centers of Brazilian settlement. All that changed quite dramatically during the 1980s as new development began to stir throughout this awakening subregion, which continues to be the scene of the world's largest migration into virgin territory as more than 200,000 new settlers arrive each year.

The North of Brazil is also widely known for its high rates of **deforestation.** Between 1995 and 2005, an area of 20,000 square

The Ponte Rio Negro, iconic symbol of the ongoing development thrust into South America's interior. This first major bridge in the Amazon Basin and Brazil's longest was completed in 2011 and spans the mighty Rio Negro just upstream from its confluence with the even greater Amazon itself. Not surprisingly, this most impressive river junction on the South American continent spawned the central Amazon's largest city, Manaus, well over a century ago. Hemmed in until recently by dense rainforests and these enormous rivers, Manaus's new bridge connects the city to its satellite, Iranduba. The latter is now booming as a result of its much-upgraded accessibility and serves as the jumping-off point on the west side of the Rio Negro, its own suburbs spearheading the outward thrust of metropolitan Manaus into an entirely new development sector

kilometers (7700 sq mi) was cleared away every year. But the worst might be over because the Brazilian government has pledged to achieve an 80 percent reduction in deforestation by 2020 and to terminate the practice by 2030. At least part of its thinking is that Brazil has enormous potential as a 'green economy' with its abundance of land, water, and sunshine. Today, nearly half of the country's energy is already obtained from renewable sources (versus less than 12 percent in the United States), and it is becoming easier to envision Brazil as a pacesetting environmental power. Although deforestation rates in Brazil have decreased more sharply than in most other countries, it is still hard to believe that the government's ambitious goals will be achieved anywhere near 2020 and 2030, respectively.

Development projects abound today in the Amazonian North. One of the most durable is the **Grande Carajás Project** in central Pará State, a huge multifaceted scheme centered on the world's largest known deposit of iron ore in the hills around Carajás (Fig. 3-16; see aerial photo in the section *Economic Geography*). Besides a vast mining complex, other new construction here includes the Tucuruí Dam on the nearby Tocantins River and an 850-kilometer (535-mi) railroad to the Atlantic port of São Luis. This ambitious development project further emphasizes the exploitation of additional minerals, cattle-raising, crop farming, and forestry. What is taking place here is a manifestation of the **growth-pole concept 26**. A growth pole is a location where a set of activities, given a start, will expand and generate widening ripples of development in the

This is what the Amazon's equatorial rainforest looks like from an orbiting satellite after the human onslaught in preparation for settlement. The colors on this Landsat image emphasize the destruction of the trees, with the dark green of the natural forest contrasted against the pale green and pinks of the leveled forest. The linear branching pattern of deforestation is the preferred approach here in Rondônia State's Highway BR-364 corridor. But farming is not likely to succeed for very long, and much of the cleared land is likely to be abandoned. Then the onslaught would resume to clear additional land—a pernicious cycle the Brazilian government has committed itself to ending by 2030.

surrounding area. According to this scenario, the stimulated hinterland could one day cover one-sixth of all Amazonia.

Understandably, tens of thousands of settlers have descended on this part of the Amazon Basin. Those seeking business opportunities have been in the vanguard, but they have been followed by masses of lower-income laborers and peasant farmers in search of jobs and land ownership. The initial stage of this colossal enterprise has boosted the fortunes of many urban centers, most of all Manaus northwest of Carajás. Here, a thriving industrial complex specializing in the production of electronic goods has emerged within the free-trade zone adjoining the city thanks to the outstanding air-freight facilities and operations at Manaus's ultramodern airport. At the same time, in and around Manaus, a new era of urban expansion has been unleashed by the opening of the spectacular Rio Negro bridge at the beginning of this decade (see photo).

But many problems have also surfaced as the tide of pioneers has rolled across central Amazonia. One of the most tragic involved the Yanomami people, whose homeland in Roraima State was overrun by thousands of claim-stakers (in search of newly discovered gold), who triggered violent confrontations that ravaged the fragile aboriginal way of life.

FUTURE PROSPECTS OF SOUTH AMERICA

South American countries have a long history of intermittent political turmoil. Dictatorial regimes ruled from one end of the realm to the other; unstable governments fell with damaging frequency. Widespread poverty, harsh regional disparities, inferior internal surface connections, limited international contact, and economic stagnation prevailed.

Today, much of that is finally being cast aside as South America enters a new era of transition. Well into the second decade of the twenty-first century, formal democracy seems to have taken hold almost everywhere. Long-isolated countries are becoming more interconnected through new transportation routes and trade agreements. New settlement frontiers are being opened. Energy resources, some long exploited and others newly discovered, are boosting national economies. Foreign states and corporations have appeared throughout the realm to buy commodities and invest in infrastructure. The pace of globalization is accelerating from Bogotá to Buenos Aires.

This exciting progress, however, still varies enormously across the realm, and the new economic map is as dominated as ever by uneven development. Furthermore, everything must be viewed against a backdrop of persistent problems of deep inequality, intensifying class divisions, growing ethnic tensions, and endemic political malfeasance. It is also apparent that this realm is overly dependent on its raw materials. Revenues increased substantially during the so-called commodity boom that marked the late 2000s and early 2010s, largely driven by voracious demand from China. By 2015, however, a new reality was emerging (even in Brazil) as foreign demand was in decline, South American revenues were dropping, and world prices were moving downward. This unexpected reversal, of course, is yet another reminder that it is never enough to rely on commodity exports, and that lasting progress requires new economic initiatives, especially local industrial development. Countries such as Brazil and Chile seem very well positioned to move forward, but others such as Venezuela, Bolivia, and Argentina continue to face much greater challenges.

POINTS TO PONDER

- Is it a coincidence that both liberation theology and *dependencia* theory originated in this realm?
- Even though the Amazon rainforest lies almost entirely within Brazil, it is often referred to as 'the lungs of the Earth.' Should its preservation be a global responsibility?
- Should China's deepening involvement in South America be a cause for concern in the United States?
- Is the United States partially responsible for the problems caused by cocaine production and trafficking in such countries as Colombia and Mexico?

© MAYA VIDON-WHITE/epa /Corbis Images

More than one million people poured into the streets of Paris on January 11, 2015 to protest the killing, four days earlier, of 11 journalists, cartoonists, and co-workers at the offices of the satirical magazine, *Charlie Hebdo*. This publication is widely known for its irreverent, nonconformist criticism of politics and religions, including—but certainly not limited to—Islam. The perpetrators, self-declared Muslim jihadists associated with al-Qaeda, claimed to be acting in the name of Allah as well as the Prophet Muhammad. This terribly tragic event sent shockwaves through Europe, triggering demonstrations in Paris and far beyond that often included Muslims who denounced the savage terrorist act. The banner above translates as 'We are Charlie,' an expression of solidarity with the victims. The ongoing intensification of Islamist militancy in Europe is deeply problematic and increasingly drives national political agendas and debates.

IN THIS CHAPTER

- The question of Europe's eastern boundary
- The specter of Islamist terrorism in Europe
- Anti-EU sentiments in Europe's heartland
- Whither Greece—in or out of the EU?
- Crisis in Ukraine: Causes and consequences

CONCEPTS, IDEAS, AND TERMS

4

THE EUROPEAN REALM

FIGURE 4-1

A strong case can be made that, over the past five centuries, Europeans have influenced and changed the rest of the world more than the people of any other realm. For good or bad, much of the world would look quite different today if it had not been for Europe. Europe's colonial empires spanned the globe and transformed societies far and near. Millions of Europeans migrated from their homelands to the Old World as well as the New, changing (and sometimes nearly obliterating) traditional communities and creating new societies from Australia to North America. Colonial power and economic incentive combined to propel the movement of millions of imperial subjects from their ancestral homes to distant lands: Africans to the Americas, Indians to Africa, Chinese to Southeast Asia. In agriculture, industry, politics, and other spheres, Europe generated revolutions—and then, through imperialism and colonialism, exported those revolutions across the world.

However, during the seven decades since World War II (1939–1945), and particularly over the last four, Europe has decidedly become a more 'ordinary' realm. The days of empire are now long gone. National economies face the same global challenges as their counterparts elsewhere. And Europe no longer plays a leading role in the world economy, even though it continues to rank among the most highly developed geographic realms. If the claim can still be made today that Europe is 'special,' it is with an eye on the remarkable process of economic and political integration that has shaped the evolution of the *European Union (EU)* over the past half-century. Although the European realm is highly diverse in a cultural-geographical sense, most of it is also tightly integrated with considerable economic powers devolved from national governments to 'Brussels,' the EU headquarters city. This integration itself, it should be noted, varies across Europe as well—making for an especially complex geographic realm.

DEFINING THE REALM

GEOGRAPHICAL FEATURES

Europe's Eastern Boundary

As Figure 4-1 shows, Europe is a realm of peninsulas and islands on the western margin of the world's largest landmass, Eurasia. It is a realm containing almost 600 million people and 40 countries, but it is territorially quite small. The European realm is bounded on the west, north, and south by Atlantic, Arctic, and Mediterranean waters, respectively. But where is Europe's eastern limit?

Each episode in the historical geography of Eastern Europe has left its particular legacy in the cultural landscape.

Twenty centuries ago, the Roman Empire ruled much of Eastern Europe (Romania is a cartographic reminder of that period); for most of the second half of the twentieth century, the Soviet Empire controlled nearly all of it. During the intervening two millennia, Christian Orthodox church doctrines spread from the southeast, and Roman Catholicism advanced from the northwest. Turkish (Ottoman) Muslims invaded and created an empire that reached the vicinity of Vienna. By the time the Austro-Hungarian

MAJOR GEOGRAPHIC QUALITIES

EUROPE

1. The European geographic realm lies on the western flank of the Eurasian landmass.

2. Though territorially small, Europe is heavily populated and fragmented into 40 states.

3. European natural environments are highly varied, and Europe's resource base is rich and diverse.

4. Europe's geographic diversity, cultural as well as physical, created strong local identities, specializations, and opportunities for trade and commerce.

5. The European Union (EU) is a historic and unique effort to achieve multinational economic integration and, to a lesser degree, political coordination.

6. Europe's relatively prosperous population is highly urbanized and aging rapidly.

7. Local demands for greater autonomy as well as cultural challenges posed by immigration are straining the European social fabric.

8. Despite Europe's momentous unification efforts, east-west contrasts still mark the realm's regional geography.

9. The effects of the 'euro crisis' pose a threat to the EU and suggests that the single currency was introduced too soon for some members. It is possible that certain countries will either choose or be forced to exit the Union.

Empire ousted the Turks more than a century ago, millions of eastern Europeans had been converted to Islam (Albania and Kosovo still remain predominantly Muslim countries). Meanwhile, it is often said that Western Europe's civilization had its cradle in ancient Greece, but that lies farther still to the southeast, beyond the former Yugoslavia.

As we shall see, the geographic extent of Europe has always been debatable. In this decade, it is a highly contentious issue in terms of European Union expansion and relations with Russia—as underscored by the turmoil that continues to plague eastern Ukraine. Europe's eastern boundary is a dynamic one, and it has changed with history. Some would say that there really is no clear boundary, that Europe's atmosphere, so to speak, just thins out toward the east. Our present definition (Fig. 4-1) places Europe's eastern boundary between Russia and its six European neighbors to the west. That definition is based on several geographic criteria that include European-Russian contrasts in territorial dimensions, geopolitical developments, cultural properties, and history.

Figure 4-1, as you may already have observed, places the Crimean Peninsula inside Russia as a consequence of its de facto annexation from Ukraine in 2014. This does not mean that we as geographers endorse that forcible annexation. But it does show that state boundaries frequently materialize through violence and the will of the victors. Thus the global map of states is in many ways the product of power struggles, territorial conflicts, and war—especially in Europe. In fact, it can be argued that much of Eastern Europe constitutes a **transition zone** [1] between the European and Russian-dominated realms (which is particularly notable in Belarus and eastern Ukraine). That argument could be made on the basis of language, religion, and/or functional spatial interaction—but it does not translate into simple claims of territorial control.

Climate and Resources

From the balmy shores of the Mediterranean Sea to the icy peaks of the Alps, and from the moist woodlands and moors of the Atlantic fringe to the semiarid prairies north of the Black Sea, Europe presents an almost infinite range of natural environments (Figs. 4-2 and 4-3).

Europe's peoples have benefited from a large and varied store of raw materials. Whenever the opportunity or need arose, the realm proved to contain what was required. Early on, these requirements included cultivable soils, rich fishing waters, and wild animals that could be domesticated; in addition, extensive forests provided wood for houses and boats. Later, coal and mineral ores propelled industrialization. More recently, Europe proved to contain sizeable deposits of oil and natural gas.

FIGURE 4-2

© J. Nijman, P. O. Muller, H. J. de Blij, and John Wiley & Sons, Inc.

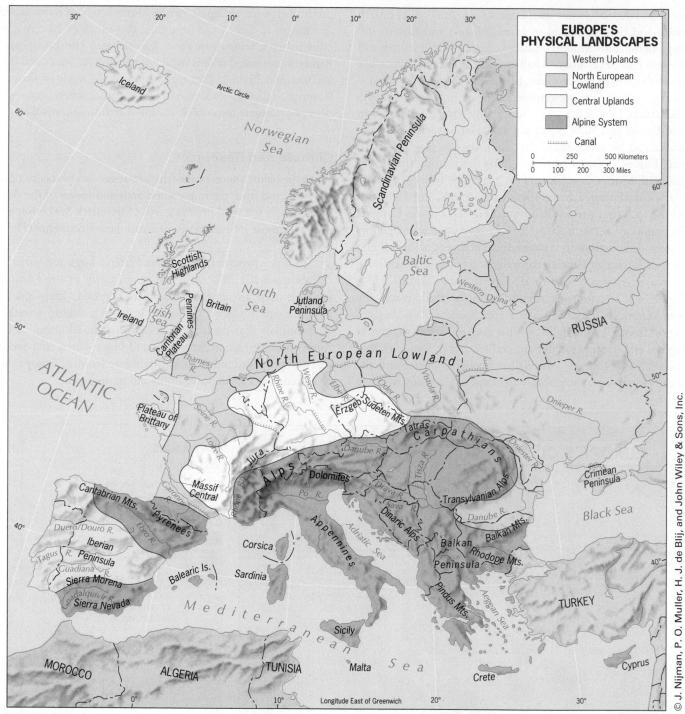

FIGURE 4-3

Landforms and Opportunities

Europe's area may be small, but its physical landscapes are varied and complex. Regionally, we identify four broad units: the Central Uplands, the southern Alpine Mountains, the Western Uplands, and the North European Lowland (Fig. 4-3).

The **Central Uplands** form the heart of Europe. It is a region of hills and low plateaus loaded with raw materials whose farm villages grew into towns and cities when the Industrial Revolution transformed this realm.

The **Alpine Mountains**, a highland region named after the Alps, extend eastward from the Pyrenees on the French-Spanish border to the Balkan Mountains near the Black Sea, and include Italy's Appennines as well as the Carpathians of Eastern Europe.

The **Western Uplands**, geologically older, lower, and more stable than the Alpine Mountains, extend from northern Scandinavia through western Britain and Ireland to the heart of the Iberian Peninsula in Spain.

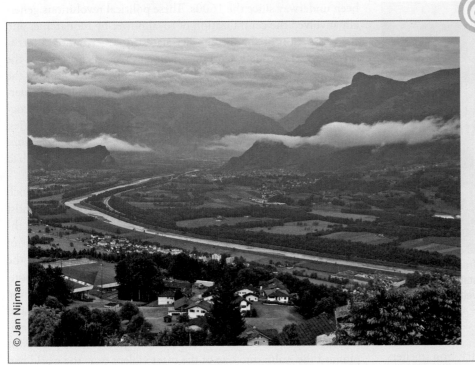

@ FROM THE FIELD NOTES...

"It has been a long day, driving north from Italy through the Swiss Alps. I just crossed the border into the mini-state of Liechtenstein to look for a place to spend the night. From my hotel in the town of Triesenberg I have a panoramic view of an upper stretch of the Rhine River. It flows from left to right in the picture. This is Western Europe's leading river and one of its defining geographic features—as a means of transportation, sometimes as a barrier, and frequently as a natural boundary. Here it forms the mini-state's border with Switzerland, located on the far side. Even though we are looking at what is a political boundary, the physical and cultural landscapes on both sides of the Rhine have much in common."

The ***North European Lowland*** stretches in a lengthy arc from southeastern Britain and central France across Germany and Denmark into Poland and Ukraine, from where it continues well into Russia. Also known as the Great European Plain, this has been an avenue for human migration time after time, so that complicated cultural and economic mosaics developed here and together produced a jigsaw-like political map. As Figure 4-3 shows, many of Europe's major rivers and connecting waterways serve this populous region, where a number of Europe's leading cities (London, Paris, Amsterdam, Copenhagen, Berlin, Warsaw) are located.

Locational Advantages

Europe's ***relative location***, at the crossroads of the land hemisphere **2**, creates considerable efficiency for contact with much of the rest of the world (Fig. 4-4). A 'peninsula of peninsulas,' Europe is nowhere far from the ocean and its avenues of seaborne trade and conquest. Hundreds of kilometers of navigable rivers, augmented by an unmatched system of canals, open the interior of Europe to its neighboring seas and to the shipping lanes of the world. The Mediterranean and Baltic seas, in particular, were critical in the development of trade in early modern times, and in the emergence of Europe's early city-states **3** such as Venice, Italy (see photo) in the south and Lübeck, Germany in the north.

And do not overlook the scale of the maps in this chapter. Europe is a realm of moderate distances and close proximities. Short distances and large cultural differences make for intense interaction, the constant circulation of goods and ideas. That has been the hallmark of Europe's geography since early modern times.

MODERN HISTORICAL GEOGRAPHY

The Industrial Revolution

From the late eighteenth century onward, Europe experienced a steady process of intensifying local functional specialization **4**. The Industrial Revolution **5**, the beginning of which is often attributed to the invention of the steam engine by Scotsman James Watt (1736–1819) in 1780, was essential in this process, and it would provide Europe with a critical edge in world affairs for a long time to come. The Industrial Revolution gathered momentum throughout the nineteenth century, and Britain was at its epicenter.

EUROPE IN THE NORTHERN HEMISPHERE

3000 kilometers=ca. 1900 miles

© J. Nijman, P. O. Muller, H. J. de Blij, and John Wiley & Sons, Inc.

FIGURE 4-4

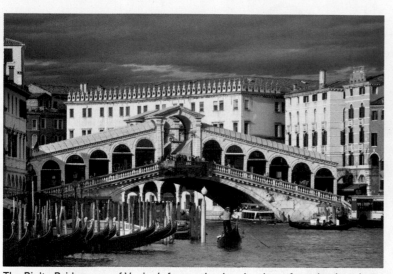

Visions Of Our Land/The ImageBank/Getty Images, Inc.

The Rialto Bridge, one of Venice's famous landmarks, dates from the time that Venice was one of Europe's most powerful and richest city-states. The first bridge over the Grand Canal at this spot dates back to the twelfth century; the present span was completed in 1591.

The power loom revolutionized the weaving industry. Iron smelters, long dependent on Europe's dwindling forests for fuel, could now be concentrated near coalfields. Engines could move locomotives as well as power looms. Ocean shipping entered a new age. The British had an enormous advantage because they controlled the flow of raw materials and held a monopoly over products that were in demand globally. In Britain, manufacturing regions developed near coalfields in the English Midlands, at Newcastle to the northeast, in southern Wales, and along Scotland's Clyde River around Glasgow.

After 1850, the Industrial Revolution spread eastward from Britain into continental Europe and by 1890 had reached as far as western Russia. In mainland Europe, the axis of industrialization followed a west-to-east belt of major coalfields extending roughly along the southern margins of the North European Lowlands; each gave rise to manufacturing complexes that can still be seen on the map today in northern France and Belgium, Germany's Ruhr Basin, the Czech Republic's Bohemia, southern Poland's Silesia, and eastern Ukraine's Donets Basin (Donbas). Abundant iron ore deposits lie in a similar west-east belt and together with coal provide the key raw materials for manufacturing steel.

As in Britain, this cornerstone heavy-metal industry spawned new concentrations of economic activity, which grew steadily as millions migrated in from the countryside to fill expanding employment opportunities. Densely populated and heavily urbanized, these industrial agglomerations became the backbone of Europe's world-scale population cluster (labeled **D** in Fig. G-7) as well as an enduring major feature on the realm-scale map of Europe's population distribution (Fig. 4-5).

Political Revolutions

The Industrial Revolution unfolded against the backdrop of the ongoing formation of Europe's national states, a process that had

been underway since the 1600s. These political revolutions generally involved the centralization of royal power by monarchies and the usurpation of the former city-states in larger national territories—which frequently produced violent reconfigurations of the realm's evolving political geography. Most historians point to the Peace (Treaty) of Westphalia in 1648 as a key step in the emergence of Europe's state system, ending decades of war and recognizing territories, boundaries, and the sovereignty of countries. But this was only the beginning, and these so-called *absolutist states*, in which monarchs held all the power and the people had few if any rights, would not last.

One of the most dramatic episodes to follow was the French Revolution (1789–1795), which soon ended the era of absolutist states. More gradual political transformations occurred in the Netherlands, Britain, and the Scandinavian countries. But a number of other components of Europe would remain much longer under the tight control of authoritarian (dictatorial) regimes headed by monarchs or despots. By the late nineteenth century, Europe became the arena for the competing ideologies of **liberalism, socialism**, and **nationalism** (national spirit, pride, patriotism). However, a generation later the rise of extreme nationalism (**fascism**) in the 1930s plunged the realm into one of the most violent wars the world had ever seen.

A Fractured Map

Europe's political map has always been one of intriguing complexity. As a geographic realm, Europe occupies only about 5 percent of the Earth's land area; but that minuscule territory is fragmented into some 40 countries—more than one-fifth of all the states in the world today. Therefore, when you look at Europe's political map, the question that arises is how did so small a geographic realm come to be divided into so many political entities? Europe's map is a legacy of its feudal and monarchic periods, when powerful kings, barons, dukes, and other rulers, rich enough to fund armies and powerful enough to extract taxes and tribute from their domains, created bounded territories in which they reigned supreme. Royal marriages, alliances, and conquests actually simplified Europe's political map. In the early nineteenth century, there still were 39 German States; a unified Germany as we know it today would not emerge until the 1870s.

State and Nation

Europe's political revolutions produced a prototype of politico-geographical organization known as the **nation-state** 6, a territorial state embodied by its culturally distinctive population. The term **nation** 7 refers to a people with a single language, a common history, and a similar ethnic background. In the sense of **nationality**, it relates to citizenship (legal membership) in the state. Only a handful of states today are so homogeneous culturally that their culture is practically conterminous with the state. Europe's prominent nation-states of a century ago—France, Spain, the United Kingdom, Italy—have become multicultural

© J. Nijman, P. O. Muller, H. J. de Blij, and John Wiley & Sons, Inc.

FIGURE 4-5

societies, with *nation* now defined more by an intangible spirit of nationalism and emotional commitment than by cultural or ethnic homogeneity.

CONTEMPORARY EUROPE: A DYNAMIC REALM

Cultural Diversity

The European realm is home to peoples of numerous cultural-linguistic stocks, including not only Latins, Germanics, and Slavs but also minorities such as Finns, Magyars (Hungarians), Basques, and Celts. This diversity of ancestries continues to be an asset as well as a liability. It has generated not only interaction and exchange, but also conflict and war.

It is worth remembering that Europe's territory is just over 60 percent the size of the United States, but that the population of Europe's 40 countries is almost twice as large as America's. This population of just under 600 million speaks numerous languages, almost all of which belong to the Indo-European language family 8 (see Figs. G-8 and 4-6). Nonetheless, most of those languages are not mutually understandable; some, such as Finnish and Hungarian, are not even members of the Indo-European family. When Europe's unification efforts began after World War II, one major problem was to determine which languages to recognize as 'official.' That problem still prevails, although English has become the realm's unofficial *lingua franca* (common language). During a visit to Europe, though, you would find that English is more commonly usable in the big cities than in the countryside and more in Western Europe than farther east. Europe's multilingualism remains a rich cultural legacy, but it is also a barrier to further integration.

Another divisive force confronting Europeans involves religion. Europe's cultural heritage is steeped in Christian traditions, but sectarian strife between Catholics and Protestants, which plunged parts of the realm into bitter and widespread conflict, still divides communities and, as until recently in Northern Ireland, can still arouse violence. Some political parties still carry the name 'Christian'—for instance, Germany's Christian Democrats.

More generally, Christianity has gradually lost adherents since the 1960s as secularization gathered momentum, especially in Western Europe. The Roman Catholic Church, long powerful in much of Europe, has steadily been losing its grip on society, and many church institutions—schools, universities, unions, political parties, charities, and clubs—have been hollowed out. Moreover, many churches have closed down or have been converted into art galleries, public meeting halls, and even corporate offices.

Today, a new force roils the religious landscape: the rise of Islam. In southeastern Europe, this takes the form of new Islamic assertiveness in an old Muslim bastion: the (Turkish) Ottoman Empire left behind millions of converts from Bosnia to Bulgaria, many of whom are now demanding greater political representation and power. In the west, this Islamic resurgence results from the relatively recent infusion of millions of Muslim immigrants from former colonies in North Africa and other parts of the far-flung Islamic World. Here, whereas mosques overflow with the faithful, churches stand nearly empty as a witness to secularism among Europeans.

Spatial Interaction

If not a single culture, then what does unify Europe? The European realm is best understood as an enormous *functional region*, a fully interdependent realm held together by highly developed, spatial economic and political networks. Contemporary Europe has seized on the realm's abundant geographic opportunities to create a huge, intensively used network of spatial interaction linking places, communities, and countries in countless ways. This interaction operates on the basis of two key principles.

First, regional complementarity 9 means that one area produces a surplus of a commodity required by another area. The mere existence of a particular resource or product is no guarantee of trade: it must be needed elsewhere. When two areas each require the other's products, we speak of *double complementarity*. Europe exhibits countless examples of such complementarity, from local communities to entire countries. Industrial Italy needs coal from Western Europe; Western Europe needs Italy's farm products.

Second, the ease with which a commodity can be transported by producer to consumer defines its transferability 10. Distance and physical obstacles can raise the cost of a product to the point of unprofitability. But Europe is small, distances are short, and the Europeans have built an efficient transport system of roads, railroads, and canals linking navigable rivers. Taken together, Europe's enormously diverse economic regions, its ultra-efficient transportation infrastructure, and its far-reaching free-trade agreements within the European Union make for a supremely interdependent geographic realm.

A Highly Urbanized Realm

About three of every four Europeans live in towns and cities (in Western Europe the share of the urban population is much higher still). Large cities are production centers as well as marketplaces, and they also form the crucibles of their national cultures. Europe's major cities tend to have long histories. They are compact, and in general the European cityscape looks quite different from its North American counterpart.

Seemingly haphazard inner-city street systems impede traffic; central cities may be picturesque, but they are also cramped. European city centers tend to be more vibrant today than cities in the United States. They offer a mix of businesses, government functions, shopping facilities, educational and arts institutions, and entertainment as well as housing for upper-income residents.

Wide residential sectors radiate outward from the *central business district (CBD)* across the rest of the central city, often inhabited by particular income groups. Beyond the central city

LANGUAGES OF EUROPE

0   200   400   600 Kilometers
0   100   200   300 Miles

MAJOR INDO-EUROPEAN BRANCHES

GERMANIC GROUP

WESTERN GERMANIC	NORTHERN GERMANIC
1 Dutch	5 Danish 8 Icelandic
2 German	6 Swedish 9 Faeroese
3 Frysian	7 Norwegian
4 English	

ROMANCE GROUP

10 Portuguese	16 Rhaeto-Romansch
11 Spanish	17 Romanian
12 Catalan	18 Corsican-Italian
13 Provencal	19 Sardinian-Italian
14 French	20 Walloon
15 Italian	

SLAVIC GROUP

WEST SLAVONIC	EAST SLAVONIC	SOUTH SLAVONIC
21 Polish	25 Russian	28 Slovene
22 Slovak	26 Ukrainian	29 Serbo-Croatian
23 Czech	27 Belarussian	30 Macedonian
24 Lusatian		31 Bulgarian

OTHER INDO-EUROPEAN BRANCHES

CELTIC GROUP

BRITANNIC	GAELISH
32 Breton	34 Irish Gaelic
33 Welsh	35 Scots Gaelic

BALTIC GROUP

36 Latvian   37 Lithuanian

HELLENIC GROUP

38 Greek

THRACIAN/ILLYRIAN GROUP

39 Albanian

INDO-IRANIAN GROUP

40 Romani (dispersed)

URALIC LANGUAGE FAMILY

FINNO-UGRIC GROUP

41 Finnish	44 Estonian
42 Karelian	45 Hungarian
43 Saami	46 Komi

SAMOYEDIC GROUP

47 Samoyedic

ALTAIC LANGUAGE FAMILY

TURKIC GROUP

48 Turkish

OTHER LANGUAGES

BASQUE

49 Basque

Areas with significant concentrations of other languages (usually adjacent national languages)

Boundary between languages

After Murphy, 1998.

FIGURE 4-6

© J. Nijman, P. O. Muller, H. J. de Blij, and John Wiley & Sons, Inc.

© Jan Nijman

@ **FROM THE FIELD NOTES...**

"If you were to be asked what city is shown here, would Paris spring to mind? Most images of the French capital show venerable landmarks such as the Eiffel Tower or Notre Dame Cathedral. I am taking this photo from the top of the famous Arc de Triomphe, looking toward *another* Paris, the burgeoning northwestern business district just beyond the city line named La Défense. There, Paris escapes the height restrictions and architectural limitations of the historic center and displays an ultramodern complexion. Glass-box skyscrapers reflect the vibrant global metropolis this is, and the central landmark is the 'Cube,' a huge open structure admired as well as reviled (as was the Eiffel Tower in its time). La Défense is ingeniously incorporated into Paris's urban design, with a straight-line, broad avenue connecting it to the Arc de Triomphe and then connecting, via the Champs Élysées to the Place de la Concorde. In terms of a grand geographic layout, Paris is hard to beat."

lies a sizeable suburban ring, but even here residential densities are much higher than those in the United States because the European preference has been to live in apartments rather than detached single-family houses and to set aside recreational spaces in 'greenbelts.' There also is a greater reliance on public transportation, which further concentrates the realm's urban and suburban development pattern. However, traditions may now be starting to change. One indication is the recent growth of cutting-edge suburban business centers that increasingly compete with the CBD in many parts of urban Europe (see the accompanying Field Note).

A CHANGING POPULATION MOSAIC

Negative Natural Population Growth

There was a time when Europe's population was (in the terminology of population geographers) exploding, sending millions to the New World and other colonies while still growing at home. But these days Europe's native population, unlike most of the rest of the world's, is actually shrinking. To keep a given population from declining, the (statistically) average woman must bear 2.1 children. For Europe as a whole that figure was 1.6 in 2013. Several countries recorded numbers at or below 1.3, including Poland, Spain, Portugal, and Greece. Such negative population growth poses serious challenges for any nation. When the population pyramid becomes top-heavy, the number of workers whose taxes pay for the social services of the elderly goes down, leading to reduced pensions and dwindling funds for healthcare.

The Growing Multicultural Challenge

Meanwhile, immigration is partially offsetting Europe's population deficit. But it may be a mixed blessing in that demographic stability comes at the risk of social disruption. Millions of Turks, Turkish Kurds, Algerians, Moroccans, West Africans, Pakistanis, and West Indians are transforming the sociocultural fabric of what once were homogeneous nation-states, especially France, Germany, the United Kingdom, Belgium, and the Netherlands. As noted earlier, a key dimension of this change is the spread of Islam in Europe. Muslim populations in Eastern Europe (such as those in Albania, Kosovo, and Bosnia) are indigenous communities that were converted during the period of Ottoman rule. The Muslim sectors of Western European countries in contrast reflect more recent immigrant infusions.

The majority of these immigrants are generally more religious than the Christian natives. They arrived in a Europe where homegrown populations are stagnant or declining, where religious institutions are weakening, and where certain cultural norms are incompatible with Islamic traditions. Integration of Muslim communities into the national fabric has been slow, their education and income levels considerably lower than those of the host population. In Western European countries particularly, Islamic immigrants are heavily concentrated in metropolitan areas. Thus, in major cities such as Hamburg, London, Paris, and Brussels, the local proportion of the Muslim population is considerably higher than the national average.

The social and political implications of Europe's cultural transformation are numerous and far-reaching. Not surprisingly, European societies are now trying to restrict immigration in various ways, and political parties with anti-immigrant agendas (especially on the extreme right) are gaining ground in several countries.

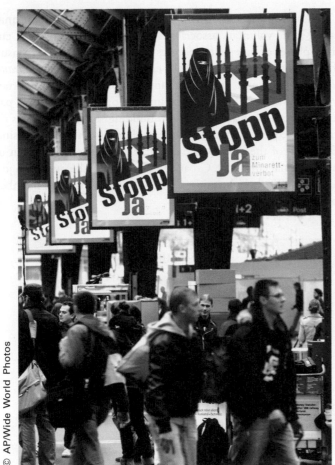

© AP/Wide World Photos

A sign of the popular backlash against the growth of Islam in Europe: a 2009 referendum in Switzerland on new minaret construction produced some ugly symbolism (note the rocket-like depiction of minarets on the Swiss flag). The Swiss electorate approved a law that prohibits the construction of minarets (read: mosques). A majority of the Swiss seemed to feel that their culture was under threat from Islam and that this type of mosque architecture gave expression to an overly assertive minority— even though there were only four minarets in the entire country! No doubt, Switzerland's nearly 500,000 Muslims (just over 6 percent of the population, mainly Turks, and not known to be particularly zealous) must have felt uneasy about the vote.

EUROPEAN UNIFICATION

Realms, regions, and countries can all be subject to dividing and unifying forces that cause them to become more or less cohesive and stable political units. Where the region or country is home to diverse populations with different political goals, it may prove difficult to avoid divergence and territorial fragmentation. Political geographers use the term **centrifugal forces 11** to identify and measure the strength of such division, which may result from religious, racial, linguistic, political, economic, or other regional factors.

Centrifugal forces are measured against **centripetal forces 12**, the binding, unifying glue of the state or region. General satisfaction with the system of government and administration, shared identities, legal institutions, and other functions of the state (notably including its treatment of minorities) can ensure stability at the state level.

Since World War II ended in 1945, Europe has witnessed a steady process of integration and unification at a much broader geographic scale. A majority of European states and their leaders recognized that closer association and regional coordination formed the key to a more stable, prosperous, and secure future. Centripetal forces have thus far prevailed, but, as the EU has expanded to include ever more countries, recent years have witnessed growing dissent and criticism, particularly in the countries where it all began in the late 1950s.

Background

At the end of World War II, much of Europe lay shattered, its cities and towns devastated, its infrastructure wrecked, its economies ravaged. If this was one of the world's most developed realms at the beginning of the twentieth century, its economic prowess had been almost completely destroyed by 1945. One of the primary motives for integration and collaboration among Western European countries, therefore, was rapid economic recovery.

The integration process was also driven by political considerations. From the perspective of countries such as France, the Netherlands, and Belgium, one of the key issues was to control defeated Germany in the postwar years, and this could only be done through close political cooperation. Thus European integration was from the start both an economic and a political affair, and concerns about Germany played a major part.

From afar, the United States was a strong supporter of (west) European integration. The ending of World War II had witnessed accelerating tensions with the Soviet Union, which had taken control over the bulk of Eastern Europe. Furthermore, communist parties seemed poised to dominate the political life of major Western European countries. The United States was intent, therefore, to have a firm hand in that part of Europe and to keep communist influences at bay. Under the so-called **Marshall Plan**, massive U.S. financial support, exceeding $100 billion in today's money, poured in to 16 European countries, including the western half of Germany not controlled by the Soviets as well as Turkey.

As these economies recovered, Germany (i.e., West Germany until the Cold War ended in 1989) became firmly embedded in a pan-European structure, the United States assumed a more distant role, and the motivations driving European unification developed a new emphasis. Economically, the process has increasingly been driven by European corporate interests: the need to facilitate an ever larger and more efficient open market that could compete globally with the United States, China, and Japan. Politically, the goals today are more and more about stabilizing a much larger and diverse Europe that now approaches the Russian frontier. It is not just about Germany anymore but rather about a much bigger and more complicated geopolitical zone that has to accommodate a large number of (still) sovereign states and at the same time maintain satisfactory relations with the Russian giant next door.

The Unification Process

European integration has been a complex, gradual, and step-by-step process aimed at the creation of a single territorial economy across Europe. It has involved (1) agreements on free trade across borders without tariffs or border controls; (2) the abolition of protectionist economic policies; (3) harmonization of national laws (e.g., on taxation) to establish a level playing field; (4) political cooperation and the creation of pan-European executive, judicial, and legislative bodies; (5) the introduction of a single currency, the *euro*; and (6) agreements to transfer specific powers from national governments to EU institutions.

European integration is a classic example of **supranationalism** 13, the voluntary association in economic, political, or cultural spheres of independent states willing to yield some measure of sovereignty for their mutual benefit. The box titled *Supranationalism in Europe* provides a timeline of the most important events and shows that European integration has in no small part been a matter of steady institution building. Figure 4-7 maps the territorial expansion of the European Union (so named only since 1993) and of the **eurozone** (not all EU countries have given up their national currency). Croatia became the 28th country to join the EU in 2013; Lithuania, in 2015, became the 19th member-country to adopt the euro.

© J. Nijman, P. O. Muller, H. J. de Blij, and John Wiley & Sons, Inc.

FIGURE 4-7

SUPRANATIONALISM IN EUROPE

1944 Benelux Agreement signed; this forerunner association aimed to fully integrate the economies of Belgium, the Netherlands, and Luxembourg.

1948 Organization for European Economic Cooperation (OEEC) established.

1949 Council of Europe created.

1951 European Coal and Steel Community (ECSC) Agreement signed (effective 1952).

1957 Treaty of Rome signed, establishing European Economic Community (EEC) (effective 1958); also known as the Common Market and 'The Inner Six,' the founding member-countries were France, (then) West Germany, Italy, Belgium, the Netherlands, and Luxembourg. European Atomic Energy Community (EURATOM) Treaty signed (effective 1958).

1967 European Community (EC) inaugurated.

1968 All customs duties removed for intra-EC trade; common external tariff established.

1973 United Kingdom, Denmark, and Ireland admitted as members of EC, creating 'The Nine.'

1979 First general elections for a European Parliament held; new 410-member legislature convenes in Strasbourg. European Monetary System established.

1981 Greece admitted as member of EC, creating 'The Ten.'

1986 Spain and Portugal admitted as members of EC, creating 'The Twelve.' Single European Act ratified, targeting a functioning European Union in the 1990s.

1993 Single European Market goes into effect. Modified European Union Treaty ratified, transforming EC into EU.

1995 Austria, Finland, and Sweden admitted into EU, creating 'The Fifteen.'

1999 European Monetary Union (EMU) goes into effect.

2002 The euro is introduced in 12 countries; the eurozone expands gradually over time (see Fig. 4-7) but not all EU members choose to give up their national currency.

2004 Historic expansion of EU from 15 to 25 countries with the admission of Cyprus, the Czech Republic, Estonia, Hungary, Latvia, Lithuania, Malta, Poland, Slovakia, and Slovenia.

2005 Proposed EU Constitution is rejected by voters in France and the Netherlands.

2007 Romania and Bulgaria are admitted, bringing total EU membership to 27 countries. Slovenia adopts the euro.

2010 Financial crisis strikes heavily indebted Greece, requiring massive EU bailout. Later in the year, Ireland follows and raises fears of similar crises in Portugal, Spain, and Italy. The future of the EMU is clouded; the value of the euro declines after a long rise against the dollar.

2012 Fiscal Compact agreement (effective 2013) signed by 20 member-states (and initially ratified-by 16) stipulates their tight adherence to rules of fiscal responsibility. The agreement sparked intense debate in various countries, and it was not until mid-2014 that all EU members had ratified the pact.

2013 Croatia gains entry as the 28th EU member-state. Politicians in several member-countries, including the United Kingdom and the Netherlands, demand a national referendum on exiting the EU.

2014 Latvia adopts the euro.

2015 In a hastily called referendum, Greece's population voted by a large majority to reject continued EU bailout terms requiring severe austerity measures—bringing a "Grexit" from the eurozone ever closer. Lithuania adopts the euro, increasing eurozone membership to 19 countries.

The EU's administrative, economic, and even political framework has become so advanced that the organization now has a headquarters with many of the trappings of a capital city. Early on, EU planners chose Brussels (already the national capital of Belgium) as the organization's center of governance. But in order to avoid giving Brussels too much prominence, they chose Strasbourg (located near the German border in the northeastern corner of France) as the seat of the European Parliament, whose elected membership represents all EU countries.

CONSEQUENCES OF UNIFICATION

The European Union is not just a paper organization. It has a profound impact on national economies, on the role of individual states, and on the daily lives of its member-countries' 503 million citizens.

One Market

EU directives are aimed at the creation of a single market for producers and consumers, businesses, and workers. Corporations should be able to produce and sell anywhere within the Union without legal impediments, whereas workers should be able to move anywhere in the EU and find employment without legal restrictions. In order to make this happen and to keep things manageable, member-states have had to harmonize a wide range of national laws from taxation to the protection of the environment to educational standards.

One major step toward union came with the introduction of a single central bank (with considerable power over, for example, interest rates) and a single currency, the euro. The euro was also meant to symbolize Europe's strengthening unity and to establish a counterweight to the no-longer-almighty American dollar. In 2002, twelve of the (then) 15 EU member-countries withdrew their own currencies and began using the euro, with only the United Kingdom, Denmark, and Sweden opting to stay out (Fig. 4-7). More

The European Council is the EU body constituted by the heads of state of all participating countries. The Council President chairs this body, a leadership position that changes every six months to allow all 28 members to have their turn within a reasonable time period. During the second half of 2014, it was Italy's turn to preside over the Council. This photo was taken on July 2 of that year, and shows the insignia of Italy displayed above the bank of European national flags at the main entrance to the Council's headquarters in Brussels. The frequent rotation of leadership among the constituent countries also extends to a number of other EU bodies, a politically unavoidable practice that is costly and can hamper efficient governance.

recently, Slovakia (2009), Estonia (2011), and Latvia (2014) have joined the eurozone, and in 2015 Lithuania became the nineteenth adopting country. The single currency, hailed as a major triumph at its inception, has evolved into a major bone of contention because it has significantly reduced the freedom of member-states to formulate fiscal policies. We will return to the continuing 'euro crisis' shortly.

A New Economic Geography

The establishment and continuing development of the EU have generated a new economic landscape that today not only transcends the old but is fundamentally reshaping the realm's regional geography. By investing heavily in new infrastructure and by smoothing the flows of money, labor, and products, European planners have dramatically reduced the divisive effects of their national boundaries. And by acknowledging demands for greater freedom of action by their provinces, States, departments, and other administrative units of their countries, European leaders have unleashed a wave of economic energy that has transformed some of these subnational units into powerful engines of growth (Fig. 4-8).

Four of these burgeoning hubs are especially noteworthy, to the point that some geographers refer to them as the **Four Motors of Europe** **14**: (1) France's southeastern *Rhône-Alpes Region*, centered on the country's second-largest city, Lyon; (2) *Lombardy* in north-central Italy, focused on the industrial metropolis of Milan; (3) *Catalonia* in northeastern Spain, anchored by the cultural and manufacturing nucleus of Barcelona; and (4) *Baden-Württemberg* in southwestern Germany, headquartered by the high-tech city of Stuttgart.

Despite economic integration and the harmonization of policies, Figure 4-8 also shows that major inequalities persist among Europe's regions as well as within EU member-states. This is an important reminder that, as noted in the introductory chapter, uneven development is a pervasive phenomenon that plays out at every geographic scale.

Diminished State Power and New Regionalism

As states relinquish some of their power to Brussels and express agreement with the ideal of realmwide integration, some of the provinces and regions *within* states have seized the opportunity to assert their cultural identity and particular economic interests. Often, the local governments in these subregions simply bypass the central government in their national capital, dealing not only with each other but also directly with foreign governments as their expanding business networks span the globe.

Thus, even as Europe's states have been working to join forces in the EU, many of those same states are confronting severe centrifugal stresses. The term **devolution** **15** has come into use to describe the powerful centrifugal forces whereby regions or peoples within a state, through negotiation or active rebellion, demand and gain political strength and sometimes autonomy at the expense of the center. Most states exhibit some level of internal regionalism, but the process of devolution is set into motion when a key centripetal binding force—the nationally accepted idea of what a country stands for—erodes to the point that a regional drive for autonomy, or for outright secession, is launched. As Figure 4-9 shows, several European countries are affected by devolutionary pressures of varying extent.

States respond to such pressures in a number of ways, ranging from accommodation to suppression. One way to deal with these centrifugal forces is to give historic regions (such as Scotland or Catalonia) certain rights and privileges formerly held exclusively by the national government. Another answer is for EU member-states to create new administrative divisions that will allow the state to meet regional demands, often in consultation with, or under pressure of, the EU. Later in this chapter, we highlight subnational autonomy movements in the United Kingdom and Spain.

REDEFINING EUROPE: DIFFICULT CHOICES

Widening or Deepening?

Expansion has always been a EU objective, and the subject never fails to arouse passionate debate. Will the incorporation of weaker economies undermine the strength of the overall organization? Should the ties and cooperation among existing members be deepened and solidified before other, less prepared countries are invited to join? Remember that member-states must adhere to strict economic policies and harmonize their political systems. This is much

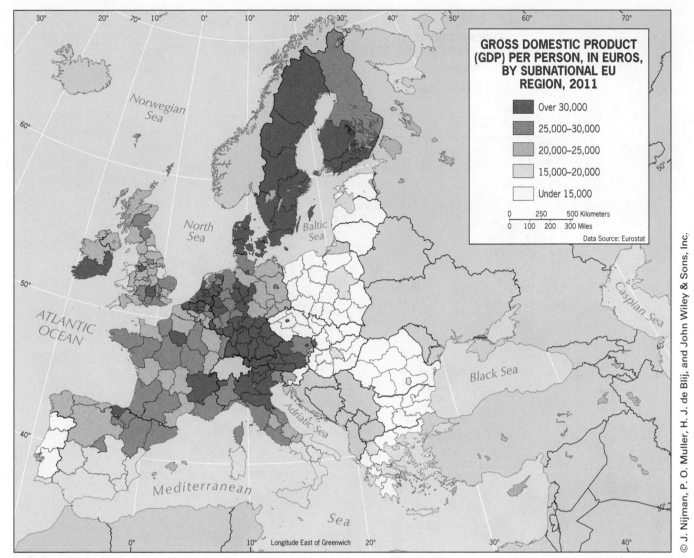

FIGURE 4-8

easier for prosperous countries with longstanding democratic traditions than for poorer nations with a volatile political past. Despite such misgivings, negotiations to expand the EU have long been in progress, and the gains of the twenty-first century are mapped in Figure 4-7.

Numerous structural implications arise from this expansion, affecting all member-countries. Some of the pre-2004 EU's less affluent countries, which used to be on the receiving end of EU subsidy programs, now had to pay up to support the much poorer new eastern member-states. Disputes also intensified over representation at EU's Brussels headquarters. And, most of all, the incorporation of weaker national economies tends to compromise the ability of the EU to move forward in unison because the weaker members are unable to keep their fiscal deficits within limits, and sometimes their industries are hurt by the more developed, competitive industries of the wealthier members. This kind of economic incompatibility among EU countries has come into sharper focus since the arrival of the single currency, and we now turn to the euro crisis that has resulted.

The Euro Crisis

Since the late 2000s, the 'euro crisis' has been an almost standard news item throughout the realm as it laid bare the limitations that accompanied the shift to a single currency and a common monetary policy. The evolving problems are not just about the euro itself, but about the inability or unwillingness of EU members to comply with fiscal standards (e.g., avoiding government deficits) and apply them as solutions within their individual economies. Keep in mind that each EU country still has its own budget as well as its own revenues and expenditures. But even though the EU budget itself remains quite small (about 1 percent of the EU's GDP), the budget rules governing the Union are in crucial ways determined by Brussels.

The economic difficulties faced by several European national governments in this decade were heightened because they had only recently introduced the single European currency—which significantly constrained their policy options. Some of the EU's weaker member-states, most notably Greece, accumulated enormous debts that violated terms in the agreement establishing the eurozone. Moreover, having a single currency implied that countries could no

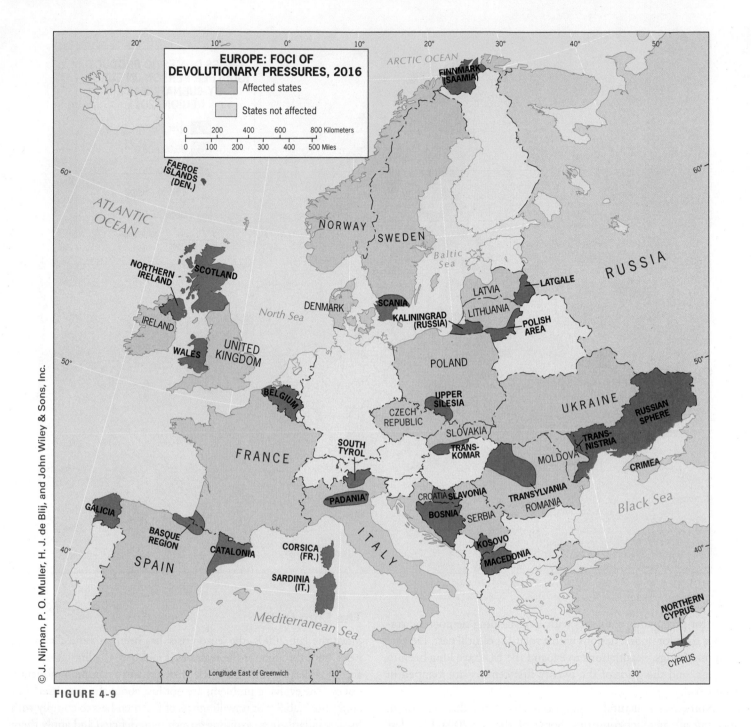

© J. Nijman, P. O. Muller, H. J. de Blij, and John Wiley & Sons, Inc.

FIGURE 4-9

longer independently pursue monetary policies such as devaluations (a common approach to promoting exports) or the lowering of interest rates (used to stimulate the economy). In other words, as countries such as Greece faced deepening economic problems, their hands were tied by their dependence on the European Central Bank—whose primary mission is to maintain the purchasing power (price stability) of the euro.

The euro crisis began in earnest in 2010 when it came to light that Greece had accumulated so massive a government deficit that an emergency bailout from Brussels was essential to avoid immediate bankruptcy and economic chaos. A year later, it was Ireland's turn: the country that not long ago had been hailed as one of Europe's major success stories had to be rescued from imminent

financial collapse. Portugal required a similar bailout in 2011, and Greece received its second in 2012. By then, fiscal problems had also emerged in Italy (forcing its prime minister to resign), in France (where austerity measures triggered widespread public protests), and even in the Netherlands (where a coalition government fell when it could not agree on severe budget cuts). Given the integrated financial system that the EU has become, the contagion of failing banks and faltering governments is now an overriding concern—as if Europe constituted a precariously balanced row of dominoes. Not surprisingly, the future of the euro as a stable currency has been thrown into uncertainty.

By mid-2015, the persistent Greek crisis had worsened. As a result, its government was not only forced to negotiate yet another

FROM THE FIELD NOTES...

"The Grasshopper, one of Amsterdam's many 'coffee shops.' For the past four decades, the Netherlands has permitted legalized sales and consumption of modest amounts of marijuana. The general consensus is that marijuana is not as harmful as hard drugs and that legalization helps to decriminalize its commerce and use. This policy is at odds with legislation in most of the rest of the EU, especially with bordering Germany, Belgium, and nearby France. It used to be that coffee shops were found only in Amsterdam and other major cities, but in recent years they have opened in small towns along the border to attract a foreign clientele (another byproduct of heightened cross-border interaction). As a result, once quiet and conservative rural villages (where lifestyles differ enormously from those in Amsterdam) started to experience heavy traffic of buyers and users from abroad—not something the townfolk were likely to appreciate. Largely for this reason, the national government has turned coffee shops in the border areas into membership-only establishments whereby only Dutch citizens and residents are eligible to join. Initially, there were plans to make this change across the entire country, including the big cities. As you might expect, reactions to this proposal varied geographically: it was supported in the rural border areas, but most people in Amsterdam and other cities viewed it as an infringement on traditional liberties and a bad business decision to boot. Said one Amsterdammer and steady customer of The Grasshopper with a tone of sarcasm: 'I never thought I'd experience this: some provincials trying to decide what happens in Amsterdam. Not in my lifetime!' In Amsterdam's coffee shops, it is business as usual."

bailout (now totaling hundreds of billions of euros since 2009), but also intensify its ongoing, severe austerity measures (see photo). Fears of Greece's exit from the EU, forced or voluntary, remained at a high level, particularly concerning the consequences of such a departure for the integrity of the EU as a whole.

The Remaining Outsiders

As Figure 4-7 shows, a number of countries and territories still remain outside the European Union. The first group includes the five remaining states that have emerged out of former Yugoslavia, plus Albania. In this cluster of new western Balkan countries, most are troubled politically as well as economically. Only Slovenia and Croatia have achieved full EU membership. The rest—Serbia, Bosnia and Herzegovina, Macedonia, Montenegro, and Kosovo—rank among Europe's poorest and most ethnically fractured states, yet they do contain more than 15 million people whose circumstances could worsen outside the 'EU club.' Better, EU leaders reasoned, to move them toward membership by demanding political, social, and economic reforms, even if the process could take several years. As of mid-2015, negotiations were proceeding with Serbia, Montenegro, and Macedonia.

The second group of outsiders is composed of Ukraine and its neighbors. Four former Soviet republics in Europe's 'far east' could someday join the EU (even Belarus, until recently the most disinterested, may now be reconsidering). The government of Ukraine itself has expressed interest, but its electorate is strongly (and regionally) divided—as well as overshadowed by a threat of Ukrainian civil war in 2015 that has put everything on hold (see Map Analysis box). Moldova views the EU as a potential supporter in the struggle against devolution along its eastern border, and a country not even on the map—Georgia, clear across the Black Sea from Ukraine—has proclaimed its interest in future membership for a similar reason: to find an ally against Russian intervention in its internal affairs (including a brief 2008 military incursion).

Finally, we should take special note of another long-mentioned candidate for EU membership—Turkey. Some EU leaders would like to include Turkey mainly for geostrategic reasons (Turkey is a member of NATO), but others are convinced that Turkey is just not 'European' enough. The Turkish candidacy for admission is currently stalled as that country deals with multiple internal problems as well as a major refugee crisis in its south-central borderland adjoining war-torn Syria.

A Greek pensioner raises his fist as he marches through central Athens as part of a demonstration involving tens of thousands on April 1, 2015. This huge crowd of protesters is shouting slogans against budget cuts in pensions and healthcare services—mandated by the EU as part of its continuing bailout of Greece's floundering economy.

FIGURE 4-10

NATO, THE UNITED STATES, AND EUROPEAN SECURITY

The emphasis on EU developments tends to overshadow another important pan-European institution—the **North Atlantic Treaty Organization (NATO)**—a far-reaching security alliance that partially overlaps with the EU (Fig. 4-10). NATO, however, is a very different organization. First, it is primarily concerned with politics and security and not with economics; NATO's purpose is to provide military protection for its members. Second, NATO is led by the United States, which provides the bulk of the alliance's military muscle, and also includes Canada. And third, NATO includes non-EU member Turkey, a country that is hugely important because of its strategic location on the southeastern edge of Europe nearest to the Middle East.

NATO was created in 1949 to provide a security umbrella for Western Europe under U.S. leadership (in response, the Soviets and their satellites created the Warsaw Pact). When the Cold War ended in 1991 with the disintegration of the Soviet Union, NATO became a more general security organization that would provide stability across Europe. It also became more concerned with emerging, twenty-first-century threats to its member-states: terrorism, attacks by 'rogue' states with weapons of mass destruction, and cyber-warfare (a very real and growing danger, with critical global implications). However, as NATO opened its doors to membership for Eastern European countries (Fig. 4-10), it initiated an eastward expansion—not unlike that of the EU—that is already producing another stand-off with Russia.

REGIONS OF EUROPE

EUROPE'S REGIONAL COMPLEXITY

In the past, Europe divided rather easily into Western, Northern (Nordic), Mediterranean (Southern), and Eastern regions, and

countries were grouped accordingly. This traditional configuration remains relevant today, but in the era of European unification, a **core-periphery** framework has been superimposed on that regional scheme, defined in terms of how central—or peripheral—countries are to the workings of the EU. In other words, Europe can still be understood as a set of formal, more or less homogeneous, cultural regions; yet at the same time, the entire realm increasingly functions as a tightly interdependent economic and political regional system, a spatial network that revolves around the EU. We distinguish four major regions that constitute the European realm (Fig. 4-11): (1) *Western Europe*, including the Britain and Ireland; (2) *Northern Europe*; (3) *Mediterranean Europe*, including Slovenia and Croatia; and (4) *Eastern Europe*, the broad eastern tier of the realm that reaches to the Russian border.

Take a close look at Figure 4-11 and note that the European Core (bounded by the red dashed line) in many places extends beyond Western Europe and rarely coincides with national borders. Thus the European Core is primarily defined on the basis of regional economic performance, not political territorialization. For example, northern Italy (but not the Italian south) and southern Sweden (but not the Swedish north) are both part of this Core; also note the inclusion of Estonia in the Core, despite its relatively remote location. This regionalization underscores how complicated and challenging it can be to understand the complexity of spatial organization. Even though Europe typifies that challenge like no other realm, the goal of this chapter is to bring clarity to what at first glance appears to be a chaotic regional arena.

WESTERN EUROPE

Western Europe consists of 11 countries including the **microstate 16** of Liechtenstein. This region also lies almost entirely within the European Core that forms the heart of the EU. The large offshore islands of Britain and Ireland (sometimes referred to as the British Isles) have been closely involved in Western Europe's history as well as its ongoing economic and political development. Let us begin on the mainland with Germany, which is both the realm's most populous country and largest economy.

Germany

Twice during the twentieth century Germany plunged Europe and the world into war. At the end of World War II in 1945, the defeated and devastated German state was divided into two parts: West Germany and East Germany (their former border is delimited by the red line in Fig. 4-13), the latter of which was forced to surrender some important territory to Poland.

West and East Germany then set forth on widely different economic and political trajectories. The Soviet governance of East Germany was established on the USSR's communist model, and as repayment for the extreme hardships the Soviet Union had suffered at German hands during the war, this rule was unremittingly punitive. The U.S.-led authority in West Germany was far less strict and aimed at rehabilitation rather than retribution. West Germany was included when the Marshall Plan was instituted in 1948, and the massive infusion of aid stimulated rapid economic recovery. At the same time, West Germany was reorganized politically into a modern federal state based on democratic foundations.

West Germany's economy soon began to thrive. Between 1949 and 1964, its gross national income (GNI) tripled as industrial output rose by 60 percent. Simultaneously, its political leaders participated enthusiastically in the negotiations that created the six-member Common Market in 1957 (and thereafter evolved to become today's EU). Geography worked in West Germany's favor: it had common borders with all but one of the original member-states. Its swiftly rebuilt transportation infrastructure was second to none in Europe. The reinvigorated spatial economy focused on heavy industrial production in the Ruhr Basin (in the hinterland of the gigantic Dutch port of Rotterdam) as well as the leading-edge manufacturing complexes centered on Hamburg in the north, Frankfurt (the leading financial and air transport hub as well) in the center, and Stuttgart in the south. West Germany now exported huge quantities of iron, steel, motor vehicles, machinery, textiles, and farm products. Until just a few years ago, Germany had the largest export economy by value in the world; although it has now been surpassed by China and the United States, Germany's exports per capita are still by far the highest on Earth.

At the time of Germany's reunification in 1990, West Germany had a population of about 62 million and East Germany 17 million. By 2015, former East Germany's population was down to less than 13 million. Decades of communist misrule in the East had yielded outdated factories, crumbling infrastructures, polluted environments, drab cities, inefficient farming, and inadequate legal and other institutions. Thus the 1990 reunification was more of a rescue than a merger, and the cost to the reconstituted German state since then has been gargantuan. Figure 4-13 maps the 16 States (or *Länder*) that make up today's reunified Federal Republic of Germany and also makes a key point: regional disparity in terms of gross domestic product (GDP) per person remains a serious problem. Note that all of former East Germany's six States (except Berlin) still rank in the lowest income category, whereas nine of the ten former West German States rank in the two higher-income categories.

Another challenge facing Germany, along with other Western European countries, is the integration of sizeable immigrant minorities. Germany has nearly 16 million immigrants, about 20 percent of the total population of 80.7 million. They not only are culturally different but also worse off economically: more than 40 percent of immigrant households live below the official poverty line, largely because the unemployment rate of immigrants far exceeds that of native Germans.

Nevertheless, Germany navigated the recent global recession better than any other European country and retains its position as the realm's economic powerhouse. Today, this country is developing closer ties with Eastern Europe, and, from the Balkans to the Baltic Sea, the Germans are regarded as the region's most desirable trading partner. Germany has also played a leading role in EU bailouts of Greece and other countries, and—in an ironic historical twist, given the initial reasons for European integration—has been a vital force in maintaining a sufficient degree of EU unity.

EUROPEAN REGIONS

- - - European Core Boundary

Western Europe

Northern (Nordic) Europe

Mediterranean Europe

Eastern Europe

0   250   500 Kilometers
0   100   200   300 Miles

FIGURE 4-11

France

Territorially, France is larger than Germany, and the map suggests that France has a superior relative location, with coastlines on the Mediterranean, the Atlantic Ocean, and, at Calais, even a window on the North Sea. But France does not have any good natural harbors, and oceangoing ships cannot navigate its rivers and other waterways very far inland.

The map of the mainland Western Europe (Fig. 4-12) reveals a significant contrast between France and Germany in terms of their **urban systems** [17], the spatial distribution and interrelationships among cities of various sizes. France has one dominant city, Paris, and no other city comes close either in size or centrality: Paris has 10.9 million residents, whereas its closest rival, Lyon, contains only 1.6 million. Germany has no city to match Paris, but it does have a wider range of medium-sized cities: in the 100,000-plus category, Germany has more than 80, whereas France has less than 40.

Paris is an example of a **primate city** [18], one that is disproportionately large compared to all others in the national urban system, exceptionally expressive of its country's culture, and (almost) always the capital. Paris grew so large and dominant within the French urban system for reasons of geography and history. Geographically, Paris has advantages in terms of both **site** [19] (the physical attributes of the place it occupies) and **situation** [20] (its location relative to surrounding areas). The *site* of the original settlement at Paris lay on an island in the Seine River, a defensible spot where the river was often crossed. This island, the Île de la Cité, was a Roman outpost 2000 years ago; for centuries its security ensured continuity. Eventually the island became overcrowded, and the city expanded along the banks of the river (Fig. 4-14A). Paris's advantageous *situation*

MAINLAND WESTERN EUROPE

POPULATION
- Under 50,000
- 50,000–250,000
- 250,000–1,000,000
- 1,000,000–5,000,000
- Over 5,000,000

- ▪▪▪ European Core Boundary
- —— Railroad
- —— Road
- —— Canal

National capitals are underlined

FIGURE 4-12

© J. Nijman, P. O. Muller, and John Wiley & Sons, Inc.

FIGURE 4-13

lay in its fertile agricultural hinterland and the presence of key waterways. The Seine River is joined near Paris by several navigable tributaries (the Oise, Marne, and Yonne). When canals extended these waterways even farther, Paris was linked to the Loire Valley of central France; the Rhône-Saône Basin of the southeast; Lorraine, a key industrial area in the northeast; and the northern border zone with Belgium. When Napoleon reorganized France and built a radial system of roads (duplicated later by railroads) that focused on Paris from all parts of the country, the city's primacy was assured (Fig. 4-14B).

Historically, France early on developed the strong tradition of a highly centralized state in which the central government in Paris maintained tight control over 96 subnational *départements*. It remained so for nearly two centuries, but France is now decentralizing. A new layer of governance, consisting of 26 larger *régions*, has been inserted at the level between Paris and the *départements* both to accommodate devolutionary forces and to seize the opportunities for regional and local growth throughout the country (Fig. 4-15).

France has long had one of the most diversified economies in the world. Northern French agriculture remains intensive and varied, exploiting the country's wide range of soils and microclimates and enjoying state subsidies and protections. In the southeast, Lyon, France's second city and headquarters of the *région* named

Leipzig is eastern Germany's biggest city after Berlin. It has a rich cultural history and has been home to, among others, Johann Sebastian Bach, Johann Wolfgang von Goethe, Richard Wagner, Friedrich Nietzsche, and Felix Mendelssohn. But during communist times Leipzig was dragged down along with the rest of East Germany. Although this city is now finally experiencing something of a revival, there are still many struggling neighborhoods and outlying communities.

away as China and Chile. More broadly, France's economic geography is marked by new high-tech industries, and it is a leading producer of high-speed trains, aircraft, fiber-optic communications systems, and space-related technologies. France is also the world leader in nuclear power, which currently supplies about 75 percent of its electricity and thereby reduces dependence on foreign oil imports; but rising public opposition today all but prohibits the construction of new generating facilities, given the national goal to reduce nuclear reliance to 50 percent by 2025.

France is another one of those European states with a rapidly aging population (totaling just over 65 million), and that poses some major challenges for the years ahead. Natural population growth is close to zero and must be compensated for through immigration. However, most immigrants hail from poor Islamic countries with very different cultures, and their integration and acceptance in French society has been a painful process. The graying of the French population means that in the future more and more elderly people must be cared for with the incomes earned by a shrinking productive workforce.

The French economy has already exhibited signs that it might encounter problems of the kind that have bedeviled countries such as Greece and Spain—which would raise serious alarms because France's economy is far bigger and of much greater importance to the EU as a whole. Government debt is high and rising; the public sector is very large (nearly twice the size as Germany's), draining expenditures while youth unemployment hovers at almost 25 percent. That the French electorate does not take kindly to

Rhône-Alpes, has become a focus for growth industries and multinational firms. This *région* is evolving into a stand-alone economic powerhouse that is becoming a driving force in the European economy. Indeed, it is one of the *Four Motors of Europe* (discussed earlier) with its own global business connections to countries as far

SITE AND SITUATION OF PARIS, FRANCE

- ▪▪▪▪ 12th Century Wall
- ═══ Wall of 1840s
- ─── Roads
- ─── Railroads
- ▨ Forests, Parks
- ▨ Built-up area

FIGURE 4-14

FIGURE 4-15

austerity measures has already been demonstrated in their 2012 election of a president from the Socialist Party. And the clearest sign that all is not well came in the response to the horrific 2015 terrorist attack on the Paris office of *Charlie Hebdo* (see chapter-opening photo). In its wake, France's right-wing *Front National* became the fastest growing French political party, espousing an extremist, ultranational agenda that is highly critical of Muslim and other immigration as well as EU policies in general.

Benelux

Three small countries are crowded into the northwestern corner of mainland Western Europe—Belgium, the Netherlands, and

Luxembourg—and are collectively referred to by their first syllables (**Be-Ne-Lux**). Their total population, 28.5 million, reminds us that this is one of the most densely peopled corners of our planet. Not coincidentally, the *Low Countries*, as they are also sometimes called, are situated at the **estuary** 21 (mouth) of the great Rhine and Scheldt rivers and have access to the seas that, throughout modern history, has been the envy of the Germans and the French. They are a highly productive trio: both the Netherlands and Belgium generally rank among the top 20 economies on Earth, and tiny Luxembourg has one of the world's highest per-capita gross national incomes.

The Netherlands, one of Europe's oldest democracies and a constitutional monarchy today, has for centuries been expanding its territory—not by warring with its neighbors but by wresting

© Jan Nijman

"Amsterdam has a latitude and a near-coastal location similar to that of Seattle—and it has comparable weather. It takes an extreme cold wave for the canals to freeze over, and on average that occurs only once every six or seven years. When it does, people cannot wait to put on their skates. It gets dark early here in winter, but there is just enough daylight after school or work to get out on the ice and experience the city in a different way."

land from the sea. About a quarter of the country lies below sea level. The urban, demographic, and economic center of gravity lies in a **conurbation** 22 (a large urban complex formed by the coalescence of two or more cities) in the west of the country. This so-called *Randstad* ('Ring City'), roughly triangular in shape, is anchored by Amsterdam, the constitutional capital; Rotterdam at the mouth of the Rhine, Europe's largest port; and The Hague, the national seat of government.

Belgium also has a thriving economy as well as a leading port in the city of Antwerp. With 11.2 million people, Belgium's regional geography is dominated by a cultural fault line that cuts diagonally across the country, separating a Flemish-speaking majority (59 percent) centered on Flanders in the northwest from a French-speaking minority in southeastern Wallonia (40 percent). Brussels, the predominantly French-speaking capital, forms a linguistic island in the Flemish-speaking sector; but the city also is one of Belgium's greatest assets because it serves as the headquarters, and in many ways as the de facto capital, of the European Union (though it is important to note that the EU does not have an official capital city). Nonetheless, devolution is a persistent problem for Belgium, with political parties that agitate for Flemish separatism constantly roiling the social landscape.

Luxembourg, one of the realm's numerous ministates, lies between Germany, Belgium, and France, with a Grand Duke as head of state (its official name is the Grand Duchy of Luxembourg), a territory of only 2600 square kilometers (1000 sq mi), and a population of just 540,000. Luxembourg has translated sovereignty, relative location, and stability into a haven for financial, service, and information-technology industries. There are more than 150 banks in this tiny country and about 14,000 holding companies (corporations that hold controlling stock in other businesses in Europe and worldwide). With its unmatched GDP per capita (more than U.S. $110,000), Luxembourg is in some ways the greatest beneficiary of the advent of the European Union. And in no country in Europe is support for the EU stronger than it is here.

The Alpine States

Switzerland, Austria, and the microstate of Liechtenstein on their border share an absence of coasts and the mountainous terrain of the Alps and little else (Fig. 4-12). Austria speaks one language; the Swiss speak German in their north, French in the west, Italian in the southeast, and even a bit of Rhaeto-Romansch in the remote central highlands (Fig. 4-6). Austria has a large primate city; multicultural Switzerland does not. Austria has a substantial range of domestic raw materials; Switzerland does not. Austria is twice the size of Switzerland and has a larger population, but far more trade moves across the Swiss Alps between western and Mediterranean Europe than crosses Austria.

Switzerland, not Austria, is in most ways the leading state in the Alpine subregion of Western Europe. Mountainous terrain and **landlocked location** 23 can constitute crucial barriers to economic development but, as is so often the case, geography poses constraints in one way while offering opportunities in another. That is why Switzerland is such an important lesson in the complexities of human geography. Through the skillful maximization of their limited opportunities (including the transfer needs of their neighbors), the Swiss have transformed their seemingly restrictive environment into a prosperous state. They deftly utilized the waters cascading from their mountains to generate hydroelectric power to develop highly specialized industries. Swiss farmers perfected methods to optimize the productivity of mountain pastures and valley soils. Swiss leaders converted their country's isolation into stability, security, and neutrality, making it a world banking behemoth, a global magnet for money. Zürich, in the German-speaking sector, is the financial center; Geneva, in the French-speaking sector, is one of the world's most internationalized cities. The Swiss feel that they do not need to join the EU—and they have no plans to apply in the foreseeable future.

Austria, which joined the EU in 1995, is a remnant of the old Austro-Hungarian Empire and has a historical geography that is far more reminiscent of unstable Eastern Europe than that of Switzerland. In fact, even Austria's physical geography seems to demand that the country look eastward: it is at its widest, lowest, and most productive in the east, where the River Danube links it to Hungary, its old ally in the anti-Muslim wars of the past.

Vienna, by far the Alpine subregion's largest city, also lies on the country's eastern perimeter. One of the world's most expressive primate cities with magnificent architecture and monumental art, Vienna today is Western Europe's easternmost major city, but its relative location has changed dramatically with EU enlargement since 2000. Even though Vienna found itself in a far more centralized position when the EU border shifted eastward, most Austrians doubted that their neighbors to the east could ever become EU members of potentially equal standing. One result is that Austrian public support for the European Union today has fallen to a record low level.

The Czech Republic

The Czech Republic is focused on the historic Bohemian Basin, the mountain-encircled national core area centered on the capital, Prague. This is another classic primate city, its cultural landscape highly reflective of Czech traditions. Prague is also an important industrial hub and collection center, surrounded by mountains that contain many valleys with small towns that specialize, Swiss-style, in fabricating high-quality manufactures. In the old Eastern Europe, even during the communist period, the Czechs always were leaders in technology and engineering; their products could be found on markets in foreign countries near and far. Moreover, Bohemia always was cosmopolitan and Western in its exposure, outlook, development, and linkages; Prague lies in the upper basin of the Elbe River, its traditional outlet through northern Germany to the North Sea. Not long ago, the Czech Republic would have been considered part of 'Central Europe,' but today it has shifted its gaze entirely to the west, with its solid economic development fully warranting inclusion in the European Core (Fig. 4-11).

The United Kingdom (UK)

It has often been observed that the geographical position of the United Kingdom, just off the European mainland, reflects its traditional hesitation to consider itself part of 'Europe.' It did join the EU in 1973 but has been reluctant to move toward tighter integration. When most member-states in 2002 adopted the new euro in favor of their national currencies, the British kept their pound sterling and delayed their participation in the European Monetary Union (EMU)—a decision they did not regret when the 'euro crisis' unfolded in the late 2000s. Inside the UK there is substantial, and possibly rising, opposition to continued EU membership. With an area about the size of Oregon and a population of 64.1 million, the UK is by European standards a fairly large country. Based on a combination of physiographic, historical, cultural, economic, and political criteria,

the United Kingdom can be subdivided into four subregions (numbered in Fig. 4-16):

1. *England*. So dominant is this subregion of the United Kingdom that the entire country is sometimes referred to by this name. England is anchored by the sprawling London metropolitan area, which by itself contains more than one-seventh of the UK's total population. Indeed, along with New York, London is regarded as one of the top-tier **world-cities** **24**, with financial, high-technology, communications, engineering, and associated industries reflecting the momentum of its long-term growth and agglomeration. Within England, there is a notable difference between the crowded, fast-paced, and globally connected southeast centered on London, and the far less prosperous north and west (see inset map in Fig. 4-16); the latter areas, once the global hearth of the Industrial Revolution, have now completely lost their dynamism as well as segments of their population in our age of deindustrialization. Spatial inequality has also been rising, and today's widening gap between rich and poor within England's residential mosaic shows no sign of abating (see photo).

2. *Wales*. This nearly rectangular, rugged territory was a refuge for ancient Celtic peoples, and in its western counties more than half the inhabitants still speak Welsh. Because of the high-quality coal reserves in its southern tier, Wales, too, was engulfed by the Industrial Revolution, and Cardiff, the Welsh capital, was once

In both 2014 and 2015, London was ranked (by the high-end real estate company, Knight Frank) as the best place for the world's wealthiest people to live and/or own a home. Unsurprisingly, it is also one of the most expensive cities in the world. To reside here at ultra-exclusive Lowndes Square in Chelsea, a high-cachet neighborhood in central London close to Buckingham Palace, be prepared to pay house prices that regularly exceed U.S. $5 million. Foreign investors, especially oligarchs from Russia and the oil-rich Persian Gulf States, are most prominent among the buyers. London may be an extreme case, but it is indicative of a trend among a select few 'hot' cities (New York, Paris, and San Francisco are others) that are experiencing skyrocketing real estate values due to foreign interest—increasingly to the detriment of established middle-class residents who can no longer afford to remain in their long-time neighborhoods.

BRITAIN AND IRELAND

POPULATION

— Railroad
— Road
—|—| Canal
•—•—• Gas pipeline
•—•—• Oil pipeline
◆ Gasfield
⚓ Oilfield
▨ Coalfield

● Under 50,000
● 50,000–250,000
● 250,000–1,000,000
● 1,000,000–5,000,000
● Over 5,000,000

National capitals are underlined
—·—· Maritime Boundaries
▨ Area of Catholic majority in N. Ireland
▨ Area with a majority of Welsh speakers
▬▬▬ European Core Boundary

0   50   100 150 200 250 Kilometers
0 25 50 75 100 125 150 Miles

GDP PER PERSON, IN EUROS, BY SUBNATIONAL EU REGION, 2011

Over 30,000
25,000–30,000
20,000–25,000
15,000–20,000

Data Source: Eurostat

© J. Nijman, P. O. Muller, and John Wiley & Sons, Inc.

FIGURE 4-16

the world's leading coal exporter. But the fortunes of Wales also declined, and many Welsh emigrated. In several cultural and economic spheres, Wales is closely tied to England and independence is not in the cards. In the Welsh Assembly, the nationalist party claims only 11 of the 60 seats. The situation in Scotland, however, is entirely different.

3. *Scotland*. With a population of 5.4 million, Scotland is a major component of the UK—but perhaps for not much longer. It is also well removed from Britain's thriving southeastern core area.

Most of the population resides in the Scottish Lowlands, anchored by Edinburgh, the capital, in the east and Glasgow in the west. Attracted there by the labor demands of the Industrial Revolution (coal and iron reserves lay nearby), the Scots developed a world-class shipbuilding industry. Subsequent decline and obsolescence were followed by high-tech development (notably around Glasgow) and Scottish participation, beginning in the 1970s, in the exploitation of oil and gas reserves beneath the adjacent North Sea. Nevertheless, a growing number of Scots feel that they are increasingly disadvantaged within the UK and should play a bigger role in the EU. In 2014, a referendum on the question of full independence won the support of 45 percent of the electorate. In 2015, separatist candidates won a landslide victory in national parliamentary elections—perhaps threatening the very survival of the UK.

4. *Northern Ireland*. Northern Ireland was home not only to Protestants from Britain, but also to a substantial population of Irish Catholics who found themselves on the wrong side of the border when Ireland became independent nearly a century ago. Ever since, conflict has intermittently engulfed Northern Ireland and spilled over into Britain. A declining share of the overall population of 1.8 million is Protestant and traces its ancestry to Scotland or England; meanwhile, the proportion of Roman Catholics is rising, and in the capital (Belfast) they achieved majority status in 2014. Catholics share their religion with the Irish Republic on the southern side of the border. Although Figure 4-16 suggests that there are majority areas of Protestants and Catholics in Northern Ireland, no such clear separation exists; mostly they reside in clusters throughout the territory, including the walled-off neighborhoods that still mark the major cities of Belfast and Londonderry.

Republic of Ireland

Hard-won independence from Britain in 1921 did not bring real economic prosperity until the final decades of the twentieth century, when Ireland became known as the **Celtic Tiger** (likening it to the miraculous, Hong Kong-style economic development of the *Asian Tigers* discussed in Chapter 9). Participation in the EU (since 1973), adoption of the euro (in 2002), business-friendly tax policies, comparatively low wages, an English-speaking workforce, and an advantageous relative location combined for a time to produce the highest rate of economic growth in the EU. During the 1990s, Ireland's booming, service-based economy, accompanied by burgeoning towns and cities (especially the capital, Dublin), skyrocketing real estate prices, mushrooming industrial parks, and bustling traffic, transformed a country long known for emigration into a magnet for industrial workers, producing new social challenges for a closely knit, long-isolated society. In 2007, however, Ireland's economic boom began to falter as global recession loomed, and a year later everything fell apart: the real estate market crashed, service industries found more favorable conditions in Eastern Europe, and unemployment soared. By 2011, the government had amassed such huge deficits that an EU bailout was required. In the mid-2010s, even though a slow recovery was taking hold, the prosperity of the Celtic Tiger days remained a fading memory.

NORTHERN EUROPE

Northern Europe's remoteness, isolation, and challenging environment, especially at latitudes higher than 60°N, have had positive binding effects for much of this domain and seem to have fostered similar cultures. The Scandinavian Peninsula carrying Sweden and Norway lay removed from most of the wars of mainland Europe and developed in relatively tranquil circumstances (even though Norway was overrun by Nazi Germany during World War II). Three of its major languages (Swedish, Norwegian, and Danish) are mutually intelligible, and in terms of religion there is overwhelming adherence to the same Lutheran church. Democratic and representative governments emerged early on, and individual rights and social welfare have been carefully protected for centuries. Women participate more fully in government and politics here than in any other part of the world.

But consider the implications of Figures 4-11 and 4-17: the southern, coastal, and more urbanized areas of the region form part of the European Core, but the sparsely populated remainder does not. In the aggregate, economic indicators for most of this region are quite strong, but almost all development is concentrated within that smaller Core zone. Territorially, northern Norway and Sweden as well as most of Finland belong to the European Periphery. The five countries of this higher-latitude domain of Europe have a combined population of only 26 million, yet their core areas make up in prosperity and external linkages what they lack in size. Interestingly, a few of the countries in the region have stopped short of full participation in the EU: Norway is not even a member, and only Finland has joined the eurozone to date.

Sweden

Sweden is Northern Europe's largest country in terms of both population (9.6 million) and territory. Most Swedes live south of latitude 60°N (which passes through Uppsala just north of the capital, Stockholm) in what is climatically the most moderate part of the country (Fig. 4-17). Here lie the primate city, core area, and the main industrial districts; here, too, are the main agricultural areas that benefit from the lower relief, better soils, and somewhat milder climate.

Sweden long exported raw or semifinished materials to industrial countries, but today the Swedes make finished products themselves, including motor vehicles, electronics, stainless steel, furniture, and glassware. Much of this production is based on local resources, including a major iron ore reserve at Kiruna in the far north (there is a steel mill at Luleå). Swedish manufacturing, in contrast to that of several Western European countries, is based in dozens of small and medium-sized towns specializing in particular products. Energy-poor Sweden was a pioneer in the development of nuclear power, but a national debate over the risks involved has reversed that course. Sweden joined the EU in 1995 but chose to stay out of the eurozone in 2002.

As Nordic Europe's biggest and most influential country, Sweden has historically dominated both Norway and Denmark. Culturally, Sweden has time and again made its presence known on the international scene, from popular music to films to literary works to the annual awarding of the Nobel Prizes.

FIGURE 4-17

Norway

Norway does not need a nuclear power industry to supply its energy needs. It has found its economic opportunities on, in, and beneath the sea. Norway's fishing industry, now augmented by highly efficient fish farms, has long been a cornerstone of the economy, and its merchant marine spans the world. But over the past four decades, Norway's economic life has been transformed by the bounty of oil and natural gas discovered beneath the floor of its sector of the North Sea.

With its limited patches of cultivable soil, high relief, extensive forests, and spectacular fjords, Norway has nothing to compare to Sweden's agricultural or industrial development. Its cities, from the capital Oslo and the North Sea port of Bergen to the historic national focus of Trondheim as well as Arctic Hammerfest, lie on

© Jan Nijman

@ FROM THE FIELD NOTES...

"From the water, outer Stockholm's suburban landscape looks very prosperous. I am cruising along Lilla Värtan, the narrow strait that separates the Swedish capital from the adjacent island of Lidingö. Stockholm is home to 1.5 million people, and its long history dates back to the 13th century. Its layout is quite unusual in that this metropolis has been built on a cluster of no less than 14 islands. Sweden is a wealthy country, and much of that affluence is concentrated in and around this capital city. Which is not to say that everybody lives like this. The myriad waterways and islands have allowed for considerable segregation, with drab, low-income, high-rise residential areas located well away from these obviously prospering suburbs."

narrow coastal lowlands and have difficult overland connections. The isolated northern province of Finnmark has even become the scene of a devolutionary movement among the reindeer-herding indigenous Saami (Fig. 4-9). The distribution of Norway's population of 5.1 million has been described as a necklace, its beads linked by the thinnest of strands, but this has not constrained national development. Economically, Norway has one of the lowest unemployment rates in Europe, and in terms of income per capita it is one of the world's richest countries.

Norwegians have a strong national consciousness and spirit of independence. In 1995, when Sweden and Finland voted to join the European Union, the Norwegians again said no. They had little desire to exchange their economic independence for the regulations of a broader, even possibly safer, Europe.

Denmark

Territorially small by Scandinavian standards, Denmark has a population of 5.6 million, the North's second-largest country after Sweden. It consists of the Jutland Peninsula and several islands to the east at the gateway to the Baltic Sea; it is on one of these islands, Sjaelland, that the capital of Copenhagen is located. Copenhagen has long been a port that collects, stores, and transships large quantities of goods. This **break-of-bulk** 25 function exists because many oceangoing vessels cannot enter the shallow Baltic Sea, making the city an *entrepôt* 26 where transfer facilities and activities prevail. The completion of the Øresund bridge-tunnel link to southern Sweden in 2000 further enhanced Copenhagen's situation (Fig. 4-17).

Denmark remains a kingdom, and in centuries past Danish influence extended far beyond its present confines. Remnants of that period continue to challenge Denmark's governance. Greenland

came under Danish rule following union with Norway (1380) and continued as a Danish possession when that union ended in 1814. In 1953, Greenland's status changed from colony to province, and in 1979 its 56,000 inhabitants were granted home rule under the new Inuit name of *Kalaallit Nunaat*. They promptly exercised their rights by withdrawing from the EU, of which they had become a part when Denmark joined in 1973. Greenland was dependent on financial support from Denmark, but this has changed in recent years following the discovery of oil off Greenland's west coast, north of the Arctic Circle.

Another restive Danish dependency is the Faroe Islands, located between Scotland and Iceland (Fig. 4-11). These 17 small islands and their 50,000 inhabitants were awarded self-governance in 1948, complete with their own flag and currency, but Denmark remains sovereign and debates continue about full independence.

Finland

Finland, territorially almost as large as Germany, has only 5.4 million residents, most of them concentrated in the southern triangle formed by the capital, Helsinki, the textile-producing complex of Tampere, and the shipbuilding center of Turku (Fig. 4-17). A land of evergreen forests and glacial lakes, Finland's national income has long been sustained by wood and wood-product exports. But the Finns have now developed a more diversified economy with staple agricultural crops and the manufacture of precision machinery and telecommunications equipment. Finland's environmental challenges and relative location have created cultural landscapes similar to those of northern Sweden and Norway, but the Finns are not a Scandinavian people—their linguistic and historical affinities are with the Estonians across the Gulf of Finland to the south (Fig. 4-6). In fact, other ethnic groups speaking these Finno-Ugric languages are also widely dispersed across what is today western Russia. Finland shares a long border with Russia, and during the Cold War it had to tread very carefully in international affairs. But after the collapse of the Soviet Union it joined the EU in 1995 and was one of the first countries to adopt the euro in 2002. In the twenty-first century, Finland has decidedly positioned itself in Europe.

Iceland

Iceland is a volcanic, glacier-studded island in the frigid waters of the North Atlantic just south of the Arctic Circle. Inhabited by people with Scandinavian ancestries (population: 325,000), Iceland is of special scientific interest because it lies astride the Mid-Atlantic Ridge. Here, the Eurasian and North American tectonic plates of the Earth's crust are diverging (see Fig. G-4), new land can be seen forming, and spectacular volcanic eruptions are periodically on display (as occurred memorably in 2010 when the Eyjafjallajökull volcano brought trans-Atlantic and domestic air traffic to a standstill for days on end).

Iceland's population is almost completely urban, and the capital, Reykjavik, is home to fully half of the country's inhabitants. The nation's economic geography is traditionally oriented to the frigid surrounding waters, whose seafood harvests have given Iceland a high standard of living. In the 1990s, Iceland embarked on economic liberalization policies, and its financial industries and banks grew rapidly. For a while, Iceland was referred to as the *Nordic Tiger*; but then, not unlike Ireland which experienced a similar boom, the economy floundered with the arrival of the global recession in 2008. Iceland's government, closely involved with some of the troubled banks, had to be bailed out by the International Monetary Fund. Subsequently, plans were initiated to join the European Union, but in 2015 the government withdrew its bid, stating that its interests were better served by remaining outside the EU.

▌MEDITERRANEAN EUROPE

As Figure 4-11 reveals, the northern sectors of two major countries of Southern Europe (Italy and Spain) form parts of the European Core. Portugal, on the western flank of the Iberian Peninsula, remains outside the Core, far less urbanized, much more agrarian, and not strongly integrated with Western Europe. Other Mediterranean countries include Greece, the island-states of Cyprus and Malta, Croatia, and Slovenia, of which only the last can be considered part of the European Core in recognition of its rapid economic development. Slovenia's recent upward trajectory is similar to that of the Czech Republic (and, as we shall presently see, Estonia).

Spain and Portugal

At the western end of Southern Europe lies the broad Iberian Peninsula, separated from France and Western Europe by the rugged mountain wall of the Pyrenees and from North Africa by the narrow Strait of Gibraltar. Spain (population: 46.4 million) occupies most of this compact Mediterranean landmass, and peripheral Portugal lies in its southwestern corner.

Spain followed the leads of Germany and France and decentralized its administrative structure as a concession to devolutionary pressures. These pressures were especially strong in Catalonia, the Basque Country, and Galicia; in response, the Madrid government created a subnational regional framework of 17 so-called *Autonomous Communities (ACs)* (Fig. 4-18). Even so, the new AC mosaic has not defused any secessionist-minded movements. Relations between Madrid and the Basque Country remain tense, and in Catalonia the drive toward independence is strengthening as the Spanish economy struggles while Catalonia continues to forge ahead. In late 2014, the Catalan AC government held a referendum in which, reportedly, about 80 percent of the population voted in favor of outright independence from Spain; however, the Spanish High Court swiftly voided this vote, declaring it to be in violation of Spain's constitution.

Centered on the dynamic, productive coastal city of Barcelona, the Catalonia AC is not only Spain's leading industrial agglomeration but has also become one of Europe's Four Motors. Catalonia is endowed with its own distinctive language and culture that find vivid expression in Barcelona's vibrant urban landscape. In recent years, Catalonia—with 6 percent of Spain's territory and 17 percent of its population—has annually produced 25 percent of all Spanish exports and more than one-third of its industrial exports.

As Figure 4-18 shows, Spain's capital and largest city, Madrid, lies near the geographic center of the state. It also lies astride an economic-geographic transition zone. In terms of annual GDP per capita, the Spanish north is far more affluent than the south, a reflection of the country's inequitable distribution of resources, development opportunities, and environmental assets (the south suffers from drought as well as inferior soils). The most prosperous ACs, apart from Madrid, are in the northeastern tier closest to the French border.

Economically, Spain has been going through a difficult stretch. Economic decline has plagued most of the country since 2008 when it was hit hard by a recession-triggered housing market crash and a major crisis in the banking sector. Unemployment in mid-2015 still seemed to be stuck near an astonishing 25 percent (more than 50 percent among youths), and Spain's government debt ranked among the highest in the EU. There is hope today that things have bottomed out because in 2014 Spain recorded (modest) economic growth for the first time since 2007.

Portugal (10.5 million) occupies the southwestern corner of the Iberian Peninsula. Because EU agricultural policies entail transfers from rich to poor areas in the Union and because the EU funds major infrastructural projects in more isolated areas, Portugal at first benefited substantially from its 1986 admission to the EU. Nevertheless, it still remains far behind Europe's leading economies. Unlike Spain, which has major population clusters on its interior plateau as well as its coastal lowlands, the Portuguese

© J. Nijman, P. O. Muller, and John Wiley & Sons, Inc.

FIGURE 4-18

are mainly concentrated within their Atlantic coastal zone. The country has several excellent natural harbors, including those of primate-city Lisbon and Sines (Fig. 4-18), with considerable potential to attract a larger share of container shipping (especially for transfer onto freight trains bound for Spain), but these ports need to become more efficient and competitive. The best farmlands lie in the moister western and northern zones of the country; but farms here are small and inefficient, and even though Portugal remains dominantly rural, it must import as much as half of its food. Like Spain, Portugal has suffered a major setback since the onset of global recession, and signs of recovery have so far been minimal.

Italy

Centrally located within Mediterranean Europe, most populous of the realm's southern states, best connected to the European Core, and economically most advanced is Italy (population: 60.2 million), a charter member of what is now the EU. The northern half of Italy stands in strong social, economic, and political contrast to such southern regions as Calabria (the 'toe' of the Italian 'boot') and Italy's two major Mediterranean islands, Sicily and Sardinia. Not surprisingly, Italy is often described as two countries—a

progressive north and a stagnant south (known as the ***Mezzogiorno***). The urbanized, industrialized north is very much a part of the European Core.

North and south are bound by the ancient (and current) capital, Rome, which lies within the narrow transition zone between Italy's contrasting halves. This clear manifestation of Europe's core-periphery distinction is referred to in Italy as the ***Ancona Line***, named after the city on the Adriatic coast where it reaches the eastern side of the peninsula (Fig. 4-19, blue line). Whereas Rome remains Italy's governmental and cultural focus, the functional core area of the country has shifted northward into Lombardy in the basin of the Po River. Here lies Mediterranean Europe's leading industrial complex, the Milan–Turin–Genoa triangle. Metropolitan Milan itself personifies the new, modern Italy—not only as its primary manufacturing center (making Lombardy one of Europe's Four Motors) but also as the headquarters of the country's financial and service industries. Although the Milan area contains just 7 percent of Italy's population, it now accounts for more than one-third of the entire country's national income.

The Italian economy also has been slow to recover from the global recession, but here too austerity measures have been a hard sell. Beyond these challenges, Italy is now increasingly burdened by the accelerating flow of hordes of migrants who illegally enter southern Italy from the north coast of Africa. In early 2015, these boat people, desperately seeking political and/or economic refuge in Europe, were streaming by the thousands across the Mediterranean in the flimsiest of vessels toward Sicily. Many of these clandestine voyages are doomed to end tragically, and the 2013 drowning of 360 Africans off the coast of Lampedusa (the small Italian island, located halfway between Libya and Sicily, shown in Fig. 4-19) first brought this crisis to the world's attention. In 2015, propelled by the disintegration of Libya, a tremendous upsurge in this immigration sharply intensified the crisis, overwhelming the EU's belated Triton rescue operation (see photo) as well as unleashing heightened chaos at intake centers on European shores and larger-scale disasters at sea (in mid-April of 2015 as many as 700 people drowned when their overloaded boat capsized just north of Libyan waters).

Slovenia and Croatia

These two countries, both lying in close proximity to Italy (Fig. 4-19), were the most progressive to emerge from the 1990s disintegration of former Yugoslavia. **Slovenia**, wedged against Austria and Italy in the hilly terrain near the head of the Adriatic Sea, has just over 2 million inhabitants, a nearly homogeneous ethnic complexion, and a productive economy. Slovenia joined the EU in 2004 and entered the eurozone three years later. **Croatia**, admitted to the EU in 2013, has a long coastline on the Adriatic that has recently become a major European tourist destination. In the long and bitter aftermath of the collapse of Yugoslavia, relations with neighboring Serbia are still strained, and the presence of a Croatian minority of about 600,000 in neighboring Bosnia continues to stir a range of political sensitivities.

FIGURE 4-19

Greece, Cyprus, and Malta

Greece is the birthplace of European civilization, a major historic presence in the Mediterranean region, and a longtime member of the EU and NATO. Geographically, it is also relatively far removed from the European Core. Seeming to dangle from the southern end of Eastern Europe, Greece was the wellspring of one of the ancient world's greatest civilizations, its scientists and philosophers still cited to this day, its famous tragedies still staged in the amphitheaters built more than 2000 years ago. Its familiar peninsulas and islands are bounded by Turkey to the east and by

Bulgaria, Macedonia, and Albania to the north (Fig. 4-11). Note that some of Greece's islands in the Aegean Sea lie on the very doorstep of Turkey; in addition, Greeks represent the majority of the population of Cyprus in the remote northeastern corner of the Mediterranean Sea.

But EU membership had other, less favorable consequences. When the Union expanded to include poorer states such as Bulgaria and Romania, Greece lost much of its EU subsidy as Brussels had to divert its equalization funds to needier members. Then the global financial crisis hit Greece especially hard in the late 2000s, and it became increasingly difficult for the government to continue to pay for such personal frills as guaranteed

Francesco Malavolta / AP Images

Illegal African migrants wait to disembark from an Italian Coast Guard patrol boat after being rescued near Sicily on February 14, 2015. During the first four months alone of that year, more than 23,000 refugees were rescued by Italy's seagoing forces. The surge of early 2015 was undoubtedly reinforced by the simultaneous political unraveling of Libya, the main springboard of this undocumented migration flow that has long been (and continues to be) controlled by smugglers. The smugglers reacted by expanding their operations to the southern Mediterranean shores of other countries, ensuring that not even a breakup of the Libyan state is going to disrupt this population movement. As Libya edges toward failed-state status, the expectation is that rising numbers of fleeing Libyan civilians will be joining the illegal northbound exodus.

lifetime employment in the public sector and generous (early) retirement, and to reward businesses with protection against competition and the loose enforcement of tax regulations. Suddenly, this proud and historic nation found itself at the center of an unfolding economic and social calamity as the over-matched Greek government—mired in mountainous debt and facing massive street riots that greeted every effort at urgently needed reform (see photo)—was forced to conform to the painful bailout terms imposed by the EU agencies providing the essential loans.

This southeastern corner of Europe also contains the island country of **Cyprus**, an EU member since 2004, but one whose fractious political geography continues to cause problems for the Union. As Figure 4-11 shows, Cyprus lies closer to Turkey than it does to any part of Europe, and for more than three centuries it was ruled by the Turkish Ottoman Empire. But today's population of 1.1 million is predominantly Greek, the people who have been there the longest. In 1983, the 40 percent of Cyprus under Turkish control, with about 100,000 inhabitants (plus more than 30,000 Turkish soldiers), declared itself the independent *Turkish Republic of Northern Cyprus*. Only Turkey recognizes this mini-state (which today contains a population of less than 300,000), whereas the international community recognizes the government on the Greek side as legitimate. Things got even more complicated when only the Greek side of the island was admitted to the EU in 2004. Resentment was high on both the Turkish side and in

Turkey itself, a reminder that Cyprus's 'Green Line' separating its Greek and Turkish communities constitutes not just a regional border but a boundary between geographic realms.

The final component of Mediterranean Europe is the mini-state of **Malta**, a cluster of five islands located just to the south of Sicily (Fig. 4-19). Although Malta contains only about 420,000 inhabitants, its history is that of a crossroads at the very center of the Mediterranean Sea, a cultural nexus that received infusions from Britain (as a former dependency), Italy, North Africa, and the Middle East. As an EU member since 2004, Malta enjoys a comparatively high standard of living reinforced by a thriving tourist industry.

EASTERN EUROPE

Europe has changed fundamentally in the quarter-century since the Cold War ended. The old Eastern Europe, its western edge sharply defined by the Iron Curtain, no longer exists, and to a certain extent we have seen the revival of an older cultural region known as Central Europe or, in its prevailing German terminology, *Mitteleuropa*. Many of these changes came about as NATO and the European Union expanded eastward. As we saw, the Czech Republic and Slovenia are have already been integrated into the European Core. By 2015, several Eastern European states had already joined the EU, and more are knocking on the door. But this has been a decidedly uneven process because of the disparate conditions and variable geographies that mark so many countries in the region. Some are unlikely to be admitted because of their very poor economic circumstances (e.g., Albania), whereas others (especially Ukraine) would pose serious geopolitical challenges.

The region defined as Eastern Europe covers roughly half of the realm's territory and encompasses no fewer than 17 countries. Our coverage is organized within the framework of four subregions: the Balkans, East-Central Europe, the Baltic States, and the Eastern Periphery.

The Balkans

The subregion of Eastern Europe was deeply impacted by the violent dismantling of communist Yugoslavia during the final decade of the twentieth century. However, this was but the latest chapter in a particularly volatile history because the Balkans constitute a classic example of a **shatter belt** **27**—a zone of persistent splintering and fracturing. Geographers employ many terms to describe the breakup of established order, and one of the best is **balkanization** **28**—the recurrent division and fragmentation of a region. That term is rooted in this part of Europe, derived from the east-west trending Balkan Mountains that stretch across Bulgaria. Today, none of the Balkan subregion's countries can be considered as part of the European Core and none are EU members.

The key state on the new map is **Serbia**, the name of what is left of a much larger domain once ruled by the Serbs, who were dominant in the former Yugoslavia. Centered on the historic capital of Belgrade on the Danube River, Serbia (with a population of 7.2 million) is the largest and potentially most important new country in the Balkans subregion. But Serbia faces some major challenges. First, more than one million Serbs live in neighboring Bosnia and Herzegovina, where they have an uneasy relationship with the local Muslims and Croats (Fig. 4-20). Second, the coastal province named **Montenegro** broke away in 2006, when voters there decided to convert it into an independent state. Third, the Muslim-majority, southernmost province of **Kosovo** declared independence in 2008, its sovereignty immediately recognized by the United States and a majority of (but not all) European governments. And fourth, Serbia still incorporates a Hungarian minority of more than 250,000 in its northern province of *Vojvodina*. Clearly, the Balkans contain a virtual kaleidoscope of ethnic identities that severely complicates any kind of orderly territorial arrangement (Fig. 4-20).

When former Yugoslavia collapsed, **Bosnia and Herzegovina** (*Bosnia* in our shorthand form) was the cauldron of calamity. No ethnic group was numerically dominant here, and this multicultural, effectively landlocked triangle of territory, lying between the Serbian stronghold to the east and the Croatian republic to the west and north, fell victim to disastrous conflict among Serbs,

Croats, and **Bosniaks** (now the official name for Bosnia's Muslims, who comprise almost half of the population of 3.8 million). At least a quarter-million civilians perished in concentration camps associated with *ethnic cleansing* practices; in 1995, a U.S. diplomatic effort resulted in a truce that partitioned the country as shown in Figure 4-20. In 2015, Bosnia reached an initial understanding with the EU that might result in a future candidacy for EU admission.

The southernmost 'republic' of the former Yugoslavia was **Macedonia**, which in 1991 emerged from the wreckage as an independent state with a mere 2 million inhabitants, of whom about two-thirds are Macedonian Slavs. As the map shows, Macedonia adjoins Muslim Albania and Kosovo, and its northwestern corner is home to the 35 percent of the Macedonians who are nominally Muslims. The remainder of this culturally diverse population are Turks, Serbs, and Roma (Gypsies). Macedonia is one of Europe's poorest countries, landlocked and powerless. Even its very name caused it problems: Greece, Macedonia's neighbor to the south, argued that this name was Greek property and therefore would not recognize it. Macedonia also confronts a long-running autonomy movement among its Albanian citizens, requiring the allocation of scarce resources to hold this fragile state together.

© J. Nijman, P. O. Muller, and John Wiley & Sons, Inc.

FIGURE 4-20

The only other dominantly Muslim country in Europe is **Albania**, where 83 percent of the population of 2.9 million adhere to Islam. Albania shares with one other country—Moldova—the status of being Europe's poorest state. It has one of Europe's higher rates of natural population growth, and many Albanians try to emigrate to the EU by crossing the Adriatic Sea to Italy. Most Albanians subsist on livestock herding and farming. They also are eager to join the EU—which is not in the cards for the foreseeable future.

East-Central Europe

As Figures 4-11 and 4-21 show, the three states of this subregion—Poland, Hungary, and Slovakia—all adjoin or are partially integrated into the European Core. All are EU and NATO

members, and Slovakia has also joined the eurozone Considering that these countries were tightly bound into the Soviet bloc until 25 years ago, these changes have been truly remarkable.

The largest and most populous of these countries is **Poland**. With 38.5 million people and lying between two historic enemies, Poland's borders have shifted time and again; but its current, post–1945 reconfiguration appears to be more durable than previous territorial layouts. As the map shows, the historic and once-central capital of Warsaw now lies closer to Russia in the east than to Germany in the west, but the country today has overwhelmingly shifted its gaze westward. After entering the EU in 2004, Poland saw hundreds of thousands of workers depart for jobs in the European Core. But many have since returned as the national economy has grown robustly in this decade (Poland may now be in its best shape ever) and economic relations with Germany are at their closest in history.

Hungary's postcommunist economic transition went surprisingly smoothly. By the mid-1990s, most government-owned companies had been privatized, and Hungary was attracting significant foreign investment. Ten years later, in 2004, Hungary joined the EU. But soon thereafter its situation began to deteriorate as the Hungarian economy was hit especially hard by the global recession of the late 2000s. Politically, heightening discord has centered on Hungary's persistent problem of **irredentism 29**, which refers to a government's open support for fellow ethnics residing in nearby countries. Ethnic Magyars (Hungarians) are found in relatively large numbers in parts of neighboring Romania, Slovakia, and Serbia (Fig. 4-22). As for domestic politics, Hungary has recently taken a turn to the far right, unleashing both nationalistic and anti-Semitic tendencies that have been roundly criticized by EU leaders.

Slovakia has followed a trajectory similar to that of Hungary: a relatively smooth postcommunist transition and rapid accession to the EU, followed by a severe economic downturn and rising political controversies that also drew rebukes from Brussels. In Slovakia, those controversies revolved around the alleged poor treatment of ethnic Germans, Hungarians, Roma (Gypsies), and LGBTQ populations.

The Baltic States

Estonia, Latvia, and Lithuania are positioned at the northeastern extremity of Eastern Europe (Figs. 4-11, 4-17). They could be grouped with the Northern European region or just with Finland, around the Baltic Sea. In our regional scheme of the European realm, we consider the Baltic States a subregion of Eastern Europe because their historical geography is so closely tied to Russia, with which they share a border and a common history as Soviet Republics. The

EAST-CENTRAL EUROPE

POPULATION

- Under 50,000
- 50,000–250,000
- 250,000–1,000,000
- 1,000,000–5,000,000

National capitals are underlined

Road
Railroad
Gas Pipeline
Oil Pipeline
European Core Boundary

0   100   200 Kilometers
0   50   100 Miles

FIGURE 4-21

FIGURE 4-22

geopolitical sensitivities regarding Russia are palpable here, and the pro-European orientation is crystal clear. All three countries joined the EU and NATO.

Estonia is the northernmost of the three Baltic states, and it has longstanding ethnic and linguistic ties to Finland. During the period of Soviet control from 1940 to 1991, Estonia's demographic structure transformed drastically: today about 27 percent of its 1.3 million inhabitants are Russians, most of whom came there as colonizers. After an initially difficult period of adjustment, Estonia has forged ahead of its Baltic neighbors, Latvia and Lithuania. Busy traffic links Tallinn, the capital, with its counterpart in Finland, Helsinki, and a free-trade zone at Muuga

Harbor facilitates commerce with Russia. But more important for this country's future was its entry into the European Union in 2004. Since then, Estonia has gained attention as a result of its economic advances, particularly its creative entrepreneurship in the high-tech and software industries (Skype, for instance, was originally an Estonian company). Having entered the eurozone in 2011, Estonia has developed an economy that has remained vigorous and it is here considered a stable part of the European Core.

As Figure 4-17 reminds us, the boundary of the European Core that traverses Europe's North includes southern Norway, Sweden, and Estonia but excludes the other two Baltic states.

Latvia, the middle Baltic state centered on the port and capital city of Riga, was tightly integrated into the Soviet system throughout Moscow's long domination. Here, as in Estonia, about 27 percent is Russian. Consider this: 25 years ago, virtually all of Latvia's trade was with the Soviet Union. Today its principal trading partners are Germany, the United Kingdom, and Sweden. Russia figures in only one category: Latvia's import of oil and gas. In 2014, Latvia once more underscored its European orientation by joining the eurozone.

Lithuania is the southernmost Baltic state, with a population of 3.2 million and a residual Russian minority of only about 5 percent. But relations with this gigantic neighbor are tense—despite a continuing dependence on Russia as a trading partner. One reason for this strained relationship has to do with neighboring *Kaliningrad*, Russia's **exclave** **30** facing the Baltic Sea (Fig. 4-17). When Kaliningrad became a Russian territory after World War II (Russia still regards this exclave as an important outpost), Lithuania was left with only about 80 kilometers (50 mi) of Baltic coastline and a small port that was not even connected by rail to the interior capital of Vilnius. Significantly, Lithuania joined NATO in 2004, and a year later it called for the demilitarization of Kaliningrad as a matter of national security. Russia did not oblige. Following in the footsteps of Estonia and Latvia, Lithuania joined the eurozone in 2015.

The Eastern Periphery

This subregion consists of the five remaining countries of Eastern Europe: Romania, Bulgaria, Ukraine, Moldova, and Belarus (Fig. 4-23). The first two joined NATO in 2004 and were admitted to the EU three years later. For the other three, these memberships are not even on the horizon. Ukraine and Belarus find themselves on Russia's doorstep, and separatist-minded eastern Ukraine (with its substantial population of ethnic Russians) descended into a *de facto* state of civil war in 2014.

Just how **Romania** managed to persuade EU leaders to endorse its 2007 accession remains a mystery to many Europeans. Romania exhibits some of Europe's worst social indicators: its economy is weak, its incomes are low, its governance has not been sufficiently upgraded from communist times, and political infighting and corruption are endemic. But this is an important country, located in the lower basin of the Danube River and occupying much of the heart of Eastern Europe. And like a number of other countries in this region, it was incorporated into NATO before it was allowed to join the EU. With 20 million inhabitants and a pivotal location on the Black Sea, Romania is a bridge between central Europe and the realm's southeastern corner, where EU-member Greece faces its historical adversary, Turkey, across both land and water (Fig. 4-23).

Across the Danube River lies Romania's southern neighbor, **Bulgaria**. The rugged Balkan Mountains form Bulgaria's physiographic backbone, separating the Danube and Maritsa basins. As Figure 4-23 shows, Bulgaria has five neighbors, several of which are in political turmoil. In 2007, Bulgaria also became an EU member once required judicial and other social reforms were finalized. But today it remains one of the Union's poorest countries, and development prospects are not encouraging. Bulgaria fronts the Black Sea and has a major port at Varna, but the main advantage it derives from its seaside location is the tourism its beaches generate. The capital, Sofia, lies near the opposite, western border with Serbia, and in its core-area hinterland modest foreign investment is slowly modernizing parts of the economic landscape.

Territorially, **Ukraine** is the realm's largest country. Demographically, however, its population of 45.5 million only qualifies Ukraine for second-tier ranking among European states, on a par with Poland and Spain. As Figure 4-23 underscores, Ukraine's relative location is crucial. Not only does it link the core of Russia to the periphery of the European Union: it also forms the northern shore of the Black Sea from Russia to Romania, in the center of which lies the strategically prominent Crimea Peninsula. Perhaps most important of all, oil and gas pipelines connect Russian fields to European markets across Ukrainian territory. Once a land of farmers tilling its famously fertile soils, Ukraine emerged from the Soviet era with the huge (and decaying) Donbas heavy-industrial complex in its east as well as with a substantial (17 percent) Russian minority population. A majority of the Russians are concentrated in the zone that stretches eastward from the Dnieper Valley—which is why that part of the country is mapped as a transition zone in Figure 4-11.

The vital Dnieper River forms a valuable geographic reference for comprehending Ukraine's division that creates the transition zone (Fig. 4-23). To its west lies agrarian, rural, mainly Roman Catholic Ukraine; in the Dnieper's great southern bend and eastward lies industrial, urban, Russified (and Russian Orthodox) Ukraine. For the most part, Ukraine's electorate is divided between a pro-Western (and pro-EU) north and west and a pro-Russian south and east. Beginning in March 2014 right after Russia's invasion and occupation of Crimea, the Russian border zone of eastern Ukraine was engulfed by an escalating conflict that pitted Ukrainian government forces against those of the separatists, now aided directly by Russian military power; effectively, the country had descended into a civil war (see Map Analysis box).

Moldova, Ukraine's tiny impoverished southwestern neighbor, was a Romanian province seized by the Soviets in 1940 and converted into a landlocked 'Soviet Socialist Republic.' A half-century later, along with the USSR's 14 other 'republics,' Moldova gained independence when the Soviet Union disintegrated. Romanians remain in the majority among its 3.6 million people, but many of the Russians and Ukrainians (each about 13 percent) have migrated across the Dniester River to an industrialized strip of land between that waterway and the Ukrainian border, proclaiming there a 'Republic of Transnistria' (see Fig. 4-23). This separatist movement constitutes only one of Moldova's multiple problems. Its economy, dominated by farming, is in decline; an estimated

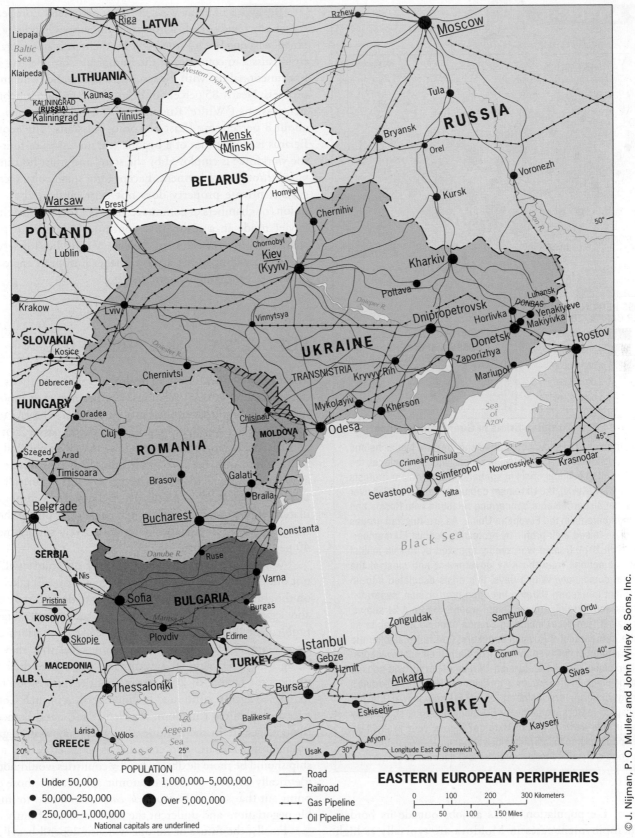

POPULATION

- Under 50,000
- 50,000–250,000
- 250,000–1,000,000
- 1,000,000–5,000,000
- Over 5,000,000

National capitals are underlined

Road
Railroad
Gas Pipeline
Oil Pipeline

EASTERN EUROPEAN PERIPHERIES

0   100   200   300 Kilometers
0   50   100   150 Miles

© J. Nijman, P. O. Muller, and John Wiley & Sons, Inc.

FIGURE 4-23

MAP ANALYSIS

FIGURE 4-24 The Ukrainian Crisis in Geographic Context

Maps are often used to show boundaries and they are just as helpful in demarcating and analyzing transition zones. Figure 4-24 places the ongoing Ukrainian political crisis in spatial context. Violent protests broke out in Kiev (Kyyiv), the Ukrainian capital, in February 2014 over major political disagreements concerning relations with Russia and possible membership in the European Union. As pro-Russian groups inside Ukraine raised their profile by protesting against EU membership, in March 2014, Crimea was swiftly annexed by Russia amidst vociferous objections from the Kiev government and most of the international community. Within days, this crisis escalated into an armed conflict in eastern Ukraine between pro-Russian, separatist rebel forces and Ukrainian government troops. Figure 4-24 shows, for each oblast (province) including Crimea, income per capita as well as the percentage of Russian speakers. Carefully examine the spatial distribution of both variables, particularly their east-west gradients. Then closely track the Russian-Ukrainian border and consult the relevant maps in this chapter to discover which of Ukraine's neighboring countries are members of the EU and/or NATO. What key geographic factors explain why EU membership is so strongly supported in the western half of Ukraine? Acess your WileyPLUS Learning Space course to interact with a dynamic version of this map and to engage with online map exercises and questions.

40 percent of the population works at jobs outside its borders because unemployment rates in Moldova are typically as high as 30 percent; smuggling and illegal arms trafficking are rife. These many misfortunes translate into weakness, and Russia's influence in the form of support for Transnistria's separatists keeps the country in turmoil.

Belarus is in many ways the most peripheral European country of all. Landlocked, autocratic, and heavy-handedly misgoverned, sustained in no small way by the transit of energy supplies from Russia to countries of the European Core, Belarus itself has few functional ties with the rest of the realm and little prospect of progress. Nearly 85 percent of Belarus's 9.5 million inhabitants are Belarussians ('White' Russians), a West Slavonic people; only about 8 percent are (East Slavonic) Russians. The economy of Belarus is almost that of a bygone era: no less than four-fifths of the workforce is employed by the state, and most housing is provided through employment. Except for a small contingent of oligarchs, private property does not exist, and neither do labor unions or chambers of commerce. This is a society stuck in the old Soviet mode in a realm that continues to undergo profound change. In effect, Belarus trades energy subsidies for serving as a stable and willing buffer against further European encroachment on Russia's shrinking western doorstep.

EUROPE'S FUTURE PROSPECTS

The future of the European realm, involving both its internal dynamics and external relations, has become ever more tightly intertwined with that of the European Union. The EU today has penetrated deeply into Eastern Europe; encompasses 28 members; and has facilitated economic progress as well as political stability as it evolved over the past six decades. This remarkable advance amounts to a tremendous achievement in this historically volatile and fractious part of the world. The EU now has a combined population of just over 500 million that constitutes one of the world's richest markets; its member-states collectively supply more than 40 percent of all exports.

Whether European integration stays on course will depend not only on the direction the realm's economy will take, but also on the ability of Europe's political leaders to demonstrate the value and legitimize the costs of integration to their electorates. The problem of the so-called *democratic deficit* is a serious obstacle: from Germany to Greece, citizens often feel that they are not being heard, yet national governments must have their votes to support the Union. By 2015, strong political opposition had been marshaled against continued EU membership in such Core countries as the United Kingdom, the Netherlands, and France. So far these voices are in the minority, but they may well be gathering momentum. Debates also continue as to whether EU membership should be more flexible, whereby countries would not have to identically commit in every economic sphere. But those opposed point out that such a 'Europe *à la carte*' would lead to interminable negotiations and undercut the basic concept of unity.

Finally, the future of Europe will also depend heavily on the resolution of its 'eastern boundary.' More than anything else, it is the eastward expansion of the European Union that is putting this question to a critical test, and the outcome will at least in part be shaped by the response from Russia.

POINTS TO PONDER **?**

- Unemployment rates among youths in Greece and Spain exceed 50 percent, and about half of those under the age of 30 still live with their parents.

- Scotland held a referendum on independence from the UK in 2014; the UK government says it wants to hold a referendum on continuing membership in the European Union in 2017.

- Brussels is the main seat of the European Union, yet it is the capital of a country faced with strong divisive forces that at times threaten its national unity.

- In Northern European countries, the south is more developed than the north; in Southern European countries, the north is more developed than the south.

© Geodakyan Artyom/ITAR-TASS Photo/Corbis

On March 18, 2015, Russia celebrated the first anniversary of its "reunification" with Crimea. This pivotal peninsula on the northern coast of the Black Sea, which contains Russia's major naval base at Sevastopol, is of the utmost geostrategic importance. It was forcibly annexed from Ukraine in 2014 despite the strenuous objections of the Ukrainian government and most of the international community led by the United States. These Russian Black Sea Fleet officers in Sevastopol's Nakhimov Square are gearing up for a high-profile parade through the city's streets. The Crimean Peninsula was "given" to the Ukrainian Soviet Socialist Republic in 1954 by the USSR's leader, Nikita Khrushchev, as a gesture of good will and friendship—at a time when the Russians were firmly in control. It was never imagined that an independent Ukraine might one day turn its back on Russia. But that is precisely what has been happening over the past few years, culminating in President Vladimir Putin's 2014 decision to invade, occupy, and forcibly reattach Crimea to the Russian Federation.

5

THE RUSSIAN/ CENTRAL ASIAN REALM

IN THIS CHAPTER

- Global warming in Eurasia's Arctic
- The forgotten promise of democracy in Russia
- Russian revanchism in the Near Abroad
- Oil, natural gas, and economic power
- Russia's population decline and its consequences
- A New Silk Road takes shape across Central Asia

CONCEPTS, IDEAS, AND TERMS

Monocentric geographic realm	1	Distance decay	12
Continentality	2	Federal system	13
Permafrost	3	Unitary state system	14
Tundra	4	Population implosion	15
Taiga	5	BRICs	16
Forward capital	6	Centrality	17
Federation	7	Shi'ite Islam	18
Russification	8	Sunni Islam	19
Command economy	9	New Silk Road	20
Satellite state	10	Eurasian Customs Union	21
Near Abroad	11	Revanchism	22

FIGURE 5-1

Every geographic realm has a dominant, distinguishing feature: Europe's jigsaw-puzzle map of 40 countries, South America's familiar triangle, Southeast Asia's thousands of islands and peninsulas. The Russian/Central Asian realm stands out because of its vast areal extent, particularly its enormous, nine-time-zone-wide, east-west dimension. The country of Russia, by itself, is twice the territorial size of Canada, the world's second-largest country, and three times as big as neighboring Europe. Not surprisingly, Russia dominates this part of the world, making this a classic example of a **monocentric realm 1**. But not all of this realm is Russian: it also includes the regions of Central Asia and Transcaucasia, which now consist of independent states that were once part of Czarist Russia and its successor Soviet Union.

DEFINING THE REALM

As Figure 5-1 reveals and a globe shows even better, Russia's entire northern coast—from its border with Norway in the west to the Bering Strait across from Alaska in the east—faces the Arctic Ocean, with all the environmental consequences you would expect. And for all of its stupendous dimensions, much of this realm lies at higher latitudes, shut off by mountains and deserts from the warmer and moister parts of Eurasia. That has posed a geographic challenge for Russian rulers and governments for as long as Russia has existed as a state. Russia's ports on the Pacific Ocean in the east lie about as far from where most Russians live as they could be. And in the west, where Russia's core area is situated, every maritime outlet is restricted in some way. The Arctic ports operate on seasonal access. The Baltic ports face a shallow sea far from the open ocean. The Black Sea ports (some leased from

Ukraine) require maneuvering through narrow straits across Turkish territory—simply to reach the eastern Mediterranean Sea.

The Russia/Central Asia realm may be gigantic in size, but its human numbers (that total 227 million) are modest and not much larger than those of such individual countries as Nigeria and Pakistan. Moreover, much of the realm's population has been declining since the 1991 collapse of the Soviet Union in response to changed political, economic, and social conditions that even now continue to produce negative demographic effects.

This realm, as defined by our global framework, consists of nine political entities: the giant Russian state itself; three small nations in Transcaucasia; and five countries in Central Asia. Transcaucasia, the mountainous zone between the Black and Caspian seas, contains Georgia, Armenia, and Azerbaijan. All are small in

MAJOR GEOGRAPHIC QUALITIES

RUSSIA/CENTRAL ASIA

1. Russia/Central Asia is an enormous realm that is almost entirely lacking in warm-water access to the world ocean.

2. This is by far the "widest" geographic realm on Earth, stretching from the Baltic Sea in the west to the Pacific Ocean in the east for some 10,000 kilometers (ca. 6200 mi) and spanning no less than nine time zones.

3. Much of Russia/Central Asia is cold and/or dry. Extensive rugged mountain zones separate the realm from warmer subtropical air, and most of it lies open to Arctic air masses.

4. Russia/Central Asia borders four other realms: Europe, North Africa/Southwest Asia, South Asia, and East Asia. These boundaries have shifted throughout history, and today some of them must be mapped as transition zones.

5. For so large a territory, the population size of Russia/Central Asia (227 million) is relatively small. In Russia itself, the population has been declining in recent decades, especially in the country's east and in most of the rural areas.

6. The core area of this realm is located in western Russia. New development is now mostly concentrated west of the

Ural Mountains; here are most of the major cities, leading industrial complexes, densest transport networks, and most productive farming areas. Outside Russia, certain parts of Central Asia (most notably Kazakhstan) have experienced rapid development based on their rich natural resources.

7. For a monocentric realm, this one is decidedly multicultural. Russian domination has long inflamed tensions within this ethnic mosaic, especially Muslim concentrations along the realm's southern periphery.

8. Central Asia is a region that highly interests China, for both its natural resources and its geopolitical role on China's western frontier. There is now much talk here about forging a 'New Silk Road.'

9. Prospects of a post-Soviet democratic transition in Russia have been effectively extinguished as Russia's leadership becomes increasingly dictatorial. The country's powerful position as a leading exporter of oil and natural gas has been a springboard in Russia's quest to reclaim its Soviet-era role as a major player on the international stage.

size, with a combined population of 17 million (7.5 percent of the realm's total); they also lie in a historically turbulent arena of conflict between Russian and numerous non-Russian peoples. The five Central Asian countries are Kazakhstan, Uzbekistan, Turkmenistan, Kyrgyzstan, and Tajikistan—sometimes referred to collectively as the "Stans" (although they do not include Afghanistan, which we assign to the South Asian realm in Chapter 8). In contrast to Transcaucasia, Central Asia covers a comparatively large area. Territorially, Kazakhstan is the biggest Central Asian country (about five times the size of France) and also has a lengthy border with Russia. Central Asia's overall population is 66 million, accounting for just below 30 percent of the realm's total.

What Transcaucasia and Central Asia have in common is that both were part of Czarist Russia and the Soviet Union. Today they remain firmly within Russia's sphere of influence (the same could be said for parts of Eastern Europe, such as Belarus and eastern Ukraine). It is also worth noting that these southern peripheral components of the realm display characteristics of a *transition zone*: Transcaucasia is positioned in a historically volatile corridor between Russia and Southwest Asia, whereas the "Stans" of Central Asia form a broad arena that has for centuries functioned as a crossroads linking Russia, Southwest Asia, South Asia, and East Asia.

PHYSICAL GEOGRAPHY OF RUSSIA/CENTRAL ASIA

Physiographic Regions

One of the first features you will notice when looking at the physiographic map of the Russian realm is a prominent north-south trending mountain range that extends from the Arctic Ocean southward to Kazakhstan (Figs. 5-1 and 5-2), dividing Russia into two unequal halves: the Russian Plain to the west and the subdivisions of Siberia to the east. This range, the ***Ural Mountains*** (numbered ① in Fig. 5-2), is sometimes designated as the eastern boundary of Europe. That is a mistake because Russia is not Europe: with both sides of the range equally Russian, the Urals cannot form a cultural boundary. Note, too, that the island of Novaya Zemlya (Fig. 5-1) is the northern extension of the Urals, which at their opposite end also extend southward into Kazakhstan.

The ***Russian Plain*** ② to the west of the Ural Mountains, is the eastward continuation of the North European Lowland, and here lies Russia's *core area*. Travel northward from the centrally located capital of Moscow, and the countryside soon changes to

RUSSIA/CENTRAL ASIA: PHYSIOGRAPHY

- - - - Limit of permafrost   ☐ Continental shelf

| 0 | 400 | 800 | 1200 | 1600 Kilometers |
| 0 | 200 | 400 | 600 | 800 | 1000 Miles |

© J. Nijman, P. O. Muller, and John Wiley & Sons, Inc.

FIGURE 5-2

coniferous (needleleaf) forests like those of Canada. The Volga River, Russia's Mississippi, flows southward in a wide arc across this plain from the forested north—but unlike the Mississippi, the Volga drains into a landlocked lake, the Caspian "Sea." Travel southward from Moscow today, and the land is draped in grain fields and pastures. Eastward, the Urals are not high enough to impede surface transportation, but they retain their prominence because this range separates two vast lowlands. At the southern end of the Russian Plain, the *Caucasus Mountains* ③ rise like a wall between Russia and the lands beyond—a multitiered barrier over which Russians have fought with their southern neighbors for centuries.

The *West Siberian Plain* ④ to the east of the Urals is often described as the world's largest unbroken lowland, and here all the rivers flow northward. In fact, this plain is so flat that over the last 1600 kilometers (1000 mi) of its course to the Arctic Ocean, the Ob River (the lowland's largest) falls less than 90 meters (300 ft) in huge meanders flanked by forests until, in the near-Arctic cold, the trees yield to mosses and lichens. Follow an imaginary trail eastward, and in just about the middle of Russia, at the Yenisey River, the West Siberian Plain gives way to the higher relief of the massive *Central Siberian Plateau* ⑤, one of the most sparsely populated areas in the habitable world.

To the south, beyond the Russia-Kazakhstan border, the West Siberian Plain transitions into the *Caspian-Aral Basin* ⑥. This physiographic region contains almost all of Central Asia and is dominated by windswept plains of moisture-starved steppes and desert. Poorly executed irrigation schemes along the rivers draining into the Aral Sea marked much of the Soviet era, and the price was paid after 1990 as an environmental disaster unfolded that has now resulted in the nearly complete drying up of the Aral Sea and the waterways that supply it. To the southeast, lining the long border with China and Mongolia, lie the *Central Asian Ranges* ⑦ that stretch northeastward from Tajikistan and Kyrgyzstan to Lake Baykal.

Continuing our eastward transect of Russia toward the Pacific, we first encounter the *Yakutsk Basin* ⑧ drained by the Lena River, and then we approach the realm's mountainous eastern perimeter (look again at Figs. 5-1 and 5-2), where the few roads and railroads must wend their way through tunnels and valleys full of hairpin turns. We map this jumble of ranges collectively as the *Eastern Highlands* ⑨ which blanket one of the most spectacular, diverse, and still-remote landscapes on the face of the Earth.

Finally, our survey of the realm's physiographic regions ends at the farthest extremity of Russia, the partially offshore *Pacific Rimland* ⑩ that consists of the huge Kamchatka Peninsula and the long, narrow island of Sakhalin to its southwest across the Sea of Okhotsk. Here, this immense geographic realm makes contact with the Pacific Ring of Fire (see Fig. G-4). There are no erupting volcanoes in Siberia and earthquakes are rare, but the Kamchatka Peninsula has plenty of both. This is one of the most volatile segments of the Pacific Ring of Fire, with more than 60 volcanoes, many of them active, forming the spine of the peninsula. Nothing here resembles Siberia (the climate is moderated by the Pacific waters offshore; the vegetation is mixed, not coniferous), and Kamchatka is one of the world's most difficult places to inhabit, effectively separated from the rest of Russia by the absence of overland connections. On the island of Sakhalin, earthquakes rather than volcanism are the leading environmental hazard. Long a battleground between Russians and Japanese, Sakhalin has remained in Russian hands since the end of World War II. In subsequent decades, the island and its maritime surroundings proved to contain substantial reserves of oil and natural gas, making it a most valuable asset in Russia's energy-based export economy.

Climate and Vegetation

As the northernmost populous country on Earth, Russia has virtually no natural barriers against the onslaught of Arctic air masses. Moscow lies farther north than Edmonton, Canada, and St.

Contrasting landscapes of the Russia/Central Asia realm. On the left is a coniferous forest in early autumn in the Tyumen Oblast (Province) near the center of the West Siberian Plain; to the right are the Tian Shan Mountains that parallel the Chinese border in northern Kyrgyzstan.

Petersburg lies at the latitude of the southern tip of Greenland. Winters are long, dark, and bitterly cold in most of Russia; summers are short and growing seasons limited. Precipitation, even in western Russia, ranges from modest to minimal because the warm, moist air carried eastward across Europe from the North Atlantic Ocean loses both heat and moisture by the time it reaches Russia. Figure 5-3 reveals the consequences. Russia's climatic **continentality 2** (inland climatic environment remote from moderating and moistening maritime influences) is expressed by its prevailing *Dfb* and *Dfc* conditions. Compare the Russian map to that of North America (Fig. G-6), and you will note that, except for a small corner off the Black Sea, Russia's climatic conditions resemble those of the U.S. Upper Midwest and interior Canada. Along its entire northern edge, Russia has a zone of *E* climates, the most frigid on the planet. In these Arctic latitudes originate the polar air masses that dominate its environments.

The Russian realm's harsh northern climates affect people, animals, plants—and even the soil. In Figure 5-2 you can find a direct consequence of what you see in Figure 5-3: a dashed blue line that starts at the shore of the Barents Sea, crosses Siberia eastward, and reaches the coast of the Pacific Ocean north of the neck of the Kamchatka Peninsula. North of this blue line, water in the ground is permanently frozen, creating an even more formidable obstacle to settlement and infrastructure than the severe weather alone. It is referred to as **permafrost 3**, and it affects other high-latitude environments as well (Alaska, for example). In the soil layer above the permafrost, seasonal temperature changes cause alternate thawing and freezing, with predictably destructive impacts on buildings, roads, railroad tracks, and pipelines.

Examine Figure 5-3 carefully, and you will see two ecological terms: *tundra* and *taiga*. **Tundra 4**, as the map suggests, refers to both climate and vegetation. The blue area mapped as *E* marks the coldest and ice-affected environmental zone in Russia and elsewhere in the high Arctic: this is frigid, treeless, windswept, low-elevation terrain where bare ground and rock prevail and mosses, lichens, patches of low grass, and a few hardy shrubs are all that grows. The stippled area on the map is the **taiga 5**—a Russian word meaning "snowforest" (also called boreal forest)—which extends across vast reaches of Eurasia as well as northern North America and is dominated by coniferous (as in pine cone) trees. As the map shows, the taiga extends southward into more moderate environments, and there it becomes a mixed forest of coniferous and deciduous trees. Although the taiga prevails from northern Scandinavia to the Russian Far East, this is the vegetative landscape most often associated with Siberia: endless expanses of rolling countryside draped in dense stands of evergreen pine trees (see left frame of photo pair).

Climates and Peoples

Climate and weather have always challenged the peoples of this realm. The high winds that drive the bitter Arctic cold southward

RUSSIA/CENTRAL ASIA: CLIMATE

B DRY CLIMATE

BS	Semiarid
BW	Arid

h=hot
k=cold

C HUMID TEMPERATE CLIMATE

Cf	No dry season
Cw	Dry winter
Cs	Dry summer

a=hot summer
b=cool summer

D HUMID COLD CLIMATE

Df	No dry season
Dw	Dry winter

c=short, cool summer
d=very cold winter

E COLD POLAR CLIMATE

E	Tundra and ice

Taiga (snowforest)

H HIGHLAND CLIMATE

H	Unclassified highlands

0 500 1000 1500 Kilometers
0 500 1000 Miles

© J. Nijman, P. O. Muller, and John Wiley & Sons, Inc.

FIGURE 5-3

deep into Eurasia, the blizzards of Siberia, the aridity of Central Asia, the temperature extremes, rainfall variability, and short and undependable growing seasons even in the more moderate western parts of the realm have always made farming difficult, to the point that grain often had to be imported into Russia.

As noted in the Introduction, and will repeatedly be observed throughout this book, humankind's long-term dependence on agriculture remains etched on the population map, even as our planet becomes ever more urbanized and its economy more globalized. By examining the climates of the Russian/Central Asian realm, we can begin to understand what the map of population distribution communicates (Fig. 5-4). The overwhelming majority of the realm's 227 million inhabitants remain concentrated in western and southwestern Russia, where environmental conditions were least difficult at a time when farming was the mainstay of most of the people. To the east of the Urals, the population is sparser and tends to cluster like beads on a string along the southern margin of the realm, becoming even more thinly distributed east of Lake Baykal. Central Asia's population distribution is very closely tied to local water supplies, which are most readily available in the hilly terrain that lines the rugged mountains of the southeast. And let us also keep in mind that nearly three-quarters of this realm's population now resides in cities and towns, so it is

not difficult to imagine just how empty vast stretches of countryside must be—especially in those frigid northerly latitudes.

Climate Change and Arctic Prospects

As every map of this high-latitude realm shows, its northern coast lies entirely on the polar side of the Arctic Circle. Such a lengthy coastline on the Arctic Ocean does not exactly constitute an advantage: most of that ocean is frozen much of the year, and only limited warmth from the North Atlantic Drift ocean current keeps the western ports of Murmansk and Arkhangelsk open a bit longer (Fig. 5-1). Ports like this (and even St. Petersburg on the Gulf of Finland, an arm of the Baltic Sea) would never have developed the way they did if Russia had better access to the world ocean.

These days it looks as though nature may be about to give Russia a helping hand. If, as most climatologists anticipate, global warming causes long-term melting of large portions of the Arctic Ocean's ice cover, that body of water may come to play a new and different role in this realm's future. Milder atmospheric conditions may shrink the area of permafrost mapped in Figure 5-2; moister air masses may improve agriculture on the Russian Plain; warmer water may keep Arctic ports open year-round; and the so-called Northern Sea Route (also known as the Northeast

FIGURE 5-4

© J. Nijman, P. O. Muller, and John Wiley & Sons, Inc.

Passage) might even open up to directly connect the Bering Strait with the North Sea (see Fig. 12-5), shortening some international sea routes by thousands of kilometers and ushering in a new era of long-distance trade.

That, at least, is how many Russians like to envision global warming—as a potential advantage nature has long denied them. And indeed, the Arctic provides much evidence for such warming: the melting of parts of the Greenland Ice Sheet, the shrinking of the average extent of permanent Arctic Ocean ice, and the reduced incidence of icebergs. Russian economic planners look at the shallow offshore waters along the Arctic Ocean coastline (see the white- and light-blue areas in Fig. 5-1) and hope that oil and gas reserves beneath those waters will come within reach of exploitation. These developments may even have the effect of expanding the realm far into the Arctic. In 2007, the Russian government even placed a metal Russian flag at the North Pole on the seafloor, under the permanent ice of the Arctic Ocean, symbolizing its intentions (see final photo in Chapter 12).

Ecologies at Risk

If, as computer models predict, global warming continues and perhaps intensifies in the polar latitudes, these ecologically sensitive environments are sure to be severely affected. Animals as well as humans have long adapted to prevailing climatic conditions, and such adaptations would be significantly disrupted by environmental change. The intricate web of relationships among species and their environments on the one hand, and among species themselves on the other, can be quickly damaged by temperature shifts. The polar bear is just one prominent example: it depends on ample floating sea ice to hunt as well as raise cubs. When sea ice contracts and wider stretches of open water force polar bears to swim greater distances during the cub-rearing season, fewer infants make it to adulthood. If ice-free Arctic summers are indeed in the offing, the polar bear could become extinct in your lifetime. Rapid ecosystem change would also endanger seal, bird, fish, and other high-latitude wildlife populations.

Such changes affect human populations as well. For example, indigenous communities still pursuing traditional lives in parts of the Arctic domain are similarly adapted to the harsher environments of the past. Their traditions, already under pressure from political and economic forces resulting from their incorporation into modern states, would be further imperiled by environmental changes that are likely to irreversibly modify the ways of life they developed over thousands of years.

And if a new era of oil and natural gas exploration and exploitation is indeed about to open, already-fragile offshore environments face even greater danger. As technologies of recovery put ever more of these reserves within reach, oil platforms, drills, pumps, and pipelines will materialize—along with risks of oil spills, pollution, and ecological damage of the kind we have seen all over the lower-latitude world. Climate change and technological development are allowing the economic penetration of a part of the world long protected from it by distance and nature.

THE NATURAL RICHES OF RUSSIA AND CENTRAL ASIA

Given the gigantic territorial layout of the realm, it is not surprising that its natural resource base is vast and varied. Indeed, it is already one of the world's leading producers and exporters of oil and natural gas. Figure 5-5 maps the growing number of producing fields dispersed across the realm: from western and central Russia eastward to Sakhalin in the Pacific Rimland, and from the Arctic shore of Siberia in the north to the Caspian Basin in the far south (Fig. 5-5).

Petroleum and natural gas are not the only underground riches of this realm. In fact, it is so well endowed with minerals that virtually all of the raw materials required by modern industry are extractable. Nor are oil and gas the only energy reserves. Major coalfields lie both east and west of the Urals as well as in Siberia's more southerly latitudes (the string of mines producing high-quality coal along the corridor of the all-important Trans-Siberian Railroad, completed in the early twentieth century, were essential to Russia's industrial development and its successful war against Nazi Germany). Large deposits of iron ore, too, are widely scattered across Russia, from the so-called Kursk Magnetic Anomaly near the border with Ukraine to the remotest corners of Siberia's Arctic north. And when it comes to other metallic ores, this realm has everything from gold to lead and from aluminum to zinc. In Central Asia, Kazakhstan has led the way in expanding the exploitation of its rich deposits of oil, natural gas, uranium, chromium, and titanium.

RUSSIA'S CZARIST ROOTS

Grand Duchy of Muscovy

During the fourteenth century, the Grand Duchy of Muscovy rose to preeminence under the rule of princes and dukes (Fig. 5-6). They extended Moscow's trade connections from the Baltic to the shores of the Black Sea and forged enduring religious ties with the leadership of the Eastern Orthodox Church in what is now Istanbul, Turkey. Then, beginning around 1450, there began a period of more than three centuries during which powerful Russian rulers—now called czars—etched Muscovy's imprints ever more widely on the map of Russia, Eurasia, and, by extension, the world.

By the sixteenth century, during the reign of Czar Ivan IV (better known as Ivan the Terrible), the Grand Duchy had been transformed into a military powerhouse and an imperial state. Ivan expanded Moscow's empire first by conquering the Islamic areas of what is now western Russia and then by gaining control over present-day Estonia and Latvia on the Baltic Sea. It was a time of almost constant warfare, and Ivan acquired his reputation from the reign of terror he unleashed in his pursuit of military discipline, centralized administrative control, and retaliation against many members of the nobility.

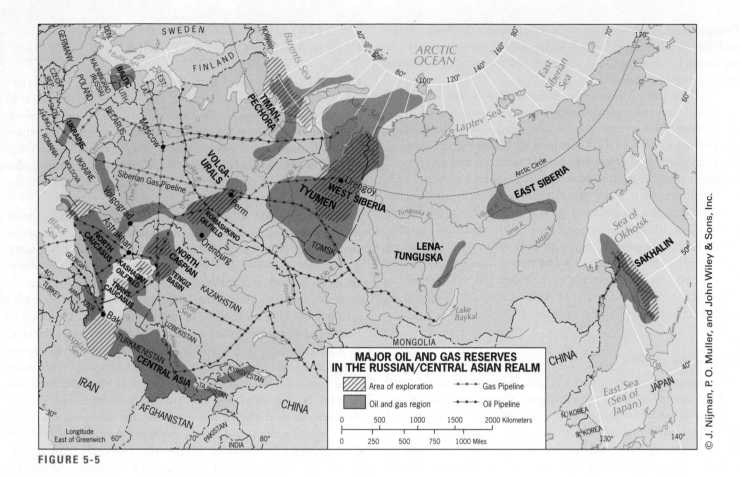

FIGURE 5-5

Eighteenth-Century Czarist Russia

When Peter the Great became czar (he ruled from 1682 to 1725), Moscow already lay at the center of a great empire—great, at least, in terms of the territories it controlled. The Islamic threat had been ended, and the growing influence of the Russian Orthodox Church was represented by its distinctive religious architecture and powerful bishops. Peter consolidated Russia's gains and endeavored to make a modern, European-style state out of his loosely knit country. He built St. Petersburg as a **forward capital** 6 on the doorstep of Swedish-held Finland, fortified it with major military installations, and made it Russia's leading port (Fig. 5-6).

Peter the Great was an extraordinary leader, in many ways the founder of modern Russia. In his desire to remake the country—to pull it from the forests of the interior to the seas of the west, to open it to outside influences, and to relocate its population—he left nothing to chance. Prominent merchant families were forced to move from other cities to St. Petersburg. Ships and wagons entering the city had to bring building stones as an entry toll. The czar himself, aware that to become a leading power Russia had to be strong at sea as well as on land, went to the Netherlands to work as a laborer in the famed Dutch shipyards to learn the most efficient methods for building them. Meanwhile, the czar's forces continued to conquer peoples and territory: Estonia was incorporated in 1721, widening Russia's window to the west; and in Siberia steady expansion soon began in the area south of the city of Tomsk (Fig. 5-6).

Under Czarina Catherine the Great, who ruled from 1760 to 1796, Russia's empire in the Black Sea Basin grew at the expense of the Ottoman Turks. The Crimean Peninsula, the port city of Odessa, and the entire northern coastal zone of the Black Sea fell under Russian control. Also during this period, the Russians made a fateful move: they penetrated the corridor between the Black and Caspian seas, whose spine is the mountainous Caucasus with its dozens of ethnic and cultural homelands, many of which were Islamized. The cities of Tbilisi (now in Georgia), Baku (Azerbaijan), and Yerevan (Armenia) were captured. Eventually, the Russian push toward an Indian Ocean outlet was thwarted by the British, who held sway in Persia (modern Iran), and also by the Turks.

Around the same time, Russian colonists were advancing far to the east, settling along the southeastern frontier, crossing the Bering Strait, and even entering Alaska in 1784. But in less than a century the Russians gave up on their North American outposts (which at their height had reached as far south as central California). The sea-otter pelts that had attracted the early pioneers were running out, European and white American hunters were cutting into the profits, and indigenous American resistance was growing. When the United States offered to purchase Russia's Alaskan holdings for $7.2 million (more than $100 million in today's dollars) in 1867, the Russian government quickly agreed, setting Alaska on a new course that culminated in its joining the United States as the forty-ninth State in 1959.

FIGURE 5-6

Nineteenth-Century Expansion

Although Russia had withdrawn from North America, Russian expansionism during the nineteenth century continued in Eurasia. As they were extending their empire southward, the Russians also took on the Poles, old enemies to the west, and succeeded in conquering most of what is now Poland, including the capital of Warsaw. To the northwest, Russia took over Finland from the Swedes in 1809.

Throughout most of the nineteenth century, however, the Russians were preoccupied with Central Asia, where Tashkent and Samarqand (Samarkand) came under St. Petersburg's control (Fig. 5-6). The Russians here were still bothered by raids of nomadic horsemen, and they sought to establish their authority over the Central Asian drylands as far southeast as the high mountains that marked China's western frontier. As it made headway here, Russia gained many Muslim subjects because it was now penetrating Islamic Asia. Nonetheless, under czarist rule local peoples retained some autonomy.

Thus the Russian state, like the states of Britain, France, and other European powers, expanded through *imperialism*. Whereas those European mother countries expanded their domains overseas, Russian influence spread overland into Central Asia, Siberia, China's inland peripheries, and the far eastern Pacific coastlands. What emerged in the end was not the greatest empire but the largest territorially contiguous empire in the world.

The czars embarked on their imperial conquests in no small part because of Russia's relative location—most importantly, the

lack of warm-water ports. Had the early-twentieth-century revolution not intervened, their southward push might have reached the Indian Ocean or even the Mediterranean Sea. Czar Peter the Great envisaged a Russia open to trading with the entire world, and he made sure his new capital of St. Petersburg on the Baltic Sea developed into Russia's leading port. But in truth, throughout the long arc of its history, Russia's geography has mostly been one of remoteness from international mainstreams of change and progress, repeatedly punctuated by episodes of self-imposed isolation.

A Multinational Empire

Centuries of Russian expansionism did not confine itself to empty land or unclaimed frontiers. The Russian state became an imperial power that annexed and incorporated numerous nationalities and cultures. This was done by employing force of arms, by overthrowing uncooperative rulers, by seizing and annexing territory, and by stoking the fires of ethnic conflict. By the time the ruthless Russian regime had begun to confront revolution among its own citizens, Czarist Russia controlled peoples representing more than 100 nationalities. The winners in the revolutionary struggle that ensued—the communists who forged the Soviet Union—did not liberate these subjugated peoples. Rather, they changed the empire's framework, binding the peoples colonized by the czars into a new system that would in theory give them autonomy as well as identity. In practice, however, it doomed those peoples to bondage and, in certain cases, extinction. The Soviet Union, which arose in the

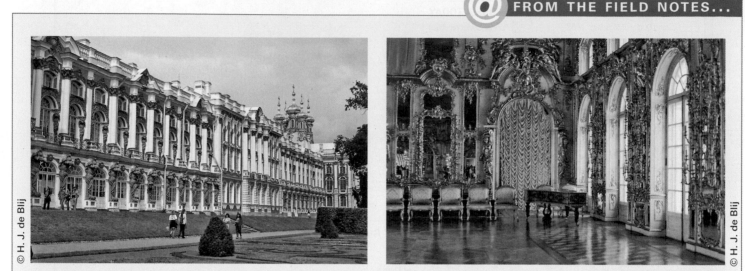

"Not only the city of St. Petersburg itself, but also its surrounding suburbs display the architectural and artistic splendor of Czarist Russia. The czars built opulent palaces in these outlying districts (then some distance from the built-up center), among which the Catherine Palace, begun in 1717 and completed in 1723 followed by several expansions, was especially majestic. These photos show the ornate sculptures that decorate the magnificent exterior (left) as well as the interior detail of a set of rooms known as the Golden Suite, of which the ballroom (right) exemplifies eighteenth-century Russian Baroque at its height. The 'glorious past' of Czarist Russia (see Fig. 5-6) seems to be reviving in the hearts and minds of today's most patriotic Russians."

wake of the communist revolution of 1917, expeditiously undertook another effort to integrate all these peoples and regions into this vast multinational state—notwithstanding the mantra-like socialist condemnations of imperialism.

THE SOVIET UNION (1922–1991)

Communism found fertile ground in the Russia of the 1910s and 1920s. In those days, Russia was infamous for the wretched serfdom of its peasants, the cruel exploitation of its workers, the excesses of its nobility, and the ostentatious palaces and riches of the czars. Ripples from Europe's Industrial Revolution during the 1890s had introduced a new age of misery for those laboring in factories. There were workers' strikes and ugly retributions, but when the czars finally tried to better the lot of the poor, it was too little too late.

There was no democracy, and the people had no way to express or channel their grievances. Europe's democratic revolution had bypassed Russia, and its economic revolution touched the czars' domain only slightly. Most Russians, as well as tens of millions of non-Russians under czarist control, faced exploitation, corruption, starvation, and harsh subjugation. When the people finally began to rebel in 1905, there was no hint of what lay in store. Over the next two decades, it would result in nothing less than a full-scale revolution that transformed Czarist Russia into a behemoth communist state. In the process, the lives of nearly 150 million now-Soviet citizens were fundamentally reorganized, and

the seeds sown for what would eventually become one of the twentieth century's superpowers on the global political stage.

The Soviet Territorial Framework

Russia's great expansion had brought myriad nationalities under czarist control; now the revolutionary government set out to restructure this heterogeneous ethnic mosaic into a smoothly functioning state, the ***Union of Soviet Socialist Republics (USSR)***—or ***Soviet Union*** in short form. The czars had conquered, but they had done little to bring Russian culture to the peoples they ruled. When the old order was overthrown in 1917, ethnic Russians themselves constituted only about half of the population of their empire.

The political framework for the new Soviet Union was based on the ethnic identities of its numerous incorporated peoples. Given the size and cultural complexity of the now Soviet Empire, it was impossible to allocate territory of equal political standing to each of the nationalities (keep in mind that the communists now controlled the destinies of well over 100 peoples, ranging from large nations to small isolated ethnic pockets). It was decided to subdivide this gargantuan realm into 15 *Soviet Socialist Republics (SSRs)*, each of which was delimited to correspond broadly to one of the major nationalities. (As Figure 5-6 shows, much later these SSR territories morphed into the framework of the national state borders that emerged after the 1991 dissolution of the USSR.) Although Russians constituted barely 50 percent of the nascent Soviet Union's population, they were (and still are) the most widely dispersed ethnic group in the realm (see Fig. 5-7). Thus the Russian Republic was by far the largest designated SSR, encompassing just over three-quarters of the Soviet Empire's territory.

As designed by Vladimir Lenin, the Soviet Union's founder, the new political system claimed to be a **federation** `7`, which suggests power sharing among the various republics—but the reality was quite different. The Russian leadership in Moscow maintained absolute control over the SSRs, and they made it Soviet policy to relocate entire peoples from their homelands in order to better fit the grand design as well as rewarding or punishing localities—sometimes capriciously. The overall effect, however, was to move minority peoples eastward and replace them with Russians. In time, this **Russification** `8` of the Soviet Empire produced substantial ethnic Russian minorities in each of the 14 other republics.

The centerpiece of the tightly controlled Soviet "federation" was the Russian Soviet Federative Socialist Republic. Containing half the USSR's population, the capital city, the realm's core area, and 76 percent of all Soviet territory, Russia was unmistakably the Empire's nucleus. Outside the Russian Republic, "Soviet" almost always meant "Russian"—the reality with which the lesser republics lived. Russians came to these 14 republics to teach (Russian was taught in the colonial schools), to organize (and frequently dominate) the local Communist Party, and to implement Moscow's economic decisions. Clearly, this was colonialism, but somehow the communist disguise—how could socialists, as the communists called themselves, be colonialists?—and the contiguous spatial nature of the Empire made it appear to the rest of the world as something else. Indeed, on the international stage the Soviet Union became a champion of oppressed peoples, a force in the post-1950 decolonization process. It was an astonishing contradiction that would, over time, be fully exposed.

The Soviet Economic Framework

The geopolitical changes that resulted from the formation of the Soviet Union were accompanied by a gigantic economic experiment: the conversion of the Empire from a czarist autocracy with a capitalist veneer to communism. From the early 1920s onward, the country's economy would now be *centrally planned*—the communist leadership in Moscow would make all decisions regarding economic planning and development. Soviet planners had two principal objectives: (1) to accelerate industrialization and (2) to collectivize agriculture. For the first time ever on such a scale, an entire country was organized to work toward national goals prescribed by a central government.

But Soviet agriculture never came close to achieving its intended level of productivity because farmers were inadequately motivated. Most opposed collectivization, but any who obstructed the communists' grand design suffered a dreadful fate. It has been estimated that between 30 and 60 million people lost their lives through imposed starvation, constant political purges, Siberian exile, and/or forced relocation. Even though this historic Soviet "experiment" produced an incalculable human tragedy, the secretive nature of

© Velar Grant/Demotix / CorbisImages

The spirit of the communist revolution found expression in 'socialist-realist' art and political propaganda. This famous landmark at the entrance to Moscow's Mosfilm Studios dates from 1937 and glorifies the working class in typical fashion: a heroic statue of individual male and female workers hoisting the iconic hammer and sickle above their heads. Russians call this sculpture *Rabochiy i Kolkhoznitsa (Worker and Kolkhoz Woman*—a kolkhoz being a collectivized farm). Even today, statues like this stud urban landscapes all across Russia.

Soviet officialdom made it possible to hide this disaster from the world.

The USSR operated a **command economy** `9`, in which state planners assigned the production of particular manufactures to particular places, often disregarding the basic rules of economic geography. For instance, the manufacture of railroad cars might be assigned (as indeed it was) to a factory in Latvia. No other factory anywhere else would be permitted to produce this equipment—even if supplies of raw materials would make it cheaper to build them near, say, Volgograd 2000 kilometers (1250 mi) away. Yet, despite an expanded and improved transportation network (see Fig. 5-11), such policies made manufacturing in the USSR extremely expensive, and the absence of competition made managers complacent and workers far less productive than they might have been.

The Downfall of the Soviet Union

On December 25, 1991, the inevitable came to pass: the Soviet Union expired. The centrally planned economy went into terminal structural failure; the arms race with the United States during the Cold War had drained too many resources; and the long effort at Russification had finally backfired, fueling the drive for independence among most of the peoples that had been forcibly incorporated into the USSR. The last Soviet president, Mikhail Gorbachev, resigned, and the Soviet hammer-and-sickle flag flying atop the Kremlin in Moscow was lowered for the final time and immediately replaced by the white, red, and blue Russian tricolor. Yet even though a new and turbulent era had opened literally overnight, deeply entrenched Soviet institutions and systems will long continue to cast their shadows over transforming Russia.

When the Soviet system collapsed and the 15 Soviet Socialist Republics became independent states in 1991, Russia was left without the empire that had taken centuries to build and consolidate—and that contained crucial agricultural as well as industrial resources. No longer did Moscow control the farms of Ukraine or the oil and natural gas reserves of the Caspian Basin and Central Asia. The former Russian Soviet Federative Socialist Republic had suddenly become the newborn country of Russia, still a state of enormous territorial size but no longer able to claim sovereignty over the valuable lands beyond its new borders.

POST-SOVIET RUSSIA AND CENTRAL ASIA

The Soviet Union's implosion signaled a major reorientation of the Russian Republic as well as the other 14 former SSRs that became independent countries. Central planning came to an end and free markets had to be introduced; national industries and corporations had to be privatized; and the communists lost their political monopoly to a new multiparty system that (at the outset) offered real elections. Although individual freedoms increased, state support in a variety of forms, from pensions to rent subsidies, mostly came to a halt. As all the components of the former Soviet Union went through an initial phase of shock, the overriding hope was that all this would be transitional, and that temporary dislocation and occasional chaos would soon be followed by a better life for most citizens. Before we examine the new political and economic forces unleashed at the regional level, we need to establish their broader context by reviewing two important developments reshaping the realm's human geography in the twenty-first century: (1) the continuing evolution of its complex cultural mosaic, and (2) Russia's geopolitical repositioning in its 'Near Abroad' (the former non-Russian SSRs) as well as the world beyond.

A Complex Cultural Mosaic

For centuries, this realm has been a culturally and ethnically diverse part of the world whose traditions and customs spill over into neighboring realms even as neighbors have come to live here. Ethnic Russians still form the overall majority, but sizeable parts of this realm are home to a growing number of non-Russians—and not just in Transcaucasia and Central Asia. As Figure 5-7 shows, the realm contains clusters of Finnish, Turkic, Armenian, Kazakh, and dozens of other minorities, but the scale of this map is too small to reflect the complexity of the total ethnic mosaic (especially within Russia itself). Here is an interesting example: in the southwest, facing the Caspian Sea, lies a small "republic" named Dagestan (which achieved notoriety in conjunction with the 2013 Boston Marathon bombings) that is part of modern Russia. It is half the size of Maine, has a population of 3 million, and contains nearly 30 ethnic communities that speak their own languages, most of which are variants of languages spoken in the nearby Caucasus, Turkey, and Iran.

As Figure 5-7 shows, the Slavic people collectively known as the Russians not only form the majority of the realm's population but also are the most widely dispersed across it. Although the large lowland west of the Urals is the historic hearth and modern core area of the Russian state, the distribution of Russians today still extends from the shores of the Arctic Sea southward to the Black Sea coast and from St. Petersburg on the Gulf of Finland all the way east to Vladivostok on the Pacific's Sea of Japan. Discontinuous nuclei and ribbons of Russian settlement are even scattered across parts of Siberia. Nonetheless, another glance at Figure 5-4 reminds us that the population as a whole decidedly concentrates in the realm's southwestern sector.

Traces of the penetration of the Islamic Ottomans hundreds of years ago can clearly be discerned on today's cultural map (Fig. 5-7): the large swath of Turkic peoples (shaded red) extending from near Nizhniy Novgorod southeastward across Kazakhstan to the fringes of Central Asia is a solid piece of evidence. Contemporary Russia has a relatively higher percentage of Muslims in its population (roughly 13 percent) than Western European countries do. As for Central Asia—whose total population is 46 percent of Russia's—the combined percentage of Muslims living in its five countries today is almost 80 percent.

Long before Islam appeared on the scene, Slavic peoples from one end of their domain to the other had accepted the teachings of the Eastern Orthodox Church. It stayed that way until the communists overthrew the czar and promptly began the disestablishment of the Church. For the next seven decades atheism was official policy; but following the demise of the USSR in the early 1990s, the Russian Orthodox Church has made a very strong comeback, today increasingly accompanied by nationalistic propaganda.

The Near Abroad

The ending of the Cold War upon the dissolution of the USSR implied an abrupt termination of Soviet geopolitical strategies that had been aimed at confronting U.S.-led alliances in Europe, the Middle East, and South Asia as well as Southeast Asia. Soviet foreign policy and power had been projected well beyond Soviet

RUSSIA/CENTRAL ASIA:
ETHNOLINGUISTIC AREAS

Armenian · Azerbaijani (Azeri) · Buryat · Caucasian · Finno-Ugric · Kalmyk · Paleo-Siberian · Russian · Tajik · Turkic · Ukrainian · Sparsely populated or uninhabited

FIGURE 5-7

© J. Nijman, P. O. Muller, and John Wiley & Sons, Inc.

borders to other geographic realms from Africa to Middle America. The downfall of the Soviet Union meant that the new Russia, shorn of its **satellite states** 10 in Eastern Europe and abandoned by all the former SSRs, had to completely reinvent its position in the world, and most immediately within Eurasia. Moreover, it had to meet this challenge in the knowledge that a total makeover of Russia's economy was the foremost priority.

In 1992, the new Russian government introduced the term **Near Abroad** 11 to refer to the newly formed countries that surround Russia today. For the most part, these encompass the 14 former Soviet republics, stretching from Estonia to Tajikistan. Those states may now be located "abroad," but seen from Moscow they are so "near"—and collectively constitute such a pronounced encirclement of Russia—that the Russian government feels entitled to take a special interest in them. What this really means is that Russia will not permit any of these countries to mount a threat of any kind, and that it reserves the right to intervene if it deems that is necessary. It is not difficult to envision this sense of territorial encirclement in Russia, particularly with the eastward expansion of the EU and NATO (now almost at Russia's western doorstep), heightened Islamic militancy in the southern periphery, and the swift rise and growing assertiveness of China in the Far East as well as Central Asia.

And then there is the issue of ethnic Russian minorities who now find themselves living "abroad." Although many Russian nationals have returned from the former SSRs since 1991, millions more stayed on. A large proportion of these expatriates reside in areas where Russians were in the majority (for example, in parts of eastern Ukraine, northern Kazakhstan, and eastern Moldova) and where post-Soviet life was generally unchanged from what it was during Moscow's rule. But elsewhere, Russians who remained in the erstwhile Soviet republics were instantly reduced to minority status; many were mistreated in various ways, and they often appealed to Moscow for assistance. Thus the concept of a Russian "Near Abroad" found its way into Russia's national discourse. Beyond Russia's borders, but of growing concern to its leadership and citizens, the "Near Abroad" became geographic shorthand for an enduring Russian sphere of influence, whereby Moscow stood ready to help its fellow ethnics in case of trouble. So far, the most acute and dangerous confrontation continues to unfold in Ukraine following Russia's unexpected, forcible annexation of the Crimean Peninsula in 2014. Within weeks, parts of eastern Ukraine had descended into full-scale warfare as separatist Russian rebels, strongly supported by Moscow, battled the Kiev-based Ukrainian government in their struggle to secede from Ukraine and achieve annexation by Russia (see box containing Fig. 4-24).

At a soccer match between Russia and Poland during the 2012 Euro Tournament, Russian fans 'greeted' the Polish team with a gigantic banner that left little doubt about their convictions—the game was played in Poland's capital, Warsaw, and those fans knew very well that it was being televised around the globe, along with their message. Today there is much fertile breeding ground in Russia for ever more assertive foreign policy claims in the Near Abroad and beyond.

Playing Russia's Energy Card

During Soviet times, the non-Russian republics all became dependent to some extent on Russia's prodigious oil, natural gas, and mineral supplies. Today, this dependency on Russian energy

reserves in particular plays a critical role in the evolving geopolitics of the realm. In parts of Central Asia, however, Russia has actually been losing ground because a number of countries contain their own abundant fuel supplies. Most of all, Kazakhstan has begun to exploit its huge reserves of oil and gas—attracting growing interest as well as substantial investments from around the world. Between 2000 and 2012, Kazakhstan's exports to Russia were almost halved from 17 to 9 percent while exports to China practically tripled from 7 to 20 percent. Oil constitutes the bulk of these exports, and the Chinese are now working with the Kazakhs to complete a pipeline clear across western China and Kazakhstan in order to tap the latter's high-yielding northern Caspian Sea oilfields. Similarly, oil- and gas-rich Turkmenistan has in recent years intensified relations with China at Russia's expense: 81 percent of Turkmen exports (mostly natural gas) now flow to China. The remaining (energy-poorer) Central Asian countries—Uzbekistan, Kyrgyzstan, and Tajikistan—are still tightly tethered to the Russian economy, and Moscow's political influence in all three remains considerable.

As for the European countries to Russia's west, dependence on Russian energy supplies is enormous, implying significant leverage for Moscow in its political dealings with this realm. Consider these current percentages of oil and gas imports from Russia: Belarus, 100; Slovakia, 98; Lithuania, 92; Bulgaria, 90; Hungary, 86;

FIGURE 5-8

Finland, 76; Latvia, 72; and Estonia, 69. Farther west, Russian imports are also substantial: Sweden, 40; Germany, 30; the Netherlands, 30; and Italy, 28. Russia also is the single biggest fuel supplier to the European Union as a whole, providing 35 percent of its oil and 30 percent of its natural gas. In the mid-2010s, and as criticism of Russia accelerated in Europe during the deepening crisis in Ukraine, there has been much talk about the need for Europe to reduce its energy dependence on Russia. But this cannot happen overnight because, quite apart from the economic and political issues involved, energy infrastructures are complex, very expensive, and can take years to build. Figure 5-8 maps both existing and proposed gas pipelines connecting Russia and Europe. Note that a great deal of Russian gas being pumped to Europe passes through Ukraine—yet another potentially critical factor in the evolving crisis there. Ukraine itself is hurriedly trying to decrease its reliance on Russia by seeking new suppliers in the West: in 2013, 95 percent of its gas came from Russia; but in 2015, so the Ukrainian government claims, that quantity was reduced to 33 percent.

REGIONS OF RUSSIA AND CENTRAL ASIA

No other single state dominates its geographic realm to the extent that Russia does, but we also need to keep in mind all those recent changes that have occurred within and beyond the realm's peripheries. After the USSR imploded a quarter of a century ago, and all the former SSRs had gone the way of independent statehood, Russia was left by itself—still of unmatched proportions territorially, but one that had lost control of bordering regions in Europe and Southwest Asia, from the Baltic Sea to the Tian Shan Mountains lining China's westernmost boundary. Today, two decades later, we refer to the Russian/Central Asian realm, which consists of Russia, the three comparatively tiny countries of Transcaucasia, and, to the east of the Caspian Sea, the five Central Asian countries that were all once part of the Soviet Union.

As we know, Russia remains a gigantic country, one so diverse in terms of its human and physical geography that it could be regionally subdivided in a variety of ways and at different scales. We will keep things simple by delimiting what can be called six first-order regions, recognizing that each of them contains its own notable subdivisions. Four of these six regions to be discussed, which are mapped in Figure 5-9, lie entirely inside Russia, and the

FIGURE 5-9

other two are located in separate sectors of Russia's southern frontier zone that is almost transcontinental in east-west length:

1. The Russian Core
2. The Southeastern Frontier
3. Siberia
4. The Russian Far East
5. Transcaucasia
6. Central Asia

THE NEW RUSSIA

Russia's Political-Administrative Structure

When Russia's new government was installed after the disintegration of the USSR, it immediately faced an old Soviet problem in the form of an exceptional geographical challenge—the sheer size of the country, its vast distances, and the remoteness of many of its far-flung spatial components. Geographers refer to the principle of **distance decay 12** in order to explain how increasing distances between places tend to reduce interactions among them. Because Russia is the world's largest country, distance is a significant factor in the relationships between the capital and outlying areas. Furthermore, Moscow lies in the far west of this behemoth country, literally half a world away from Russia's Pacific coast. Still another problem is the astonishing size variation among the country's myriad administrative units, as can be seen in Figure 5-10. Whereas the (territorially) smallest divisions are concentrated in the Russian core area, the largest lie far to the east, where the Republic of Sakha is nearly 1000 times the size of the Republic of Ingushetiya. On the other hand, the populations of the enormous eastern divisions are minuscule compared to those of the smaller ones in the Russian Core. Such diversity spells administrative difficulty.

Figure 5-10 reflects this especially complex Russian administrative system, which in large part is a legacy of the Soviet Union. There are eight Federal Districts (newly created in 2000) that are subdivided into 21 Ethnic Republics and Autonomous Regions, and 35 Russian Regions. These administrative units function with varying degrees of power and autonomy from the federal government. The 21 Republics, established to recognize and accommodate the largest ethnic minorities, include a cluster in the far south, another lying to the east of Moscow, and a third that parallels the far-off Mongolian border.

In 1992, most of Russia's administrative units signed a document known as the Russian Federation Treaty, which committed them to cooperate within the new **federal system 13**. At first, a few holdouts refused to sign, including a Republic in the remote Caucasus periphery, Chechnya-Ingushetiya, where Muslim rebels had launched a campaign for independence (Fig. 5-10). Chechnya-Ingushetiya subsequently split into two separate Republics. Eventually, only Chechnya refused to sign the Russian Federation Treaty; Russian military intervention soon followed and triggered a prolonged, violent conflict, producing disastrous consequences for Chechnya's people and infrastructure (the capital, Groznyy, was completely destroyed).

In 2000, the newly elected administration of President Vladimir Putin moved to diminish the influence of the Regions by creating a new geographic framework that combined the 83 Regions, Republics, and other entities into a set of eight new Federal Districts—not to enhance their clout in Moscow, but to increase Moscow's authority over them. As Figure 5-10 shows, each of these new Federal Districts has a capital, elevating cities such as Rostov and Novosibirsk to a status secondary only to Moscow. In a related move supported by Russia's parliament, Putin decreed that Regional governors would be appointed rather than elected—thereby concentrating even more power in Moscow. Thus Russia's federal system began its steady shift toward a **unitary state system 14**—one dominated by a centralized government that exercises power equally throughout the country. This administrative reorganization not only reinforced the position of western Russia as the unrivaled political focus of the state: this heartland is also the demographic center of gravity (Fig. 5-4) and is by far Russia's most accessible as well as urbanized region (Fig. 5-11).

Russia's Shrinking Population

Some of these administrative problems are exacerbated by an ominous trend: the Russian population has been shrinking steadily since 1990. Certain parts of the country have witnessed a veritable **population implosion 15**. When the Soviet Union disintegrated in 1991, Russia's population was about 149 million. Today, that total has dropped by 4 percent and is down to 143.5 million—despite the in-migration of several million ethnic Russians from the old non-Russian SSRs.

Undoubtedly, the transition away from Soviet rule is one reason: uncertainty is known to cause families to have fewer children; moreover, abortion is widespread in Russia. Although the birth rate has stabilized at 13 per thousand, it is Russia's death rate that has been rising and now exceeds 13 per thousand. According to the latest data, this produces an annual net change of population that hovers around zero.

Russian males are the ones most affected. Male life expectancy dropped from 71 in 1991 to 65 in 2013 (over the same period, the rate for females dipped only slightly to 76—a life expectancy fully 17 percent higher). Males are more likely to be afflicted by alcoholism and related diseases, by AIDS (chronically underreported in Russia), by heavy smoking, and by suicide, accidents, and murder. On average, a Russian male is nine times more likely to die a violent or an accidental death than his European counterpart. Shockingly, fewer than half of today's male Russian teenagers will survive to age 60.

In 2012 and 2013, according to its own demographic statistics, Russia seemed to be bucking the trend: for the first time in 20 years the overall population expanded, albeit minimally so. It is too early to tell if this turnaround will last, and it may well prove

RUSSIA'S ADMINISTRATIVE STRUCTURE

Ethnic Republics and Autonomous Regions

Russian Regions

National capital is underlined

38.4 Population, in millions, 2010 census

● Federal District capital

— Federal District boundary

0 100 200 300 400 500 Miles
0 200 400 600 800 Kilometers

RUSSIAN ADMINISTRATIVE UNITS
(named after their capitals)

1. Astrakhan	10. Kurgan	19. Penza	28. Tambov
2. Belgorod	11. Kursk	20. Pskov	29. Tula
3. Bryansk	12. Lipetsk	21. Rostov	30. Tver
4. Chelyabinsk	13. Mariy-el	22. Ryazan	31. Ulyanovsk
5. Ivanovo	14. Moscow	23. St. Petersburg	32. Vladimir
6. Kaluga	15. Novosibirsk	24. Samara	33. Volgograd
7. Kemerovo	16. Omsk	25. Saratov	34. Voronezh
8. Kostroma	17. Orenburg	26. Smolensk	35. Yaroslavl
9. Krasnodar	18. Orel	27. Stavropol	

FIGURE 5-10

J. Nijman, P. O. Muller, and John Wiley & Sons, Inc.

181

RUSSIA/CENTRAL ASIA: CITIES AND SURFACE LINKS

•	Under 50,000	Railroad built
		before 1917
•	50,000–250,000	Railroad built
		1917–1991
●	250,000–1,000,000	Road
●	1,000,000–5,000,000	
⬤	Over 5,000,000	National capitals are underlined

0 100 200 300 400 500 600 800 1000 Kilometers
0 200 400 600 800 1000 Miles

J. Nijman, P. O. Muller, and John Wiley & Sons, Inc.

FIGURE 5-11

MAP ANALYSIS

FIGURE 5-12

Between the dismantling of the Soviet Union in 1991 and 2014, Russia's total population dropped from 149 million to less than 144 million. Because this population decline has varied markedly from region to region, its geographic impact needs to be more closely examined. Figure 5-12 shows the percentage of population loss from 1989 to 2010 in Russia's eight Federal Districts. Can you identify the spatial pattern of population shrinkage? This map also shows the size of the economy (GDP) in each District for 2010. Is it possible to determine which is the better predictor of population decline in a District, geographic remoteness or economic robustness? Access your WileyPLUS Learning Space course to interact with a dynamic version of this map and to engage with online map exercises and questions.

to be but a momentary reprieve. Whatever the outcome, the situation is grimmest at the regional level, where a few Federal Districts have been hit hard by very substantial population losses. Out-migration unquestionably plays a role because people are no longer forced to stay in place by a totalitarian regime. Villages in particular continue to empty out all across Russia, and at least 20,000 have turned into "ghost villages"—more than 15 percent of all such settlements in the country two decades ago. The Map Analysis box and especially Figure 5-12 provide further insights into the spatial dimension of Russia's ongoing population decline.

A VOLATILE ECONOMY

At the beginning of this century, it looked like Russia's economy had been turned completely around from what it was under its communist predecessor. Private property, start-up companies, trade, foreign investment, and functioning stock exchanges—all the things we regard as vital components of a free market economy—seemed to have found their way into post-Soviet Russia. The city of Moscow changed beyond recognition with a new high-rise skyline and unprecedented linkages to the global economy. After some major (and for many people painful) adjustments were made, Russia's GDP grew robustly as the price of shares on the Moscow stock exchange skyrocketed. In common parlance, this performance catapulted Russia into membership in the exclusive club of the so-called BRICs **16**, the world's four biggest emerging market economies (Brazil, Russia, India, China). Russia's *nouveaux riche* became known for their extravagant consumption on shopping trips to Paris, Milan, and New York; for sending their children to the highest-cachet Swiss and British boarding schools; and even for purchasing premier soccer-team franchises in Europe.

By 2015, however, it became apparent that Russia's advancement had been precarious at best. Too much of it was based primarily on exports of oil and gas; incoming foreign direct investment was anemic in comparison with the other emerging markets; the get-rich-quick schemes of a handful of oligarchs (highly successful businessmen—but almost never women) did not benefit the masses; and the gargantuan state apparatus left by the Soviets was responsible for widespread corruption as well as close connections to organized crime syndicates. The value of the "bribes market" was consistently estimated to be an alarming 20 percent of the national economy. Thus in 2014 when world oil prices came crashing down and stayed there, the nosediving Russian ruble ignited a severe economic downturn.

Even though statistics on the Russian economy since 2000 suggested abundant growth, ample evidence has also accumulated to indicate that most citizens were unable to better their standard of living. Frustrations with the economic situation have boiled over into the political arena. The (rigged) parliamentary elections of 2011 that handed victory to Putin's United Russia Party triggered massive demonstrations on Moscow's streets (see photo).

A portion of the huge crowd on the streets of downtown Moscow in February 2012, braving the –22°C (–4°F) cold to challenge the bid of Vladimir Putin to reclaim the presidency of the Russian Federation. The efforts of these several thousand protesters notwithstanding, Putin coasted to victory a month later proclaiming that he had received 63 percent of the vote.

Nonetheless, a few months later Putin himself easily won election to return as president until 2018—when his previously engineered constitutional change kicks in to allow him to run for reelection for yet another six-year term (his fourth). Prospects for true democracy in Russia, therefore, seem likely to be deferred until at least the mid-2020s.

THE RUSSIAN CORE

The Russian Core, broadly defined, extends from the western border of the Russian realm to the Ural Mountains in the east (Fig. 5-9). This is the Russia of Moscow and St. Petersburg, of the Volga River and its industrial cities. This is the most developed part of this enormous realm and the focus of its transportation network (Fig. 5-11).

Moscow is the megacity hub of a wider area encompassing some 50 million inhabitants (no less than one-fifth of the realm's entire population), most of them concentrated in cities that, during the communist period, specialized in assigned industries such as automaking in Nizhniy Novgorod (the "Soviet Detroit") and textile fabrication in Ivanovo. But today, Russia is less a diversified manufacturing country than a commodity producer and exporter (dominated by oil and natural gas). The tall steel-and-glass towers you see rising above the CBDs of the biggest cities mainly belong to energy companies such as Gazprom. And the cars being built are more likely to display such foreign brand names such Fiat or Volkswagen.

Moscow itself has grown rapidly in recent years and today has a population of just over 12 million, of which more than 2 million are undocumented migrants. It is the economic and business opportunities that continue to draw people in, but steadily worsening congestion and an increasingly brutal social environment mean that many dream of leaving the city as soon as they can afford to. The Moscow metro system (now the world's fifth largest) carries more than nine million passengers daily, and new route extensions and metro stations are built nearly every year. To handle all this overcrowding, the government has announced plans for a massive expansion of the capital to the southwest over the next two decades (Fig. 5-13).

St. Petersburg (formerly Leningrad) remains Russia's second city, with a population of 5.0 million. In the late seventeenth and early eighteenth century, Czar Peter the Great and his architects transformed this city at the head of the Baltic Sea's Gulf of Finland into the "Venice of the North." St. Petersburg's palaces, churches, waterfront façades, bridges, and monuments still give the city a decidedly European look unlike that of any other city in Russia. It was proclaimed the Russian capital in 1712, with Moscow serving as the distant second city. The Imperial Winter Palace and the adjoining Hermitage (one of the biggest art museums on Earth) at the heart of the city are among a host of surviving architectural treasures. Today, St. Petersburg is again important to Russia's trade with the rest of the world, and the city has become a major destination for many cruise ships. But it possesses none of Moscow's

© J. Nijman, P. O. Muller, H. J. de Blij, and John Wiley & Sons, Inc.

FIGURE 5-13

locational advantages, at least not with respect to the domestic market. With Moscow these days increasingly defined in terms of global finance and producer services, St. Petersburg's niche involves trade, culture, and tourism.

Central Industrial Region

At the heart of the Russian Core lies the Central Industrial Region (Fig. 5-14). The precise definition of this subregion varies because all regional definitions are subject to debate. Some geographers prefer to call this the Moscow Region, thereby emphasizing that for over 400 kilometers (250 mi) in all directions from the capital, everything is oriented toward this historic focus of the state. As Figure 5-11 underscores, Moscow has maintained its decisive **centrality** **17**: roads and railroads converge in all directions—from Ukraine in the south; from Minsk (Belarus) and the rest of Eastern Europe in the west; from St. Petersburg and the Baltic coast in the northwest; from Nizhniy Novgorod and the Urals in the east; from the cities and waterways of the Volga Basin in the southeast (a canal links Moscow to the Volga, the country's single most important navigable river); and even to the subarctic northern periphery that faces the Barents Sea, where the strategic naval port of Murmansk and the lumber-exporting outpost of Arkhangelsk lie.

Povolzhye: The Volga Region

Another major subregion within the Russian Core is the ***Povolzhye***, the Russian name for an area that extends along the middle and lower valley of the Volga River as well as its tributary, the Kama, above the city of Kazan. It would be appropriate to call this the Volga Region, for this greatest of Russia's rivers is its lifeline and most of the Povolzhye's cities lie on its banks (Fig. 5-14). The city of Volgograd is centrally located within this subregion, on the west bank of the Volga. It was named Stalingrad from 1925 to 1961, and that name will always be associated with the decisive battle here that became the turning point of World War II on Nazi Germany's Eastern Front.

The elongated Povolzhye (the *vol* refers to the Volga) forms the eastern flank of the Central Industrial Region, its cities and farms sustained by this Mississippi River of Russia. In the 1950s, the crucial Volga-Don Canal was opened, creating new linkages and opportunities. Next, the full dimensions of the oil- and gasfields in and near the Povolzhye became clear (see Fig. 5-5), and for some time these reserves were the largest in the entire Soviet Union. And finally, the Volga Region's northwestward linkages, via the Moscow and Mariinsk canals to the Baltic Sea, were improved and augmented by upgraded rail and road connections. Today, the industrial composition of the Volga's riverside cities continues to evolve—but if the Central Industrial Region is Russia's heart, then the Povolzhye is its aorta.

The Urals Region

The Ural Mountains form the eastern edge of the Russian Core. They are not particularly high; in the north they consist of a single range, but southward they broaden into a hilly zone. Nowhere are they an obstacle to east-west surface transportation. An enormous storehouse of metallic mineral resources located in and near the Urals has made this subregion a natural place for industrial development. Today, the Urals Region, well connected to the nearby Volga and Central Industrial Regions, extends from Serov in the north to Orsk in the south (Fig. 5-14).

The heart of central Moscow, with the triangular Kremlin (right of center) fronting on adjacent Red Square.

© Travelwide/Alamy

© J. Nijman, P. O. Muller, H. J. de Blij, and John Wiley & Sons, Inc.

RUSSIA'S MANUFACTURING REGIONS

▨ Manufacturing regions

— Major railroad

FIGURE 5-14

THE SOUTHEASTERN FRONTIER

From the southeastern flanks of the Ural Mountains to the headwaters of the Amur River, and from the latitude of Tyumen to the northern zone of neighboring Kazakhstan, lies Russia's vast Southeastern Frontier region, the product of a gigantic Soviet experiment in the eastward extension of the Russian Core along the axial Trans-Siberian Railroad (Fig. 5-9). As the maps of Russia's cities and transportation (Fig. 5-11) as well as population distribution (Fig. 5-4) suggest, this region is more densely peopled and more fully developed in the west than in the east. To the east of the Yenisey River, linear settlement prevails, marked by ribbons and clusters along the Trans-Siberian Railroad and the newer Baykal-Amur Mainline Railroad that parallels it to the north. Two subregions dominate the human geography: the Kuznetsk Basin in the west and the Lake Baykal area in the east.

The Kuznetsk Basin (*Kuzbas*)

Some 1500 kilometers (950 mi) east of the Urals lies another of Russia's primary regions of heavy manufacturing resulting from the communist period's national planning: the Kuznetsk Basin, or *Kuzbas* (Fig. 5-14). The leading city, located just outside the Kuzbas, is Novosibirsk, which lies at the intersection of the

Trans-Siberian Railroad and the Ob River as the symbol of Russian enterprise in the vast eastern interior. To the northeast lies Tomsk, one of the oldest Russian towns east of the Urals,

Wolfgang Kaehler/SuperStock

The river port of Krasnoyarsk, where the south-to-north-flowing Yenisey River meets the Trans-Siberian Railroad, Russia's east-west transport axis, near the center of the Southeastern Frontier region (see Fig. 5-9). Supplies brought by train from the west are shipped northward to settlements in the Siberian interior; raw materials from Siberia arrive by boat and barge for dispatch to Russia's factories and markets. Large storage facilities stand across the river.

founded in the seventeenth century and now caught up in the modern development of the Kuzbas. Southeast of Novosibirsk lies Novokuznetsk, a city that produces steel for this subregion's machine and metal-working plants as well as aluminum products from Urals bauxite.

The Lake Baykal Area (*Baykaliya*)

To the east of the Kuzbas, development becomes more insular, and distance becomes a stronger adversary. North of the central segment of the Mongolian border zone and eastward around Lake Baykal, larger and smaller settlements cluster along the two railroads to the Pacific coast (Fig. 5-14). West of the lake, these rail corridors lie in the headwater zone of the Yenisey River and its tributaries. A number of dams and hydroelectric projects serve the Angara Valley, particularly the city of Bratsk. Mining, lumbering, and some farming sustain life here, but isolation dominates it. The city of Irkutsk, near the southern end of Lake Baykal, is the principal service center for the enormous Siberian region to the north as well as for a lengthy east-west stretch of southeastern Russia.

Beyond Lake Baykal, the Southeastern Frontier really lives up to its name: this is southern Russia's most rugged, remote, and forbidding country. Settlements are rare, with many being mere camps. The Buryat Republic (see Fig. 5-10) is part of this zone; the territory bordering it to the east was taken from China by the czars and may yet become an issue in the future. Where the Russian-Chinese boundary turns southeastward along the Amur River, the region labeled the Southeastern Frontier peters out and Russia's Far East begins.

Oymyakon is known as the coldest town on Earth. It is located north of the city of Yakutsk in eastern Siberia. The average daily temperature during winter hovers around −46° Celsius (−51° Fahrenheit), and the record low recorded here in 1933 was −68°C (−90°F). A few hundred people live here, and there is even a school—which does not open if the temperature drops below −52°C (−62°F). Cellphones would not work here, even if there was satellite coverage, and visitors are advised not to wear glasses—they freeze to your face.

SIBERIA

Before we assess the potential of Russia's Pacific Rim, we should remember that the ribbons of settlement just discussed hug the southern perimeter of this gargantuan country, avoiding the huge Siberian region to the north (Fig. 5-9). Siberia extends from the Ural Mountains to the Kamchatka Peninsula—a vast, frigid, forbidding land. Larger than the conterminous United States but inhabited by fewer than 20 million people, Siberia quintessentially symbolizes the Russian environmental plight: huge distances, cold temperatures aggravated by fierce Arctic winds, difficult terrain, poor soils, and limited options for survival.

But Siberia does possess valuable resources. From the days of the first Russian explorers, Siberia's riches have beckoned. Gold, diamonds, and many other precious minerals were found. Later, metallic ores including iron and bauxite were discovered. Still more recently, the Siberian interior proved to contain massive deposits of oil and natural gas (Fig. 5-5) that now contribute significantly to Russia's energy supply and exports.

As the physiographic map (Fig. 5-2) reveals, major rivers—the Ob, Yenisey, and Lena—flow gently northward across Siberia and the Arctic Lowland into the Arctic Ocean. Hydroelectric power development in the basins of these rivers has generated electricity used to extract and refine local ores, and to run the lumber mills that have been set up to exploit the vast Siberian forests.

Understandably, the population geography of Siberia is highly fragmented. Most of the region is devoid of inhabitants, with hundreds of square kilometers of empty territory separating the relatively few islands of settlement (Fig. 5-4). Nonetheless, some small clusters of ethnic Russians have developed. Accordingly, the Yenisey River can be traced on the map of the realm's peoples (Fig. 5-7) as a thin ribbon of Russian settlements north of Krasnoyarsk; farther east, the upper Lena Valley can similarly be recognized.

THE RUSSIAN FAR EAST

Officially, the huge Far Eastern Federal District incorporates six Russian administrative regions and four residual ethnic republics (Fig. 5-10). But many Russian citizens have a different view of what constitutes the geographic region they call the Far East. Justifiably, they see most of the northern zone of the official Far East as a continuation of Siberia. To them, the real Far East is constituted by the territory beyond the Southeastern Frontier to the Pacific coast, the island of Sakhalin, the Kamchatka Peninsula, and a narrow strip of land along the northern shore of the Sea of Okhotsk (Figs. 5-9 and 5-15).

In Soviet times, citizens who agreed to move to this most distant region to work in factories or for the administration were rewarded with special privileges to compensate for their hardship. Make no mistake: even though the Pacific Ocean's proximity tends

© H. J. de Blij

© H. J. de Blij

"The Russian Far East is losing people at the fastest rate among all Russian Federal Districts, and my 2009 field trip to Petropavlovsk-Kamchatsky on the Kamchatka Peninsula only reinforced what I had learned in Vladivostok some years earlier. During Soviet times, Russians willing to move to the Far East were given special privileges (an older man told me that 'being as far away from Moscow as possible was reward enough'), but today locals here feel abandoned, even repressed. In 1989, Petropavlovsk had about 270,000 residents; today that total is estimated to be fully 100,000 fewer. It is not just the bleak environment—the ubiquitous ash and soot from dozens of active volcanoes on Kamchatka, covering recent snowfalls and creating a pervasive black mud—it is also a loss of purpose that people feel here. There was a time when the Soviet windows on the Pacific were valuable and worthy; now the people scramble to make a living, and when they find a way, for example, by selling used cars bought cheaply in Japan and brought to the city by boat, Moscow sends law enforcement to halt the illegal trade. Petropavlovsk has no surface links to the rest of the world, and this springtime view of the city at the foot of the volcano Koryakskaya (left) and a street scene near the World War II museum (right) suggest why the Far East is depopulating."

to temper natural environments here, life anywhere in the Russian Far East is tough. Moreover, Moscow's favors were withdrawn early in the postcommunist era, so under the new democracy hundreds of thousands of inhabitants voted with their feet and simply departed. Petropavlovsk, the largest urban center on the Kamchatka

© SmityukYuri/ITAR-TASS Photo /Corbis Images

A 2014 naval exercise of the Russian Pacific Fleet near its home base of Vladivostok, the port city that serves as the eastern terminus of the Trans-Siberian Railroad. Russia's growing visibility and military strength in the Far East over the past few years suggests a widening geostrategic posture in the East Asian arena—an impression reinforced in 2015 by the rapid warming of relations with China.

Peninsula, exemplifies the devastating consequences of this exodus (see *Field Note*).

The Far East, too, has significant reserves of what underpins today's Russian economy—oil and natural gas, most importantly, the deposits located on and around the island of Sakhalin. And, as Figure 5-15 also reminds us, the southern sector of this region lies between additional reserves and potential markets: China next door and Japan across a narrow sea. We are not used to thinking about Russia as an Asia-Pacific power, but it certainly qualifies. The distance to Moscow notwithstanding, the Far East provides Russia with highly valuable geopolitical as well as economic assets. In 2012, Russia hosted the ***APEC*** (Asia-Pacific Economic Cooperation) summit in Vladivostok, spotlighting its strengthening military presence in this part of the world. By choosing to hold the summit at this largest Russian port on the Pacific—which is also the home base of the Russian Pacific Fleet—the newly-restored Putin regime signaled the high priority it placed on safeguarding its strategic interests on these easternmost coasts and adjacent waters (see photo).

TRANSCAUCASIA

Now let us focus on the smallest but most complicated region of all—the area where Russia's leaders, communists as well as czars, pushed the expanding state's influence across and beyond the

FIGURE 5-15

Caucasus Mountains. On regional maps (Figs. 5-9 and 5-16), it looks as though the Russian Core extends to the international boundary (which coincides, more or less, with the crestline of this towering range). However, things are not that simple. When you travel southward from Volgograd or Rostov, the landscape—physical as well as cultural—does not change; the Russian Plain's gently rolling countryside with its villages and farms is very much like that south of Moscow. But taking the train south from Astrakhan in the Volga River Delta, you will soon see the Caucasus wall looming in the distance, but you may not realize that you have entered a different world until you get off near its base at Groznyy, the capital of Chechnya, one of Russia's 21 ethnic Republics

(Fig. 5-16). Just observing the mix of peoples makes you aware that you have left the heartland and reached Russia's periphery.

Figure 5-16 is important enough to spend some serious time studying it. Here the communist leaders who succeeded the czars confronted multiple challenges. First, inside the Russian state itself, there were many minority peoples (such as the Chechens of Chechnya) who had fought the czars' armies and who now faced new rulers. Second, there were peoples who had sided with the czars and who wanted to cooperate with the Soviet regime, such as the Ossetians. To recognize the aspirations of these peoples, some (but not all) of whom were Muslim, the communist administration designated ethnic 'republics' for them. On our map, these are

SOUTHERN RUSSIA AND TRANSCAUCASIA

POPULATION

- Under 50,000
- 50,000–250,000
- 250,000–1,000,000
- 1,000,000–5,000,000
- Over 5,000,000

National capitals are underlined

———	Railroad
———	Road
··········	Canal
•─•─•─•	Oil pipeline
○─○─○	Proposed oil pipeline

0   50   100   150   200   250   300   350 Kilometers
0   50   100   150   200 Miles

KAZAKHSTAN

Volgograd

Tsimlyansk Res.

Don River

UKR.
Mariupol
Rostov

Sea of Azov

Tikhoretsk

KALMYKIYA

Elista

Astrakhan

Caspian Sea

RUSSIA

Kropotkin

Novorossiysk
Krasnodar
Armavir
Stavropol
Maykop
ADYGEYA
Budennovsk
Cherkessk
Mineral'nyye Vody
KARACHAYEVO-CHERKESSIYA
Pyatigorsk
Sochi
KABARDINO-BALKARIYA
Nalchik
Groznyy
Makhachkala
ABKHAZIA
Vladikavkaz
INGUSHETIYA
CHECHNYA
Sokhumi
NORTH OSSETIA
Nazran
DAGESTAN

Black Sea

CAUCASUS MOUNTAINS

Samtredia
Kutaisi
SOUTH OSSETIA
Tskhinval
Pankisi Gorge
Supsa
Gori
Tbilisi
Kura R.
Batumi
AJARIA
GEORGIA
Rustavi
Borcka
Lake Mingacevir
TURKEY
AZERBAIJAN
Trabzon
Ceyhan
Sumgait
Gyumri
Gänjä
Baki (Baku)
ARMENIA
NAGORNO-KARABAKH
Yerevan
Xankändi (Stepanakert)
L. Sevan
NAXÇIVAN
AZER.
Armenian controlled
Agri
AZERBAIJAN PROVINCE
Van
Marand
Lake Van
Ardabil
Tabriz
IRAN
Lake Urmia

CHECHNYA
Kargalinskaja
Russian North (Plains)
Iscerskaja
Selkovskaja
Terek R.
Cervlënnaja
INGUSHETIYA
Groznyy
Gudermes
NORTH OSSETIA
Argun
Sunzha R.
Shali
Urban-Industrial Middle Zone
CAUCASUS
Vedeno
Chechen South (Mountains)
Shatoy
Argun R.
DAGESTAN
MTS.
GEORGIA

0   20   40 Kilometers
0   10   20   30 Miles

Longitude East of Greenwich

FIGURE 5-16

colored pink to distinguish them from the rest of Russia, which is shown in tan.

Russia's Internal Periphery

From Buddhist-infused Kalmykiya on the perimeter of the Caspian Sea westward to mainly Orthodox Christian Adygeya near the Black Sea, the Internal Periphery of southern Russia is defined by eight ethnic Republics, each with its own identity (Fig. 5-16).

Physiographically as well as culturally divided Chechnya, as the inset map suggests, lies in a transition zone between the plains and mountains, with the mountains serving as a refuge for those opposed to Moscow's rule. When the post-Soviet Russian government asked all of the country's Regions and Republics to sign its proposed Russian Federation Treaty in 1992, Chechnya's leaders refused, believing that they now had the opportunity to terminate Russian control. Its western neighbor, Ingushetiya, got caught up in the anti-Russian activism, and the ensuing drawn-out conflict also took a terrible toll in adjacent North Ossetia where, in 2004, more than 350 students, teachers, and parents died in a terrorist attack on a school. Still farther west, things remained calmer, and no problems at all arose in generally pro-Moscow Adygeya.

This tier of ethnic Republics along Russia's southernmost border displays all the properties of a disadvantaged periphery: not only do they share a history of subjugation by the powerful Russian Core to the north, but they also lag far behind the rest of the Federation by almost any measure of social progress. In terms of health, education, income, opportunity, and other such indicators, the peoples of the Internal Periphery, whether they are pro-Russian or not, have a long way to go to achieve current Russian living standards. And, as noted above for Chechnya, there is much simmering resentment of Russian rule, past and present.

In Chechnya's case, such anger has its most recent roots in the Second World War, when Soviet leaders accused the Chechens of sympathizing, and even collaborating, with the Nazi invaders. Soviet dictator Josef Stalin then ordered the entire Chechen population loaded onto trains and exiled to the desert of Kazakhstan across the Caspian Sea. Tens of thousands died along the way or after arriving there, and even though Stalin's successor, Nikita Khrushchev, in 1957 allowed the survivors to return to their homeland, the Chechens never forgave their tormentors. Then in 1992, just after the Soviet Union disbanded, they saw their opportunity: they refused to sign the Russian Federation Treaty that would make their republic an integral component of the new Russian state, and they mounted a campaign for independence that led to a bitter and costly war that destabilized a number of neighboring republics as well. Chechen terrorists eventually carried this campaign to the very heart of Moscow, killing hundreds of civilians there and forever altering Russia's political landscape. When presidential candidate Vladimir Putin promised victory, the Russians elected him overwhelmingly in 2000. Subsequent military intervention did overpower the Chechens, and the war came to an end in 2007;

Chechnya now has a pro-Russian regime that receives substantial economic support from Moscow in exchange for fierce military loyalty. Nonetheless, the Internal Periphery as a whole remains restive and barely stable.

Russia's External Periphery

Russian (and later Soviet) manipulation of the ethnic map did not terminate at the international boundary we see in Figure 5-16, but it did take a different turn. Beyond the Russian border, the Soviet Empire extended its control over the three adjoining Transcaucasian entities where the Russians had for centuries exerted influence and waged war. These three political units were Georgia (mapped in yellow in Fig. 5-16), facing the Black Sea; Azerbaijan (green), bordering on the Caspian Sea; and Armenia (lavender), landlocked in the middle. During Soviet times, Georgia was a loyal partner (Stalin was a Georgian by birth). Christian Armenia appreciated membership in the Soviet Union because of the security it provided next door to hostile Islamic Turkey. And Islamic Azerbaijan, whose ethnic majority comprises a people known as the *Azeris* (who have close kinship with Iranian Azeris directly across their southern boundary with Iran), was a valued member of the USSR because of its rich oil reserves.

But note in Figure 5-16 that Azerbaijan has a large exclave on the Iranian border (Naxcivan), separated from the main part of the country by Armenian territory—and that the Armenians hold another exclave named Nagorno-Karabakh inside what would appear to be Azerbaijan's territory (which is home to more than 130,000 Christian Armenian citizens who were placed under Islamic Azerbaijan's jurisdiction by Soviet mandate).

Soviet rule kept the lid on potential conflict in the three countries that now constitute Transcaucasia. After the demise of the USSR, Russia sought to retain its influence and maintain stability in this vital sector of the Near Abroad. It wanted to keep Azerbaijan's oil flowing to and through Russia (see the pipeline connecting Baki [Baku] to the Black Sea terminal at Novorossiysk in Fig. 5-16); Russia also wanted to keep Georgia in the fold, and it looked for ways to mitigate the long-running clash between Azerbaijan and Armenia over Nagorno-Karabakh.

But things have not gone well for Moscow in post-Soviet Transcaucasia. Propelled by Western investments, Azerbaijan in 2006 began exporting oil via a new pipeline across Georgia and Turkey to the latter's Mediterranean port of Ceyhan, providing it with a more favorable economic as well as political alternative to Russian transit. Meanwhile, Armenia and Azerbaijan have not been able to resolve their sometimes violent quarrel over Nagorno-Karabakh. And things really fell apart in Georgia, where the government pursued pro-European policies and allegedly mistreated its pro-Russian Ossetian minority, and where Russia openly supported devolutionary initiatives in the northwestern corner of the country, a small ethnic entity known as Abkhazia (Fig. 5-16; see *Field Note*).

Soon after the USSR disbanded in 1991, the disagreement between Russia and Georgia escalated into serious discord. The

Russians closed their border with Georgia, shutting Georgian farmers out of their crucial market. Then the Russians started issuing Russian passports to Abkhazian citizens of Georgia. Next, the Russians signaled even stronger support for their Ossetian allies, and in a still-disputed sequence of events during 2008 Russian forces entered South Ossetia and engaged the Georgian military in a brief but costly conflict. After a cease-fire was brokered by international mediators, Moscow took the extraordinary step of recognizing the two pro-Russian autonomous regions of Georgia—Abkhazia and South Ossetia—as independent states! Russian military intervention here sent shudders across Europe and much of the rest of the world—and turned out to be the prelude to similar Russian aggression in 2014, which saw the forcible annexation of Crimea followed by the outbreak of hostilities in nearby eastern Ukraine.

Unlike Azerbaijan, predominantly agricultural Georgia has no energy card to play—although it did agree to allow the new pipeline from Azerbaijan to Turkey to traverse its territory, thereby angering the Russians even further. In addition, the Georgian government has courted the European Union and even NATO, seeking the security Georgians have lacked for centuries. The Azeris of Azerbaijan, whose prevailing religion is **Shi'ite Islam** 18 like that of neighboring Iran, are caught in the global web (and political geography) of energy demand and supply, but most ordinary people have benefited little from their country's petroleum wealth. The landlocked 3 million Armenians live in a world of unsettled scores: with the Turks over their decimation by Turkish Ottoman Muslims during World War One, with the Azeris over their exclave

of Nagorno-Karabakh, and with the international community over the proper recognition of their plight. Transcaucasia may be the smallest region of the Russian/Central Asian realm, but it is densely populated by 17 million people who inhabit one of the most volatile peripheries in this part of the world.

CENTRAL ASIA

The sixth and final region of the realm is Central Asia, which comprises five states that simultaneously became independent upon the expiration of the USSR in 1991: Kazakhstan, Uzbekistan, Turkmenistan, Kyrgyzstan, and Tajikistan. Regarded until quite recently as both isolated and remote, Central Asia's relative location in the very heart of Eurasia is now being translated into an increasingly strategic regional position. Geopolitically, this region lies directly between Russia and China as well as bordering Transcaucasia, Iran, and parts of South Asia. In terms of economic geography, Central Asia contains huge deposits of oil and natural gas, and key parts of it are being developed as a bridge between China and western Eurasia in the form of major new overland transport corridors that are being likened to a twenty-first-century version of the historic Silk Road.

In Central Asia today, Turkic cultural influences blend with Islam and Soviet legacies. During its hegemony over this region, the USSR tried to suppress Islam and install secular regimes. Nevertheless, Islam reasserted itself strongly after the Soviet withdrawal and has now become one of the defining qualities of

© John O'Loughlin

@ FROM THE FIELD NOTES...

"Upon first entering Sukhum, the capital of the 'de facto' state of Abkhazia on the Black Sea coast (see Fig. 5-16), I visited the former government house—now an eerie and unintended war memorial. It was the scene of an intense battle in 1993 when the Georgian government unsuccessfully tried to suppress Abkhazian separatists. The building bears the marks of bullets, bombs, and fires, and is deliberately maintained in this condition as a reminder of the war. The Abkhazians repelled the Georgians and insist on their own spelling of the city's name (Georgians use Sokhumi); also note that it is the Abkhazian "national" flag flying on the top of the burnt-out building. The legacies of war in the Caucasus are dramatically evident in the landscape, with the omnipresence of ruined buildings, war memorials, empty houses of displaced persons, refugee camps, military checkpoints, minefields, abandoned farms, ruined roads, and destroyed statues such as the one of Lenin that used to stand on the now empty base in front of the building. Like many other structures (over half by some estimates) in Sukhum, this one remains as it was when the fighting ended, but its considerable size and high-rise prominence in a city of small, low buildings make it visible from everywhere in town and a daily reminder of the war's brutality."

Central Asia. From Almaty to Samarqand (the region's most densely populated area), mosques are being restored and reactivated, and Islamic dress is again part of the cultural landscape. National leaders make high-profile visits to Mecca, and most are sworn into office with a hand on the holy Quran.

Central Asia is marked by enormous ethnic complexity and diversity. Figure 5-7, however, is but a generalization of the highly intricate mosaic of peoples and cultures: every cultural territory on this map contains clusters of minorities that cannot be captured at this scale. Thus people of different faiths, languages, and lifestyles rub shoulders everywhere, and this often produces friction that can escalate into ethnic conflict. And it should not come as a surprise, in this post-Soviet geographic context, that democracy is hard to come by: each of these five countries is either autocratic or has flawed representative governments (Tajikistan, in fact, is already being labeled by some as a failed state).

The Five States of Central Asia

Kazakhstan is the region's giant but lies between two even greater territorial titans: Russia and China (Fig. 5-17). During the Soviet era, the northern tier of Kazakhstan was heavily Russified, and in the 2010s ethnic Russians still account for 22 percent of the total national population of 17 million. To stay on good terms with post-Soviet Russia, the Kazakhs carefully maintain the railway and highway links that connect their north to Russia—and they even

FIGURE 5-17

Astana, Kazakhstan's capital since 1998, exhibits spectacular architecture to symbolize the Kazakh nation, its ongoing economic development and modernization, and its aspiration to become the key urban center for all of Central Asia. This is the Bayterek Tower, which depicts a mythical giant poplar tree containing a golden egg in its branches.

moved their capital to Astana in the heart of Russified Kazakhstan (see photo). Today, this country is in the forefront of Central Asia's energy boom: it already ranks among the world's leading producers of uranium, and in 2014 it ranked 15th among the oil-producing countries. Figure 5-17 further reveals Kazakhstan's intermediate position between the huge oil reserves of the northern Caspian Basin and China. New oil and gas pipelines operating across Kazakhstan are serving to reduce China's dependence on oil imports carried by tankers along distant international sea lanes. Kazakhstan's economic growth has averaged 8 percent annually over the past decade, a development thrust that is certain to continue as the country becomes even more of an oil power as well as a leading beneficiary of China's New Silk Road project (described in the next section).

Uzbekistan occupies the heart of the region and borders every state within it. Uzbeks not only make up more than 80 percent of the population of 30.2 million, but also form substantial minorities in a number of neighboring countries. The capital, Tashkent, lies in the eastern core area of the country, where most of the people live in towns and farm villages, and where the crowded Fergana Valley is the focus. In the desert northwest lies what is left of the still-shrinking Aral Sea, whose feeder rivers were diverted into cotton fields and croplands during the Soviet occupation; heavy use of pesticides contaminated the groundwater, and countless thousands of local people continue

to suffer from cancers and other catastrophic medical problems as a result.

Turkmenistan, the autocratic desert republic that extends all the way from the Caspian Sea to the borders of Afghanistan, has a population of 5.2 million, of which nearly 90 percent are ethnic Turkmen. During Soviet times, communist planners initiated work on a massive project—the Garagum (Kara Kum) Canal designed to transfer water from Turkestan's eastern mountains into the heart of the desert. Today the canal is 1100 kilometers (700 mi) long, and it has enabled the cultivation of more than one million hectares (ca. 3 million acres) of cotton, vegetables, and fruits. The plan is to extend the canal all the way to the Caspian Sea. Meanwhile, Turkmenistan has greatly expanded its oil and gas extraction from Caspian coastal and offshore reserves, most of which is transported to China via a new pipeline that runs through Kazakhstan to northwestern China.

Kyrgyzstan's topography and political geography are reminiscent of the Caucasus. The agricultural economy is generally weak, consisting of pastoralism in the highlands and farming in the valleys. Mapped in yellow in Figure 5-17, Kyrgyzstan lies intertwined with Uzbekistan and Tajikistan to the point of having exclaves and enclaves along their common borders. Ethnic Kyrgyz constitute about 70 percent of the population of 5.7 million. More than three-fourths of the people profess allegiance to Islam, and *Wahhabism*, an orthodox form of **Sunni Islam 19**, has achieved a

The first train (on February 25, 2015) pulls out of the new portside container terminal in Lianyungang on China's Yellow Sea coast to begin its 3750-kilometer (2300-mile) westbound run to Almaty, Kazakhstan's capital. The freight-rail corridor it will traverse is the pioneering component of the New Silk Road, the Chinese high-profile Eurasian transportation initiative that may fundamentally change relations between China, Russia, and Central Asia.

strong foothold here as well; the town of Osh is often referred to as the Central Asian headquarters of this movement. In 2010, ethnic rioting erupted in Osh between the Kyrgyz majority and the Uzbek minority, killing hundreds and putting the country in danger of breaking apart. Tens of thousands of Uzbeks, who make up roughly 14 percent of the total population but form a prosperous merchant class, fled their homes.

Tajikistan's mountainous scenery is even more spectacular than Kyrgyzstan's, and here too topography is a barrier to the integration of a multicultural society. The Tajiks, who constitute more than 80 percent of the population of 8.2 million, are ethnically Persian (Iranian), not Turkic, and speak an Indo-European language related to Persian. Most Tajiks, despite these cultural affinities, are Sunni Muslims, not Shi'ites. Territorially minute though it is, regionalism to the point of state failure plagues Tajikistan: the government in Dushanbe is repeatedly at odds with the barely connected northern part of the country (Fig. 5-17), a boiling cauldron not only of Islamic revivalism but also of anti-Tajik, Uzbek activism.

A New Silk Road?

Central Asia is a still very much a region in evolution—redefining its relations with Russia, searching for new ways to tame its ever-turbulent ethnic mosaic, and working to lure the interest of investors worldwide in the quest to more fully develop its bountiful resource base. A noted above, once-dormant Central Asia has now become a key crossroads linking the different realms and powerful countries that surround it. Of the utmost importance today is China's growing challenge to Russia's longstanding regional influence here. The centerpiece of that effort is an enormous project the Chinese announced in 2013: the **New Silk Road 20**. The historic Silk Road, of course, was the ancient linear trade network that connected Mediterranean Europe to Asia, stretching eastward from Constantinople (modern Istanbul) to Xian and beyond in eastern China—and named after the prized commodity that had been perfected by the Chinese: silk.

The geographic idea behind the New Silk Road is the Chinese initiative to construct an ultramodern overland routeway extending from East Asia to Europe via Central Asia—a new 'Eurasian land bridge.' The Chinese claim that it will benefit everyone, particularly the Central Asian countries hungry for investment and cutting-edge industrial technology. China's funding of the project will mainly go into building infrastructure—railroads, highways, and freight terminals—but Chinese companies would also become a driving force in the industrialization of Central Asia. Because of its relative location, Kazakhstan is of prime importance, and in 2015 the Kazakh and Chinese governments completed the first set of deals said to be worth U.S. $23 billion.

China's motives are fourfold. The first entails the insatiable Chinese appetite for raw materials and energy supplies, and Central Asia is richly endowed with both. Second, as industrial production in China shifts westward (discussed in Chapter 9), Europe-bound goods become costlier to ship if they must be first sent east to the Pacific coast and then transshipped via the circuitous oceanic route around the entire southern edge of Eurasia. Besides shipping costs, significant time would be saved as well: an existing railroad that moves freight from China to western Kazakhstan and connects to a rail line that crosses Russia and Belarus on its way to Germany is said to require only 14 days versus the 60 days needed to sail the much slower and indirect oceanic route. The third reason for the New Silk Road involves a geostrategic imperative: the need to secure open overland access in case of maritime blockades (especially in the hotly disputed South China Sea [discussed in Chapter 10]). And fourth, the New Silk Road will propel the economic development of China's far western periphery, particularly the restive Autonomous Region of Xinjiang, through which all routes to Kazakhstan are channeled.

As new freight-rail lines open (see photo) along with oil and gas pipelines that connect Central Asia to China, trade between the two—at the expense of Russia—is steadily increasing. If China and Russia are seen to be competing for influence in Kazakhstan, then China has taken a major step forward. Figure 5-18 is a map redrawn from a Chinese government publication that appeared in 2014, and reveals a most interesting perspective on this contest: the proposed main routeway of the New Silk Road veers southward from Kazakhstan to Uzbekistan before turning westward toward Europe—thereby avoiding Russian territory altogether.

Russia's latest response is a half-hearted attempt to strengthen the so-called **Eurasian Customs Union** **21**, a supranational organization it created in 2010 with Belarus and Kazakhstan that was subsequently joined by Armenia and Kyrgyzstan in 2015. Even so, Russia may already have lost out to China's initiative in Central Asia—perhaps providing another reason why Russia continues to intensify its presence in the Far East as it keeps an ever watchful eye on China all around their common, 4200-kilometer (2600-mile)-long border.

A REALM IN FLUX

A quarter-century ago, one of the most remarkable politico-geographical events in history took place: the sudden, peaceful expiration and dismemberment of the Soviet Union, global superpower and anchor of the world's next-to-last colonial empire. What remains today is the centerpiece of the Soviet Empire, recast as the Russian Federation—still the world's biggest territorial state, but fringed with a string of countries along many of its edges that were once part of the old USSR. To the south, most of these states are still considered to be part of this geographic realm, while others, to the west, have gravitated into Europe's orbit. Russia's relations with this 'Near Abroad' are not stable. The 1991 breakup of the USSR may seem like a long time ago, yet the current situation still bears many imprints of the Soviet era and ensures that this realm will remain in flux. Soviet times were repressive and economically stagnant, whereas the

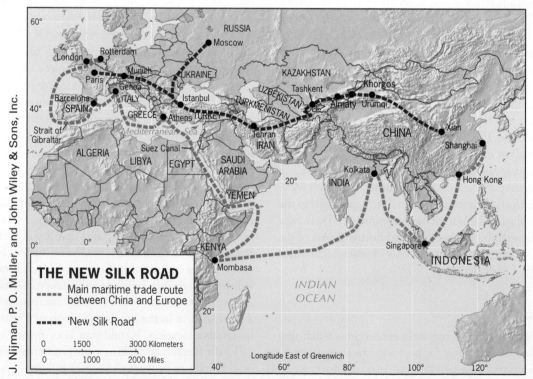

THE NEW SILK ROAD

- - - - - Main maritime trade route between China and Europe

- - - - - 'New Silk Road'

0   1500   3000 Kilometers
0   1000   2000 Miles

J. Nijman, P. O. Muller, and John Wiley & Sons, Inc.

FIGURE 5-18

new Russia has offered greater freedom and business opportunities. Geopolitically, however, the Putin years (he has served as president since 2000, interrupted by a four-year term as prime minister) bear all the hallmarks of **revanchism** 22—retaliatory policies aimed at recovering lost territory. Unmistakably, the eastward advance of the European Union as well as NATO, along with the anti-Russian disposition of former SSRs such as Ukraine and the Baltic states, have fueled reactionary Russian policies.

It is hard to separate Russia's foreign relations from its internal political developments. Inside Russia, the promise of democracy has mostly faded away, whereas the benefits of the post-Soviet era have almost exclusively been conferred on the favored few as income inequality steadily widened. Over the past few years, the masses have been demonstrating their discontent, most notably in the 2012 antiregime street protests in Moscow and other cities. They are disgusted with the pervasive corruption as well as a postcommunist economy that is not really a free market (some still derisively refer to it as mafia-capitalism) and fails to provide any of the securities of the old Soviet system. Although many long for a better future, others harbor a persistent nostalgia for the past. Not surprisingly, half the Russians surveyed in a recent poll agreed that it was "a great misfortune that the Soviet Union no longer exists."

POINTS TO PONDER

- Russia has land borders with 14 countries and three other geographic realms. Compare that to the conterminous United States, which borders only two countries and only one other realm.

- High-speed rail travel, so urgently needed in this far-flung realm, has been extremely slow to develop. The first such route opened in 2010, linking the capital, Moscow, with the leading port and second city, St. Petersburg; but since then, not a kilometer of new track has been laid.

- One out of seven Russian villages has been abandoned, with many turning into ghost towns.

- According to a poll marking the twentieth anniversary of the Soviet Union's demise, 70 percent of Russians agreed that the changes since 1991 had not benefited ordinary people.

- Landlocked, once-remote Central Asia could soon become a first-order global geopolitical hotspot involving Russia, China, India, and a host of regional powers.

© Amr Sayed/ZUMA Press / Corbis Images

Although the Arab Spring of 2011 did not produce the democratic reforms that so many had hoped for, popular protests continue throughout most of the realm—with women playing a prominent role in countless demonstrations. On March 1, 2015, these students at the Egyptian capital's Cairo University raised their banners aloft that pictured a detained fellow protester and demanded her release. Since the 2011 overthrow of long-entrenched, autocratic President Hosni Mubarak, two other presidents have come and gone. But the military retains tight control, and the prospects for democracy seem as far-off as ever.

6

THE NORTH AFRICAN/ SOUTHWEST ASIAN REALM

IN THIS CHAPTER

- Escalating terrorism unleashed by ISIS
- Egypt embarks on megaprojects
- The steady rise of the Gulf States
- Libya's disintegration
- The deepening of Syria's tragedy
- An Iranian Spring?

CONCEPTS, IDEAS, AND TERMS

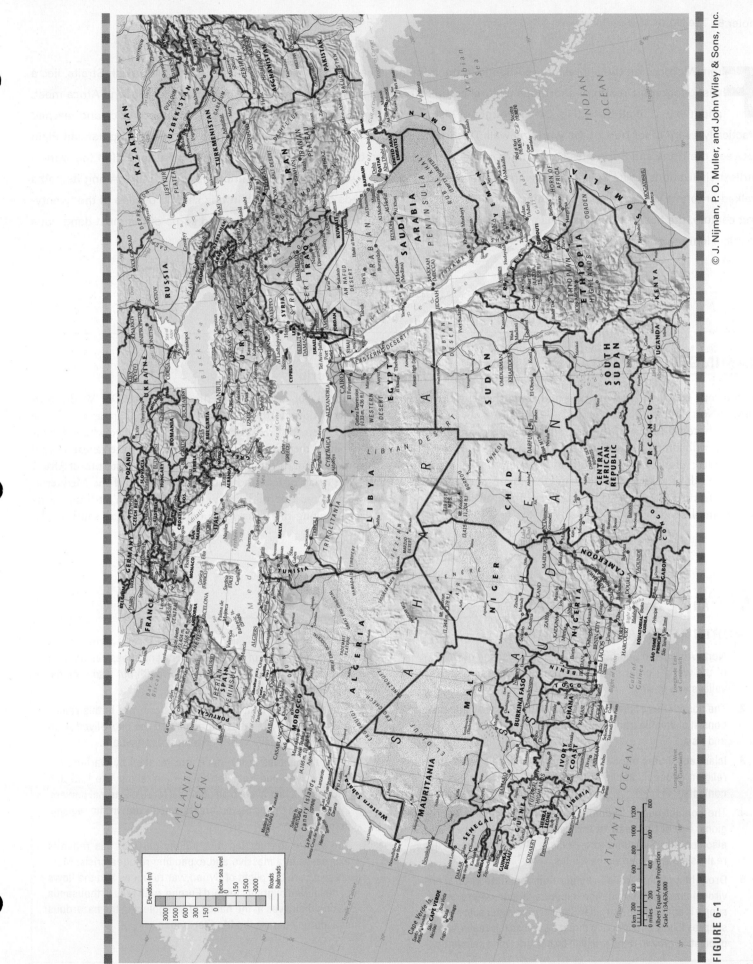

FIGURE 6-1

From Morocco on the shores of the Atlantic to the Iranian Plateau, and from the Horn of Africa to the Turkish Straits, lies a vast geographic realm of enormous cultural complexity. It stands at the crossroads where Europe, Asia, and Africa meet, and it is part of all three (Fig. 6-1). Throughout history, its influences have radiated across Eurasia and Africa and reached practically every other part of the world as well. This is one of humankind's primary source areas. On the Mesopotamian Plain between the Tigris and Euphrates rivers (in modern-day Iraq) and on the banks of the Egyptian Nile arose several of the world's earliest civilizations. In its soils, plants were domesticated that are now grown from the Americas to Australia. Along its paths walked prophets whose religious teachings are still followed by hundreds of millions. And in this second decade of the twenty-first century, the heart of this realm has been engulfed by political turbulence, religious strife, and some of the most dangerous conflicts in the world.

DEFINING THE REALM

NAMING THIS PIVOTAL REALM

As you may conclude from its long and somewhat cumbersome name—North Africa/Southwest Asia—this is a sprawling, geographically complex realm. In our era of high-speed communication you will sometimes see it referred to as *NASWA*, the first letters of its regional deployment. And it is tempting to refer to it in other kinds of geographic shorthand, based on some of its outstanding features. But such generalizations can be misleading, and some examples follow.

A 'Dry World'?

This realm is, for instance, sometimes called the Dry World, containing as it does northern Africa's immense Sahara as well as the Arabian Desert. But most of the realm's people live where there is water—along the Nile River, along the hilly Mediterranean coastal strip or *tell* (meaning mound in Arabic) of northwesternmost Africa, along the Asian eastern and northeastern shores of the Mediterranean Sea, in most of the Tigris-Euphrates Basin, in far-flung desert oases, and along the lower mountain slopes of Iran south of the

▮▮▮ MAJOR GEOGRAPHIC QUALITIES

NORTH AFRICA/SOUTHWEST ASIA

1. North Africa and Southwest Asia were the scene of several of the world's great ancient civilizations, based in its river valleys and basins.

2. The boundaries of the North African/Southwest Asian realm consist of volatile transition zones in several places in Africa and Asia.

3. Islam dominates the cultural geography of this realm, and religion is deeply implicated in political developments and conflicts.

4. The conflict-ridden Middle East, one of six constituent regions, lies at the heart of this realm. It includes Israel, but also contains Syria and Iraq—where ISIS has triggered the realm's most dangerous hostilities.

5. Drought and unreliable precipitation dominate natural environments, and population clusters occur where water supply is adequate to marginal. Conflict over water sources and supplies looms as a growing threat all across this realm, where population growth rates are high by current world standards.

6. Certain countries of the realm have enormous reserves of oil and natural gas, creating great wealth but also regional disparities.

7. Democracy and individual freedoms are rare in this realm, and the high hopes that accompanied the short-lived Arab Spring of 2011 have vanished almost everywhere.

8. Several countries in the realm are ravaged by warfare, especially Syria and Iraq. Terrorist organizations like ISIS and al-Qaeda have proven to be powerful forces. External powers, including the United States, are repeatedly drawn into regional conflicts.

9. The severe dysfunction that afflicts parts of this realm is expressed in a massive and expanding refugee crisis. Most pressing are the plight of Syrian war refugees and the flows of desperate migrants toward Europe, with tens of thousands risking their lives at the hands of smugglers in the hazardous sea journey across the Mediterranean.

NORTH AFRICA/SOUTHWEST ASIA:
POPULATION DISTRIBUTION, 2016
One dot represents 50,000 persons

FIGURE 6-2

Caspian Sea. Figure 6-2 reflects this co-varying relationship between population distribution and available water. Quite a few of this realm's prominent physiographic features—river valleys, basins, and deltas; moist coastlines; adequately watered mountain basins—are virtually defined by the clusters you see on this map.

So we know this geographic realm as a place where water is almost always at a premium; where consumers (such as Israelis and their Arab neighbors) compete for limited supplies; where peasants often struggle to make soil and moisture yield a small harvest; where nomadic peoples and their livestock still migrate across dust-blown flatlands; where oases are green islands in the desert that sustain local farming, supply weary travelers, and support trade across vast expanses of aridity.

But this is also the land of the Nile Valley and Delta as well as the Tigris-Euphrates Basin—respectively, the lifelines of Egypt and Iraq; the crop-covered *tell* of coastal Algeria; and the verdant shores of western Turkey. Compare Figure 6-1 to Figure G-6, and the dominance of **B** climates becomes evident, underscoring just how water dependent this realm's highly clustered population is.

Is *This* the 'Middle East'?

This realm is commonly referred to as the Middle East. That must sound odd to someone in, say, India, who might think of a Middle West rather than a Middle East! The name, of course, reflects the biases of its source—the Western world—which saw a Near East in Turkey; a Middle East in Egypt, Arabia, and Iraq; and a Far East in China and Japan. Still, the term has taken hold, and it can be seen

and heard in everyday usage by scholars, journalists, and members of the United Nations. Even so, we feel that it should only be applied to one of the regions of this far-flung realm, not to NASWA as a whole.

An 'Arab World'?

Another shorthand label often used for North Africa/Southwest Asia is the Arab World. This term implies a uniformity that does not actually exist. First, the name *Arab* is applied loosely to the peoples of this area who speak Arabic and related languages, but ethnologists normally restrict it to certain occupants of the Arabian Peninsula—the Arab 'source.' In any case, Turks are not Arabs, and neither are most Iranians or Israelis. Moreover, although the Arabic language prevails from Mauritania in the west across all of North Africa to the Arabian Peninsula, Syria, and Iraq in the east, it is not spoken in certain parts of this realm. In Turkey, for example, Turkish is the major language, and it has Ural-Altaic rather than Arabic's Semitic or Hamitic roots; the Iranian language is a member of the Indo-European linguistic family (see Fig. G-8). Other 'Arab World' languages that have separate ethnological identities are spoken by the Jews of Israel, the Tuareg people of the western Sahara, the Berbers of northwestern Africa, and the myriad peoples of the wide transition zone between North Africa and Subsaharan Africa to the south.

An 'Islamic World'?

Finally, and perhaps most significantly, this realm is routinely referred to as the Islamic World. Indeed, the Prophet Muhammad was born on the Arabian Peninsula in AD 571, and Islam, the religion to which he gave rise, spread across this realm and beyond after his death in 632. This great saga of Arab conquest saw Islamic armies, caravans, and fleets carry Muhammad's teachings across deserts, mountains, and oceans, penetrating Europe, converting the kings of West African states, threatening the Christian strongholds of Ethiopia, invading Central Asia, conquering most of what is now India, and even reaching the peninsulas and islands of Southeast Asia. Today, the world's largest Muslim state is Indonesia, and the Islamic faith extends far outside the realm under discussion (Fig. G-9). In this context, the term *Islamic World* is also misleading in that it suggests that there is no Islam beyond NASWA's borders.

In any case, this 'World of Islam' is not entirely Muslim either. Christian minorities continue to survive in all of the regions and many of the countries of the North Africa/Southwest Asia realm. Judaism has its base in NASWA's Middle East region; and smaller religious communities, such as Lebanon's Druze and Iran's Bahais, many of them under pressure from Islamic regimes, continue to diversify the realm's religious mosaic.

HEARTHS OF CULTURES

This geographic realm occupies a pivotal part of the world: here Africa, the source of humankind, meets Eurasia, crucible of

human cultures. Two million years ago, the ancestors of our species walked from East Africa into North Africa and Arabia and then spread all across Asia. Less than one hundred thousand years ago, *Homo sapiens* crossed these lands on their way to Europe, Australia, and, eventually, the Americas. Ten thousand years ago, human communities in what we now call the Middle East began to domesticate plants and animals, learned to irrigate their fields, enlarged some of their settlements into towns, and formed the earliest states. The world's dominant monotheistic religions originated here: first came Judaism, followed by the rise of Christianity, and most recently the heart of the realm was stirred and mobilized by the teachings of Muhammad and the **Quran** (Koran), and Islam arose. Today, this realm is a cauldron of religious and political turmoil—complicated by foreign interventions; tormented by internal conflict; empowered and enriched by oil in selected places; and plagued by grinding poverty in many others.

Dimensions of Culture

In the introductory chapter, we discussed the concept of culture and its regional expression in the cultural landscape. **Cultural geography** **1**, we observed, is a wide-ranging and comprehensive field that studies spatial aspects of human cultures, focusing not only on cultural landscapes but also on **culture hearths** **2**— the crucibles of civilization, the sources of ideas, innovations, and

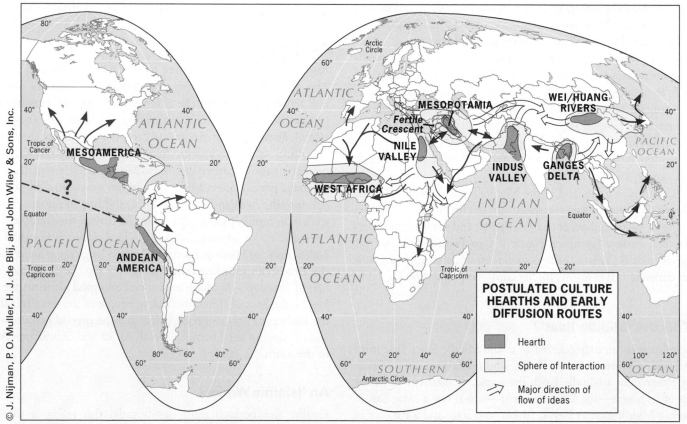

© J. Nijman, P. O. Muller, H. J. de Blij, and John Wiley & Sons, Inc.

POSTULATED CULTURE HEARTHS AND EARLY DIFFUSION ROUTES

Hearth

Sphere of Interaction

Major direction of flow of ideas

FIGURE 6-3

ideologies that transformed regions and realms. Those ideas and innovations spread far and wide through a set of processes that we study under the rubric of **cultural diffusion** **3**. Because we understand these processes better today, we can reconstruct many of the ancient routes by which the knowledge and achievements of culture hearths spread—that is, *diffused*—to other areas.

Another aspect of cultural geography, particularly noteworthy in the context of the North Africa/Southwest Asia realm, is the study of **cultural landscapes** **4** that a dominant culture creates. Human cultures exist in long-term accommodation with (and adaptation to) their natural environments, exploiting opportunities that these habitats present and coping with the extremes they can impose. In the process, they fuse their physical landscape and cultural landscape into an interacting unity between nature and society.

Rivers and Communities

In the basins of the major rivers of this realm (the Tigris-Euphrates system of modern-day Turkey, Syria, and Iraq as well as the Nile of Egypt) lay two of the world's earliest culture hearths (Fig. 6-3). Mesopotamia (which means land between the rivers) possessed fertile alluvial soils, abundant sunshine, ample water, and animals and plants that could be domesticated. Here, in the Tigris-Euphrates lowland between the Persian Gulf and the uplands of present-day Turkey, arose one of humanity's first culture hearths, a cluster of communities that grew into larger societies and, eventually, into the world's first states.

Mesopotamia

Mesopotamians were innovative farmers who domesticated cereal grains such as wheat and knew when to sow and harvest various other crops, water their fields, and store their surpluses. Their knowledge spread to villages near and far, and a *Fertile Crescent*, a region of significant agricultural productivity, evolved, extending from Mesopotamia across southeastern Turkey into Syria and the eastern Mediterranean coast beyond (Fig. 6-3).

Irrigation was the key to prosperity and power in Mesopotamia, and urbanization was its major reward. Among many settlements in the Fertile Crescent, some thrived, grew, enlarged their hinterlands, and diversified socially and occupationally; others failed. What determined success? One theory, the **hydraulic civilization theory** **5**, holds that cities that could control irrigated farming over large hinterlands held power over others, used food as a weapon, and prospered. One of these cities, Babylon on the Euphrates River, endured for nearly 4000 years (from 4100 BC). A busy port, its walled and fortified center endowed with temples, towers, and palaces, Babylon for a time was the world's largest city.

Egypt and the Nile

Egypt's cultural evolution may have started even earlier than Mesopotamia's, and its focus lay upstream from (south of) the Nile Delta and downstream from (north of) the first of the Nile's series of rapids, or cataracts. This part of the Nile Valley is surrounded

© Jan Nijman

@ FROM THE FIELD NOTES...

"Flying north over central Egypt from Luxor's ruins toward Cairo along the Nile's ribbon of green, I pondered the famous remark by the ancient Greek historian Herodotus that 'Egypt is the gift of the River Nile.' The Nile brings water from the interior of Africa that crosses the vast northern deserts and is crucially important to a country with as little rainfall as Egypt. About 95 percent of the Egyptian population lives within 20 kilometers (12 mi) of the river's banks. Almost all of Egypt's agriculture has always depended on irrigation, which is how this narrow strip of desert turned fertile. Notice the razor-sharp boundaries between irrigated and nonirrigated land. Undoubtedly, the Nile was the lifeline that supported one of the ancient world's major civilizations."

by inhospitable desert, and unlike Mesopotamia (which lay open to all comers), the Nile provided a natural fortress here. The ancient Egyptians converted their security into progress. The Nile was their leading avenue of trade and interaction; it also supported agriculture through irrigation (see photo). The Nile's cyclical ebb and flow was much more predictable than that of the Tigris-Euphrates river system. By the time Egypt fell victim to outside invaders (around 1700 BC), a full-scale urban civilization had emerged. Ancient Egypt's artist-engineers left a magnificent legacy of huge stone monuments, some of them containing treasure-filled crypts of god-kings called Pharaohs. These tombs have enabled archeologists to reconstruct the ancient history of this culture hearth.

Today, the world continues to benefit from the accomplishments of the ancient Mesopotamians and Egyptians. They domesticated cereals (wheat, rye, barley), vegetables (peas, beans), fruits (grapes, apples, peaches), and several animals (horses, pigs, sheep). They also advanced the study of the calendar, mathematics, astronomy, government, engineering, metallurgy, and a host of other skills and technologies. In time, many of their innovations were adopted and then modified by other cultures in the Old World and eventually in the New World as well. Europe was the greatest beneficiary of these legacies of Mesopotamia and ancient Egypt, whose achievements constituted the foundations of Western civilization.

STAGE FOR ISLAM

In a remote place on the Arabian Peninsula, where the foreign invasions of the Middle East had had little effect on the Arab communities, an event occurred early in the seventh century that was to change history and affect the destinies of people in many parts of the world. In a town called Mecca (Makkah), about 70 kilometers (45 mi) from the Red Sea coast in the Jabal Mountains, a man named Muhammad in the year AD 611 began to receive revelations from Allah (God). Muhammad (571–632) was then in his early forties. Arab society was in social and cultural disarray, but Muhammad forcefully taught Allah's lessons and began to transform his culture. His personal power soon attracted enemies, and in 622 he fled from Mecca to the safer haven of Medina (Al Madinah), where he continued his work until his death. This moment, the *hejira* (meaning migration), marks the starting date of the Muslim era, Year 1 on Islam's calendar. Mecca, of course, later became Islam's holiest place.

The Faith

The precepts of Islam in many ways constituted a revision and embellishment of Judaic and Christian beliefs and traditions. All of these faiths have but one god, who occasionally communicates with humankind through prophets; Islam acknowledges that Moses and Jesus were such prophets but considers Muhammad to be the final and greatest prophet.

Islam brought to Arabian culture not only the unifying religious faith it had lacked but also a new set of values, a new way of life, a new individual and collective dignity. Islam dictated observance of the Five Pillars: (1) repeated expressions of the basic creed, (2) daily prayer, (3) a month each year of daytime fasting (Ramadan), (4) the giving of alms, and (5) at least one pilgrimage to Mecca in each Muslim's lifetime (the *hajj*). Islam prescribed and proscribed in other spheres of life as well. It forbade alcohol, smoking, and gambling. It tolerated polygamy, even though it acknowledged the virtues of monogamy. Mosques appeared in Arab settlements, not only for the (Friday) sabbath prayer, but also as social gathering places to knit communities closer together. Mecca became the spiritual center for a divided, widely dispersed people for whom a collective focus was something new.

The Arab-Islamic Empire

Muhammad provided such a powerful stimulus that Arab society was mobilized almost overnight. The prophet died in 632, but his faith and fame spread like wildfire. Arab armies carrying the banner of Islam formed, invaded, conquered, and converted wherever they went. As Figure 6-4 shows, by AD 700 Islam had reached far into North Africa, into Transcaucasia, and into most of Southwest Asia. In the centuries that followed, it penetrated both southern and eastern Europe, Central Asia, West Africa, East Africa, South Asia, and Southeast Asia, even reaching into China by AD 1000.

The spread of Islam provides a good illustration of a set of processes known as **spatial diffusion** **6** that focus on the way ideas, inventions, and cultural practices propagate through a population in space and time. Diffusion takes place in two forms: **expansion diffusion** **7**, when propagation waves originate in a strong and durable source area and spread outward, affecting an ever larger region and population; and **relocation diffusion** **8**, in which migrants carry an innovation or an idea (or belief). Islam spread through expansion as well as relocation diffusion, reaching such faraway places as the Ganges Delta in South Asia, what is now Indonesia in offshore Southeast Asia, and East Africa.

The heart of Islam, however, remained in Southwest Asia. There, Islam became the cornerstone of an Arab Empire with Medina as its first capital. As the Empire expanded, its headquarters shifted from Medina to Damascus (in what is now Syria) and later to Baghdad on the Tigris River (in modern-day Iraq). And it prospered. In architecture, mathematics, and science, the Arabs overshadowed their European contemporaries (now mired in their medieval standstill). The Arabs established institutions of higher learning in many cities including Cairo, Baghdad, and Toledo (Spain), and their distinctive cultural landscapes united their far-flung domain. Non-Arab societies in the path of the Muslim drive were not only Islamized, but also

FIGURE 6-4

© J. Nijman, P. O. Muller, H. J. de Blij, and John Wiley & Sons, Inc.

Arabized, adopting other Arab traditions as well. Islam had spawned a culture; it still continues to function at the heart of that culture today.

Islam and Other Religions

Two other major faiths also originated in the *Levant* (the lands bordering the Mediterranean's eastern coast, extending inland for about 300 kilometers [200 mi])—Judaism and Christianity—with both being older than Islam. Islam's rise submerged many smaller Jewish communities, but the Christians, not the Jews, waged centuries of holy war against the Muslims, seeking, through the Crusades, not only to drive Islam back but to reestablish Christian communities where they had dominated before the Islamic conquest. The aftermath of that lengthy campaign still marks the realm's cultural landscape today. A substantial Christian community remains in Lebanon, and Christian minorities also survive in Israel, Egypt, Syria, and Jordan. Since World War II, the most intense conflict to date has pitted Israel, the Jewish state formed in 1948, against its Islamic neighbors near and far. Jerusalem—holy city for Judaism, Christianity, and Islam—lies squarely in the middle of this confrontation.

Islam Divided

For all its vigor and success, Islam is still fragmented into sects. The earliest and most consequential division arose after Muhammad's death between those who believed that only a blood relative should follow the prophet as leader of Islam (*Shi'ites*) and others who felt that any devout follower of Muhammad was qualified to take over (*Sunnis*). Overall, the Sunnis prevailed, and, in the centuries that followed, the great expansion of Islam was mainly propelled by Sunnis; the Shi'ites survived as minorities scattered throughout the realm. Today, even though roughly 90 percent of all Muslims are Sunnis, the Shia–Sunni schism has strongly resurfaced to become the root of a number of the realm's ongoing conflicts.

The minority Shi'ites have always vigorously promoted their version of the faith. In the early sixteenth century, their work finally paid off: the royal house of Persia (present-day Iran) made Shi'ism the only legal religion throughout its vast empire. That domain extended from Persia into lower Mesopotamia, into Azerbaijan, and into western Afghanistan and Pakistan. As the world map of religions (Fig. G-9) shows, this created for Shi'ism a large culture region and gave the faith unprecedented strength. Iran remains the bastion of Shi'ism in the realm today, and its appeal continues to radiate into neighboring countries and even farther afield.

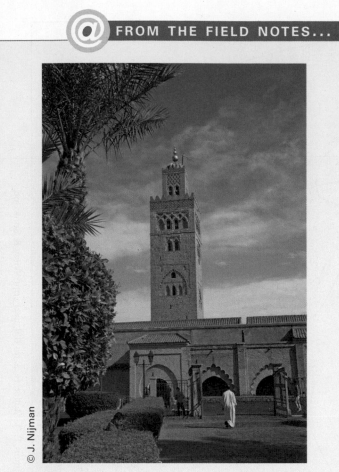

© J. Nijman

"As I walked through the medina (old city) of Marrakech (Morocco) in early 2015, I came upon the famous Koutoubia Mosque. Its minaret is the city's signature landmark, standing 77 meters (250 ft) tall. It was built in the twelfth century and reminds me of the masterful architecture and cosmopolitanism that flourished in parts of the Arab-Islamic Empire while European culture lay dormant during the depths of the so-called Dark Ages. In fact, this Islamic cultural imprint also extended into southern Europe, and comparable architectural wonders still stand in the Spanish cities of Seville and Granada. When, in the late fifteenth century, Spain expelled its Jewish population, some found refuge across the Strait of Gibraltar in Morocco. I tried to take a look inside the Mosque, but non-Muslims were refused entry (has that always been the case?). Today, Morocco is a relatively stable outpost on the western edge of a decidedly turbulent and deeply challenged realm."

THE OTTOMAN EMPIRE AND ITS AFTERMATH

If we want to understand this sprawling realm today, we must appreciate the historical role of the Ottoman Empire, its enormous expansion, and the ways it was undone by other great powers. It is ironic that Islam's last great advance into Europe eventually led to European occupation of Islam's very heartland. The Ottomans

(named after their leader, Osman I), based in what is today Turkey, conquered Constantinople (now Istanbul) in 1453 and then pushed into southeastern Europe. Soon Ottoman forces were on the doorstep of Vienna; they also invaded Persia, Mesopotamia, and North Africa. At its height, the Ottoman Empire under Suleyman the Magnificent (ruled 1522–1560) was the most powerful state in western Eurasia.

Although the Ottoman Empire survived for more than four centuries until its dissolution in 1923, it lost territory as time went on. First came losses to the Hungarians, then to the Russians, and later to the Greeks and Serbs until, after World War I, the European powers took over its provinces and turned them into colonies—colonies we now know by the names of Syria, Iraq, Lebanon, and Yemen (Fig. 6-5). As the map shows, the French and the British took several possessions; even the Italians annexed the part of the Ottoman domain that is now Libya.

The boundary framework that the colonial powers created after World War I to delimit their new holdings was unsatisfactory. As Figure 6-2 illustrates, this realm's population is clustered, fragmented, and strung out in river valleys, coastal zones, and crowded oases. The colonial powers laid out long stretches of boundary as ruler-straight lines across uninhabited territory; they saw no need to adjust those boundaries to cultural or physical features in the landscape. Other borders, even some in desert zones, were poorly defined and never marked on the ground. Later, when the colonies had become independent states, such boundaries led to quarrels, even armed conflicts, among neighboring Muslim states. In some instances, they left entire nations, such as the Kurds and the Palestinians, without any territory of their own, thereby rendering them **stateless nations** 9.

Population and the New Political Framework

Although Islam and its cultural expressions overspread this geographic realm, its political and social geographies are divided and often fractious. We encounter states with strong internal divisions (Iraq, Lebanon), stateless nations (Palestinians, Kurds, Berbers), and territories still in the process of integration (Western Sahara, Palestine). Of its two dozen states and territories, the three largest and in many ways the most important are Egypt in North Africa, Turkey on the threshold of Europe, and Iran at the margins of Central Asia. All three have populations between 75 and 82 million, modest by world standards. But at the other end of this continuum are ministates such as Bahrain, Qatar, and Djibouti, each with fewer than 2.5 million inhabitants. It is quite important to study Figure 6-1 attentively because the boundary framework—much of it inherited from the colonial era—now lies at the root of so many of the realm's most difficult problems.

It is also useful to take another careful look at Figure 6-2 because it shows how differently populations are distributed within the countries we will examine. Compare the three largest countries just mentioned: Egypt's population forms a river-valley ribbon and delta cluster; Turkey's, between the Black Sea and the Mediterranean Sea, is far more evenly dispersed, although its west

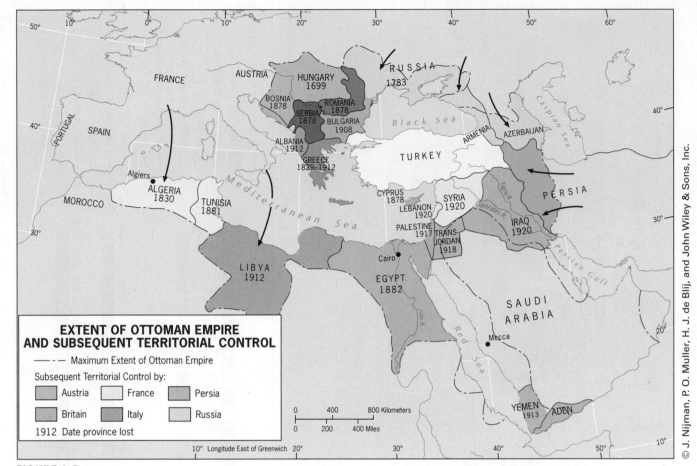

FIGURE 6-5

is more populous than the east; and in Iran you can draw a line down its center from the Caspian Sea to the Persian Gulf that separates the comparatively crowded west, where the capital is located, from the drier and far more sparsely peopled east. In North Africa, almost everyone seems to live along or near the Mediterranean coast, but no such coastal concentrations mark the Arabian Peninsula except in its far south, where Yemen's inhabitants cluster near the Red Sea.

THE POWER AND PERIL OF OIL

About 30 of the world's countries have significant oil reserves, with six of the nine leading countries located in this realm: Saudi Arabia, Iran, Iraq, Kuwait, the United Arab Emirates, and Libya. In general terms, NASWA's deposits of oil (and associated natural gas) are concentrated within two discontinuous zones (Fig. 6-6). The most productive of these zones extends from the southern and southeastern part of the Arabian Peninsula northwestward around the rim of the Persian Gulf, reaching into Iran and continuing northward into Iraq, Syria, and southeastern Turkey, where it peters out. The second zone lies across North Africa and extends from north-central Algeria eastward across northern Libya to Egypt's Sinai Peninsula.

Winners and Losers

Saudi Arabia is the world's largest oil exporter, and the combined oil production by the countries of the North Africa/Southwest Asia realm far exceeds that from all other sources. Throughout recent times, the United States has been the world's leading importer, followed closely by China (the United States is by far the biggest consumer of oil but is itself also the top producer, making it less dependent on imports than it would otherwise be).

As Figure G-10 indicates, the production and export of oil and gas has vaulted a number of this realm's countries into the higher-income categories. But petroleum wealth also has entangled these Islamic societies and their governments in global strategic affairs. When regional conflicts create instability in producing countries that have the potential to disrupt international supply lines, powerful consumers are tempted to intervene and have done so.

When the colonial powers laid down the boundaries that partitioned this realm among themselves, no one knew about the riches that lay beneath the ground. A few wells had been drilled, and production in Iran had begun as early as 1908 and in Egypt's Sinai Peninsula in 1913. But the major discoveries came later, in some cases after the colonial powers had already withdrawn. Some of the newly independent countries, such as Libya, Iraq, and Kuwait, found themselves endowed with wealth undreamed of when the Ottoman Empire collapsed. As Figure 6-6 shows,

NORTH AFRICA/SOUTHWEST ASIA: OIL AND NATURAL GAS

0 250 500 750 Kilometers
0 250 500 Miles

BARRELS OF OIL PRODUCED ANNUALLY

Over one billion
100 million–one billion
Under 100 million
Little or no oil production

— Oil and gas production region
◣ Oilfield
◗ Gasfield
Oil pipeline
Proposed oil pipeline
Gas pipeline
IRAN OPEC member
Algiers Oil shipping terminal

© J. Nijman, P. O. Muller, and John Wiley & Sons, Inc.

WORLD OIL PRODUCTION
0% 10% 20% 30% 40% 50% 60% 70% 80% 90% 100%
U.S. SAUDI ARABIA RUSSIA CHINA CAN. U.A.E. IRAN IRAQ MEXICO KUWAIT VENEZ. NIGERIA NORWAY ANGOLA OTHER N.A./S.W. ASIA OTHER

WORLD OIL RESERVES
0% 10% 20% 30% 40% 50% 60% 70% 80% 90% 100%
VENEZUELA SAUDI ARABIA CANADA IRAN IRAQ KUWAIT U.A.E. RUS. LIBYA OTHER N.A./S.W. ASIA OTHER

WORLD NATURAL GAS PRODUCTION
0% 10% 20% 30% 40% 50% 60% 70% 80% 90% 100%
UNITED STATES RUSSIA IRAN CAN. QATAR SAUDI ARABIA CHINA NORWAY NETH. TURK. NIGERIA MALAYSIA VENEZ. INDON. OTHER N.A./S.W. ASIA OTHER

WORLD CONVENTIONAL NATURAL GAS RESERVES
0% 10% 20% 30% 40% 50% 60% 70% 80% 90% 100%
RUSSIA IRAN QATAR SAUDI ARABIA U.S. TURK. U.A.E. VENEZ. NIGERIA OTHER N.A./S.W. ASIA OTHER

FIGURE 6-6

however, others were less fortunate. A few countries had (and still have) potential. The smaller, weaker emirates and sheikdoms on the Arabian Peninsula always feared that powerful neighbors would try to annex them. The unevenly distributed oil wealth, therefore, created yet another source of division and distrust among Islamic neighbors.

A Foreign Invasion

The oil-rich countries of the realm found themselves with a coveted energy source, but lacking in the skills, capital, or equipment to exploit it. These resources had to come from the Western world and entailed what many tradition-bound Muslims feared most: a strong foreign presence on Islamic soil, outside intervention in political as well as economic affairs, and penetration of Islamic societies by the vulgarities of Western ways. The transport of oil from these countries to destinations in other geographic realms also raised the importance of strategic arteries that connect NASWA to the rest of the world. Most significant are **choke points** 10, the narrowing of international waterways that increase the risk of ship collisions and vulnerability to piracy and other forms of attack. This realm contains no less than five of the world's most critical choke points: the Suez Canal and Bab el Mandeb Strait at opposite ends of the Red Sea; the Strait of Hormuz at the mouth of the Persian Gulf; the Turkish Straits that connect the Black and Mediterranean seas; and the Strait of Gibraltar at the Mediterranean's outlet to the Atlantic Ocean (Fig. 6-1).

Foreign intervention often aggravated clashes between the traditional and the modern. Oil revenues created cultural landscapes in which gleaming skyscrapers towered over ornate mosques and centuries-old communities. The social chasms between the rich, well-connected, Westernized elites and less fortunate local citizenries bred resentment. In Iran, these issues played a role in the alienation of the ruling shah (king) from his own people, fueling a religious revival that ultimately led to the monarch's ouster in 1979. Thus oil, foreign involvement, economics, and religion are inextricably intertwined in the realm's petroleum-producing countries.

The Geography of Oil's Impact

As is true of every exploited natural resource, oil has completely transformed cultural landscapes in certain parts of the North Africa/Southwest Asia realm and left others virtually unchanged. For hundreds of millions of this realm's inhabitants, a patch of tillable soil and a source of water still mean more to daily life than all the oil in OPEC.* The countryside in Arab as well as non-Arab regions in this

---

*OPEC, the Organization of Petroleum Exporting Countries, is the international oil cartel (syndicate) formed by 12 producing countries to promote their common economic interests through the setting of joint pricing policies and the limitation of market options for consumers. Its eight NASWA members are Algeria, Iran, Iraq, Kuwait, Libya, Qatar, Saudi Arabia, and the United Arab Emirates.

Jose Fuste Raga/Corbis

The Burj Khalifa, centerpiece of the ultramodern city of Dubai on the Persian Gulf in the UAE. Completed in 2010, the 163-story Burj is the world's tallest building, topping out at 828 meters (2716 ft)—just over half a mile high, almost twice the height of New York's Empire State Building.

part of the world continues to carry the imprints of centuries of cultural tradition, not decades of oil-driven modernization. Nonetheless, oil and natural gas—their location, production, transportation, marketing, and sale—have produced fundamental changes that include the following:

1. ***Urban Transformation.*** Undoubtedly, the most visible manifestation of oil wealth is the modernization of cities. The tallest building on Earth rises above Dubai (see photo) in a ministate named the United Arab Emirates (UAE), but it is only one in a forest of new glass-encased skyscrapers, many of which test the limits of design and engineering. Besides the capitals (such as Saudi Arabia's Riyadh), the landscapes of a number of other cities on the Arabian Peninsula also reflect the riches oil has brought—yes, even Mecca's.

2. ***Variable Incomes.*** Oil and natural gas prices fluctuate on world markets. In 2008, oil sold for more than $140 per barrel before plunging to less than $40; by 2013, the price of a barrel had climbed back to nearly $120, only to drop to less than $50 in mid-2015. When energy prices are high, several states in this realm rank among the highest-income societies in the world, but the countries without oil have to pay dearly for their energy. Even when prices decline, many petroleum-exporting countries manage to remain at least in the upper-middle-income category (Fig. G-10).

3. ***Infrastructure.*** Massive spending on airports, seaports, bridges, tunnels, four-lane highways, public buildings, shopping malls, recreational facilities, and other components of national infrastructure creates an image of comfort and affluence quite unlike that prevailing in countries without significant oil or gas revenues. Saudi Arabia, for instance, is now engaged in a gigantic modernization project extending from coast to coast.

4. ***Industrialization.*** A number of far-sighted governments among those with oil wealth, realizing that reserves will not last forever,

are investing some of their income in industries that will survive the oil-exporting era. Petrochemical manufacturing using domestic supplies, aluminum, steel, and fertilizers are among those industries, although others potentially more promising, especially in high-technology fields, are only slowly becoming a part of this important initiative.

5. *Regional Disparities.* Oil wealth, like other high-value resources, tends to create strong regional contrasts both within and among countries. Saudi Arabia's ultramodern east coast is a world apart from most of its interior, which remains a land of barren deserts, widely scattered oases, enormous distances, and isolated settlements. Among countries the contrasts can also be striking, for example, between Qatar and Yemen.

6. *Foreign Investment.* Governments and private entrepreneurs have invested enormous amounts of oil-generated wealth in foreign countries, buying financial securities as well as acquiring prestigious hotels, upscale retailers, and other high-cachet properties. These investments have created a network of international linkages that not only connects NASWA states and individuals to the national economies of other realms but also links them to the growing Islamic communities in those countries.

7. *Foreign Involvement.* To many inhabitants of this realm, particularly those with strong Islamic-revivalist convictions (a topic to be discussed shortly), the inevitable presence of foreigners (including businesspeople, politicians, architects, engineers, and even armed forces) on Islamic soil is an unwelcome byproduct of the energy era. In fact, public opinion in Saudi Arabia forced its ruling regime to negotiate the 2003 departure of U.S. troops from the country's territory.

8. *Intra-Realm Migration.* Petroleum wealth enables governments, industrialists, and private individuals to hire workers from less-favored parts of the realm to labor in the oilfields, ports, and numerous menial service occupations. This has brought many Shi'ites to the countries of the eastern Arabian Peninsula, where they now form significant sectors of national populations; hundreds of thousands of Palestinian Arabs also sought temporary employment in local industries here. In 2015, it was estimated that Saudi Arabia, with a permanent population of just below 29 million, hosted 8.2 million foreign workers, about 5 million of whom come from elsewhere in the NASWA realm.

9. *Migration from Other Realms.* The labor market in such burgeoning places as the UAE's Dubai and Abu Dhabi has also attracted hordes of workers from beyond the realm's borders during recent periods of rapid growth. Wages in South Asian and African countries are considerably lower than those paid by the building industry or private employers in oil-boom-driven NASWA states. Besides low-level construction jobs, these workers serve as domestics, gardeners, trash collectors, and the like. In 2015, almost 90 percent of the total population of the United Arab Emirates and Qatar consisted of foreign employees! Over the past few years, substandard working conditions and wage issues have also led to protests that have exposed the harsh conditions faced by many guest workers throughout this realm.

10. *Diffusion of Revivalism.* Islamist regimes utilize oil and gas revenues to support Muslim communities as well as their mosques and cultural centers throughout the world. No NASWA country has spent more money on such causes than Saudi Arabia, and thousands of mosques from England to Indonesia prosper as a result. This example of relocation diffusion creates myriad nodes of recruitment for the faith—and ensures the dissemination of revivalist principles.

FRAGMENTED MODERNIZATION

More than any other realm, this one is marked by a human spatial mosaic of **fragmented modernization 11**. To a certain degree NASWA is highly modern and prosperous, but in many more respects it remains traditional, stagnant, and quite poor. These disparities that characterize so many of the realm's societies are best explained by the uneven impact of oil, the lack of democracy, and the mercurial role of religion.

Oil-Generated Inequalities

Oil brought this realm into contact with the outside world in ways unforeseen just a century ago. Oil has strengthened and empowered some of its peoples, but has dislocated and imperiled others. It has truly been a double-edged sword.

We should remind ourselves that the great majority of NASWA's inhabitants are not, in their daily lives, directly affected by the changes the energy era has brought. Most Moroccans, Tunisians, Egyptians, Jordanians, Yemenis, Lebanese, and countless millions of others—Kurds, Palestinians, Berbers, Tuaregs—make ends meet by trading or farming or working at jobs their parents and grandparents performed. Take the case of Iran, which as a country earns about two-thirds of its income from oil and natural gas. Only one-half of 1 percent of that country's workforce (just over 100,000 out of more than 20 million workers) is employed in the energy-related sector. By far the largest number engaged in any single occupation are Iran's five-million-plus farmers. And for all their oil, Iranians in 2014 earned an average monthly income of only (U.S.) $448 per month (less than 60 percent of the monthly earnings in neighboring, energy-poor Turkey). As we shall discover, cultural-geographic rather than economic-geographic forces mainly shape the regionalization of this realm.

The two dozen states that constitute NASWA display an often stunning degree of variety and diversity. This difference is accentuated by the severe disparities that result from the uneven distribution of the realm's oil wealth. And even within those countries possessing large energy reserves, glaring inequalities have arisen between their have and have-not citizens. Cutting-edge urban skylines on the Arabian Peninsula stand in sharp contrast to Nile Valley villages virtually unchanged for millennia. Modern expressways cross deserts still traversed by nomadic traders riding camels. Social geographies vary just as greatly—for instance, the female literacy rate in Lebanon is 86 percent, yet only 48 percent in Yemen.

MAP ANALYSIS

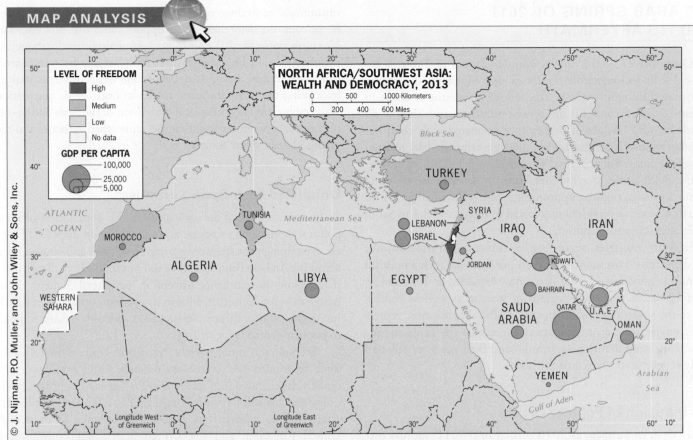

FIGURE 6-7

Worldwide, high levels of national development tend to be associated with democratic political systems that offer a full range of personal freedoms. Conversely, it is thought that the poorer the country the less likely it is to be a democracy (for example, think of North America versus Subsaharan Africa). But all this can vary considerably from one realm to another or within realms. Figure 6-7 maps income per capita against level of democracy for selected NASWA countries. Can a clear relationship between wealth and democracy be observed in this realm? Can you discern any regional patterns? Comparing this map with the one in Figure 6-6, do you think oil and gas production is associated with the spatial variations you observe? Access your WileyPLUS Learning Space course to interact with a dynamic version of this map and to engage with online map exercises and questions.

The Absence of Democratic Traditions

Democracy has been a rarity in this realm. Although specific reasons vary, the majority of states share a history of autocratic, conservative regimes, usually backed strongly by their military forces. Most countries were part of the Ottoman Empire before World War I (see Fig. 6-5) and only became independent after a period of (attempted) European rule that was, in some cases, sanctioned by the League of Nations (the predecessor of the United Nations that functioned from 1920 to 1946). Accordingly, Egypt, Iraq, Palestine, Transjordan, and parts of Yemen were administered by Britain; Lebanon, Syria, Tunisia, Morocco, and Algeria were ruled by France; and Libya was controlled by Italy.

In most cases, the Europeans faced persistently tenacious opposition from local populations but were reluctant to let go. Independence was usually preceded by intense conflict or even war; at any rate, none of these countries can be said to have been prepared to govern themselves as European-style democracies.

Indeed, virtually everywhere the newly independent countries fell into the hands of the military (such as Egypt) or restored monarchs with strong military support (such as Yemen).

After World War II, these autocratic regimes were often cemented into place through foreign involvement. Mostly, this was about oil: foreign powers needed oil, showed little interest in domestic politics, and provided incumbent regimes with abundant revenues to build up their military forces, secure their political position, and maintain order. Saudi Arabia, the world's richest petro-state, is the quintessential example. In other cases, it was Cold War politics that effectively supported the realm's nondemocratic governments.

Political regimes in this realm vary, but at least until the Arab Spring movement emerged in late 2010, most government leaders had been in power for several years, and their top priority was to preserve their domination and the status quo. Overall, their rule had been generally conservative and based on top-down control with minimal concern for the 'people on the street.'

THE ARAB SPRING OF 2011 AND ITS AFTERMATH

On December 17, 2010, in the Tunisian capital of Tunis, a 26-year-old fruit-and-vegetable vendor named Mohammed Bouazizi became so distraught when police, following months of constant harassment, confiscated his unlicensed produce stand, that he set himself on fire. His act of desperation resonated with countless Tunisians whose frustration with their government had reached the boiling point. Bouazizi died soon afterward, but his self-immolation quickly came to symbolize Tunisia's 'Jasmine Revolution.' Within a month, popular opposition to the regime had grown so fierce that Tunisia's president was forced to resign and flee to Saudi Arabia after 23 years in power.

In Arab countries, according to conventional wisdom, not only democratic governance but even the *desire* for such a political system was nonexistent. That myth now exploded as Tunisians of all backgrounds, including legions of women as well as youths, took to the streets to demand an end to official corruption, repression, cronyism, and economic mismanagement. The Tunisian people were insisting that they be heard—and their voices swiftly and loudly echoed across much of the realm.

The Diffusion of Popular Unrest

What began in Tunisia, as 2010 turned into 2011, rapidly diffused eastward across North Africa and into adjacent parts of Southwest Asia. Within a few months, popular uprisings had spread to Egypt, Libya, Syria, Yemen, and Bahrain. The conditions in all these countries were similar: they were ruled by long-established autocratic regimes that failed to bring economic progress, repressed their own people, and had lost touch (most especially with the younger generation). Throughout the realm, people now closely followed these events via television and the Internet, and grassroots revolutionary fervor spread like wildfire. It may well have been the most spectacular **domino effect 12** in recent history.

The pivotal tile in this row of falling dominoes was Egypt. The most heavily populated, historically prominent, and centrally positioned of the Arab countries, Egypt had for decades suffered from stunted economic growth and the stagnation of an out-of-touch, repressive government. In Cairo's Tahrir Square and elsewhere, the masses boldly took to the streets and soon forced the ouster of President Mubarak, even though protesters were repeatedly confronted by harsh, bloody encounters with the police and military forces; nearly 1000 demonstrators were killed in these clashes. The events in Egypt generated an even more powerful round of shockwaves that reverberated throughout the realm. If the tide could be turned here at the Arab epicenter, reasoned the fearless throngs of protesters in several other countries, it could happen anywhere.

In most NASWA countries, more than half the population is under 25 years of age. This enormous demographic cohort has grown up in a world in which information flows almost uncontrollably, so these young people know only too well that their home countries deny them many opportunities that exist elsewhere. Not surprisingly, the 2011 uprisings were predominantly (though not exclusively) led by the younger generation, ranging from teenagers to thirty-somethings. In Tunisia, Egypt, and Syria, students took the lead, using the Internet's social media to organize demonstrations. This emphasized another critical dimension of the Arab Spring—the disconnect between the realm's aging, entrenched autocrats and their young, globally aware subjects.

Unfulfilled Arab-Spring Promises

If the very term *Arab Spring* expressed the hopefulness of a new political dawn, such romantic notions were soon dispelled. First of all, even though the spate of uprisings resonated and jangled nerves across the Arabian Peninsula, they did not seriously challenge any of that region's governments (except in peripheral Yemen). It was here—as well as in Jordan, Morocco, and Iran—that the established monarchies and/or theocracies prevailed with only modest responses to popular demands for reform.

Second, the more overtly tyrannical regimes in Libya and Syria responded with stubborn resistance and iron fists. Libya quickly degenerated into civil war as President Muammar Qadhafi unleashed his military forces against his unarmed people. That attack was so vicious that a makeshift alliance of France, the United Kingdom, and the United States (later subsumed by NATO) intervened with aerial bombing raids to help protect the Libyans from annihilation by their own government. By October 2011, the tide had turned, the capital of Tripoli fell, and Qadhafi was cornered and killed in his hometown of Sirte on the central coast. However, that did not end the conflict. The country's power structure soon imploded, the civil war intensified, and by 2015 Libya was well on its way to becoming a failed state. In Syria, protesters were met with nothing less than state-sanctioned violence perpetrated by the army and the air force, aimed at all civilians who opposed the dictatorial regime of President Bashar al Assad. This ruthless campaign of repression not only triggered a brutal civil war in 2011 that still continues, but also fractured the country in a manner that lured in some of the most radical extremists, who quickly turned the conflict into a regional war. By mid-2015, the death toll resulting from this unrelenting carnage was estimated to total more than 300,000.

And third, where it had appeared that a transformation of government had been achieved—most notably in Egypt—democratic prospects steadily faded as the military conspired to retain *de facto* control over the country. In retrospect, the Arab Spring was fueled in large part by the aspirations of masses of young people yearning for economic progress and democracy, however they faced an uphill struggle without the institutions for a successful transition (such as an independent judiciary, a free press, and a multiple-party electoral system). Five years after it began, the short-lived Arab Spring had fully run its course—its initial promise and excitement gradually eroding into crushed hopes with the inexorable restoration of the business-as-usual autocratic regimes.

POLITICS AND RELIGION

Another facet of the Arab Spring is that it was driven by populist grievances concerning the suppression of religious freedoms. Keep in mind that almost all of the governments in this realm were (and are) secular militaristic regimes that view Islamic parties and organizations with suspicion and as threats to political stability. Religious leaders and activists saw their opportunity as part of the wider revolt that was primarily aimed at the removal of repressive regimes. Thus the protest movement, overall, was both complex and potentially conflicted. In fact, the notions of democracy that stirred so many of the young protesters ran counter to the fundamentalist views of reactionary Islamists, a divide that added yet one more obstacle to much awaited social change in this realm.

Religious Revivalism

To the extent that the Arab Spring exposed the vulnerability of some long-entrenched secular regimes, it may have unintentionally fueled Islamic fundamentalism—or as Muslims prefer to call it, religious revivalism **13**. That resurgence had already been in the making for decades. During the 1970s, the *imams* (mosque worship leaders) in Shi'ite Iran sought to reverse the shah's moves toward liberalization and secularization: they wanted to (and soon did) recast society in a traditional, revivalist Islamic mold. An *ayatollah* (leader under Allah) replaced the deposed shah in 1979; Islamic rules and punishments were instituted. Urban women, many of whom had been considerably liberated and educated during the shah's regime, were forced to return to more traditional roles.

The rise of Islamic revivalism was neither limited to Iran nor confined to Shi'ite communities. Many Muslims—Sunnis as well as Shi'ites—in various parts of the realm disapproved of the erosion of traditional Islamic values, the corruption of society by European colonialists and later by Western modernizers (which brought economic benefits only to the elites), and the declining power of the faith within the secular state. As noted above, most governments in the NASWA realm continue to be both secular *and* repressive, and they do not provide the economic progress their citizens yearn for. Put in those terms, it is not that difficult to understand why many people seek solace in religion as a way to regain hope and dignity.

Some have taken their return to the faith much farther and imbued it with a fanaticism that is alien to most Muslims—and this has set Muslim against Muslim in many parts of the realm. Revivalists fired the faith with a new militancy, aggressively challenging the status quo from Iran to Algeria. The militants forced their governments to ban 'blasphemous' books, to re-segregate schools according to gender, to enforce traditional dress codes, to legitimize religion-based political parties, and to heed the wishes of the *mullahs* (teachers of Islamic ways). Militant Muslims also proclaimed that Western ideas about modernization, inherited from colonialists and adopted by Arab nationalists, were incompatible with the dictates of the Quran.

Terrorism in the Name of Islam

The great majority of Muslims are not fundamentalists; not all fundamentalists are militants; and not all militants are terrorists. Most religious scholars point out that the Islamic faith is no more predisposed toward militancy than other religions. And we should not forget that terrorism is a tool of war that has been used around the world, and indeed throughout modern history, by people of many different beliefs and ethnic backgrounds.

But nowadays the religion of Islam has proven to be a significant vehicle for political mobilization and a powerful source of inspiration for militants with terrorist agendas. To be sure, many Muslim adherents contest this extreme interpretation of the Quran; on the other hand, there is little doubt that militants justify their actions as expressions of their particular fundamentalist beliefs. Theirs is a 'holy war'—a *jihad* **14**—and as such it is a deeply reactionary movement that looks to the past rather than the future.

Broadly speaking, two extremist Islamic organizations are dominant today in spreading terrorism within this realm. The first is **al-Qaeda** (the name means 'base' or 'foundation'), an internationally networked organization whose goal is to establish Islamic rule across NASWA. They originated in Afghanistan in the late 1980s and have since masterminded terrorist acts in many other countries, from Kenya to Indonesia to the United States (where al-Qaeda was responsible for the 9/11 attacks). After their leader, Osama Bin Laden, was killed by U.S. Special Forces in Pakistan in 2011, the organization was weakened and forced to decentralize. Currently, al-Qaeda's main 'branch' is constituted by **AQAP**–shorthand for Al-Qaeda in the Arabian Peninsula. It is especially active in Yemen, where it controls key territory and worked to destabilize the country to the point of its 2015 collapse into chaotic civil war.

The other leading Islamist terrorist organization is **ISIS**, which has rapidly gained importance since 2013. The acronym ISIS stands for *Islamic State of Iraq and Syria*; the organization is also known by the acronyms *IS* (*Islamic State*) and *ISIL* (*Islamic State of Iraq and the Levant*). ISIS is notoriously violent and employs unprecedented tactics of intimidation, its videos of decapitations of 'infidels' sending shockwaves around the globe. This highly militarized organization emerged in Iraq after the U.S. invasion of 2003, and initially declared allegiance to al-Qaeda. Thereafter, it steadily assumed a more independent role in Iraq and then expanded into unguarded, sparsely populated eastern Syria following the outbreak of civil war in 2011. Exploiting the chaos of that war and Sunni-Shi'ite schisms in both countries, ISIS ever more effectively rallied Sunnis against secular governments that were dominated by Shi'ites.

The threat posed by ISIS in Syria intensified so rapidly that it essentially dissipated much of the international opposition to the Assad regime and its war crimes against countless thousands of Syrian civilians. By the end of 2013, ISIS came to be regarded as the greater danger, and the U.S.-led allied coalition (which included Saudi Arabia, the UAE, the UK, and France) launched a campaign of air strikes on ISIS strongholds that, despite its limited success, was still continuing at press time.

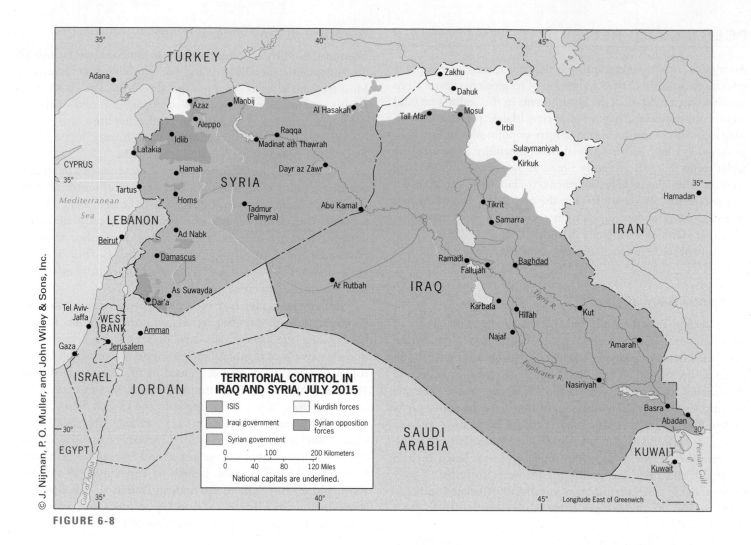

© J. Nijman, P. O. Muller, and John Wiley & Sons, Inc.

FIGURE 6-8

By mid-2015, ISIS had become an established force that controlled large swaths of territory in eastern Syria as well as in northern and western Iraq (Fig. 6-8). Arguably, ISIS evolved over a short period of time from a dispersed terrorist organization into a territorial quasi-state (perhaps even a formative *insurgent state*, a concept discussed in Chapter 3), with the Syrian city of Raqqa serving as its capital. Its armed forces are estimated by the U.S. government to total more than 30,000 fighters (ISIS claims it has 200,000); strict Islamic (*sharia*) law prevails in its territories; common people and shopkeepers are taxed; and control of oilfields and infrastructure provides sufficient revenues to fund its military operations. As ISIS tries to hold and consolidate its territorial gains, its ambitions in this realm are clearly not limited to parts of Iraq and Syria. Most notably, in March 2015 the organization claimed responsibility for a terrorist attack at a museum in Tunisia's capital that left 21 people dead, 18 of them foreign tourists (or in ISIS parlance, 'citizens of Crusader countries'). And even more ominously, the designs of ISIS reach far beyond NASWA: it has declared itself to be nothing less than a **caliphate 15**—an imperial-scale Islamic government led by a direct successor to the Prophet Muhammad who rules and exerts moral authority over Muslims worldwide. The most recent

caliphate was the Ottoman Empire, whose caliphs ruled from 1299 until abolition in 1922. Were the ISIS caliphate to succeed, it would first be governed by 'Caliph Ibrahim'—better known today as the vicious terrorist leader, Abu Bakr al-Baghdadi.

A REALM IN TURMOIL

Although the Arab Spring movement failed to achieve the diverse goals of the protesting masses in the streets of Cairo and other cities, it did destabilize those parts of the NASWA realm in and around the countries that experienced the greatest violence. A good example is Libya, whose political breakdown unleashed a wide spectrum of radical Islamist groups across northern Africa, especially in Mali. And what began as an internal conflict in Syria, soon escalated into a dangerous regional war when it spread into neighboring Iraq—drawing in a wide range of parties that now include ISIS, Turkey, Iraqi Kurdistan, Iran (traditional supporter of the Shi'ite cause in Syria and Iraq), bordering states overwhelmed by refugee flows (such as Lebanon and Jordan), and the more than 60 outside powers of the U.S.-led, anti-ISIS coalition. NASWA, and its Middle East region in particular, have a long history of conflict, and the present turmoil is as intractable as ever. As one Arab author recently observed, history in this part of the world does not move forward: it slips sideways. That insight is deeply etched on the realm's ever-changing political map.

NASWA's political instabilities, raging conflicts, social dislocations, and associated economic problems have generated a rising tide of refugee flows in recent years. No longer limited to internal and short-distance, cross-border population shifts, these migratory movements are now increasingly directed toward Europe. As Figure 6-9 indicates, a majority of the Europe-bound refugees attempt to cross the Mediterranean Sea from the coasts of Algeria, Tunisia, Libya, and Egypt, with most aiming to reach Italian (EU) soil. The distance to Lampedusa, Italy's southernmost island south of Sicily, is just over 100 kilometers (65 mi) from Tunis and about 250 kilometers (150 mi) from Tripoli. But these desperate illegal migrants are completely in the hands of smugglers, who charge exorbitant prices to cram as many as possible into their rickety boats for the dangerous voyage north. Not surprisingly, the Libyan coast serves as the primary staging area because that country has no functioning government left to control this form of human trafficking.

A closer look at Figure 6-9 shows that a large number of refugees originate in troubled areas of Subsaharan Africa. For them,

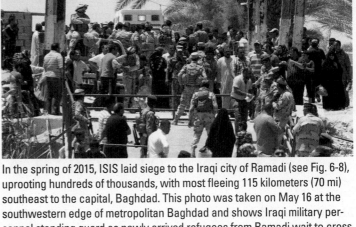

In the spring of 2015, ISIS laid siege to the Iraqi city of Ramadi (see Fig. 6-8), uprooting hundreds of thousands, with most fleeing 115 kilometers (70 mi) southeast to the capital, Baghdad. This photo was taken on May 16 at the southwestern edge of metropolitan Baghdad and shows Iraqi military personnel standing guard as newly arrived refugees from Ramadi wait to cross the Euphrates River on the Bzeibez floating bridge. The next day, ISIS captured and took full control of Ramadi, a key step in its quest to overrun Baghdad. In the meantime, Iraqi government forces were regrouping to plan the battle to retake Ramadi. But this successful advance of ISIS had showcased its considerable military capabilities and was anxiously monitored by many in Iraq and the outside world.

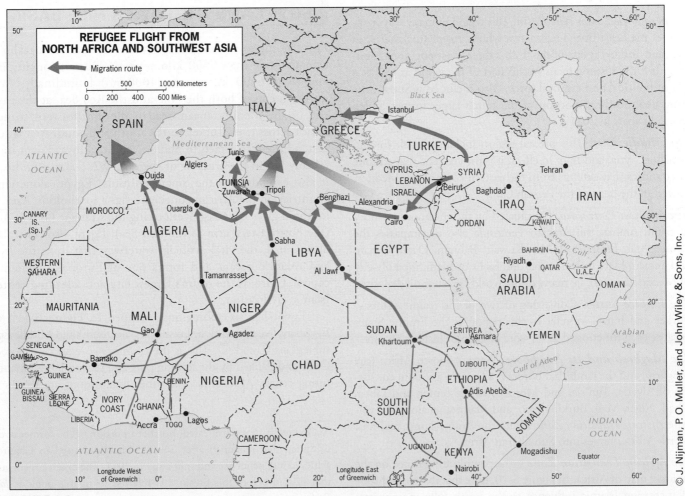

FIGURE 6-9

several northern African ports on the Mediterranean serve as entry points into Europe, the final step after a very long overland journey. To some extent, smugglers and their migrant clients who ply these trans-African routes are facilitated by geography in their efforts to evade detection: the inland Saharan components of Algeria, Libya, and Egypt lie in the vast desert that lacks natural boundaries as well as clearly demarcated, patrolled international borders.

The hardest part for most refugees, whatever their source, begins at the Mediterranean. The latest numbers are staggering:

for 2014, European authorities reported more than 200,000 detected crossings by undocumented immigrants (estimates for 2015 go much higher). Of those who complete the harrowing sea passage, thousands are refused asylum and are forced to return to their country of origin. Many others never make it. During the first four months of 2015, more than 1800 refugees drowned in Mediterranean waters (with about 700 perishing in a single mid-April incident); projections for the entire year range as high as 30,000 deaths.

REGIONS OF NORTH AFRICA/SOUTHWEST ASIA

Identifying and delimiting regions in the sprawling North Africa/Southwest Asia realm is quite a challenge. Population clusters are widely scattered in some areas, highly concentrated in others. Cultural transitions and landscapes—internal as well as peripheral—make it difficult to discern a regional framework. This, furthermore, is a highly changeable realm and always has been. Nonetheless, we can distinguish the following six regions that make up this far-flung realm today (Fig. 6-10):

1. *Egypt and the Lower Nile Basin.* This region in many ways constitutes the heart of the realm as a whole. Egypt (together with Iran and Turkey) is one of the realm's three most populous countries. It is the historic focus of this part of the world as well as a major political and cultural force. Also included is Sudan, the truncated, northern portion of the much larger former Sudan that split apart in 2011.

2. *The Middle East.* This pivotal region includes Israel, Jordan, Lebanon, Syria, and Iraq. In effect, it is the crescent-shaped zone of countries that extends from the eastern coast of the Mediterranean to the head of the Persian Gulf.

3. *The Arabian Peninsula.* Dominated by the enormous territory of Saudi Arabia, the Arabian Peninsula also encompasses the United Arab Emirates (UAE), Kuwait, Bahrain, Qatar, Oman, and Yemen. Here lie the source and focus of Islam, the holy city of Mecca; here, too, lie many of the world's greatest oil deposits.

4. *The Empire States.* Consisting of two of the realm's giants—Turkey and Iran—we refer to this region as the Empire States because each functioned as the core of a major historic empire.

5. *The Maghreb and Its Neighbors.* Western North Africa (the *Maghreb*) and the lands that border it also form a region, consisting of Algeria, Tunisia, and Morocco at the center, and Libya, Chad, Niger, Mali, Burkina Faso, and Mauritania along the volatile *African Transition Zone* where the Arab-Islamic realm of North Africa merges into Subsaharan Africa.

6. *The African Transition Zone* is the complicating factor on the regional map of Africa. In Figure 6-10, note that this striped zone of increasing Islamic influence to the north fully incorporates

certain countries (e.g., Somalia in the east and Senegal in the west) while cutting across others, in the process creating Islamized northern sectors and non-Islamic southern ones (e.g., Nigeria, Chad, Ivory Coast). Note, too, that an elongated strip of Islamic dominance extends southward from the Horn of Africa, lining the Indian Ocean coastlands of both Kenya and Tanzania.

EGYPT AND THE LOWER NILE BASIN

Egypt occupies a pivotal location in the heart of a realm that extends more than 9600 kilometers (6000 mi) longitudinally and some 6400 kilometers (4000 mi) latitudinally. At the northern end of both the Nile and the Red Sea, at the eastern end of the Mediterranean Sea, in the northeastern corner of Africa across from Turkey to the north and Saudi Arabia to the east, adjacent to Israel, to Sudan,* and to Libya, Egypt unmistakably lies in the crucible of the NASWA realm. And because it also extends into the Sinai Peninsula, Egypt, alone among countries on the African continent, has a foothold in Asia. This foothold gives it a coast overlooking the strategic Gulf of Aqaba (the northeastern arm of the Red Sea). Egypt also controls the Suez Canal, the vital artificial waterway that links the Indian and Atlantic oceans and forms one of Europe's lifelines. The capital, Cairo, is the realm's largest city and a leading center of Islamic civilization.

Egypt has six subregions, mapped in Figure 6-11. Most Egyptians live and work in Lower (i.e., northern) and Middle Egypt, subregions ① and ②. This is the country's core area, anchored by Cairo and flanked by the leading port and second-largest manufacturing center, Alexandria. The other four regions are remote and

---

Sudan refers to the truncated, northern portion of the former country of Sudan that remained after South Sudan (discussed in Chapter 7) achieved independence and split away in 2011. The much larger country that existed before this breakup is referred to in this book as *former Sudan.*

POLITICAL UNITS AND GEOGRAPHIC REGIONS OF NORTH AFRICA/SOUTHWEST ASIA

Egypt and the Lower Nile Basin

The Maghreb and Its Neighbors

Middle East

Arabian Peninsula

The Empire States

African Transition Zone

POPULATION

- Under 50,000
- 50,000–250,000
- 250,000–1,000,000
- 1,000,000–5,000,000
- Over 5,000,000

National capitals are underlined

— Railroad
— Road
---- Canal

0   200   400   600   800   1000 Miles
0   400   800   1200   1600 Kilometers

FIGURE 6-10

© J. Nijman, P. O. Muller, and John Wiley & Sons, Inc.

sparsely populated, especially Upper Egypt ③, the Western Desert ④, and the Sinai ⑥. The northern stretches of the Eastern Desert ⑤ are now in the path of Cairo's encroaching urban sprawl (see photo).

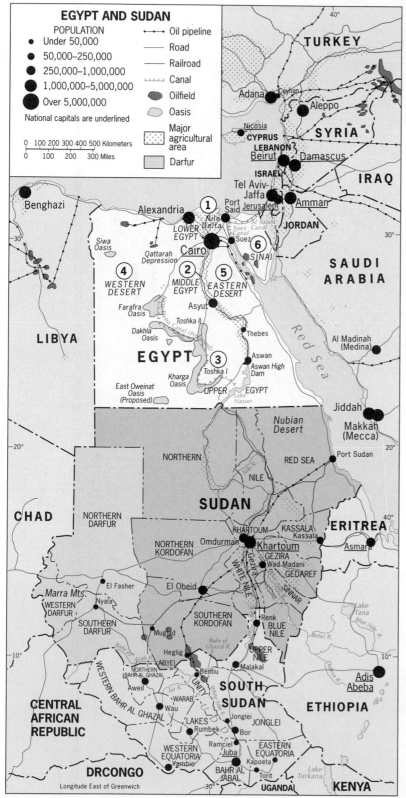

FIGURE 6-11

© Jan Nijman

Gift of the Nile

Egypt's Nile is the aggregate of the two great branches upstream: the White Nile, which originates in the streams that feed Lake Victoria in East Africa, and the Blue Nile, whose source lies in Lake Tana in Ethiopia's highlands. The two Niles converge at Khartoum, capital of present-day Sudan. About 95 percent of Egypt's 82.1 million people live within 20 kilometers (12 mi) of the great river's banks or in its delta (Figs. 6-11, 6-2).

Before dams were constructed on the Nile, the ancient Egyptians used *basin irrigation*, building fields with earthen ridges and trapping the annual floodwaters with their fertile silt, to grow their crops. That practice continued for thousands of years until, during the nineteenth century, the construction of permanent dams made it possible to irrigate Egypt's farmlands year-round. These dams, with locks for navigation, controlled the Nile's annual flood, expanded the country's cultivable area, and allowed farmers to harvest more than one crop per year on the same field. The largest of these dams, the Aswan High Dam completed in 1970, created Lake Nasser (the reservoir that extends southward 150 kilometers [100 mi] into northern Sudan) and increased Egypt's irrigable land by nearly 50 percent. Today it still provides a sizeable proportion of the country's electricity.

Much farther upstream, in the western highlands of Ethiopia where the Blue Nile originates, its government is constructing the Grand Ethiopian Renaissance Dam, with more than one-third of the funding coming from China. After completion in 2017, the dam is expected to boost electrical power production to meet all domestic demands as well as earn new revenues from sales to Ethiopia's energy-deficient neighbors. But the project has also attracted growing opposition (and raised the potential for conflict) in downstream Sudan and Egypt, who fear a reduction in water supplies flowing into the lower Nile Basin.

Egypt's elongated oasis along the Nile, just 5 to 25 kilometers (3 to 15 mi) wide, broadens north of Cairo across a delta anchored in the west by the historic city of Alexandria and in the east by Port Said, gateway to the Suez Canal. The delta contains extensive farmlands, but it is a troubled area today. The ever more intensive use of the Nile's water and silt upstream is depriving the delta of desperately needed replenishment. And the low-lying delta is also at risk due to geological subsidence and sea-level rise, increasing the danger of salt-water intrusion from the Mediterranean that can severely damage soils.

Political Upheaval and Economic Challenges

Before 2011, Egypt had experienced steady economic growth for almost a decade at an annual rate of around 5 percent. At the same time, its annual rate of population growth declined while literacy and per-capita consumption increased. Egypt became an exporter of fruits, vegetables, rice, and textiles in addition to its longtime staple, cotton. The country also nurtured a thriving tourist industry that accounted for about one-tenth of GDP. And the exploitation of (modest) oilfields in the Sinai and adjacent Red Sea (Fig. 6-11) resulted in reduced expenditures on fuel imports.

© Jan Nijman

"Taking off from Cairo International Airport at the northeastern edge of this megacity of more than 18 million, we flew over one of Cairo's newest housing developments at the edge of the northern sector of the Eastern Desert. Look at this residential density—in a desert! With a birth rate more than twice that of the United States, Egypt's critical problem is the provision of a sufficient water supply—which right now is steadily decreasing. Earlier on this day in November 2009, we took a field trip through this area led by a local geographer. These are middle- and upper-middle-class residential complexes. All those new apartment towers will need water, and so will anything green planted here. Where will these people work and how will they get there? Is this sustainable urban development?" [For some possible answers, see the concluding paragraph on Egypt that discusses plans for a new, sustainable capital city to be located somewhere out here.]

But then came the tumultuous events of the Arab Spring and its aftermath, which devastated the tourist industry and damaged much of the rest of the economy. Despite the efforts of the military-controlled Egyptian government to facilitate a return to stability, annual economic growth rates have hovered around 2 percent since 2011. Politically, in the longer run Egypt may be moving toward the so-called **Turkish Model** **16**—an espoused multiparty democracy (though marked by autocratic tendencies) that has a place for Islamic parties, yet is not dominated by them. Meanwhile, to regain economic momentum, the regime has initiated major infrastructural projects to spur employment and reduce popular discontent. In 2014, a consortium of domestic and foreign companies began work on the enlargement of the Suez Canal to accommodate two-way traffic and double the Canal's capacity; unlike the Panama Canal, no locks need to be built and the project is to be completed in late 2015. Plans have also been unveiled for the construction of a new capital by the early 2020s, to be located just to the east of Cairo. With the backing of foreign investors, this ultramodern city would be designed and built in full accordance with principles of sustainability, emphasizing state-of-the-art healthcare, educational, and recreational facilities for its planned five million residents.

Sudan

As Figure 6-11 indicates, Egypt is bordered by two countries that have recently gone through even more turbulent times than Egypt itself: Sudan to the south and Libya to the west. Sudan, the northern remnant of the 2011 breakup of former Sudan, is almost twice the size of Egypt (but with barely 45 percent of the population),

and lies centered on the confluence of the White Nile (from Uganda) and the Blue Nile (from Ethiopia). Here the capital, Khartoum, anchors a sizeable agricultural area where cotton was planted during colonial times. Almost all of Sudan is desert, with irrigation-based farming along the banks of the White Nile and the Blue Nile. The country also has a 500-kilometer (300-mi) coastline on the Red Sea, where Port Sudan lies almost directly across from Jiddah and Makkah (Mecca) in Saudi Arabia.

Note that a large part of Sudan is situated in the African Transition Zone (Figs. 6-10 and 6-19), which helps to explain some long-running regional conflicts. The triad of southwestern provinces named *Darfur* (Fig. 6-11) are inhabited by the *Fur*, a mix of Arabized pastoralists and settled farmers who in past years apparently sympathized with anti-Muslim rebels in the far south of former Sudan. This pretext led to an especially gruesome conflict with the Khartoum government's military forces in the early 2000s, and as many as 400,000 (mostly civilians) may have died. The southern part of former Sudan, in contrast to the north, was in many ways a typical equatorial African region, dominated culturally by Christianity and animist African religions. Ongoing conflicts between north and south were exacerbated by the discovery of oil on contested land. A long, bloody war ensued, with an estimated 1.5 million deaths, and finally ended in the breakup of former Sudan, with the south gaining independence in 2011 as the newborn country of South Sudan (discussed separately in Chapter 7).

Sudan today is a far more homogeneous Islamic state, with a Muslim population of no less than 95 percent (compared to the 68 percent of pre-breakup Sudan). Even though the still unofficial international boundary puts the majority of former Sudan's oil reserves in South Sudan, the pipelines for export run northward across post–2011 Sudan. Both Sudan and South Sudan remain involved in costly military operations and internal as well as cross-border conflicts with each other—while their populations remain in critical need of economic and political progress.

THE MIDDLE EAST

The regional term **Middle East**, as noted at the beginning of this chapter, is not satisfactory, but it is so common and generally used that avoiding it creates more problems than it solves. It originated when Europe was the world's dominant realm and when places were *near*, *middle*, and *far* from Europe: hence a Near East (Turkey), a Far East (China, Japan), and a Middle East (Egypt, Arabia, Iraq).

Five countries constitute the Middle East (Fig. 6-12): Iraq, largest in population and territorial size, facing the Persian Gulf; war-torn Syria, next in both categories and fronting the Mediterranean; Jordan, linked by the narrow Gulf of Aqaba to the Red Sea; Lebanon, whose survival as a unified state is a perennial question; and Israel, Jewish nation in the crucible of the Islamic domain. The complexity of this region today is heightened by the presence of two nonstate entities: Kurdistan and ISIS. Because of the extraordinary importance of the region in world affairs, we focus on its cultural, economic, and political geography.

© J. Nijman, P. O. Muller, H. J. de Blij, and John Wiley & Sons, Inc.

MIDDLE EAST

- Shi'ite area, Iraq
- Shi'ite-Sunni area, Iraq
- Sunni area, Iraq
- Kurdish majority
- ISIS-controlled area, July 2015

- Road
- Railroad
- Pipeline
- Canal
- Oil reserve

National capitals are underlined

0   50   100   150   200   250 Kilometers
0   50   100   150 Miles

Longitude East of Greenwich

FIGURE 6-12

Iraq's Enduring Importance

Even a glance at the regional map (Fig. 6-12) underscores why Iraq is crucial among the states of the Middle East. With about 60 percent of the region's total area, 45 percent of its predominantly Arab population, and most of its valuable energy and agricultural resources, Iraq is key to the Middle East's fortunes. Having no less than six neighbors, it is no surprise that Iraq is nearly landlocked, and this map shows just how narrow and congested its single outlet to the Persian Gulf is. As a result, a network of pipelines across Iraq's neighbors must carry much of Iraq's oil export volume to coastal terminals in other countries.

Iraq is divided into three cultural domains in which religion, ethnicity, and tradition form the basis of its political geography (Fig. 6-12). The largest population sector is composed of about 24 million Shi'ites (73 percent of the national total of 33.4 million) in the southeast, whose religious affinities are with neighboring Iran. The next largest is the Sunni minority of about 8 million (24 percent) in the north and west. Perhaps the most striking feature on this map is the striped zone that blankets northeastern Iraq, large parts of Turkey and Iran, and a few pockets of northern Syria. In this territory the majority of inhabitants are not Arabs but Kurds—regarded by many as the world's biggest stateless nation. More specifically, about 6 million Kurds (18 percent of the Iraqi population) live in the northern highlands of Iraq; adjacent southeastern Turkey is the traditional home of at least 15 million more; another 6 million or so inhabit northwestern Iran; and 2 million congregate in the scattered Kurdish clusters of northern Syria.

The U.S.-Led Invasion and Its Aftermath

The United States invaded Iraq in 2003, claiming that it possessed nuclear and other weapons of mass destruction and that al-Qaeda had important bases there. Neither charge proved to be true, and the invasion has been heavily criticized ever since. When American combat troops finally departed in 2011, the dictator Saddam Hussein had been eliminated, a new democratically elected government was functioning, and U.S.-trained police and military forces were in place.

But the U.S. incursion had also unleashed horrendous sectarian violence between majority Shi'ites and minority Sunnis as well as enabling al-Qaeda terrorists (who had not previously been active inside Iraq) to greatly undermine the country's security situation. Even more ominously, these events empowered the Sunni jihadists of ISIS (as discussed above), who have now established themselves in the north and west, challenged the government in Baghdad, and effectively plunged Iraq into civil war. The United States, reluctantly drawn back into the conflict in 2015, formed a large multinational coalition but confined its involvement to air strikes on ISIS military targets in northern Iraq and Syria—with little immediate effect. Take another look at the geopolitical patchwork mapped in Figure 6-12, and it appears that classifying Iraq as a national territorial state is no longer realistic.

A Kurdish State in the Making?

The Kurdish homeland in northern Iraq is now independent in all but name. It has benefited from American support (both financial and diplomatic) over the past decade; it has become politically stabilized and is functioning more or less democratically; and its economy is fueled by the significant revenues it earns from the major oilfields it exploits. By 2015, the process aimed toward achieving statehood (including a popular referendum) was well under way. However, this is a most delicate political process because 'Kurdistan' extends well beyond Iraq into Turkey and Iran—strongly organized states that have no interest in setting a precedent that might force them into territorial concessions.

The Kurdish nation is the product of diverse ethnic influences. Its language has Persian rather than Arabic roots. The majority adhere to Sunni Islam, although the Kurds are known to be rather pluralistic in their religious beliefs and practices. Like many other Sunni communities in Iraq and Syria, the Kurds have no sympathy for ISIS, whose violent jihadist offensives continue to be met with fierce resistance from Kurdish armed forces supported by the air power of the U.S.-led coalition.

Syria

Like Lebanon and Israel, Syria possesses a Mediterranean coastline where crops can be raised without irrigation. Behind this densely populated coastal zone, Syria has a much larger interior than these neighbors, but its areas of productive capacity are widely dispersed and many are unusable during the current civil war (Fig. 6-12). Damascus, the capital located in the southwestern corner of the country, was built on an oasis and is considered to be one of the world's oldest continuously inhabited cities.

Although Syria's population of 22.8 million is 73 percent Sunni Muslim, the ruling elite has long come from a small Shi'ite Islamic sect based in the country—the Alawites—who account for only about 12 percent of the population. Leaders of this powerful minority have retained control over the state for decades by ruthlessly suppressing dissent. In 2000, president-for-life Hafez al-Assad died and was succeeded by his son, Bashar.

When the Arab Spring movement reached Syria in 2011, it was immediately met with a savage military crackdown by the Assad regime. But popular resistance did not disappear, and the violence swiftly escalated into a full-fledged civil war marked by regime-inflicted atrocities on civilian populations in cities as well as the countryside. By mid-2015, at least 300,000 had died in the violence, and countless additional thousands were reported to be incarcerated or missing. This conflict has also displaced at least 11 million people—7.6 million forced to relocate within Syria plus nearly 4 million more who fled the country for the squalid refugee camps of neighboring Turkey, Jordan, and Lebanon. Internationally, even though the murderous Assad regime has been almost universally condemned for perpetrating the worst humanitarian crisis of our time, it continues to receive steady (if tacit) support from Russia, Iran, and China.

As discussed earlier, it did not take ISIS long to enter the fray in Syria, and soon its own brutalities reached a level that undercut international criticism and opposition to the Syrian regime. Today ISIS is widely regarded as the greater threat, especially since it expanded its operations to also become a leading aggressor in Iraq. Figure 6-8 starkly illustrates how Syria has fragmented into several parts since 2011, with each part controlled by a different faction—the Syrian government, opposition forces, ISIS, and Kurdish forces. But keep in mind that this map is only a snapshot of the fluid situation on the ground during July of 2015. Yet, however this political terrain eventually stabilizes, even more so than in Iraq it is extremely difficult to imagine a future restoration of Syria as an integrated national territorial state.

Jordan

With a poverty-stricken capital (Amman), lacking in oil reserves, and possessing only a small and remote outlet to the Gulf of Aqaba, Jordan (population 6.5 million) has survived with U.S., British, and other aid. It lost its West Bank territory in a 1967 war with Israel, including its sector of Jerusalem (then the kingdom's second-largest city). No third country has a greater stake in a settlement between Israel and the Palestinians than does Jordan, especially because Palestinians are the majority here.

Jordan did not experience the kind of upheaval that shook many other Arab countries in 2011 because the monarchy had enjoyed widespread support and government policies lacked the level of repression found in Syria or Tunisia. Yet there were some demonstrations to air grievances, and King Abdullah II had to dismiss his cabinet to quell the protests. Jordan's long-term challenges are primarily economic. The wider regional turmoil has not only deterred tourism but brought in more than 700,000 refugees across Jordan's northern border with war-ravaged Syria. It is also important to recognize the remarkable long-term stability of this country within this most volatile of political environments—perhaps comparable to Morocco in this regard, and suggesting that responsible NASWA monarchies have the ability to sense and respond to the pulse of the people.

Lebanon

The map suggests that Lebanon has significant geographic advantages in this region: a lengthy coastline on the Mediterranean Sea; a well-situated capital, Beirut, on its shoreline; oil terminals along its coast; and a major capital (Syria's Damascus) in its hinterland. The scale of Figure 6-12 cannot reveal yet another asset: the fertile, agriculturally productive Bekaa Valley in the eastern interior. Nevertheless, if Lebanon is well endowed from a physical-geographic perspective, its chaotic cultural geography and fragile geopolitical circumstances are problematic. As a result, the country has gone back and forth between good times and bad. Lebanon (population 4.5 million) is sometimes aptly described as a 'garden without a fence' because of its vulnerability to outside interference (especially from neighboring Syria), its latest burden being the 1.1 million

refugees it has received from Syria's civil war through mid-2015. With so many external powers and influences involved in its affairs, Lebanon is unlikely to achieve stability and significant progress in the foreseeable future.

Israel and the Palestinian Territories

Israel, the Jewish state, lies in the center of the Arab domain (Fig. 6-10). Since 1948, when Israel was created as a homeland for the Jewish people on the recommendation of a United Nations commission, the Arab-Israeli conflict has cast a deep shadow across this entire realm. Figure 6-13 helps us to understand the complex issues involved here. In 1946, the British, who had administered this area in post-Ottoman times, granted independence to what was then called Transjordan, the kingdom east of the Jordan River. In 1948, the orange-colored area became the UN-sponsored state of Israel—including, of course, land that had long belonged to Arabs in this territory called Palestine.

As soon as Israel proclaimed its independence, it was attacked by neighboring Arab states. Israel, however, not only held its own but pushed the Arab forces back beyond its borders, gaining the green-colored areas in 1949 (Fig. 6-13). Meanwhile, Transjordanian

Israel's Security Barrier takes several forms—a concrete wall, a spiked metal fence—and its effect also has multiple dimensions. This section of the barrier separates the village of Abu Dis on the outskirts of Jerusalem (left) from the West Bank (right). While Israel defines this demarcation as a security issue, the Arabs, citing its location inside Palestinian territory, argue that it is motivated primarily by politics.

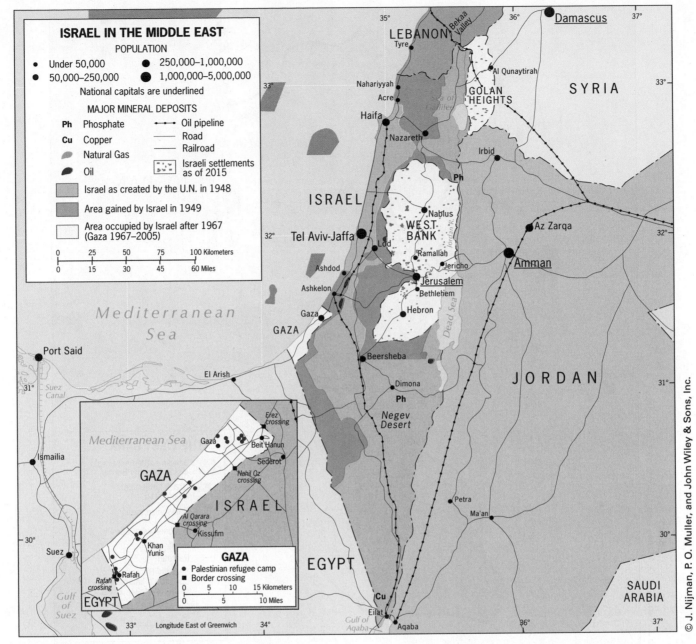

FIGURE 6-13

armies crossed the Jordan River and annexed the yellow-colored area named the West Bank (of the Jordan River), including part of the city of Jerusalem. The king called his newly enlarged country Jordan.

Further conflict was to follow. In 1967, a six-day-long war produced a major Israeli victory: Israel took the Golan Heights from Syria, the West Bank from Jordan, and the Gaza Strip from Egypt; it also conquered the entire Sinai Peninsula all the way to Egypt's Suez Canal (Fig. 6-13). In later peace agreements, Israel returned the Sinai to Egypt but not the Gaza Strip.

All this strife produced a huge outflow of Palestinian Arab refugees and displaced persons. The Palestinian Arabs, therefore, constitute another of this realm's stateless nations; about 1.7 million continue to live as Israeli citizens within the borders of Israel, but more than 2.7 million now reside in the West Bank and another 1.7 million in the Gaza Strip. Many Palestinians also live in neighboring and nearby countries, including Jordan (3.2 million), Syria (630,000), Lebanon (440,000), and Saudi Arabia (280,000); another 600,000 or so reside in Iraq, Egypt, and Kuwait; and some live in other countries around the world, including the United States (ca. 350,000). The global total of the stateless Palestinian population in 2015 was estimated to be 11.8 million.

Israel is approximately the size of Massachusetts and has a population of 8.1 million (including its 1.7 million Arab citizens), but because of its location, its powerful military, and its strong

© J. Nijman, P. O. Muller, and John Wiley & Sons, Inc.

FIGURE 6-14

international links, these data do not reflect Israel's importance. Israel has been a recipient of massive U.S. financial aid, and American foreign policy has been to seek an accommodation between Jews and Palestinians, as well as between Israel and its Arab neighbors. Washington continues to strongly pushing in the direction of a *two-state solution*, but neither side expressed the political will to pursue that route during 2015.

Geographic Obstacles to a Two-State Solution

1. *The West Bank.* This would be the centerpiece of a future Palestinian state, but its 2.7 million Palestinians are being joined by growing numbers of Israeli Jewish settlers—who in mid-2015 numbered around 370,000, dispersed across a discontinuous network of 237 settlements fortified by walls and/or security barriers (Fig. 6-14). Moreover, the West Bank is almost entirely bounded by the Israeli-built *security fence*.

2. *The Security Fence.* In response to the infiltration of suicide terrorists, Israel has now almost completed the walling-off of the West Bank along the Security Barrier border mapped in Figure 6-14. This reinforcement, which in places penetrates deeply into former Palestinian areas, imposes much inconvenience for Palestinians as well as outright hardship for many where it runs between their homes and farm fields. Palestinians demand that

Jerusalem's urban landscape is rife with cultural contrasts—and is also home to the holy sites that mean so much to Jews, Christians, and Muslims.

the wall be taken down; Israelis reply that the Palestinian National Authority (the interim self-government body) must control the terrorists.

3. **_Jerusalem._** The United Nations intended Tel Aviv to be Israel's capital, and Jerusalem an international city (see photo). But the 1948–1949 war ended with Israel holding the western part of the city and Arab forces the eastern sector. Then, in the 1967 war, Israel conquered all of the West Bank, including East Jerusalem; in 1980, the Jewish state reaffirmed Jerusalem's status as its capital, calling on all countries to relocate their embassies from Tel Aviv (some did for a while, but none are there today). Meanwhile, the government redrew the map of the ancient city, building Jewish settlements in a ring around East Jerusalem that would obliterate the old distinction between a Jewish west and an Arab east. This enraged Palestinian leaders, who still view Jerusalem as the eventual headquarters of a future Palestinian state. Renewed U.S. efforts to persuade Israel to halt the construction of new settlements in East Jerusalem have so far been futile. Also note, in Figure 6-14, that the Security Barrier runs well to the east of Jerusalem today.

4. **_The Gaza Strip._** Israel in 2005 decided to withdraw from the Gaza Strip, removing all Jewish settlements there and yielding control of this territory to the (then) Palestinian Authority (Fig. 6-13, inset map). Gaza has few natural resources, and its near-encirclement by Israel puts it in a most precarious position. Relations between Israel and Hamas (the extremist Palestinian faction that rules Gaza) are severely strained and repeatedly lapse into open conflict—usually in the form of rocket attacks from Gaza into Israel, followed by punishing Israeli retaliation.

Israel lies in the path of an impetuous geopolitical storm. The four obstacles just discussed are only part of the overall problem: others involve water rights, compensation for expropriated land, Arabs' 'right of return' to pre-refugee abodes, and the exact form a Palestinian state would take.

THE ARABIAN PENINSULA

The regional identity of the Arabian Peninsula is clear: south of Jordan and Iraq, the entire peninsula is encircled by water. This is a region of (mainly) old-style sheikdoms made fabulously wealthy by oil, modern-looking emirates, and the site where Islam originated. But it also includes, at its southern periphery on the Gulf of Adan, the economically deprived country of Yemen, which in 2014 collapsed into political chaos and war.

As a region, the Arabian Peninsula is environmentally dominated by a harsh desert habitat and politically dominated by the Kingdom of Saudi Arabia (Fig. 6-15). More than twice the territorial size of Egypt, Saudi Arabia is the realm's second-biggest state after Algeria. Saudi Arabia's neighbors (proceeding clockwise from the head of the Persian Gulf) are Kuwait, Bahrain, Qatar, the United Arab Emirates (UAE), the Sultanate of Oman, and the Republic of Yemen. Together, these countries on the eastern and southern fringes of the peninsula contain 44 million inhabitants; the largest by far is Yemen, with just under 25 million.

Saudi Arabia

Saudi Arabia itself has only 28.8 million residents (plus about 8 million expatriate workers, at least one-fourth of whom are undocumented) in its vast territory, but we are already aware of the kingdom's importance on the region's oil map (Fig. 6-6): the Arabian Peninsula contains the world's largest concentration of known petroleum reserves. As Figure 6-15 shows, these reserves lie in the eastern part of the country, particularly in the vicinity of the Persian Gulf, and extend southward into the Rub al Khali (Empty Quarter).

Figure 6-15 also reveals that most of the economic activities in Saudi Arabia are concentrated in a wide belt across the 'waist' of the peninsula, from the oil boomtown of Dhahran on the Persian Gulf through the national capital of Riyadh in the interior to the Mecca–Medina area in the far west near the Red Sea. A modern transportation and communications network has recently been completed, but in the more remote zones of the interior, Bedouin nomads still ply their ancient caravan routes across the vast deserts.

Efforts to reduce Saudi Arabia's regional economic disparities have been impeded by the enormous cost of bringing water from deep underground sources to the desert surface to stimulate agriculture, by the sheer size of the country, and by a population growth rate that is still 50 percent above the global average. Nonetheless, housing, healthcare, and education have experienced major improvements, and industrialization also has been stimulated in such newer cities as Jubail on the Persian Gulf and Yanbu on the Red Sea. In fact, the latter is part of the historic Mecca–Medina corridor, which is now modernizing rapidly.

Saudi Arabia's conservative monarchic rule, official friendship with the West, and enormous social contrasts stemming from ongoing economic growth have raised political opposition, for which no adequate channels exist. Consider this: women are still not allowed to drive and were allotted the right to vote for the first time in 2015 (in municipal elections); women cannot even travel

ARABIAN PENINSULA

POPULATION

- • Under 50,000
- • 50,000–250,000
- ● 250,000–1,000,000
- ● 1,000,000–5,000,000
- ● Over 5,000,000
- Ⓔ Economic City

National capitals are underlined

LAND USE

- Dry farming
- Desert
- Permanent crops and arable lands

	Oilfield
	Gasfield
▬▬	4-lane road
────	Major road
────	Road
•─•─•	Oil pipeline
•◆•◆•	Gas pipeline

0 100 200 300 400 500 Kilometers

0 100 200 300 Miles

Longitude East of Greenwich

FIGURE 6-15

today without a male guardian! Even though things remained relatively quiet here, the Saudi regime watched nervously as the Arab Spring uprisings reverberated across the realm in 2011. By no coincidence, the government was quick to announce a new jobs program involving billions of dollars while beefing up police presence on the streets. But Saudi Arabia's planners have also learned to think much bigger than this, and their grandiose plans now call for nothing less than a complete economic transformation of the country.

The Saudis have good reason to invest in their economic future. Today, 75 percent of the population is under the age of 30, and

unemployment among these young people may run as high as 40 percent. In order to accelerate the generation of the millions of jobs that will be needed, the Saudi royal rulers are spending their country's oil money on six new *economic cities* that by 2020 will create 1.3 million new jobs (see Fig. 6-15). Overall, in anticipation of the end of the oil era, the ultimate goal is to transform Saudi Arabia from a petrodollar-based economy into a world-class industrial power.

Saudi Arabia is an important player in the geopolitics of this realm and a longtime American ally—even if the United States and Saudi Arabia make a strange pair at times, given Saudi Arabia's highly imperfect democratic record. The Saudis are mostly orthodox Sunnis and they eye the rising regional influence of Shi'ite Iran with much trepidation. In 2015, they intervened in the escalating conflict in neighboring Yemen (see below) with air strikes to push back Houthi rebels, a Shia militia that was advancing against the Sunni forces of the beleaguered Yemeni government.

The Rise of the Gulf States

Five of Saudi Arabia's six neighbors on the Arabian Peninsula face the Persian and Oman gulfs: Kuwait, Bahrain, the United Arab Emirates, Qatar, and Oman (Fig. 6-15). All are known for their oil reserves, from which they earn a consequential foreign income. On a GDP per-capita basis, these five are all high-income countries; they are also highly urbanized and quite modern in certain ways, but traditional in others; and all are monarchies in the (Sunni) Islamic tradition. The response to popular demands during the short-lived Arab Spring was comparable to that in Saudi Arabia: overly modest political concessions combined with increased public spending to appease the masses. Another similarity is that they each contain a substantial population of foreign workers.

The Gulf states generally maintain cooperative relations with the United States. They are generally anti-jihadist, and their economies are becoming ever more globalized. The recent rise of these states is the result of a well-planned diversification away from a singular reliance on oil revenues. They seek various niches in the global economy, such as Bahrain's emergence as a center of international finance. Most spectacular is the UAE's burgeoning Dubai. This ultramodern city has turned itself into a world-scale air

Foreign laborers at work on the construction of Qatar's al-Wakrah Football Stadium in 2015. This is one of nine (!) new stadiums being built for the 2022 World Cup tournament. The decision to award this highest-profile sports event to Qatar has been widely criticized, not least because summer temperatures here are among the hottest on Earth (an average July day in Doha, the capital, records a high of 42°C [108°F] and a low of 31°C [88°F]). The Qataris have responded by promising to build covered, air-conditioned stadiums—yet another extravagance in a Gulf state that seems to have unlimited money to spend. A far more serious criticism, however, has also been leveled at Qatar: the consistently poor treatment of its (overwhelmingly foreign) workforce.

transportation hub (its airport surpassed London's Heathrow in 2014 to become the busiest on Earth) as well as tourist destination—to be crowned by Dubailand, the world's largest entertainment complex that will open in 2019. Moreover, these countries are particularly adept at promoting themselves through global public relations campaigns—think of the high visibility of the UAE's state-supported airlines (Emirates and Etihad) airlines, or the 2022 FIFA World Cup tournament being awarded to Qatar (see photo), home of the global media giant, Al Jazeera.

These five states have learned how to take advantage of their geographic position on the energy-rich Persian Gulf, whose busy oil-tanker traffic must navigate the treacherous Strait of Hormuz outlet, one of the most strategic choke points in the world ocean (Fig. 6-15). At the regional scale, the leading geographic asset of the Gulf states is their central location in Southwest Asia, providing a stable political and economic haven in an otherwise turbulent realm; moreover, none of these small countries even has a hostile neighbor today. Not a single one of these advantages, however, applies to Yemen, the country that occupies the southern corner of the Arabian Peninsula.

Yemen

With a population of 24.4 million—approximately 56 percent Sunni and 43 percent Shi'ite—Yemen is plagued by an overabundance of poverty, centrifugal political forces, and terrorist activity. In the chaotic mid-2010s, an ongoing Shi'ite rebellion against the central government has destabilized the north even as a secessionist movement continues to gain ground in the south, where oilfields encourage such ambitions.

Figure 6-15 highlights the geographic appeal of Yemen for al-Qaeda and other Islamic terrorist groups. Not only does Yemen lie at the back door of Saudi Arabia: its southern tip overlooks another of the world's most crucial and busiest choke points at the Indian Ocean entrance to the Red Sea—the Bab el Mandeb ('Gate of Grief') Strait—where ships tightly converge and run the risk of capture by pirates. Across the Gulf of Adan to the south lies another terrorist haven (and pirate stronghold), the failed state of Somalia, and just across the Red Sea to the west lies Eritrea, a likely target for extremists of the al-Qaeda stripe.

The mass demonstrations of Yemen's Arab Spring in 2011 against the 32-year-old regime of President Ali Abdullah Saleh led to his resignation a few months later. But here, too, the removal of an autocratic government was not followed by a democratic turn. On the contrary, Yemen experienced a political transition that mirrored Libya's—a brief spasm of destabilization, an ensuing power vacuum, and the swift emergence of opposing, armed factions.

By 2014, Yemen's government had been rendered powerless and the country descended into anarchy as three opposing forces began their battle for supremacy. One of the combatants is the **Houthi** tribe, based in the country's far west, that is dominated by Shi'ites; another contender is the powerful jihadist branch of **al-Qaeda of the Arabian Peninsula (AQAP)**, which is expanding its territory westward from the central Gulf of Adan coast; the third belligerent force consists of a Sunni-led secessionist movement in Yemen's southwest. The conflict is further exacerbated by foreign involvement: for example, Iran actively supports the Shi'ite Houthis, whereas Saudi Arabia has conducted air strikes against them to shore up what is left of the country's Sunni-dominated government. If Syria, Iraq, and Libya now constitute a NASWA league of failed states, by early 2015 Yemen had fully qualified to join that roster.

THE EMPIRE STATES

Two major states, both with long imperial histories, dominate the region that lies immediately to the north of the Middle East and Persian Gulf (Fig. 6-10), where Arab ethnicity gives way but Islamic culture endures—Turkey and Iran. Although they share a short border and are both Muslim countries, they display significant differences as well. Even their versions of Islam are different: Turkey is an officially secular but dominantly Sunni state, whereas Iran is the heartland of Shi'ism. Turkey's leaders have worked to establish satisfactory relations with Israel; on several occasions, Iran's leadership promised to 'wipe Israel off the map.' Turkey has

been involved in negotiations aimed at entry into the European Union. Iran's goals have been quite different as that country has defied Western efforts to constrain its nuclear ambitions—though a mid-2015 agreement may alter that direction.

These parts of Southwest Asia are a long way from North Africa, and given their location, they are geographically linked to regions outside the realm, particularly Central Asia and westernmost South Asia. For centuries, Turkey and Iran have been involved in what are now, respectively, Turkmenistan and Afghanistan. Iran's southeastern Baluchi nation extends well into southwestern Pakistan, whereas its far northwestern Azeri minority spills over the border into Azerbaijan. Most importantly, Afghanistan constitutes a transition zone between NASWA and South Asia because that war-ravaged country has been tightly enmeshed in the geopolitics and cultures of both realms (see Fig. 6-10). Today, however, Afghanistan is more firmly connected to South Asia, and as such will be discussed separately in Chapter 8.

Turkey

As Figure 6-16 indicates, Turkey is a mountainous country with generally moderate relief; it also exhibits considerable environmental diversity, ranging from steppe to highland (see Fig. G-6).

© J. Nijman, P. O. Muller, J. H. de Blij, and John Wiley & Sons, Inc.

FIGURE 6-16

On central Turkey's dry Anatolian Plateau, villages tend to be small, and subsistence farmers grow cereals and raise livestock. Coastal plains are not large, but they are productive as well as densely populated. Textiles (from home-grown cotton) and farm products dominate the export economy, but Turkey also has substantial mineral reserves, some oil in the southeast as well as major dam-building projects on the Tigris and Euphrates rivers, and a small steel industry based on domestic raw materials.

The ancient capital of Turkey was Constantinople (long renamed Istanbul), located on the Bosporus, a narrow waterway component of the strategic Turkish Straits that connect the Black and Mediterranean seas. But the struggle for Turkey's survival had been waged from the heart of the country, the Anatolian Plateau, and it was here that Kemal Atatürk (modern Turkey's founding leader who ruled from 1923 to 1938) decided to place his seat of government (Fig. 6-16). Ankara, the new capital, possessed some unique advantages: it would remind the Turks that they were (as Atatürk always said) Anatolians, it lay much closer to the center of the country than Istanbul, and it could therefore act as a stronger unifier.

Even though Atatürk moved the capital eastward and inward, his orientation was westward and outward. To implement his plans for Turkey's modernization, he initiated reforms in almost every sphere of life within the country. Islam, formerly the state religion, lost its official status, and Turkey became a secular state whose armed forces ensured that the Islamists would not take over again. The state took over most of the religious schools that had controlled education. The Roman alphabet replaced the Arabic. A modified Western code supplemented Islamic law. Symbols of old—growing beards, wearing the fez—were prohibited. Monogamy was made law, and the emancipation of women was initiated. The new regime emphasized Turkey's separateness from the Arab World, and in

many ways it has remained aloof from the affairs that engage other Islamic states.

Turkey and Its Neighbors

From before Atatürk's time, Turkey has had a history of intolerance of minorities. Soon after the outbreak of the First World War, the (pre-Atatürk) regime decided to expel the Armenians, concentrated in the country's northeast. Nearly 2 million Turkish Armenians were uprooted and brutally forced out; well over a million died in an eight-year genocidal campaign, the centennial of whose beginning was widely marked in 2015.

In modern times, Turkey also has been criticized for its harsh treatment of its large and regionally concentrated Kurdish population. About one-fourth of Turkey's population of 74.9 million is Kurdish, and successive Turkish governments have mishandled relationships with this minority nation. The historic Kurdish homeland lies in the southeast of Turkey, centered on Diyarbakir (Fig. 6-16), but millions of Kurds have moved to the shantytowns around Istanbul—and to jobs in EU countries. With Kurdish nationalism rising across the border in embattled Iraq, Turkey has responded in two principal ways: by suppressing Kurdish insurgent movements and by awarding more rights and freedoms to those Kurds not involved in armed resistance.

Turkey's overriding foreign policy issue today is Syria. By mid-2015, well over a million Syrian refugees had settled on Turkish soil, and the alleged cross-border movements of Syrian rebel forces had drawn fire from government troops inside Syria. The rapid emergence of ISIS in Syria, especially the fierce battles with Syrian Kurds near the Turkish border, has invited speculation about Turkey's real position: although it is clear that Turkey opposes

@ FROM THE FIELD NOTES...

"Having spent the night in Istanbul's Beyoğlu District on the European side, I walked across the Atatürk (Unkapani) Bridge to the Eminönü District on the south side of the Golden Horn, intending to make it to the upscale Fatih District later. But getting a clear view of the Hagia Sophia 'Museum' was not easy from this angle, standing amid the congested buildup in this hilly area. In a line of taxis at a taxi stand one had a sign saying 'Speak English,' so I asked the driver about getting to a good vantage point, and he showed me on my map how to find a lookout point on the terrace of a local restaurant. From there, I could see the bridge I had crossed to my left, the Hagia Sophia Church with its later minarets in the distance, the Sultan Haseki Hürrem Baths to the right, and in the foreground a simple, local mosque. 'You should know the rules about mosques,' he said. 'The number of minarets reveals the importance of any mosque. One minaret, and it's local, neighborhood. Two, and it's likely to be more prosperous, perhaps the work of a very successful citizen or community. Any more than two, and the builders have to have not only religious sanction but also state permission to build. And if they get that, they go for four.' Thousands of mosques, many of them architectural treasures, grace the urban landscape of Istanbul, whose population today is just over 14 million."

jihadist Islamic forces and formally participates in the U.S.-led anti-ISIS coalition, it would certainly benefit from an overall weakening of the Kurds at the hands of ISIS.

Turkey and the European Union

With its diversified and growing economy, secular government, and independent posture in international affairs, Turkey is the key predominantly-Islamic state potentially eligible for admission into the European Union. Even though formal talks on admission were launched in 2005, progress has been very slow because enthusiasm among Turks for joining the EU has decidedly cooled over the past decade. For one thing, Europe has been forced to deal with severe economic problems while Turkey's economy has performed comparatively well. For another, many Turks take offense at the unrestrained public debate in Europe concerning their candidacy, often including criticisms of Turkish customs and traditions not only in Turkey but also in existing Turkish communities already established in Europe. Therefore, if the EU should eventually extend an invitation to join, it is by no means certain that a future Turkish government would accept. Not surprisingly, polls indicate that popular support for EU membership has plunged from 75 percent in 2000 to barely 27 percent in 2014.

Turkey is a crucial state in a unique geographical position that is trying to steer an independent and singular course within one highly volatile realm while accommodating the demands of another. Its vital importance was dramatically confirmed in 2011 when Arab Spring protesters across NASWA repeatedly espoused the *Turkish Model* as the ideal future scenario for their countries—a mildly Islamist government that does not alienate the religiously devout and that prioritizes democratic reform as it pursues vigorous economic growth.

Iran

Iran has its own history of conquest and empire, collapse and revival. Oil-rich and vulnerable, Iran already was just a remnant of a once-vast regional empire that waxed and waned, from Europe's Danube Basin to the Indian subcontinent's Indus Valley, over more than two millennia. Long known to the outside world as *Persia* (although the people had called themselves Iranians for centuries), the state was officially renamed *Iran* by the reigning shah (king) in 1935. His intention was to stress his country's Indo-European cultural heritage as opposed to its Arab and Mongol infusions. But for all the rhetoric about modernization, the shahs failed to advance the lot of the common people, and by 1979 a revivalist revolution had engulfed the country. The monarchy was replaced by an Islamic republic ruled by an ayatollah (leader under Allah), and Islamists continue to tightly control this Shi'ite bastion today.

A Crucial Location and Dangerous Terrain

As Figure 6-10 shows, Iran occupies a critical position in this tempestuous realm. It controls the entire corridor between the Caspian Sea and the Persian Gulf. To the west it adjoins Turkey and Iraq, both historic enemies. To the north (west of the Caspian Sea) Iran borders both Azerbaijan and Armenia, where once again Islam confronts Christianity. To the east Iran meets Afghanistan and Pakistan, and to the northeast lies similarly volatile Turkmenistan.

As Figure 6-17 reveals, Iran is a country of mountains and deserts. The heart of it is an extensive upland, the Iranian Plateau, that lies surrounded by even higher mountains, including the Zagros in the west, the Elburz in the north along the Caspian Sea coast, and the mountains of Khorasane to the northeast. This mountainous terrain also signals geologic danger: here the Eurasian and Arabian tectonic plates actively converge, causing major and often devastating earthquakes (Fig. G-4). The Iranian Plateau is actually a huge highland basin marked by salt flats and wide expanses of sand and rock. The highlands wrest some moisture from the air, but elsewhere only oases interrupt the arid monotony—oases that for countless centuries have been stops on the area's caravan routes.

The Ethnic-Cultural Map

Today, Iran's population of 77.5 million is 71 percent urban, and the capital, Tehran (population 8.4 million), lies far to the north on the southern slopes of the Elburz Mountains, at the heart of modern Iran's core area.

Iranians dominate Iran's affairs, but the country is no more ethnically unified than Turkey. Figure 6-17 shows that three corners of Iran are inhabited by non-Iranians, two in the west and one in the southeast. Iran's northwestern corner is part of greater Kurdistan, the multistate territory in which Kurds are in the majority and of which southeastern Turkey also is a part (Fig. 6-12). Between the Kurdish area and the Caspian Sea the majority population is Azeri, who number more than 16 million in Iran and also form the ethnic majority (of about 9 million) in the neighboring state of Azerbaijan to the north. In the southwest, concentrated in Khuzestan Province at the head of the Persian Gulf, an Arab cluster predominates. A significant part of Iran's huge oil reserve is located here (Abadan is the main export terminal, shown in Fig. 6-17), but this Arab community is among Iran's poorest and most restive—in a province whose stability could not be more essential to the country's economy. The third ethnically non-Iranian corner lies in the southeast, where a coastal Arab minority focused on the key port of Bandar-e-Abbas on the Strait of Hormuz gives way eastward to a scattered and still-nomadic population of Baluchis, who have strong ties to fellow ethnics in Balochistan Province just across the border of neighboring Pakistan. The Iranian component of the Baluchi homeland is a remote and inconsequential area, but in Pakistani Balochistan that country's government faces a persistent rebellion by tribes who feel slighted by official neglect.

Iran's Choices

Iran is a crucial player in the geopolitics of Southwest Asia. Its rise as a nascent nuclear power and its regional ambitions are hardly

FIGURE 6-17

surprising. Tehran has a major stake in developments in Iraq and Afghanistan, both neighbors; it has an already-nuclear and unstable Pakistan on its southeastern border; it has a strong interest in the fate of Shi'ite majorities and minorities elsewhere in the realm; it has long been an avowed enemy of Israel and a strong supporter of Palestinian causes; and it has funded organizations labeled as terrorist whose actions have even reached across the Atlantic.

But Iran has also suffered for decades from the effects of Western economic sanctions because of its efforts to acquire nuclear capabilities. It possesses the realm's second-largest concentration of oil reserves, which amount to fully half of those in Saudi Arabia. Unlike the Gulf States, Iran is overly dependent on oil and lacks a diversified economy. Slow economic development and the traditional disposition of the Islamic state (even under the ayatollahs) have kept most Iranians in a state of poverty and backwardness. And yet, young urban Iranians can be surprisingly modern (and Western) in their outlook, and there is a palpable yearning among them for openness, democratization, and economic progress.

Iran's worsening economic plight, accentuated by those international embargoes as well as rising domestic discontent, has made the regime ever more receptive to voluntarily downsizing its nuclear ambitions in exchange for an end to sanctions. On July 14, 2015, just such a pact was agreed upon by Iran and the U.S. (partnered by China, Russia, Germany, France, and the UK).

THE MAGHREB AND ITS NEIGHBORS

The countries of northwestern Africa are collectively called the **Maghreb**, but the Arab name for them is more elaborate than that: *Jezira-al-Maghreb*, or 'Isle of the West,' in recognition of the great Atlas Mountain range rising like an enormous island from the Mediterranean Sea to the north and the sandy flatlands of the immense Sahara to the south (see Fig. 6-1).

Whereas Egypt is the gift of the Nile, the Atlas Mountains facilitate the settled Maghreb. These high ranges wrest from the rising air enough orographic rainfall to sustain life in the intervening valleys, where good soils support productive farming. From the vicinity of Algiers eastward along the coast into Tunisia, annual rainfall averages greater than 75 centimeters (30 in), a total more than three times as high as that recorded for Alexandria in Egypt's delta. Even 240 kilometers (150 mi) inland, the slopes of the Atlas still receive over 25 centimeters (10 in) of rainfall. The effect of this topography can even be read on the world climate map (Fig. G-6): where the highlands of the Atlas terminate, aridity immediately takes over.

The Atlas Mountains are aligned southwest-northeast and begin inside Morocco as the High Atlas, with elevations over 4000 meters (13,000 ft). Eastward, two major ranges dominate the landscapes of Algeria proper: the Tell Atlas to the north, facing the Mediterranean, and the Saharan Atlas to the south, overlooking the great desert. Between this pair of mountain chains, with each consisting of several parallel ranges and foothills, lies a series of intermontane basins markedly drier than the northward-facing slopes of the Tell Atlas. Within these valleys, the **rain shadow effect** 17 of the Tell Atlas is reflected not only in the semiarid, steppe-like natural vegetation but also in land-use patterns: pastoralism replaces cultivation, and stands of short grass and shrubs blanket the countryside.

The Maghreb Countries

The countries of the Maghreb include Morocco, the last remaining North African kingdom; Algeria, a secular republic besieged by religious-political problems; and Tunisia, smallest and most Westernized of the three (Fig. 6-18). **Morocco**, a relatively stable and conservative kingdom in a revolution-scarred region, is tradition-bound and economically weak. Like some of the other Arab monarchies in this realm, Morocco's government introduced some modest reforms early in this decade to preempt growing popular discontent. Its core area lies in the north, anchored by four major cities—Casablanca, Marrakech, Tangier, and the capital, Rabat. But the political attention of the Moroccan government is focused on the south, where it seeks to incorporate its neighbor, **Western Sahara** (a former Spanish dependency with more than 600,000 inhabitants, many of them immigrants from Morocco). Even if this campaign is successful, it will do little to improve the lives of

FIGURE 6-18

@ FROM THE FIELD NOTES...

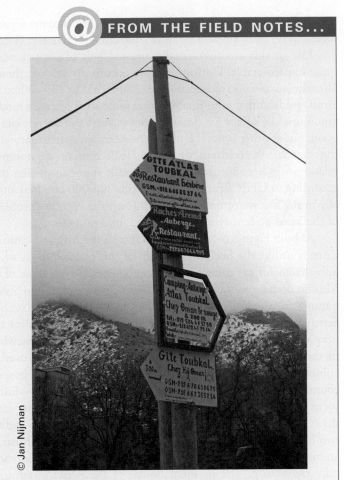

© Jan Nijman

"It is hard to believe this scene is in North Africa. I am at the village of Imlil, Morocco, near the base of Mount Toubkal, the highest peak in the Atlas Mountains at 4,167 meters (almost 14,000 ft). It is February and snow can be seen on the north-facing slopes. Most of the people residing in these small villages have simple, hard lives, and keeping warm can be a challenge. Ethnically, they are Berbers and speak a language quite different from Arabic, even though Arab cultural influences are strong throughout the country. But these directional signs, for shelter and food, are in French, a colonial vestige aimed at the few foreign tourists who come through this area. At this moment, it looks like it might soon be snowing again. What a stark contrast to the bone-dry, sizzling hot, endless desert flatlands that cover nearly all of the rest of northern Africa."

most of Morocco's 33 million people, of whom hundreds of thousands have migrated to Europe.

Algeria, France's one-time colony whose agricultural potential drew more than one million European settlers to its arable coastal strip known as the *Tell*, now has an economy based primarily on its substantial oil and natural gas reserves. The country's primate city, Algiers, is centrally situated within the densely settled corridor lining the Mediterranean that contains most of Algeria's 39.2 million inhabitants (Fig. 6-2). But a large contingent of Algerians—officially estimated at about 2 million—has emigrated to France, where they form one of Europe's largest

Muslim population sectors. Hard-nosed military control combined with generous social spending kept the Arab Spring movement at bay. But jihadists affiliated with ISIS have become active in the remote Sahara of southeastern Algeria, where they took control of some gasfields and beheaded a French hostage, arousing much indignation in Algiers, France, and beyond.

The smallest of the Maghreb states, **Tunisia**, lies at the eastern end of the region. Tunisia in many ways outranks surrounding countries: it has a higher urbanization level, higher social indicators, and a much lower growth rate (among its population of just under 11 million) than elsewhere in the Maghreb. Most of the country's productive capacity lies in the north in the hinterland of its historic capital, Tunis. As noted previously, this country took center stage as the source of the Arab Spring uprisings that diffused across NASWA. Of all the revolutions sparked by this movement, Tunisia's seems to have been the most successful so far. But in mid-2015, preserving its democratic gains remained a major challenge. And being the region's most Westernized country also attracts trouble: separate vicious terrorist attacks on a Tunis museum and a nearby beach (in March and June of 2015) killed a total of 59 international tourists.

Libya

The Maghreb's eastern neighbor, Libya, is unlike any other North African country: an oil-rich desert state whose population is almost entirely clustered in settlements along the coast—and suffering through a chaotic, multiethnic civil war that broke out following the 2011 Arab Spring revolution with its overthrow of the Qadhafi dictatorship (Fig. 6-18).

Almost rectangular in shape, Libya (population: 6.2 million) faces the Mediterranean Sea between the Maghreb states and Egypt. What limited coastal-zone agricultural possibilities exist lie in the two subregions known as Tripolitania in the northwest, centered on the capital of Tripoli, and Cyrenaica in the northeast, focused on the city of Benghazi. But it is oil, not farming, that propels the economic activity that remains in this highly urbanized country. The oilfields are located well inland from the central Gulf of Sidra, linked by pipelines to coastal terminals. Libya's two interior corners, the desert Fezzan district in the southwest and the Kufra oasis in the southeast, are connected to the coast by two-lane roads subject to frequent sandstorms.

Muammar Qadhafi had been one of the realm's most violent despots, clinging to power for more than four decades. His brutal ground and air attacks on antiregime demonstrations in several Libyan cities in early 2011 were not unexpected. Within a few months, however, advancing rebel forces captured a number of cities including Benghazi, which became their main base of operations. Assisted by NATO air support, which enforced a no-fly zone that grounded the government's military aircraft, the rebels prevailed; they soon tracked down and killed Qadhafi, who had fled Tripoli to take refuge in his hometown of Sirte.

But even though the Libyan people had rid themselves of a despised tyrant, their country collapsed into political chaos in late 2011 because a new democratic government could not be formed. Local tribes, ethnic minorities, and jihadists of various stripes, long suppressed under Qadhafi, swiftly reemerged to contest each other's turf, but only succeeded in fragmenting the country. Even ISIS had gained a foothold by early 2015, and their beheading of 21 Egyptian Coptic Christians in a jihadist-dominated eastern border town met with immediate reprisals by Egypt's air force. Bottom line: Libya today has become another of NASWA's failed states.

Adjoining Saharan Africa

Between the chain of North Africa's coastal states and the southern margins of the Sahara lies an east-west tier of five states, all but one of them landlocked, dominated by the world's greatest desert, sustained by modest ribbons of water, and thoroughly under the sway of Islam: Mauritania, Mali, Niger, Burkina Faso, and Chad (Fig. 6-10). Although their huge overall territory is sparsely populated (Fig. 6-2), the populations of these countries total 67 million, not insignificant by African standards. Furthermore, all five are at least partially located within the African Transition Zone—a region bedeviled by awesome political and environmental challenges (as we shall shortly see).

The destabilizing effects of Libya's breakdown are especially felt in Mali, territorially the size of Texas and California combined, with a population of 15.3 million that mostly inhabits the country's southern half. Until recently, Mali was cited by many as an African democratic model; but that ended in 2012 when the country descended into chaos after a military coup ousted its president. The military was said to be frustrated with the government's handling of the Tuareg insurgency in the sparsely populated far north (Fig. 6-18). But the coup sparked a rebel surge, and shortly thereafter Mali was effectively split in two. A Tuareg-Jihadist alliance took control of the fabled city of Timbuktu—located near the center of this hourglass-shaped country—and declared that the north was now an independent state named *Azawad*. In 2013, France sent thousands of its troops to help its former colony recapture Timbuktu, did so, and then turned over the effort to restore central-government rule over northern Mali to a UN-mandated, all-African military force. That approach proved ineffective, and by 2015 sporadic fighting had resumed between government and rebel forces in the country's pivotal central waist.

THE AFRICAN TRANSITION ZONE

Mali's well-populated southern half, which contains its core area, lies in the final region to be discussed in this chapter—the African Transition Zone (Fig. 6-10). From the Atlantic coast of Mauritania and Senegal in the far west to the Horn of Africa that protrudes into the Indian Ocean in the extreme east, a semiarid belt of land, about 500 kilometers (300 mi) wide, lies between the bone-dry Sahara to the north and the more humid savanna zone to the south. This is the *Sahel* (meaning 'shore'—with the bordering

Sahara analogous to an ocean), where steppe (*BS*) climates separate desert (*BW*) from tropical savanna (*Aw*) climates (see Fig. G-6).

Life in this environmentally stressed belt of steppelands is quite unpredictable. During the nine-month dry season, there is virtually no rain and temperatures rise to 50°C (122°F). When the erratic midsummer rains finally come, some areas receive a modest amount while others are flooded. In wet years the harvest can be good, but at other times it is almost nonexistent, particularly during the Sahel's recurrent droughts. Thus human–environment relationships here are always fragile, and it is not surprising that, according to UNICEF, malnutrition kills 225,000 children annually in the five westernmost Sahelian countries. But the problems of the Sahel go far beyond climate and food provision.

As both Figures 6-10 and 6-19 indicate, the African Transition Zone is unlike the five other regions of this realm. This is where Subsaharan Africa's cultures to the south intersect with those of the Muslim domain to the north. This Islamic penetration into Subsaharan Africa engulfs some countries in their entirety (such as Senegal and Burkina Faso) and divides others into Muslim and non-Muslim sectors (Nigeria, Ivory Coast, and Chad). As elsewhere in the world where certain geographic realms meet and overlap, complications abound within the African Transition Zone. In some countries, the geographic transition from Muslim to non-Muslim society is gradual, as in Sierra Leone and Ivory Coast. In other areas, the cultural divide is far more abrupt, as in eastern Ethiopia and the two halves of former Sudan, where the border between Christian/animist African cultures and Islamic African communities is practically razor-sharp. The southern border of the African Transition Zone, the religious frontier we refer to as Africa's Islamic Front 18 (mapped in Fig. 6-19), is therefore neither static nor advancing uniformly.

Conflict marks most of the African Islamic Front: longstanding in former Sudan, where a decades-long war for independence by non-Islamic Africans in the south against the Arabized, Muslim north—which finally resulted in the 2011 creation of South Sudan—cost millions of lives; intermittent in Nigeria, where violent Islamic revivalism continues to threaten that country's prospects; and recent in Ivory Coast, where political rivalries reawakened religious strife in a once stable country. Nonetheless, the most conflict-prone sector of the African Transition Zone lies in the east, involving not only now-bifurcated Sudan but also the historic Christian state of Ethiopia, which is nearly encircled by predominantly Muslim societies.

The Horn of Africa

The most potentially explosive subregion of the African Transition Zone is the Horn of Africa, which consists of Djibouti, Eritrea, Somalia, and Ethiopia (Fig. 6-20). We will focus here on the last two, the Horn's largest countries. But first note that the ministate of Djibouti, fully 97 percent Muslim, is increasingly significant because it lies directly across from Yemen and overlooks the narrow entry to the Red Sea, the Bab el Mandeb Strait (see last photo in chapter). Its coastal neighbor to the north, Eritrea, came into

FIGURE 6-19

Map legend:

AFRICAN TRANSITION ZONE: PERCENT MUSLIMS BY COUNTRY

- Over 90
- 50 – 90
- 11 – 50
- Under 11
- Islamic Front
- Armed conflict, current or recent

existence in 1993 when it separated from Ethiopia, but it still engages in boundary disputes and clashes with Ethiopia that have damaged the economies of both countries. Eritrea's early hopes for representative government and economic progress were dashed by war, dictatorial rule, and, most recently, its involvement in Islamic sectarian strife.

Somalia

The key component in the eastern sector of the African Transition Zone is Somalia, where 10.5 million people, almost all of them Muslim, live at the mercy of a desert-dominated environment that requires cross-border migration into Ethiopia's easternmost Ogaden zone in pursuit of seasonal pastures. As many as 3 to 5 million Somalis reside permanently on the Ethiopian side of the border, but this is hardly the only such division the Somali "nation" faces. Somalia's cultural geography comprises an assemblage of five major ethnic groups fragmented into hundreds of clans—all engaged in an endless struggle for power as well as survival.

In the early 2000s, Somalia fragmented into three major subdivisions (Fig. 6-20). In the north, the sector known as *Somaliland*,

FIGURE 6-20

which had proclaimed its independence in the 1990s and which remains by far the most stable of the three, essentially functions as a state even though the 'international community' will not recognize it as such. In the east, a conclave of local chiefs declared their territory to be distinct from the rest of Somalia, gave it the name of **Puntland**, and asserted varying degrees of autonomy. In the south, where the official capital, Mogadishu, is located on the Indian Ocean coast, local secular warlords and Islamic militias continued their grim contest for supremacy, the warlords supported by U.S. aid. In 2006, the

militias stormed into the capital and took control, ousting the warlords and proclaiming their determination to create an Islamic state. In this decade, the government regained control of Mogadishu in 2011, but the al-Qaeda-linked **al-Shabaab** Islamic terrorist movement it ousted reemerged strongly in southern Somalia and uses it as a base to launch its bold attacks; a U.S. drone strike killed its leader in 2014, perhaps fatally weakening this still-dangerous jihadist organization. Nonetheless, the Islamic-extremist mosaic continues to blanket and control large territorial expanses of Somalia, political

Jacques Descloitres, MODIS LandRapid Response Team, NASA/GSFC

The narrow Bab-el-Mandeb Strait connects the Red Sea (upper left) to the Gulf of Adan arm of the Indian Ocean (lower right). This classic choke point—the only one located within the African Transition Zone—accommodates one of the world's most crucial sea lanes, especially for oil shipments. With the unstable African Horn on the western shore and failed-state Yemen on the eastern side, it is not hard to envision the Strait's surrounding turmoil escalating into attempts to disrupt traffic here. That would trigger outside intervention, not only by such NASWA countries as Saudi Arabia and Egypt, but also by the United States, Europe's powers, and very likely China.

stability is as hard to imagine as ever today, and the country continues to be regarded as the world's poster-child **failed state 19**.

Ethiopia

Ethiopia, the last country covered in this chapter, allows us to end on a high note in more ways than one. The highlands of Ethiopia, where the country's core area, capital (Adis Abeba), and its historic Christian heartland lie, are virtually encircled by dominantly Muslim societies (Fig. 6-20). As Figure 6-19 reminds us, Ethiopia's eastern Ogaden is traditional Somali country and almost entirely Islamic. Today, one-third of Ethiopia's population of 94.1 million is Muslim, and here the Islamic Front is sharply defined. With the continent's second-largest population, Ethiopia is also an important African cornerstone state. It not only faces some of the usual challenges of the African Transition Zone but has been landlocked as well since it lost Eritrea in 1993. Yet Ethiopia's highland climate and soils, quite conducive for productive agriculture, strongly contrast with the less encouraging conditions that prevail in the stressed semiarid environments across most of the African Transition Zone. Indeed, Ethiopia exports fruits, vegetables, flowers, coffee, and even tea. Moreover, this country has in recent years attracted considerable investment from China to finance badly needed infrastructure, construction, and agribusiness; and in terms of economic growth Ethiopia has ranked among the world's top 25 countries for most of this decade. Politically, Ethiopia remains stable despite some questionable recent elections managed by a hard-nosed, oppressive regime. But economically, at least, Ethiopia's recent performance far exceeds that of other countries in the African Transition Zone. This not only reminds us of the importance of geography but also of the pitfalls of **environmental determinism 20**: geography explains a great deal because it poses constraints and offers opportunities—but geography never dictates.

POINTS TO PONDER ?

- The Arab Spring movement of the early 2010s did not produce major advances toward democracy, but it did contribute to political destabilization in the realm's regions and individual countries.

- In Saudi Arabia, one of America's foremost trading partners and military allies, women are forbidden to drive and in 2015 were given only limited voting privileges.

- In 2014, Dubai International Airport (DXB) in the United Arab Emirates was the world's busiest airport.

- In 2015, the North Africa/Southwest Asia realm counted no less than six failed states (Iraq, Syria, Yemen, Libya, Mali, and Somalia)—more than in any other realm.

- Ethiopia may be partially located in the African Transition Zone, but in recent years it has had one of the fastest growing economies in the world.

The Ngorongoro Crater in northern Tanzania is one of the world's most famous wildlife conservation areas. The background ridge is actually part of the rim of this largest inactive volcano crater on Earth, which formed at least two million years ago when a huge volcano exploded and caved in on itself. This UNESCO World Heritage Site is home to wildebeests and zebras (shown here) as well as dozens of other large mammals, including giraffes, elephants, leopards, lions, and hippos. Ecotourism is an economic mainstay of this immediate area.

© Jan Nijman

IN THIS CHAPTER

- The Ebola crisis
- Africa's fast-growing economies
- China's foothold in Africa
- Boko Haram in Nigeria's north
- Angola's oil boom
- Xenophobia in South Africa

CONCEPTS, IDEAS, AND TERMS

7

THE SUBSAHARAN AFRICAN REALM

FIGURE 7-1

The African continent occupies a special place in the physical as well as the human world. This is where **human evolution 1** began. In Africa we formed our first communities, spoke our first words, and created our first art. From Africa our ancestor hominins spread outward into Eurasia more than 2 million years ago. From Africa our species emigrated, beginning perhaps 95,000 years ago, northward into present-day Europe and eastward via southern Asia into Australia and, much later, farther afield into the Americas. Disperse our forebears did, but we should remember that, at the source, we are all Africans.

DEFINING THE REALM

For millions of years, Africa served as the cradle for the emergence of humankind. For tens of thousands of years, Africa served as the source of human cultures. Yet Africa has not experienced economic development on a par with other world regions. Its underdevelopment is difficult to separate from foreign involvements: from imperialism to colonialism to the geopolitical impact of the Cold War. Today, there are hopeful signs that this catastrophic era is ending as a number of states are being propelled forward by the new global economic relationships they are building.

The focus in this chapter will be on Africa south of the Sahara, also known as *Subsaharan Africa*, an unsatisfactory but popular name that applies to the extensive landmass south of the great desert (Fig. 7-1). The African continent contains two geographic realms: the African, extending from the southern margins of the Sahara to the Cape of Good Hope, and the western flank of the North African/Southwest Asian realm. Although the great desert constitutes a formidable barrier between the two, the powerful influences of Islam crossed the Sahara from North Africa centuries before the first Europeans set foot in West Africa. By that time, the African kingdoms lining the southern margins of the Sahara had been converted, creating an Islamic foothold in the African Transition Zone (the final region discussed in Chapter 6) that now marks the northern periphery of the realm we are about to investigate.

AFRICA'S PHYSIOGRAPHY

Africa accounts for about one-fifth of the Earth's entire land surface. Territorially, it is as big as China, India, the United States, Mexico, and Europe combined. The north coast of Tunisia lies 7700 kilometers (4800 mi) from the southernmost coast of South Africa. Coastal Senegal, in West Africa, lies 7200 kilometers (4500 mi) from the tip

▌▌ MAJOR GEOGRAPHIC QUALITIES

SUBSAHARAN AFRICA

1. Physiographically, Africa is a massive plateau continent without a major "spinal" mountain range but with a set of Great Lakes, several major river basins, variable rainfall, soils of generally inferior fertility, and dominantly savanna and steppe vegetation.

2. Hundreds of distinct ethnic groups make up Subsaharan Africa's culturally rich and varied population mosaic. They far outnumber the states in this realm, and rarely do state and ethnic boundaries coincide.

3. Approximately three-fifths of Subsaharan Africa's peoples still depend on farming for their livelihood.

4. The realm is rich in raw materials vital to industrialized countries, but many economies continue to rely on primary activities—the extraction of resources—and not the greater income-generating activities of manufacturing and assembly.

5. The realm is famous for its wildlife, but many species are threatened. Combining bioconservation together with sustainable development is a major challenge when the outside world covets elephant tusks and rhinoceros horns.

6. Severe dislocation still affects a number of Subsaharan African countries, from South Sudan to Zimbabwe.

7. Foreign interest in the realm's natural resources, ranging from commodities to agricultural land, has expanded steadily in this century.

8. Subsaharan Africa has recently experienced rapid economic growth, with a half-dozen countries now ranking among the fastest growing in the world.

of Africa's Horn in easternmost Somalia. These distances have critical environmental implications. Much of Africa is far from maritime sources of moisture. In addition, as Figure G-6 shows, large parts of the landmass lie in latitudes where global atmospheric circulation systems produce arid conditions. As a result, water supply is one of Africa's foremost problems.

As the physiographic map (Fig. 7-2) shows, Africa's topography is marked by several properties not replicated on other landmasses. Alone among the continents, Africa does not possess a mountain backbone; neither the northwestern Atlas nor the far southern Cape Ranges are in the same league as the Andes or Himalayas. Where Africa does have mountainous terrain, as in Ethiopia and South Africa, it takes the form of deeply eroded plateaus or, as in East Africa, high snowcapped volcanoes. Furthermore, Africa is one of only two continents containing a cluster of Great Lakes. These lakes (with the exception of Lake Victoria) lie in deep trenches called **rift valleys** **2**, which form when huge parallel fractures appear in the Earth's crust and the

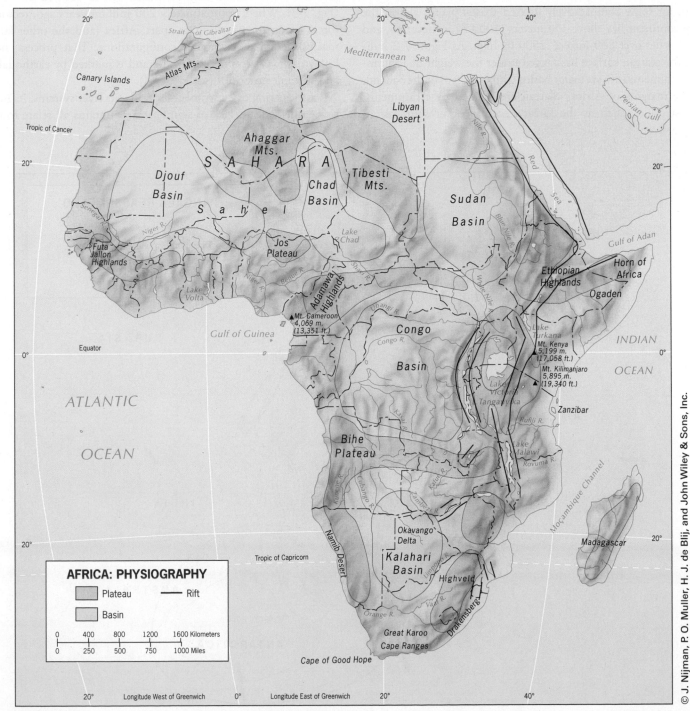

FIGURE 7-2

strips of crust between them sink to form great, steep-sided, linear valleys. These rift valleys, stretching more than 9600 kilometers (6000 mi) from the Red Sea to Swaziland, are mapped by the red lines in Figure 7-2.

Africa's rivers, too, are unusual: their upper courses frequently bear landward, seemingly unrelated to the coast toward which they eventually flow. Several rivers, such as the Nile and the Niger, have inland as well as coastal deltas. Major waterfalls, notably Victoria Falls on the Zambezi, or lengthy systems of cataracts separate the upper from the lower river courses.

Finally, Africa may be described as the plateau continent. Except for some comparatively limited coastal plains, almost the entire continent lies above 300 meters (1000 ft) in elevation, and half of it lies over 800 meters (2500 ft) high. As Figure 7-2 indicates, the plateau surface has sagged under the weight of accumulating sediments into six major basins (three of them in the Sahara), but the margins of Africa's plateau are marked by escarpments, often step-like, such as the Great Escarpment of southeastern

South Africa that marks the eastern edge of the Drakensberg Mountains.

Continental Drift and Plate Tectonics

Africa's remarkable and unusual physiography informed geographer Alfred Wegener's hypothesis of **continental drift 3**. All of the present-day continents, Wegener reasoned, lay assembled into one giant landmass called ***Pangaea*** not very long ago (ca. 220 million years) in geologic time. The southern part of this supercontinent was *Gondwana*, of which Africa formed the core (Fig. 7-3). When, approximately 200 million years ago, tectonic forces began to split Pangaea apart, Africa (and the other landmasses) acquired its present configurations. That process, now known as *plate tectonics*, continues and is marked by earthquakes and volcanic eruptions (see Fig. G-4).

Africa's ring of escarpments, its rifts, its river systems, its interior basins, and its lack of significant mountains all relate to the

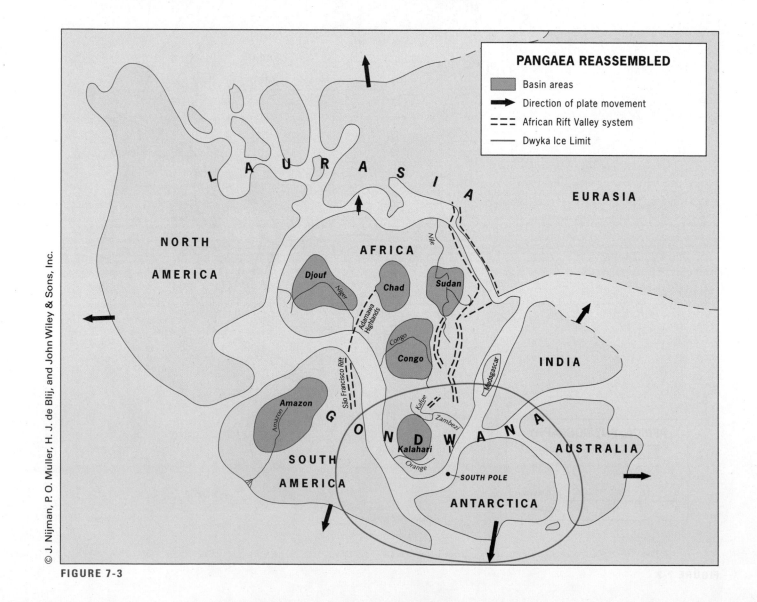

© J. Nijman, P. O. Muller, H. J. de Blij, and John Wiley & Sons, Inc.

FIGURE 7-3

continent's central position within Pangaea—all pieces of the puzzle that led to an explanation based on plate tectonics theory.

AFRICA'S HISTORICAL GEOGRAPHY

Africa is the cradle of humankind. Archeological research has chronicled 7 million years of transition from Australopithecenes to hominins to *homo sapiens*. It is, therefore, ironic that we know comparatively little about Subsaharan Africa from 5000 to 500 years ago—that is, before the onset of European colonialism. This is partly due to the colonial period itself, during which African history was neglected, numerous African traditions and artifacts were destroyed, and many misconceptions about African cultures and institutions became entrenched. It is also a result of the absence of an accurate and detailed written history about the region until the contemporary period.

African Genesis

Africa on the eve of the colonial period was far from a static realm. In interior West Africa, cities were developing; in central and southern Africa, peoples were moving, readjusting, sometimes struggling with each other for territorial supremacy. African cultures had been established in all the environmental settings shown in Figure G-6 for thousands of years and thus long before the intrusions of Islamic and/or European influences. One of these, the Nok culture, endured for over eight centuries (north of the Niger-Benue confluence in what is now central Nigeria) from about 500 BC to the third century AD. The Nok people made stone as well as iron tools, and they left behind a treasure of art in the form of clay figurines representing humans and animals.

Early Trade

West Africa, over a north-south span of a few hundred kilometers, displayed a wide range of contrasting environments, economic opportunities, modes of life, and products. The peoples of the tropical forest in the south produced and needed goods that were different from the products and requirements of the peoples of the dry, distant north. For example, salt was a prized commodity in the forests, where humidity precluded its formation, but it was plentiful in the semiarid steppe and the desert beyond. This enabled the desert-margin peoples to provide salt to the forest peoples in exchange for ivory, spices, and dried foods. Thus there evolved a degree of *regional complementarity* between the peoples of the forest and those of the drylands. And the savanna peoples—those located in between—found themselves in a position to channel and handle this profitable trade.

The markets in which these goods were exchanged prospered and grew, some into urban centers. One of these old cities, now an epitome of isolation, was once a thriving center of commerce and learning and one of the leading crossroads in the world—Timbuktu

(now located in Mali). In fact, its university is one of the oldest anywhere, with a library that holds irreplaceable documents, thankfully now being preserved. Others, predecessors as well as successors of Timbuktu declined, some of them into oblivion. Still other savanna cities, such as Kano in northern Nigeria, remain important today.

Early States

Strong and durable premodern states arose in the interior of West Africa (Fig. 7-4). The oldest known state was ancient Ghana, located to the northwest of present-day Ghana and centered on parts of what are now Mali, Burkina Faso, and Ivory Coast. Ancient Ghana managed to weld various groups of people into a stable territory between the ninth and twelfth centuries AD. Taxes were collected from its citizens, and tribute was extracted from subjugated peoples along Ghana's periphery; tolls were levied on goods entering the state, and an army maintained control. Muslims from the drylands to the north invaded Ghana in 1067, when it was in decline, hastening the state's demise as it shattered into smaller units.

In the centuries that followed, the focus of politico-territorial organization in this West African *culture hearth* (see Fig. 6-3) shifted almost continuously to the east—first to ancient Ghana's successor state of Mali, which was centered on Timbuktu and the middle Niger River Basin, and then to the state of Songhai, whose focus was Gao, a city farther down the Niger that still exists (Fig. 7-4). This eastward movement may have been the result of the growing influence and power of Islam. Traditional animist religions prevailed in ancient Ghana, but Mali and its successor states sent massive, gold-laden pilgrimages to Mecca along the savanna corridor south of the Sahara, then turning northward to follow a Nile route that passed through present-day Khartoum and Cairo.

Beyond West Africa

West Africa's savanna zone undoubtedly experienced momentous cultural, technological, and economic developments, but other parts of ancient Africa made progress as well. Early states emerged in present-day Sudan, Eritrea, and Ethiopia. Influenced by key innovations from the Egyptian culture hearth to the north, these kingdoms were stable and durable: the oldest, Kush, lasted 23 centuries (Fig. 7-4). The Kushites built elaborate irrigation systems, forged iron tools, and created impressive structures as the ruins of their long-term capital and industrial center, Meroe, reveal. Nubia, to the southeast of Kush, became Christianized until the Muslim tide overtook it in the eighth century. And Axum was the richest market in northeastern Africa, a powerful kingdom that controlled Red Sea trade and endured for more than six centuries.

The process of **state formation** 4 spread throughout Africa and was still in progress when the first European contacts occurred in the late fifteenth century. Large and effectively organized states

© J. Nijman, P. O. Muller, H. J. de Blij, and John Wiley & Sons, Inc.

FIGURE 7-4

developed on the equatorial west coast (notably Kongo) and on the southern plateau from the southern margin of the Congo River Basin southeastward to Zimbabwe. East Africa, as well, gave rise to a number of city-states, including Mogadishu, Mombasa, and Sofala.

At a broader scale, a crucial series of events commencing about 5000 years ago affected virtually all of West, Equatorial, and Southern Africa: the Great Bantu Migration from present-day Nigeria and Cameroon southward and eastward across the continent. This epic expansion appears to have occurred in waves that populated the Great Lakes area and ultimately penetrated South Africa, where it culminated in the formation of the powerful Zulu Empire in the nineteenth century (Fig. 7-4).

All this reminds us that, before European colonization, Africa was a realm of rich and varied cultures, diverse lifestyles, technological advances, circulation, and trade. But it was also a highly fragmented realm with a cultural mosaic (Fig. 7-5) constituted by

myriad ethnic groups sharing a common ancestry, culture, language, and/or religion. After the Europeans intervened and redrew the social and political map, the modern African state emerged as an amalgamation of multiple ethnic groups that almost entirely disregarded the preexisting cultural-territorial layout.

THE COLONIAL TRANSFORMATION

European involvement in Subsaharan Africa began in the fifteenth century. It disrupted the path of indigenous African development and irreversibly altered the entire cultural, economic, political, and social complexion of the continent. Starting quietly in the late 1400s, Portuguese ships groped their way all along the west coast and by 1488 had rounded the Cape of Good Hope. Their goal was

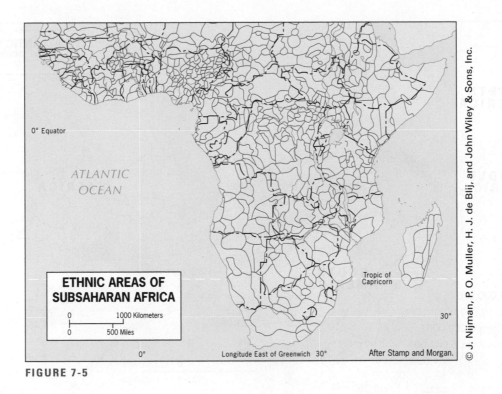

FIGURE 7-5

to find a sea route to the spices and riches of India and points east. Soon other European countries were dispatching their vessels to African waters, and a string of coastal stations and forts sprang up. In West Africa, the part of the continent closest to the European spheres in Middle and South America, the initial impact was strongest. At their coastal control points, the Europeans bartered with African intermediaries for the slaves who were destined to work the New World plantations, for the gold that had been flowing northward across the Sahara, and for highly prized ivory and spices.

Suddenly, the centers of activity lay not with the inland cities of the savanna belt but with the foreign stations on the Atlantic coast. As the interior declined, the coastal peoples thrived. Small forest states gained unprecedented wealth, transferring and selling slaves captured in the interior to the European traders on the coast. Dahomey (now called Benin) and Benin (now part of neighboring Nigeria) were two modern states built on the slave trade.

Impact of the Slave Trade

Millions of Africans were forced to migrate from their homelands to the Americas, especially Brazil, the Caribbean, and the United States. The slave trade was a catastrophe that was facilitated, in part, by the peril of proximity. The northeastern tip of Brazil, by far the largest single destination for the millions of Africans forced from their homes in bondage (Fig. 7-6), lies about as far from the nearest West African coast as South Carolina lies from Venezuela. This was a short maritime intercontinental journey indeed (it is more than twice as far from West Africa to South Carolina).

Slavery was not new to West Africa. In the interior of Africa and within city-states, kings, chiefs, and prominent families traditionally took a few slaves, but the status of these slaves was relatively benign. Indeed, large-scale slave trading was first introduced to East Africa long before the Europeans brought it to western Africa. African intermediaries from the coast raided the interior for able-bodied men and women from different ethnic groups and marched them in chains to the Arab markets on the coast (the island of Zanzibar was one such slave trading market). There, packed into specially built sailboats, they were carried off to Arabia, Persia, and India.

When the European slave trade took hold in West Africa, however, its volume and impact were far greater than anything seen before, and so were the horrors. Europeans, Arabs, and collaborating Africans now ravaged the continent, forcing perhaps as many as

South Africa's Cape Town, whose magnificent site is dominated by Table Mountain, was founded by the Dutch in 1652 as a waystation for empire building in Southeast Asia's East Indies.

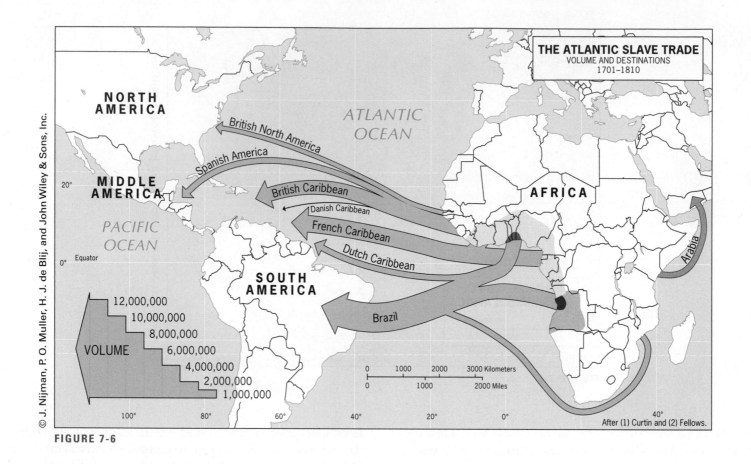

© J. Nijman, P. O. Muller, H. J. de Blij, and John Wiley & Sons, Inc.

FIGURE 7-6

30 million persons away from their homelands in captivity (Fig. 7-6). Families, entire communities, and even cultures were destroyed; those surviving in exile suffered unfathomable misery. The European presence completely reoriented West Africa's trade routes. Even though the Europeans' insatiable demand for slaves ravaged the inland population, it did not lead to any major European thrust toward the interior or produce colonies overnight. African intermediaries were well organized and strong, and they held off their European competitors for centuries.

Colonization

In the second half of the nineteenth century, the Europeans finally laid claim to virtually all of Africa. In 1884, a conference was convened in Berlin to carve up the African map among the major European powers. On maps spread across a large table, representatives from these powers drew boundaries, swapped territory, and forged a new map that would become a liability in Africa decades later. Amazingly, at no point in the proceedings were African voices heard. As Figure 7-7 indicates, when the three-month conference was in progress, most of Africa remained under traditional African rule. Then, in the early 1900s, the European colonial powers moved to take control of all the areas they had marked off and acquired on their maps.

It is important to examine Figure 7-7 carefully because the colonial powers governed their new dependencies in very different ways, and their contrasting legacies are still in evidence today.

Some colonial powers were democracies at home (Great Britain and France); others were dictatorships (Portugal and Spain). The British established a system of **indirect rule 5** over much of their domain, leaving indigenous power structures in place and making local rulers representatives of the British Crown. This was unthinkable in the Portuguese colonies, where harsh direct control predominated. The French sought to establish culturally assimilated elites that would disseminate French ideals in the colonies. King Leopold II of Belgium claimed Congo Free State as his own personal fiefdom. Subsequently, he unleashed a savagely ruthless campaign of exploitation and genocide. Leopold's reign of terror has become known as Africa's most brutal demographic disaster (after the slave trade): as many as 10 million people were exterminated.

The post-Berlin colonial map entailed a profound political reorientation—yet it lasted barely half a century. Ghana (then known as the Gold Coast) offers a good example. Its colonial ties were formalized despite the constant challenging of the mother country by indigenous ethnic groups. But in 1957, Ghana became the first African country to gain independence—triggering a movement that swept across the realm (compare the 1958 and 1970 maps in Fig. 7-7). These days, most of Subsaharan Africa has already marked 50 years of independence, and in retrospect the colonial period should now be viewed as only an interlude rather than a paramount chapter in African history. It did, however, leave a deep and enduring geographic imprint, especially on the realm's cultural landscape.

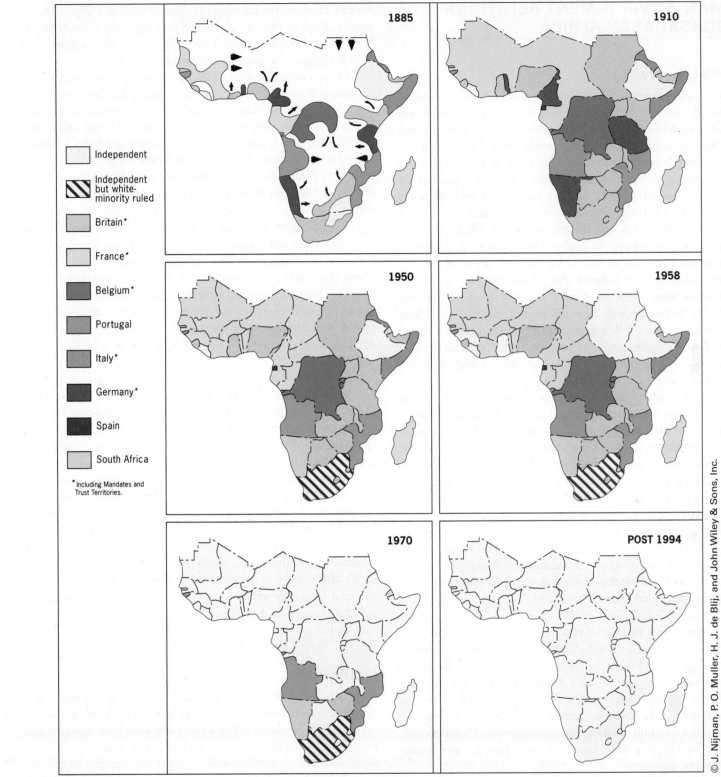

Independent

Independent
but white-
minority ruled

Britain*

France*

Belgium*

Portugal

Italy*

Germany*

Spain

South Africa

* Including Mandates and
Trust Territories.

1885

1910

1950

1958

1970

POST 1994

FIGURE 7-7

HUMAN–ENVIRONMENT RELATIONS IN SUBSAHARAN AFRICA

Colonial Legacy

Africa's tropical forests and extensive savannas form the world's last refuges for many species of wildlife, from primates and zebras to giraffes to wildebeests. But since colonial times, these natural environments have been under constantly rising threat. Early on, the Europeans introduced hunting as a "sport" (a practice that did not exist within African cultural traditions), thereby triggering the mass destruction of animals. Later, a number of colonial rulers and postcolonial national governments laid out game reserves and other conservation areas, but these were neither sufficiently large nor adequately connected to enable herd animals to adhere to their seasonal migration routes. As we shall see, the same climatic variability that plagues farmers also affects wildlife. When the rangelands wither, the animals seek better pastures; but if the fences of game reserves wall them off, they cannot survive. Where there are no fences, wildlife invasions of farmland destroy crops, and the farmers retaliate. Clearly, after thousands of years of equilibrium, the competition between humans and animals in Africa has entered a new era of intensifying stress.

It is not solely population pressure that threatens African wildlife. Today, international criminal syndicates are heavily engaged in the illegal exporting of wild animals and animal parts, and this activity has become the world's fourth most profitable illicit trade (after narcotics, human trafficking, and arms smuggling). The horn of the rhinoceros became a valued property for Arabs who fashion them into dagger handles and, in powdered form, an alleged aphrodisiac and cancer treatment for wealthy East Asians. As a result, the northern white rhinoceros is nearing extinction, and the number of all species of rhino has fallen from half a million a century ago to less than 30,000 today. A terrible tragedy is also befalling the African elephant as the value of its ivory tusks has tripled since 2010 (in 2015, a kilogram of ivory in Beijing fetched U.S. $2100); if this illicit Chinese market is not quickly terminated, this entire species will disappear in the next decade. Note that even though Africans are complicit in international smuggling and that small quantities of ivory are for sale in informal markets in Nigeria and Angola, it is external demand that drives this clandestine trade. Meanwhile, competition for land has more to do with the decline of Africa's lion population (from around 250,000 in 1975 to barely more than 25,000 today), whereas illegal logging and mining threaten a host of other species including the great ape.

The Geography of Farming

It would seem that space should be plentiful for both wildlife and humans. Figure 7-8 does not present a picture of a densely peopled realm—although there are major clusters of population in coastal West Africa, East Africa encircling Lake Victoria, and the African Horn where Ethiopia's highlands sustain a large concentration. Most of the rest of the realm is relatively sparsely populated, even though its overall natural-increase rate is more than double the global average of 1.2 percent annually.

Despite an ongoing surge in urbanization, a large majority of Africans still continue to depend on farming for their livelihood. But African environments are difficult for millions of small farmers, and the realm's highly productive areas (the Great Lakes area, the Western Rift Valley, and the cooler, moister parts of South Africa) are minuscule in comparison to the huge, fertile, alluvial basins of China or India. The great majority of African farmers face daunting challenges that include: (1) climatic variability, (2) difficulties in their national policy environments, (3) problems in reaching world markets, and (4) growing competition from Asian and Latin American agricultural producers, all of which result in unstable prices and diminishing returns for many commodities. Rhetoric about free trade in the wealthy Global Core notwithstanding, many of those governments subsidize their own farmers, and their environmentally conscious consumers favor more expensive home-grown food products. Not surprisingly, from their point of view, African farmers say they are getting a raw deal in the global marketplace.

Although adequate soil and water supplies are critical for farming, a number of political and economic factors are also influential. Among them are land tenure; commercial (for-profit) versus subsistence (household-level) agriculture; type of farming system (rotational, shifting cultivation); crop prices; government farm policies; and degree of mechanization.

With roughly 60 percent of the realm's livelihoods dependent on farming, the issue of land tenure may be the most important of all. **Land tenure** **6** refers to the way people own, occupy, and use land. African traditions of land tenure are different from those found in Europe or the Americas. In most of Subsaharan Africa, communal, not individual, rights prevail—often held in perpetuity by chiefs, large extended families, and villages. Thus users of the land have only temporary, custodial rights to it; they cannot sell that piece of land and must follow the rules of their community leaders.

Land Ownership

After the arrival of the Europeans, their colonial administrators prioritized control of the most fertile areas of each colony. In certain cases, this required military conquest; elsewhere, coercion and eviction sufficed. Before the colonial era, many traditional African livelihood systems adhered to sustainable land management practices such as letting fields lie fallow and rotational grazing of their livestock. Viewing this land as unused, however, colonial regimes began the process of **land alienation** **7** (confiscation). Much later, after the colonizers withdrew, many newly independent African countries initiated programs whereby land would revert to traditional forms of ownership and management—a land reform movement that in the end did little to improve conditions for still-marginalized small farmers.

SUBSAHARAN AFRICA:
POPULATION DISTRIBUTION, 2016
One dot represents 50,000 persons

FIGURE 7-8

© J. Nijman, P. O. Muller, H. J. de Blij, and John Wiley & Sons, Inc.

Rapid population growth since the end of the colonial era further intensified pressures on the realm's agricultural land. Traditional systems of land use, involving subsistence farming in various forms ranging from shifting cultivation to pastoralism, work best when populations are fairly stable and tenure is communal. But the African population explosion of the mid-twentieth century set into motion an accelerated demand cycle in which food producers could no longer rest soils and pasturelands—thereby unleashing new rounds of farmland degradation that steadily drove down crop yields.

Subsistence Farming

Although commercial farming prevails in parts of Africa, most farmers are engaged in subsistence agriculture that entails grain crops (corn, millet, sorghum) in drier areas and root crops (yams, cassava, sweet potatoes) where moisture is adequate. Others herd livestock, mostly cattle and goats, as they deal with the constant challenges of environmental variability. But too many farmers and pastoralists have been unable to secure regular access to markets as well as stable prices for their products because of government policies that often promote one particular type of export-oriented crop (e.g., peanuts in Gambia, tea and coffee in Kenya).

The decline of African agriculture has been disproportionally hard on women who, according to prevailing estimates, produce about two-thirds of all local food in Subsaharan Africa. Development policies frequently pay too little attention to the role of gender in agricultural production (women account for at least 15 percent of all farmland holders), so that farms operated by women are smaller, more isolated, and endowed with poorer-quality soils than those cultivated by men. Moreover, women always have a harder time than men in accessing credit.

The **Green Revolution** **8**—the development of more productive, drought-tolerant, pest-resistant, higher-yielding types of grain—had less impact in Africa than elsewhere during the latter decades of the twentieth century. Belatedly, a renewed effort has now been launched to jump-start Africa's New Green Revolution. Its twin emphases involve the adoption of more effective farming methods and the raising of higher-value crops in order to boost farmer income and enhance the realm's food production output. However, the introduction of genetically modified crops is controversial and encounters substantial resistance today. Many African

© Jan Nijman

"Flying a small propeller plane over East Africa's semiarid grasslands, we had a good view of a Maasai village near the border between Kenya and Tanzania. The Maasai are a pastoral people who herd cattle, sheep, and goats. I could clearly see the fenced 'kraals' where they keep their animals at night (during the day they are moved around) and the small dwellings that surround them. The more animals a man (!) owns, the higher his social status. But this way of life is under pressure: many have decided to move to towns and cities because they offer more and better opportunities. And this area is also located in the northern extension of Tanzania's famous Serengeti National Park, where wildlife preservation comes into direct conflict with the Maasai way of life. 'It is really impossible,' said the park guide on the 10-seater plane with us, pointing down at the settlement. 'Of course they want to protect their cattle, but they are killing lions and other animals that should be protected. And we must tell them all the time that they have to stay out of certain areas because the cattle compete for food with the wild animals.' Nodding agreeably, I thought to myself: yes, but isn't this their land? As with so many other places around the world where we want to protect nature, we must involve the local inhabitants. That is much easier said than done."

governments, influential media, and individual farmers regard biotechnological solutions with increasing hostility.

Neocolonial Land Grabs

During the postcolonial era, agricultural land in Africa was of little interest to outside investors, but over the past decade an enormous change has taken place. Many African countries now offer large tracts of land for sale or long-term leases to major investors, a number of whom are associated with multinational **agribusiness 9** corporations and private equity firms. Governments and other large landowners view the sale and leasing of food production resources they control (land, soil, water) as the quickest return on their existing investments to increase agricultural output and maximize foreign revenues. As for the investors, they are motivated by the steady rise in global food prices, brightening prospects for biofuels, and growing shortages of arable land in their own countries. They also see African smallholder agriculture as achieving only a tiny fraction of its potential productivity—which the outsiders believe can be significantly upgraded through the widespread application of contemporary agricultural strategies and technology. Although investors from other geographic realms form the majority of the participants in this land rush, certain African elites have joined in as well—either directly or in partnership with public and private institutions.

By the mid-2010s, more than 4 million hectares (9.8 million acres) of African land were under negotiation or already managed by foreign investors—an area greater than the combined agricultural land of France, Germany, Belgium, the Netherlands, Denmark, and Switzerland. Most of this land was concentrated in Ethiopia, Congo,* Moçambique, and Sudan. So far, the impact of these land grabs has been mixed. In some cases, output on large commercial farms far exceeds previous levels of production. But in many others, investors seem to have had speculative intentions and are content to let farmland lie idle. Either way, substantial numbers of small farmers are dispossessed, with adverse economic effects that reverberate throughout affected localities. Thus for countless thousands of such displaced workers, without any say of their own, traditional pathways to access land and livelihoods have been upended by these institutional changes in land use. Understandably, mistrust is rising rapidly among local populations, and the success of these land deals is increasingly variable: for example, Sudan has performed very poorly, while Moçambique's situation is said to have improved notably.

POPULATION AND HEALTH

Medical geography 10 is concerned with the tracking of disease outbreaks to identify their sources, detect their carrier agents, trace their spatial diffusion, and prevent their recurrence. Alliances between

---

Two countries in Africa have the same short-form name, *Congo*. In this book, we use **DRCongo** for the much larger Democratic Republic of the Congo, and **Congo** for the smaller Republic of Congo just to its northwest.

medical personnel, epidemiological researchers, and geographers yield significant results. Medical specialists know how a disease ravages the body; geographers know how environmental conditions (such as variations in wind direction and river flow) can affect the spread of disease carriers (*vectors*), and their application of GIS techniques to analyze spatial dimensions of diseases is making many contributions to medical research these days.

In Subsaharan Africa today, hundreds of millions of people carry one or more maladies, often without knowing exactly what ails them. A disease that infects many people (the *hosts*) in a kind of equilibrium, without causing rapid and widespread deaths (such as hepatitis or hookworm), is said to be **endemic 11** to that population. People affected may not die suddenly or dramatically, but their quality of life and productive capacity are hindered as their overall health is weakened and can deteriorate rapidly when a more acute illness strikes.

Epidemics and Pandemics

When a disease outbreak occurs at the local or regional level, it is called an **epidemic 12**. It may claim thousands, even tens of thousands, of lives, but it remains geographically confined, perhaps limited by the spatial range of its vector. Trypanosomiasis, the disease known as sleeping sickness and vectored by the tsetse fly, is a good example of a regional-scale epidemic (Fig. 7-9). The extensive herds of savanna wildlife form the *reservoir* of this disease, and the tsetse fly transmits it both to livestock and people. It is endemic to wildlife, but it also kills cattle, so Africa's herders try to keep their animals in tsetse-free zones. African sleeping sickness appears to have originated in a West African source area during the fifteenth century, and from there it spread throughout much

of tropical Africa. Its epidemic range is limited by that of the tsetse fly: where there are no tsetse flies, there is no sleeping sickness.

When a disease spreads worldwide, it is known as a **pandemic 13**. Subsaharan Africa's and the world's most deadly vectored disease is malaria, transmitted by a mosquito and the killer of at least 600,000 people each year. Because people living in the world's poorest (lower-latitude) countries are most vulnerable to malaria, that burden falls overwhelmingly on Africa—which not only accounted for 85 percent of all cases worldwide in 2013, but also 90 percent of the deaths from this disease (mostly children under 5 years of age). Ongoing eradication campaigns against this deadly mosquito vector are scoring some successes, and the UN's World Health Organization reports that better preventive and control measures have reduced malaria mortality in Africa by more than half since 2000. Yet despite these gains, the realm's long war against this scourge is far from being won and is certain to continue well into the future.

HIV/AIDS: The Continuing Struggle

AIDS stands for Acquired Immune Deficiency Syndrome, the body's failure to protect itself against the slow-acting, human immunodeficiency virus (HIV). If unchecked, this virus multiplies in the body, causing the immune system damage that produces AIDS. Globally, more than 78 million people have contracted the disease, killing about half of them. Once again, Africans are most disproportionately affected: AIDS first erupted in this realm during the 1930s, and a subsequent large-scale outbreak here in the 1970s swiftly became the springboard for a pandemic that soon reached the United States in 1981. There is mounting evidence that AIDS in Subsaharan Africa peaked around 1998 and that the rate of new infections has been slightly reduced since then. Uganda, in particular, has won wide acclaim in this effort, and its success has inspired a new wave of public health strategies and assistance programs in many AIDS-affected countries.

The geography of HIV/AIDS is complex, but the Subsaharan Africa realm stands out prominently (with fully 70 percent of the world's cases), as do its remarkable internal patterns. Most notable are the 10 countries of the Southern Africa region that bear a major brunt of the global AIDS burden: collectively, they alone account for *just over 40 percent* of the world total of AIDS cases. When standardized for population size, the dire situations of particular countries become clear: an average 24 percent of the populations of Lesotho, Swaziland, and Botswana are HIV-positive, and close behind at 19 percent is neighboring South Africa. Moreover, Swaziland now has the highest occurrence of this disease in the world, with AIDS killing 50 people each day while HIV newly infects another 55. HIV prevalence is much lower in West and Equatorial Africa, ranging from 2 to 5 percent in most countries. The highest incidence of HIV/AIDS in the region of East Africa occurs in Uganda (7 percent) and Kenya (6 percent). In absolute numbers, the largest HIV-infected national populations are found in South Africa (6.4 million) and Nigeria (5.6 million).

© J. Nijman, P. O. Muller, H. J. de Blij, and John Wiley & Sons, Inc.

WEST AFRICAN FOCUS ca. 1400

ca. 1850

1901

1907

1908

ORIGINAL TSETSE FLY VECTORS THAT TRANSMIT THE TRYPANOSOMES

→ Diffusion of African sleeping sickness

After Willett.

FIGURE 7-9

Recent studies have shown that HIV/AIDs in Africa also has important social and economic dimensions. More women than men are infected, underscoring greater female vulnerability mainly associated with inferior social and economic status. Children by the millions have been orphaned, sharply driving up the costs of caring for them. Societies have been deprived of productive workers as well as heavily burdened by related social costs, to the extent that HIV/AIDS can siphon off as much as 1.5 percent from GDP in the most afflicted countries.

Why must Africa suffer so disproportionately? First, HIV/AIDS originated in tropical African forest margins and spread rapidly through all segments of society. Second, the social stigma associated with this sexually transmitted disease makes acknowledging and treating it especially problematic. Third, life-extending medications are both expensive and very difficult to provide in remote rural areas. Fourth, governmental leadership during the AIDS crisis has varied from highly aggressive and effective (as in Uganda) to catastrophically negligent (South Africa stands out here), and there is always the risk that complacency can set in.

Although Africa's HIV/AIDS situation remains serious today, a major battle is being waged against the disease and progress is being made. Public health campaigns throughout the realm—promoting better education, condom usage, and voluntary testing—are having beneficial effects. Generic anti-HIV drugs are becoming cheaper and more widely available through international assistance. At present, more than half of Africa's HIV victims are receiving anti-retroviral therapy, and universal coverage remains the goal until a vaccine can be developed (which most experts believe will take at least another decade). But even though the number of new infections is declining across the realm, several high-risk populations remain virtually unaffected by the latest advances—particularly refugees, sex workers, prison inmates, and those living in extreme poverty.

The Ebola Crisis of 2014–2015

Another very dangerous illness that can ravage parts of Subsaharan Africa is Ebola Virus Disease (EVD)—as the world discovered during fall of 2014. That is when the largest EVD outbreak on record occurred in West Africa, resulting in thousands of deaths, devastating inadequate healthcare systems, and damaging the economies of countries still recovering from years of ruinous civil war. The hyped-up overreporting of the Western media notwithstanding, this epidemic was almost entirely limited to three countries: Sierra Leone (where the outbreak originated) and its two eastern neighbors, Liberia and Guinea. As of June 2015, the WHO had reported a total of 27, 225 Ebola cases, with all but 36 confined to these countries; 11,164 of those cases were fatalities (41 percent), but only 15 occurred outside the three countries. Of the few EVD cases beyond Sierra Leone, Liberia, and Guinea, almost all were confined to West Africa's Nigeria and Mali, as were 14 of the 15 deaths. In fact, the remaining fatality (a misdiagnosed traveler from Liberia) occurred in Dallas, Texas; the three other EVD patients in the United States all fully recovered.

By mid-2015, the West African EVD epidemic (the worst Ebola outbreak in history) appeared to be entering its final stage. The WHO had already declared Liberia to be Ebola-free, and only a limited number of virus hotspots remained in Sierra Leone and Guinea. In Guinea, new cases of EVD were in decline as public health teams strove to achieve full containment. Initial assessments of the epidemic were also emerging and show that its impacts went far beyond the healthcare arena. Ebola had caused serious disruptions to travel, transportation, and trade in this part of West Africa, including its ties to the outside world; agricultural production had diminished, undermining an already stressed food security situation; and more ominously, many major international investors, now fearful of the disease threat, had (at least temporarily) withdrawn from the afflicted countries. Unfortunately, Sierra Leone, Liberia, and Guinea will have to contend with these and other negative economic consequences long after their last new Ebola case has been recorded.

CULTURAL PATTERNS

Today, just about 60 percent of Subsaharan Africa's population still inhabit—and work in or near—Africa's hundreds of thousands of villages. They speak one of the 2000-plus languages in use in the realm. Africa's (numerically) largest groups of people constitute major nations, such as the Yoruba of Nigeria and the Zulu of South Africa. Africa's smallest groups of people number just a few thousand. Overall, this geographic realm has the most complex and fragmented cultural mosaic on Earth (see Fig. 7-5).

African Languages

Africa's linguistic geography (Fig. 7-10) is a key component of that cultural intricacy. It is clear that the Subsaharan Africa realm begins approximately where northern Africa's Afro-Asiatic language family (mapped in yellow) gives way, a changeover at its sharpest in West Africa.

The realm's dominant language family is the Niger-Congo family (mapped in purple), of which the Kordofanian subfamily is a tiny, historic, northeastern outlier (Fig. 7-10), and the Niger-Congo languages extend across the realm from West to East and Southern Africa. The Bantu language forms the largest branch of this subfamily, but Niger-Congo languages in West Africa, such as Yoruba and Akan, also have millions of speakers. Another important language family is the Nilo-Saharan family (mapped here in orange), extending from Maasai in Kenya northwest to Teda in Chad. No other language families are of similar extent or importance: the Khoisan family, of ancient origins, now survives among the dwindling Khoi and San peoples of the Kalahari in the southwest; the small white minority in South Africa speak Indo-European languages; and Malay-Polynesian languages prevail in Madagascar, which was peopled from Southeast Asia long before Africans reached it.

interregional usage within the country. On one hand, using a European, colonial language as an official medium invites criticism; on the other hand, making a dominant local language official would unleash negative reactions from ethnic minorities. One creative country today is Rwanda, which promotes its Bantu language (Kinyarwanda) in tandem with English for instruction in schools to speed the separation from its Francophone past.

Religion

Africans had their own belief systems long before Christians and Muslims arrived to convert them. Despite their diversity, all of Subsaharan Africa's cultural groups held a consistent view of their place in nature. Spiritual forces, according to African tradition, are manifest everywhere in the natural environment: in forests, rivers, mountains, and communal lands—a religious worldview known as *animism*. Gods and spirits are believed to affect people's daily lives, witnessing every move, rewarding the virtuous and punishing (through injury or crop failure) those who misbehave.

As with land tenure, the religious views of Africans clashed fundamentally with those held by the colonizers. Monotheistic **Christianity** first penetrated Africa in the northeast when Nubia and Axum were converted, and Ethiopia has been a Coptic Christian stronghold since the fourth century AD. But the Christian churches' real invasion did not commence until the onset of colonialism after the turn of the sixteenth century. Christianity's various denominations made inroads in different locales: Roman Catholicism in much of Equatorial Africa, mainly at the behest of the Belgians; the Anglican Church in British colonies; and Presbyterians and others elsewhere. And these days, as in South America, Evangelical churches steadily gain adherents.

Islam had a rather different arrival and impact. Today, about one-third of the realm's population is Muslim. By the time of the colonial invasion, Islam had advanced out of Arabia, across the Sahara, and partway down the coasts of Africa (particularly along the Indian Ocean). Muslim clerics converted the rulers of African states and commanded them to convert their subjects. They Islamized the savanna states and penetrated into present-day northern Nigeria, Ghana, and Ivory Coast. They encircled and isolated Ethiopia's Coptic Christians and Islamized the Somali people in Africa's Horn. They established beachheads on the coast of Kenya and took over offshore Zanzibar. Muslim and European-Christian proselytizers competed for African minds, but Islam proved to be a far more pervasive force. Today, tension between Islam and Christianity is especially evident all along the Islamic Front, one of the defining regional characteristics—and the dynamic southern border—of the African Transition Zone (discussed at the end of Chapter 6).

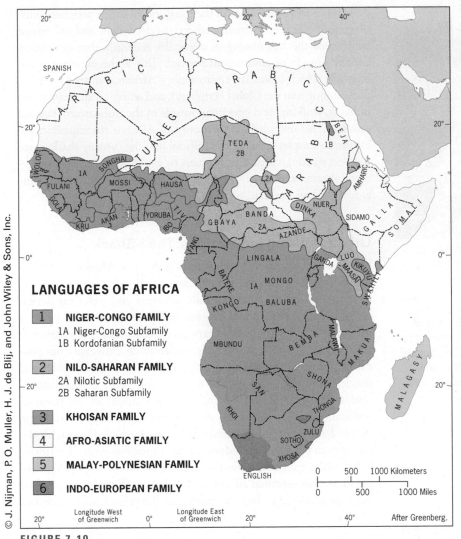

LANGUAGES OF AFRICA

1 **NIGER-CONGO FAMILY**
1A Niger-Congo Subfamily
1B Kordofanian Subfamily

2 **NILO-SAHARAN FAMILY**
2A Nilotic Subfamily
2B Saharan Subfamily

3 **KHOISAN FAMILY**

4 **AFRO-ASIATIC FAMILY**

5 **MALAY-POLYNESIAN FAMILY**

6 **INDO-EUROPEAN FAMILY**

0   500   1000 Kilometers
0   500   1000 Miles

After Greenberg.

FIGURE 7-10

© J. Nijman, P. O. Muller, H. J. de Blij, and John Wiley & Sons, Inc.

The Most Widely Used Languages

About 40 African languages are spoken by 1 million people or more, and a half-dozen by 10 million or more: Hausa (34 million), Yoruba (28 million), Ibo, Swahili, Lingala, and Zulu. Although English and French have become important *linguae francae* in countries such as Nigeria and Ivory Coast (officially *Côte d'Ivoire* in the Francophone manner), African languages also serve this purpose. Hausa is a common language across the West African savanna; Swahili is widely used in East Africa. And pidgin languages—mixtures of African and European tongues—are found along West Africa's coast; millions of Pidgin English (called *Wes Kos*) speakers use this medium in Nigeria and Ghana. However, at least 60 African languages are on the verge of extinction, and Nigeria alone has 17 endangered (near-extinct) languages.

Multilingualism **14** can be a powerful centrifugal force in society, and African governments have tried with varying success to promote national alongside local languages. Nigeria, for instance, made English its official language because none of its 500-odd languages, not even Hausa or Yoruba, had sufficient

The faithful kneel during Friday prayers outside a mosque in the city of Kano in northern Nigeria. Whereas most northerners want to practice Islam within the multireligious, multiethnic society of Nigeria, areas governed by strict Islamic (*Sharia*) law are being expanded, bringing official and *Sharia* law into confrontation throughout the north. Nigeria's viability as a unified state is now being threatened by the rise of Boko Haram, the brutally violent jihadist organization that is seeking to expand from its base in the far northeast by declaring itself a caliphate in 2014. Can this country avoid the fate of Syria and thwart the formation of another ISIS-type "Islamic State"? And how will Nigeria's moderate Muslims respond to the terrorist tactics of the extremist Boko Haram cult?

URBANIZATION AND AFRICA'S DUAL ECONOMIES

Subsaharan Africa may still be one of the least urbanized geographic realms, but people everywhere are moving to towns and cities at an accelerating pace. Today, almost 40 percent of the realm's inhabitants reside in urban settlements (up from 28 percent as recently as 2000), and in absolute numbers they total more than the entire population of the United States. Even though the growing tide of migrants is overwhelming the realm's already inadequate urban infrastructure, rapid urbanization is projected to continue far into the future—a current crisis that will soon be transforming into a social catastrophe because half of all babies born into the world between now and 2050 will be African.

The biggest African cities operate as the anchors of incipient national core areas, housing government offices and the headquarters of the most influential corporations and extractive industries. The **formal economy 15**, under governmental control that regulates the civil service, business, and industry as well as their workforces, produces a formal urban landscape featuring a modern government district, clusters of newer high-rise buildings, upscale hotels, elite gated residential communities, and even branded fast-food outlets. From a distance, such cityscapes resemble those of urban centers in the Global Core, but on closer inspection it is apparent that the formal sector is far from dominant here. In the streets, on sidewalks next to shop windows, in

downtown-area slums, and on the rough-hewn urban periphery, hustling street hawkers, garment makers, jewelry sellers, and unlicensed taxis signal the functioning of a parallel economy that lies almost entirely outside of government control. This is the **informal economy 16**, which looms large across every African country (as well as those throughout the Global Periphery), and entails an alternative way of living, working, and operating outside of the mainstream national economy. An overwhelming majority of Africans participate in the informal sector by residing in unofficial housing (usually shacks—see the next section), transacting business by avoiding banks and the payment of taxes, and by relying mainly on family, friends, and acquaintances for their survival.

Urban Housing Innovations: The I-Shack

Shacks are ubiquitous in this realm's poverty-stricken urban landscapes (see Field Note). Substandard dwellings are always unacceptable, but with some carefully planned and practical design improvements shacks can be greatly enhanced to yield a better quality of life for their disadvantaged inhabitants. The Sustainability Institute of South Africa's Stellenbosch University has produced a new prototype shack design called the *i-shack* ("improved shack"), and a pilot project to develop and apply this concept is under way at Enkanini, an informal settlement in the town of Stellenbosch that lies at the eastern edge of metropolitan Cape Town. I-shacks are rudimentary solar-powered dwellings with the capacity to support two indoor lights, a motion-sensitive outdoor light, and a mobile phone charger. Its windows are strategically positioned for optimal air circulation and sunlight heating, while the roof is sloped to harvest rainwater. Recycled cardboard boxes

A typical urban landscape in Subsaharan Africa. This aerial view of the city of Arusha in northern Tanzania (population: ca. 600,000) shows several features of the settlement layout in urban Africa. The irregular pattern of low-rise, low-density development reflects a general lack of planning, complex land ownership issues, and piecemeal as well as patchwork investment in construction.

@ **FROM THE FIELD NOTES...**

"Shacks are a conventional form of housing in the Subsaharan African city, and their construction goes on all the time. Most residents of these informally built dwellings get little more than a "roof" over their heads, and they must make do without a water supply, electricity, garbage collection, and sanitation. Quite frequently, these structures must be built in hazardous environments, such as those prone to flooding. A number of savvy informal workers learn how to specialize in shack construction and then sell pre-ordered shacks that can be fully assembled on any site in a matter of hours. In 2010, I watched this shack being put together in Soweto—the famous South African township that was the longtime home of Nelson Mandela—while engaging in field research on informal entrepreneurship in Soweto and other outlying settlements in the Johannesburg area."

and "Tetra Pak" containers are used for insulation between the exterior zinc surface and the interior, and flame-retardant paint is employed to minimize the risk of fire. I-shacks can be constructed and fitted for U.S. $660, and represent a significant and affordable upgrade on the everyday self-built shack made out of corrugated iron and discarded wood—which are too damp in winter and especially fire-prone in summer.

AFRICA RISING?

Despite colonial legacies, formidable environmental challenges, and a long history of serious adversity, twenty-first century Africa is steadily emerging on the global scene. Although conditions remain harsh in many places, the achievements of the past decade are quite encouraging, even stunning in certain cases (such as the lengthening of life expectancy in the realm by a remarkable 6 years since 2000). Gathering economic momentum is the new watchword, and the development outlook for a growing number of countries appears robust.

Economic growth in Subsaharan Africa has averaged more than 5 percent annually since 2002, outperforming all other realms (Fig. 7-11). Percentagewise, this even surpasses China's economic expansion over the same period, and in 2014 (the latest year for which data are available) several African countries—led by DRCongo, Ethiopia, and Ivory Coast—exceeded China's annual GDP growth rate. At the same time, total foreign investment in the realm dramatically rose more than fivefold (from U.S.

$11 billion in 2001 to $57 billion in 2013), much of it focused on the burgeoning resource-extraction sector. Today, that financial influx is increasingly being channeled into telecommunications, banking, infrastructure, and even manufacturing.

What underlies these far more positive economic assessments of Africa's development potential? First, the realm is demonstrably closing the gap between its economic performance and that of the developed world. Second, the management of Africa's national economies is gradually improving, and the resolution of political conflicts together with the expansion of democratic governance has led to ever more favorable business environments. Third, a new generation of policy-makers, activists, and business leaders is emerging across most of Subsaharan Africa; they are Africans to the core, but with an increasingly global mindset derived from their participation in the Internet age, easier international travel and communication, and greater exposure to current worldwide trends and thinking. Fourth, the African debt crisis is winding down, and economic relations with Western donor countries are improving. And fifth—and most important of all—China has exploded onto Africa's economic scene, influencing trade, investment, foreign aid, and diplomatic relations.

China's Role in Emerging Africa

Through much of modern history, Westerners have perceived Africa to be both remote and peripheral to international affairs. Although the subordinate, asymmetrical relationship that long tied this realm to Europe's imperial powers is still visible (in terms of foreign

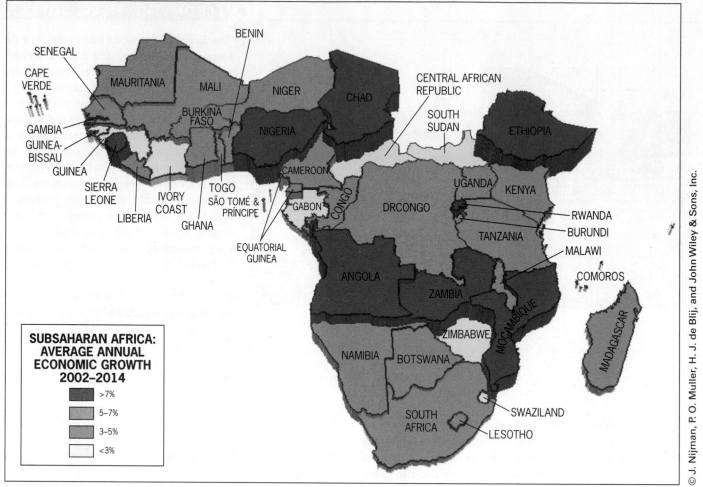

FIGURE 7-11

investment, trade, and external oversight), evidence is mounting that things have begun to change. The recent rise of emerging-market countries such as Brazil, Russia, India, and China—the so-called **BRICs**—as well as the retreat of various Western development models as the only game in town, mean that African leaders now have a wider range of options in their international relations.

Many African states are now forging political and economic partnerships of their own choosing. They are casting aside the shackles of colonial thinking and its postcolonial legacies and taking more control of their own international economic policies. Such enhanced agency enables African states to become full-fledged players in international relations rather than the objects of other countries' foreign policies—the dawning of a new era that some are calling a 'second independence' for Africa.

Nothing marks the elevation of Subsaharan Africa's stature more than China's continuing surge of investments in infrastructure, energy-producing facilities, and civic construction. Since 2010, Ghana, Nigeria, and Zambia each have more than U.S. $10 billion worth of Chinese-funded projects under development; South Africa, Tanzania, and Angola (all resource-rich countries) are also prioritized (Fig. 7-12). China has now become this realm's largest trading partner and foremost foreign investor as well as a major aid provider and a

leading source of immigrants (elaborated by Howard French in his monumental 2014 book, *China's Second Continent: How a Million Migrants Are Building a New Empire in Africa*). China also holds significant assets in the form of oil leases, joint mining ventures, and timber concessions. There is also an important difference between the Chinese and Western engagement with contemporary Africa that is worth noting: Beijing does not require African states to emulate its values. This new kind of engagement establishes an alternative enabling environment, wider options for maneuverability, and the opportunity for Africans to chart a different development course—ideally one of their own choosing.

Africa's ongoing rise as a strategic component of the global economy is based on the realm's natural resources. This has brought major new opportunities to develop infrastructure and real estate (e.g., satellite cities, high-tech growth poles, modern CBDs, and shopping complexes), provide commercial and producer services for Africa's mushrooming urban population, and accelerate the emergence of a middle class. Furthermore, many of the realm's countries are now being courted as diplomatic allies in global politics. In a world where votes matter more than ever, those of the 54 Subsaharan African states increasingly resonate in international forums that shape trade, environmental, and security-related policies.

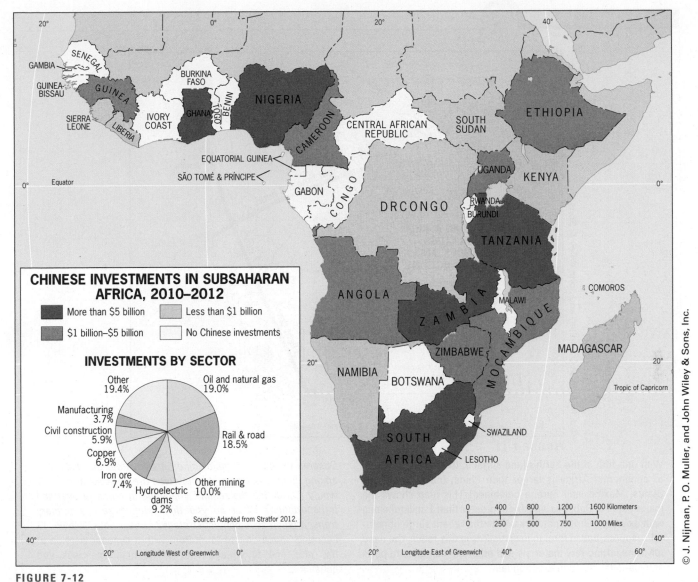

CHINESE INVESTMENTS IN SUBSAHARAN
AFRICA, 2010–2012

More than $5 billion   Less than $1 billion

$1 billion–$5 billion   No Chinese investments

INVESTMENTS BY SECTOR

Other 19.4%
Oil and natural gas 19.0%
Manufacturing 3.7%
Civil construction 5.9%
Copper 6.9%
Iron ore 7.4%
Rail & road 18.5%
Hydroelectric dams 9.2%
Other mining 10.0%

Source: Adapted from Stratfor 2012.

© J. Nijman, P. O. Muller, and John Wiley & Sons, Inc.

FIGURE 7-12

Africa's Need for Supranational Cooperation

Economic interaction among the states of Subsaharan Africa has been stifled for centuries, and geography has played a key role. The realm's huge territorial size facilitated the manipulation of its political geography throughout colonial times in order to isolate individual colonies and connect them to Europe. Accordingly, roads and rail networks were laid out to serve the interests of colonial empires: they emphasized convergence on colonial capitals, access to resource extraction sites, and, above all, direct linkages to seaports for shipping goods to the mother country. Because so few overland routes penetrated inland border areas, let alone crossing Africa's international boundaries, these colonial arteries reinforced intra-realm fragmentation. And what little cross-border commerce took place was heavily shackled by tariffs, harassment from customs officials, outrageous demands for bribes, and other unnecessary protectionist measures. At the continental scale, the shortcomings of the colonial infrastructural

legacy are even more apparent (Fig. 7-13)—a transportation development issue treated in the Map Analysis box. Consider this: even today intra-realm trade accounts for only 13 percent of total trading activity within Subsaharan Africa compared to almost 70 percent in Europe.

To overcome such disadvantages, African states must improve their international interaction by committing themselves to the kind of supranational cooperation that is strengthening the internal cohesion of geographic realms from the Americas to Europe to Southeast Asia. At the continental scale, the Organization of African Unity (OAU) was established for this purpose in 1963 and was superseded by the African Union (AU) in 2001. Efforts at the regional level include the Economic Community of West African States (ECOWAS), created by 15 countries in 1975 to promote cooperation in trade, transportation, and industry in that region; and in the early 1990s, 12 countries joined to form the Southern African Development Community (SADC) and were soon followed by the 20-odd states that

© J. Nijman, P. O. Muller, and John Wiley & Sons, Inc.

FIGURE 7-13

With one-fifth of the Earth's land surface, the territorial extent of continental Africa is larger than China, India, the United States, Mexico, and Europe combined! This map shows the major cities of Subsaharan Africa (greater than 1 million) along with existing railroad networks. With the recent surge of natural-resource-based foreign investments in this realm, the need for transporting raw materials and other commodities to ports and markets has risen exponentially.

Observe where most major cities are located, the distances among them, and where rail networks are concentrated. How would you define the realm's challenge in terms of overland transportation? What do you think should be the primary transportation development strategy? Access your WileyPLUS Learning Space course to interact with a dynamic version of this map and to engage with online map exercises and questions.

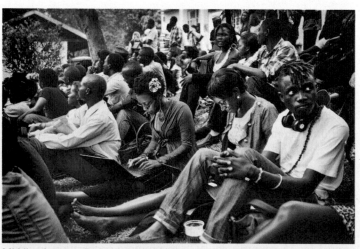

© Yannick Tylie / Corbis

Middle-class youths at a recent midsummer cultural festival in Kampala, Uganda. Demands for education, cultural goods, computer and Internet use, and access to leading entertainment media are skyrocketing, and upscale consumption is steadily on the rise in many of Subsaharan Africa's cities.

founded the Common Market for Eastern and Southern Africa (COMESA).

Although institutional structures for economic integration already exist, the various initiatives need streamlining. A promising step in that direction was taken in 2015 with the signing of the Tripartite Free Trade Agreement (TFTA) that will form a "Cape-to-Cairo" free-trade zone to be known as the Continental Free Trade Area. Coordinated by the AU, this pact involves the 26 countries participating in the amalgamation of SADC, COMESA, and the EAC (East African Community), and will create Africa's largest common market (with a total population of 630 million) when it begins operating in 2017. However, such commendable intentions aside, supranational organizations have yet to prove they can be mechanisms for meaningful economic progress. So now more than ever, the countries of Subsaharan Africa need to summon the considerable political will to build lasting regional cooperation and realmwide interdependence.

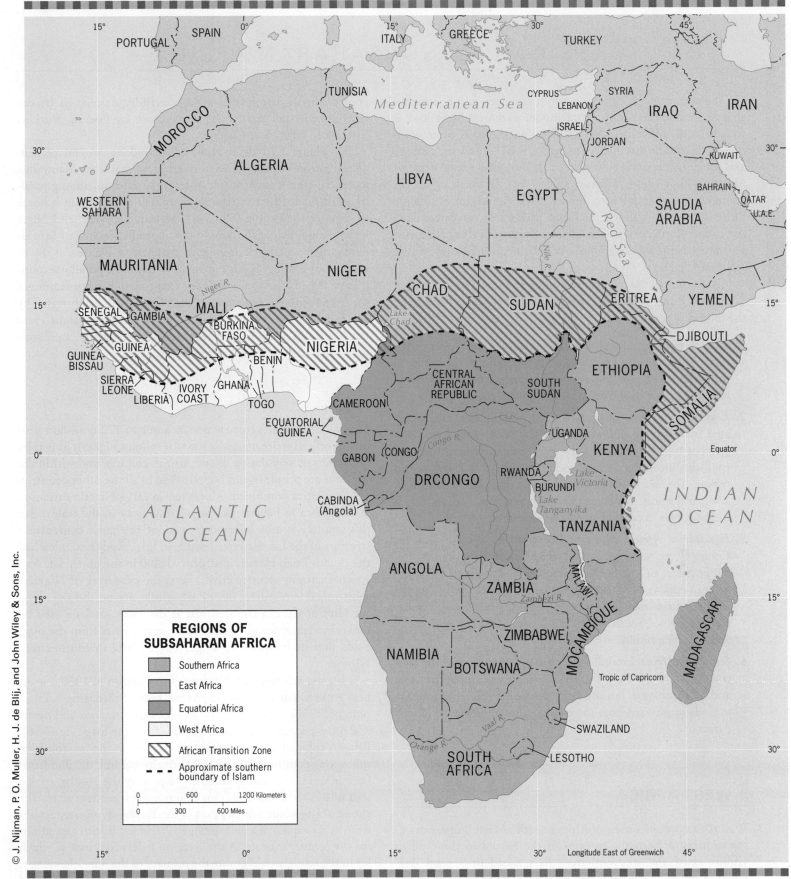

REGIONS OF SUBSAHARAN AFRICA

- Southern Africa
- East Africa
- Equatorial Africa
- West Africa
- African Transition Zone
- - - Approximate southern boundary of Islam

0 — 600 — 1200 Kilometers
0 — 300 — 600 Miles

FIGURE 7-14

REGIONS OF SUBSAHARAN AFRICA

On the face of it, Africa is so massive, compact, and continuous that any attempt to justify a contemporary regional breakdown is fraught with difficulties. Nevertheless, on the basis of environmental distributions, ethnic patterns, cultural landscapes, distance and proximity, historic culture hearths, and other spatial data, we identify a four-region structure (Fig. 7-14). By focusing on the contents of that framework, however, we should not overlook a fifth region that straddles the broad divide between Subsaharan Africa and North Africa—the *African Transition Zone*, discussed separately as the final region of the North Africa/Southwest Asia realm in Chapter 6.

Beginning in the north at the southern margin of the African Transition Zone, we work our way east and then south through the following regions:

1. *West Africa*, which includes the countries of the western coast north of the equator and those on the margins of the Sahara in the interior. This populous region of fifteen countries is anchored in the southeast by the realm's demographic giant, Nigeria.

2. *East Africa*, where equatorial natural environments are moderated by elevation and where plateaus, lakes, and mountains, some carrying permanent snow, exemplify the countryside. Six countries, including the highland part of Ethiopia, comprise this region. The island-state of Madagascar, marked by Southeast Asian influences, is neither Southern nor East African, but is included here because of its relative location.

3. *Equatorial Africa*, much of it defined by the basin of the Congo River, where elevations are lower than in East Africa, temperatures are higher and moisture more abundant, and where most of Africa's surviving tropical rainforests stand. Among the nine countries that make up this region, which now includes recently formed South Sudan, DRCongo dominates territorially and demographically.

4. *Southern Africa*, extending from the southern tip of the continent to the northern borders of Angola, Zambia, Malawi, and Moçambique. Ten countries constitute this region, which extends southward from the tropics and whose giant is South Africa.

▌ WEST AFRICA

West Africa occupies most of Africa's northwestern Bulge, extending southward from the margins of the Sahara to the Gulf of Guinea coast and westward from Lake Chad to Senegal (Fig. 7-15). A total of 17 states (also counting Chad and offshore Cape Verde) constitute this region. In addition to once-Portuguese Guinea-Bissau and long-independent Liberia, West Africa encompasses four former British and nine former French dependencies.

Four countries, comprising a huge territory dominated by dry environments but sparsely populated, are landlocked: Burkina Faso, Mali, Niger, and Chad.

As Figure 7-15 shows, political boundaries extend perpendicularly from the coast into the interior, so that from Mauritania to Nigeria, the West African habitat is parceled out among parallel, coast-oriented states. The southern half of the region is home to most of the people (Fig. 7-8). Mauritania, Mali, and Niger include too much of the *Sahel*—the unproductive belt of Sahara-bordering, semiarid environment (mapped in Figure 7-15)—to sustain populations as large as those of the most populous countries of Nigeria, Ghana, and Ivory Coast. Another key feature of the region's human geography is that West Africa is bifurcated by a religious divide in upheaval—the turmoil-ridden *Islamic Front* (see Fig. 6-20) that forms the southern edge of the African Transition Zone (the black dashed line in Fig. 7-14).

Nigeria

Nigeria, this region's cornerstone, is home to 173.6 million people (unofficial estimates go at least 10 million higher). This is by far the largest population of any African country, and within the region more people live in Nigeria than in all the other countries of West Africa combined. Moreover, in 2014 Nigeria surpassed South Africa to become the largest economy in the realm. But this country is also bisected by one of the most contentious stretches of the Islamic Front, which in this decade has unleashed the vicious Boko Haram insurgency based in the northeast. And another current security crisis affects the exporting of Nigeria's most valuable resource, oil; pirate attacks are on the rise along the Gulf of Guinea coast, aimed at the main sea lane used by tankers to transport oil to Global Core countries from the huge fields that lie beneath the Niger River Delta in southeastern Nigeria.

Nigeria achieved independence from Britain in 1960, but its new government faced the herculean task of managing a European political creation containing three major nations and nearly 250 other peoples. For reasons obvious from the map (Fig. 7-16), Britain's colonial imprint was always stronger in the two southern subregions than in the north. Christianity became the dominant faith in the south, and southerners, especially the Yoruba, took a lead role in the transition from colony to independent state. The choice of Lagos, the primary port of the Yoruba-dominated southwest, as the capital of a federal Nigeria reflected British aspirations for the country's future. A three-region federation, two of which lay in the south, would ensure the primacy of the non-Islamic part of the state. But this framework did not last very long and required frequent modification. Nigeria today has 36 States, and the capital was moved from Lagos to more centrally located Abuja more than a quarter-century ago (Fig. 7-16).

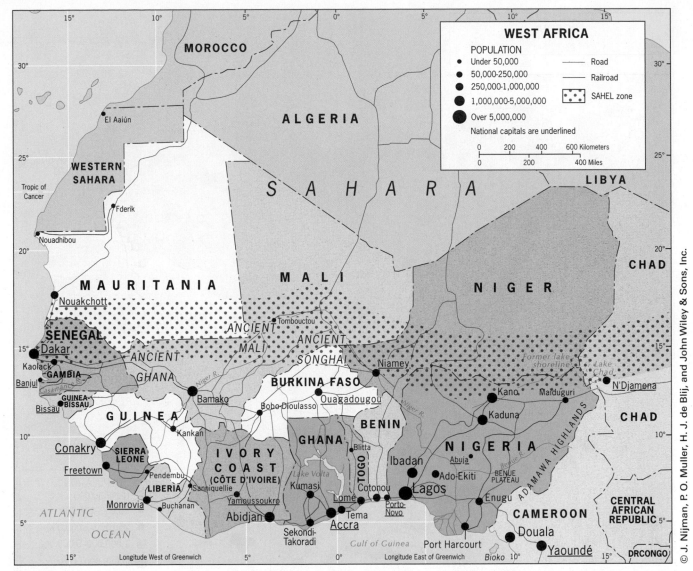

FIGURE 7-15

Oil Booms and Busts

Large oilfields were discovered beneath the Niger Delta in the 1950s, and this discovery transformed the agriculture-dominated economy into one based on petroleum exports. But before long, Nigeria's petroleum riches brought more bust than boom. New misguided development plans focused on ill-founded industrial schemes and the like, while farming, still the mainstay of most Nigerians, fell into neglect. Worse, poor management, corruption, outright theft of oil revenues during military misrule, and excessive borrowing against future oil income led to economic disaster. The country's infrastructure collapsed; in cities, basic services broke down; and in rural areas, clinics, schools, water supplies, and roads to markets crumbled.

Nonetheless, Nigeria has now become the realm's largest economy, and its GDP growth over the past decade is among the highest on the African continent. Most of that success results from holding its position as the world's twelfth-largest oil producer,

with the United States one of its main customers. But these economic benefits have yet to reach the great bulk of the fast-growing Nigerian population (birth rates here are double the global average), 80 percent of which still subsists on the equivalent of less than one U.S. dollar a day.

Islamic Revivalism and Boko Haram

In 1999, Nigeria's hopes were heightened as a democratically elected president was sworn into office. But shortly thereafter its problems magnified when northern States, beginning with Zamfara, began to promulgate strict-Islamic *Sharia* law **17**. When Kaduna State followed suit, violence between Christians and Muslims was unleashed and devastated that State's venerable capital city, Kaduna. There, and in the 11 other northern States mapped in green in Figure 7-16, the imposition of *Sharia* law led to the departure of thousands of Christians, intensifying the

FIGURE 7-16

cultural fault line that we now know is the Islamic Front. That schism continues to deepen, and the kind of Islamic revivalism now taking hold in the north has reached a level that is imperilling the stability—perhaps even the unity—of West Africa's cornerstone state.

In 2002, a jihadist Islamic organization calling itself **Boko Haram** **18** (meaning "Western education is sinful"), emerged in the northeasternmost State of Borno (Fig. 7-16), an impoverished, underrepresented corner of Nigeria where repressive central government policies had triggered growing militancy. Support for Islamic militants from the State's heavily Muslim population is mainly drawn from young, unemployed, northern Nigerian men, who have lost faith in the politicians and in their own future.

Boko Haram's "religious cleansing" campaign was initially confined to Borno, but since 2008 its barbaric activities have expanded across the entire northeastern quadrant of Nigeria, including high-profile terrorist attacks in Kano and even the federal capital, Abuja, well to the south. Targeting civilians, Boko Haram attacks during 2014 killed at least 10,000 and forced tens of thousands to flee their home villages (see photo). On April 14, 2014, the world's attention was seized by the dramatic kidnapping of 279 female adolescents from a school in the village of Chibok, to be forcibly married to Boko Haram fighters—igniting the global *#Bringback-ourgirls* social media campaign.

In early 2015, following the lead of ISIS, Boko Haram declared itself an independent caliphate consisting of the Nigerian

The aftermath of an unimaginably vicious 2013 Boko Haram attack on Kawuri in northeastern Nigeria, which resulted in the murder of 70 villagers and the destruction of every home. About a year later, the jihadist killers struck again in a nearby town as 50 gunmen stormed a school at dawn, stabbing 43 students to death to avoid attracting attention by using their firearms and homemade bombs.

territory it controlled, centered on Borno's capital, Maiduguri, a city of more than 700,000 (Fig. 7-16). Operating under the banner of its own black flag, Boko Haram today dominates an area larger than West Virginia that contains a population of almost 3 million. Yet until a newly elected regime took office in April of 2015, the Nigerian government's response to Boko Haram's

insurgency had been mystifyingly halfhearted and ineffective—even declining the assistance of a multinational African force to launch a counteroffensive. However, Africa's self-described "caliphate," which may or may not be linked to ISIS, continues to pose a massive threat to Nigeria's (as well as West Africa's) viability that cannot go unchallenged.

Coast and Interior

West Africa's other countries also exhibit distinct regional geographies. Typical of the Gulf of Guinea coastal zone west of Nigeria is neighboring **Benin**, with a population of 10.3 million and growing cultural and economic ties with northeastern Brazil's Bahía State, where many of its people were taken in bondage and where elements of West African culture survive. Benin's internal geography is comparable to that of neighboring **Togo** (population: 6.8 million): both are markedly elongated, narrow political units with savannas in the north and humid tropical lowlands near the coast, but without usable rivers. In contrast, **Burkina Faso** is representative of West Africa's much drier interior—impoverished and landlocked, but with undeveloped reserves of gold and a commercial economy that relies on exporting cotton. Importantly, with Muslims constituting 59 percent of Burkina Faso's 16.9 million inhabitants, this country—together with Mali (93 percent), Niger (99 percent), and transitional Chad (56 percent)—lies firmly within the Islamic orbit of northern West Africa.

Among the coastal states, **Ghana**, known in colonial times as the Gold Coast, was the first modern West African state to achieve independence, with a democratic government and a sound commercial

@ FROM THE FIELD NOTES...

"Even in the poorer countries of the world, you see something that has become a phenomenon of globalization: gated communities. Widening wealth differences, security concerns, and real estate markets in societies formerly characterized by traditional forms of land ownership combined to produce this new element in the cultural landscape. This is the entrance to Golden Gate, the first private gated-community development in Accra, Ghana, begun in 1993 as a joint venture between a Texas-based construction company and a Ghanaian industrial partner. Upscale gated communities are now found in major urban areas throughout the realm and are most widespread in the upper and upper middle-income neighborhoods of South Africa's suburbs and cities."

@ **FROM THE FIELD NOTES...**

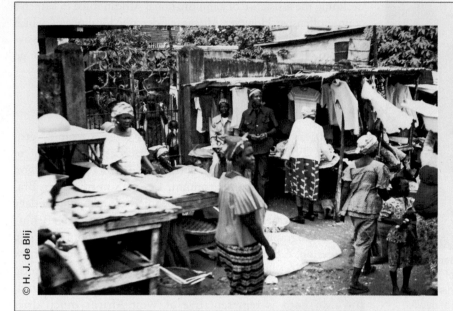

© H. J. de Blij

"Wherever you go in West Africa, you are struck by the bright colors, especially in the markets. And Freetown, Sierra Leone, on this Friday seems to be one huge market, the streets lined with stalls, young and old crowding intersections in huge numbers. Most of the goods were for local trade; compared to what you see in East Africa, where tourism plays a bigger role, these markets still have a strong domestic atmosphere. Music plays everywhere, in shops, out in the open; and the whole city seems suffused with the aromas of spicy foods."

economy based on cocoa exports. Both democracy and the economy soon failed, however in the 1990s a military regime was replaced by a stable representative government and recovery began. By 2007, Ghana celebrated 50 years of independence, its democracy maturing, its economy forging ahead, corruption persistent but declining, and its international stature rising. The same year also brought a U.S. $500-million-plus aid package from America's Millennium Challenge Corporation to expand commercial agriculture, further improve infrastructure, and combat poverty—based in no small part on Ghana's solid reputation as an African trailblazer.

Proceeding westward from Ghana along the southern coast of the African Bulge, **Ivory Coast** (officially *Côte d'Ivoire*) has had a turbulent history. Following independence from France in 1960, it translated the next three decades of autocratic but stable rule into economic progress; however, by the late 1990s the political succession had become badly entangled in the north-south, Muslim-Christian schism that degenerated into two rounds of civil war between 2002 and 2011. Conflict has now largely subsided, but this country of 20.3 million remains deeply troubled as it tries to rebuild its shattered society and cocoa-based commercial economy. Farther up the coast is another key Bulge country, **Senegal**. This democratic state demonstrates what stability can facilitate: without oil and other major income-generating resources, and with an overwhelmingly subsistence-farming population, Senegal nevertheless managed to achieve some of the region's highest GNI levels.

Between Senegal and Ivory Coast lie three other coastal countries that are just emerging from a quarter-century of civil war and horror. **Liberia**, founded in 1822 by freed American slaves who returned to Africa with the help of U.S. colonization societies, was governed by Americo-Liberians for six generations and sold rubber and iron ore abroad. A military coup in 1980 ended that era, and by 1989 full-scale civil war had embroiled virtually every ethnic group in the country. This brutal protracted conflict claimed the lives of about one-tenth of the population of 4.3 million, and hundreds of thousands of others fled the country. Liberia finally turned the corner in 2006 with the election of Ellen Johnson Sirleaf as Africa's first female president, who succeeded in stabilizing the now-democratic political system. **Sierra Leone** followed a similar path from British dependency to republic to one-party state and military dictatorship. In the 1990s, a rebel movement, financed by diamond sales, inflicted dreadful punishment on the local population, but a remarkable turnaround resulted from British intervention during the late 2000s. Diamonds are still the leading export, and now-stabilized Sierra Leone (population: 6.1 million) ranked among Africa's fastest growing economies until Ebola struck in 2014. The same cannot yet be said of neighboring **Guinea**, where dictatorial mismanagement and a violent power struggle combined to ruin economic opportunities in both agriculture and mining. As Figure 7-15 shows, Guinea borders the troubled countries just discussed plus Ivory Coast, and it has involved itself in the affairs of all of them—receiving in return a stream of refugees in its border zones. Guinea, with a population of 11.8 million, has long been under dictatorial rule and remains one of Africa's lowest-income states. This need not be because the country possesses major deposits of bauxite (aluminum ore) and gold, can produce far more coffee and cotton than it does, and controls productive offshore fishing grounds.

We should keep in mind that the numerous conflicts noted above were and are geographically concentrated in relatively few areas and that they are not representative of vast, populous West Africa as a whole. Tens of millions of farmers and herders who manage to cope with fast-changing environments in the demanding Sahel live remote from the newsmaking conflicts along the well-watered coast. Overall, to take fuller advantage of the development opportunities sweeping across the realm, West Africa's biggest challenge is to achieve both economic survival and nation-building through political stability.

▮ EAST AFRICA

To the east of the chain of Great Lakes that marks the eastern border of DRCongo, the land rises from the Congo Basin to the East African Plateau. Hills and valleys, fertile soils, and copious rains mark this transition in Rwanda and Burundi. Eastward the rainforest disappears and the open savanna cloaks the countryside. Great volcanoes tower above a rift-valley-dissected upland. At the heart of the region lies Lake Victoria. Farther north the surface rises above 3300 meters (10,000 ft), and so deep are the trenches cut by geologic faults and rivers there that the land was appropriately called Abyssinia (now Ethiopia). Five countries, in addition to the highland component of Ethiopia, constitute this East African region: Kenya, Tanzania, Uganda, Rwanda, and Burundi (Fig. 7-17). Kenya is the most important country from a Western perspective, but Tanzania and Ethiopia have larger populations. And the landlocked countries of Rwanda and Burundi may be small in size, but their fierce ethnic conflicts have spilled over into neighboring countries and contributed to the high regional level of geopolitical instability.

An aerial view, near the Tanzanian-Kenyan border, of the escarpment (right) that lines the edge of East Africa's great Eastern Rift Valley. The lush landscape of this hillside extends clear across the wide valley's well-watered floor, a savanna environment of alternating trees and grasslands, teeming with wildlife.

Tanzania

Territorially, Tanzania (its name a hybrid derived from the 1964 union of *Tan*ganyika and offshore *Zan*zibar) is the largest East African state. Its 49.3 million people are drawn from more than 100 ethnic groups. No single ethnic group is large enough to dominate the state; moreover, Muslims constitute 30 percent of the population, highly concentrated on the coast within the southern, tail end of the African Transition Zone. This is also a country without a core because its clusters of population and zones of productive capacity lie dispersed—mostly on its margins along the Indian Ocean coast, near the shores of Lake Victoria in the northwest, adjoining Lake Tanganyika in the far west, and facing Lake Malawi in the interior south. Experimenting with socialism and farm collectivization in the post-independence period scared off international investors and tourists, and Tanzania soon bottomed out as one of the world's poorest countries. Since 1990, the government has pursued a different course and prospects have improved. Tanzania today is a leading gold exporter, an emerging oil and natural gas producer, and an increasingly popular safari destination (led by well-managed Serengeti National Park in the central north), and its economy has grown at a rate of around 7 percent a year for the past decade.

Ethiopia

Ethiopia, at 94.1 million, is Subsaharan Africa's most populous country after first-place Nigeria. The highland zone of Ethiopia forms the northernmost component of East Africa, but its extreme north and northeast lie within the African Transition Zone behind the Islamic Front (Fig. 7-17, red-dashed line). Culturally, as well as physiographically, Ethiopia is part of East Africa: the two peoples that together dominate its ethnic complexion (the Oromo and the Amhara) are neither Arabized nor Muslim—they are Africans. Uniquely, deep colonial penetration was avoided here. Adis Abeba, the historic capital and headquarters of a Coptic Christian, Amharic empire, held its own against Italy's attempted colonial takeover (except for a brief period in the late 1930s). Indeed, the Ethiopians, based in their mountain fortress (Adis Abeba lies more than 2400 meters/8000 ft above sea level), became colonizers themselves, at times taking control of much of the Islamic part of the African Horn to the east.

During the past decade, Ethiopia has made considerable economic progress and ranks as one of the world's fastest-growing economies. Much of this growth has been attributed to government initiatives, such as selling enormous tracts of publicly owned arable land to foreign agribusiness corporations and welcoming a wide range of Chinese foreign investment. Today, China is investing billions of (U.S.) dollars yearly in infrastructure projects (roads, bridges, hydropower facilities), and Chinese firms are even beginning to engage in offshore manufacturing here (the Hujian Group, for instance, produced 2 million pairs of shoes in Ethiopia in 2014). This financial infusion has decidedly contributed to Ethiopia's

FIGURE 7-17

rapid economic growth (ca. 9 percent annually since 2002). These gains notwithstanding, the country's economy remains focused on such agricultural commodities as coffee, tea, spices, and livestock: no less than 85 percent of the Ethiopian workforce is still composed of subsistence farmers.

Uganda

Uganda is a landlocked, Lake Victoria-fronting country of 37.6 million, bordered by deeply troubled neighbors. To the north lies newly independent South Sudan, by many measures Africa's poorest state and plagued by ongoing civil conflict. To the west lies the heavily contested eastern frontier of DRCongo, where government forces and insurgents fight seemingly endless battles. To the south lies Rwanda, scene of recurrent mass genocides. All this violence has not left Uganda untouched, a destination for persistent refugee flows. In addition, Uganda itself had to overcome the legacies of one of Africa's most brutal dictators, Idi Amin, whose economic nationalization drive wreaked havoc on the country's agricultural, mining, and trade sectors throughout most of the 1970s. Another calamity followed in the 1980s at the height of the late-twentieth-century HIV/AIDS pandemic: more than a million Ugandans succumbed, orphaning 1.2 million children. Responding massively in the aftermath, the government launched such an effective public-health campaign that the country became a global poster-child in the struggle against this disease.

Rwanda and Burundi

Rwanda and Burundi, with respective populations of 11.8 million and 10.2 million, are Africa's most densely inhabited countries. Here, Tutsi pastoralists from the north subjugated Hutu farmers, who had themselves made serfs of the local Twa people. Belgian colonial rulers promoted the minority Tutsi population at the expense of the Hutu majority, and nothing changed following independence in the early 1960s. Not until three decades later did the oppressed majority finally erupt in the horrific genocide of 1994, when almost one million Rwandans were killed in 100 days. Although Rwanda has been stabilized and is one of the fastest-growing African economies (8 percent annually over the past decade), it still relies on a narrow export base of coffee and tea—while two-thirds of the population live in extreme poverty. Unfortunately, this longstanding conflict continues to simmer in this corner of East Africa, having spilled over into the DRCongo in the form of armed combat to control strategic minerals, particularly diamonds and coltan (a valuable mineral used in mobile phones and other electronic devices).

Kenya

Over the past half-century, Kenya has played a leading role in East Africa. This country of 44.4 million contains both the region's largest city, its capital of Nairobi, and its principal port, Mombasa. After gaining independence from Britain in 1963, Kenya chose a capitalist path of development, aligning itself with Western interests. Even though it lacked major mineral deposits, Kenya built a relatively prosperous economy based on exports of coffee, tea, and other food products as well as a thriving tourist industry keyed to its magnificent landscapes and wildlife preserves.

But the good times lasted barely a generation as increasingly serious problems arose. Kenya during the 1980s had the highest rate of population increase in the world, and pressure on farmlands and fringes of the wildlife preserves steadily mounted—leading to widespread poaching and a concomitant decline in tourism. Beginning in the mid-1990s, excessive rains triggered massive flooding that was followed by the onset of a long and disastrous drought, resulting in famine across most of the interior. Meanwhile,

The mobile phone revolution has made a dramatic impact on farmers in many parts of the developing world, especially in Subsaharan Africa where distances are lengthy, land lines are absent, and market information is scarce. Mobile phone users in the realm totaled no less than 600 million in 2014, becoming not only the leading communication device but also a primary platform for a variety of tasks normally performed on computers. Farmers and market women, such as this entrepreneurial Kenyan, now employ smartphones to access real-time commodity market prices, transfer money to suppliers and transport companies, and send funds to relatives back in their home village. Remarkably, all of this can be accomplished without maintaining a bank account!

government corruption siphoned off public funds that should have been invested, and ethnic tensions were allowed to explode into outbreaks of violence, especially after the 2008 fraudulent presidential election. Moreover, the HIV/AIDS pandemic dealt the country a devastating blow, particularly in its western provinces. Kenya's latest misfortune is the growing danger of terrorist attacks in this decade. The Somali-based jihadist group, al-Shabaab, has been in the forefront, killing dozens of upscale shoppers in its high-profile 2013 assault on Nairobi's Westgate Mall—sending shockwaves of insecurity through Kenyan society. By mid-2015, the jihadists had transformed Kenya's coastal-strip portion of the African Transition Zone (Fig. 7-17) into a territorial stronghold overrun by a Boko Haram-style reign of terror, directed at all non-Muslims.

Kenya's greatest challenge is its continuing failure to build a unified nation out of its 41 constituent ethnic groups. Geography, history, and politics have placed the Kikuyu—who account for only 20 percent of the population—in a position of dominance. That is hotly contested by the other groups, especially the Luhya, Kalenjin, Luo, and Kamba, who collectively constitute 50 percent of the population; and in Kenya's periphery, there are additional major peoples such as the Maasai, Turkana, Boran, and Galla

© Jan Nijman

"Taking an early morning stroll along the beach near the town of Maroantsetra on Madagascar's northeastern coast, I stopped to observe a group of fishing people at work. It was a very basic way of fishing, with minimal equipment. Using a couple of small rowboats, they dropped their net about 20 meters (65 ft) from shore and then pulled it in toward the beach. It was a meager haul. Most of this catch, they said, was intended for consumption by their families and the rest would be sold informally. More than 70 percent of the people in this country live below the national poverty level, and in rural and coastal areas the percentage is even higher. These economic problems stand in sharp contrast to the country's natural beauty—or what remains of it after years of severe environmental degradation."

(Fig. 7-17). Adding yet another dimension to this ethnic cauldron, large numbers of Somali refugees from that country's never-ending conflicts have concentrated in Nairobi's Eastleigh neighborhood—triggering worsening ethnic discrimination as well as police harassment, increasingly in response to the threatening stance of the Islamic Front rooted along Kenya's eastern seaboard.

Nairobi, like all major African cities, struggles daily with these and many other growing pains of accelerating urbanization. But there is also a brighter side to the continuing development of this busy skyscrapered capital that lies at the heart of a metropolis of four million. Nairobi today is widely regarded throughout the realm as an emerging information technology hub. The centerpiece is Konza Techno City, the new planned satellite city to be built on the suburban frontier south of Nairobi. This future technopole is being billed as Africa's **Silicon Savannah**, and the Kenyan government envisions the creation of an ultramodern sustainable city, driven by a burgeoning IT sector capable of generating as many as 100,000 jobs by 2030.

Madagascar

Only 400 kilometers (250 mi) off Africa's central east coast lies the world's fourth-largest island, Madagascar (Fig. 7-14). But Madagascar is part of neither East Africa nor Southern Africa. The first human settlers arrived here about 2000 years ago—not from nearby Africa but from distant Southeast Asia. Subsequently, the powerful Malay kingdom of the Merina came to flourish in the highlands, and its language, Malagasy, became the indigenous tongue of the entire island (see Fig. 7-10). Today the Merina remain the largest of nearly 20 ethnic groups in the population of 22.9 million. Like mainland Africa, Madagascar experienced colonial invasion and competition, but the Merina successfully resisted conquest for centuries. Madagascar finally was forced to become part of France's empire in 1897; it left in 1960 at the height of Africa's independence movement.

Because of its Southeast Asian imprint, Madagascar's staple food is rice. It has some minerals, but the economy is weak and

constantly bedeviled by political turmoil. The country has also been burdened for decades by its high population growth rate (still well above double the global mean), and its colonial-built infrastructure has crumbled. Madagascar's geographic isolation allowed for the evolution of unique tropical flora and fauna—but the disastrous human onslaught has destroyed almost three-quarters of the island's rainforest home of this biological treasure.

EQUATORIAL AFRICA

The term *equatorial* has both locational and environmental connotations. The equator bisects Africa, but only the western two-thirds of central Africa displays the conditions associated with the low-elevation tropics: intense heat, extreme humidity, copious rainfall, little seasonal variation, rainforest vegetation, and enormous biodiversity. To the east, where East Africa begins beyond the Western Rift Valley, elevations rise, and cooler, more seasonal climatic regimes prevail. Equatorial Africa, lying west of this regional boundary, is physiographically dominated by the gigantic bowl-shaped Congo Basin. This region consists of nine states, of which the Democratic Republic of Congo—formerly Zaïre, and for convenience labeled **DRCongo**—is by far the largest and most important country (Fig 7-18). Five of the remaining eight states—Gabon, Cameroon, Congo (Republic), Equatorial Guinea, and São Tomé and Príncipe—have Atlantic coastlines. The Central African Republic, the southern part of Chad that belongs to this region, and recently independent South Sudan, are all landlocked and therefore economically constrained. In the limited space we have, our focus here is on the first and last of these nine countries.

The Democratic Republic of Congo (DRCongo)

With a territory not much smaller than the United States east of the Mississippi, a population of 67.5 million, a rich and varied

FIGURE 7-18

mineral base, much serviceable agricultural land, abundant water supplies, and lush tropical rainforests, DRCongo would seem to have all the ingredients needed to lead this region and, indeed, all of Subsaharan Africa. But powerful centrifugal forces, arising from its physiography and cultural geography, have always pulled this country apart. The immense, heavily forested heart of basin-shaped DRCongo creates formidable barriers to overland movement between east and west as well as north and south. Many of

the country's productive areas lie along its periphery, separated from one another by vast distances. Not surprisingly, these areas tend to look across the border to one or more of DRCongo's nine neighbors for outlets, markets, and ethnic kinship ties.

DRCongo has only a tiny window (37 kilometers/23 mi) facing the Atlantic Ocean, barely enough to accommodate the wide mouth of the Congo River. Oceangoing ships can reach the port of Matadi, inland from which falls and rapids make it necessary to move goods by road or rail to Kinshasa, the megacity capital of 11.6 million. In the distant other southern corner of DRCongo (3200 kilometers/2000 mi by river and rail from Kinshasa) lies Katanga Province, which contains most of the country's riches of copper and other mineral resources (Fig. 7-18). The narrow protrusion southeast of Lubumbashi into central Zambia is the heart of the Copperbelt, from which these minerals have flowed onto world markets for over a century via rail lines that connect to ports on the Indian Ocean coast of East and Southern Africa.

DRCongo has been embroiled in conflicts since independence in 1960. Its latest cycle of civil war began in the 1990s when neighboring Rwanda's conflict spilled over into what was then still known as Zaïre. Ever since, competing networks of local rebel groups, insurgents from adjoining countries, and the militaries of various African states have combine to ravage the country. This series of interconnected conflicts has now become known as "Africa's continental war." Among its complex causes are local disputes over land and resources, the expansionist goals of rebel groups and predatory neighboring states, and myriad ethnic political grievances. By mid-2015, the estimated death toll had surpassed 5 million. As for DRCongo's economic geography, it was being reshaped by rebel groups controlling the export of copper, coltan, and other highly valuable minerals, which increasingly enter the global marketplace through a network of clandestine trade channels still in the making.

South Sudan

This newest state to appear on the world political map became independent in 2011, the outcome of a near-unanimous referendum in the southern provinces of former Sudan (a now-truncated country discussed in Chapter 6). South Sudan's birth came in the aftermath of six disastrous decades of postcolonial strife within former Sudan, a brutal conflict in which more than 1.5 million died, magnified by the bifurcation of the country along the religious divide between Islam and Christianity-animism that we know as the Islamic Front (see Fig. 6-20).

The British effectively ruled former Sudan from the 1890s until independence in 1956. One of the roots of the country's long-running internal conflict, which peaked in the resumption of its post-independence civil war between 1983 and 2005, lay in the decision of the British colonial administration to combine northern Sudan, which was heavily Arabized and Islamized, with a sizeable African/Christian-dominated area to the south. As soon as the British departed in the mid-1950s, the Khartoum-based regime in the north sought to impose its Islamic rule on southern Sudan, immediately triggering the first civil war that lasted until 1972. By the time the bitter second civil war erupted a decade later, former Sudan's strife had impoverished the country to the point that its per-capita income ranking was among the lowest on Earth. That slowly began to change in the 1990s when oil was discovered, including major deposits in the embattled south (Fig. 7-18). Eventually, leaders in the south saw oil as a ticket to self-sufficiency, and the north felt that it too would walk away with control of several oilfields.

Animosity between the two halves of former Sudan persists. Indeed, a low-grade, oil-based conflict broke out soon after the political split of 2011—and may finally be ending following a truce in 2015. Nonetheless, South Sudan continues to be plagued by grinding poverty, violent internal ethnic conflict, and ineffective governance (the country's population size can only be estimated—at 11.5 million). Meanwhile, the overwhelming majority of the people are either subsistence farmers or cattle herders—a deeply traditional livelihood in which individual and family status depends on the number of livestock owned.

SOUTHERN AFRICA

As a geographic region, Southern Africa consists of all the countries lying south of Equatorial Africa's DRCongo and East Africa's Tanzania (Fig. 7-19). Thus defined, this region extends southward from Angola and Moçambique (on the Atlantic and Indian Ocean coasts, respectively) to South Africa, and includes half a dozen landlocked states. Also marking the northern limit of the region are Zambia and Malawi. As noted above, Zambia is nearly cut in two by the elongated, mineral-rich land extension from southeastern DRCongo; similarly, Malawi penetrates deeply into Moçambique.

Africa's Resource-Rich Region

Southern Africa constitutes a geographic region in both physiographic and human terms. Its northern periphery marks the southern limit of the Congo Basin in a broad upland that stretches across Angola and into Zambia (the narrow tan lobe extending eastward from the Bihe Plateau in Fig. 7-2). Most of this region is plateau country, and much of the interior upland is framed by the prominent, near-coastal Great Escarpment. There are two consequential river systems: the Zambezi (which forms the border between Zambia and Zimbabwe) and the Orange-Vaal (South African rivers that combine to demarcate southern Namibia from South Africa).

Southern Africa has long been the continent's richest region materially. A great north-south corridor of mineral deposits extends through the eastern heart of the region from the Zambia/DRCongo Copperbelt through Zimbabwe's Great Dyke and South Africa's Bushveld Basin and Witwatersrand to the goldfields and diamond mines scattered around Bloemfontein in the geographic center of South Africa (Fig. 7-19). Ever since these minerals began to be exploited in colonial times, many migrant laborers have come to work in the mines of these countries. And in Southern Africa's far northwest, huge onshore and offshore oil reserves are propelling the economic development of Angola. The region's agricultural diversity matches its mineral wealth, especially so in South Africa: vineyards drape the slopes of South Africa's Cape Ranges and the country's relatively high latitudes and its range of altitudes create environments for apple orchards, citrus groves, banana plantations, pineapple farms, and myriad other crops.

© J. Nijman, P. O. Muller, H. J. de Blij, and John Wiley & Sons, Inc.

FIGURE 7-19

Despite this considerable wealth and potential, only a few of the ten countries of Southern Africa have prospered, and the gap between them is steadily widening (see Fig. 7-11). As Figure G-10 indicates, four countries—South Africa, Angola, Botswana, and Namibia—now rank in the World Bank's upper-middle-income category; in contrast, three remain mired in the low-income category (Malawi, Moçambique, and Zimbabwe). The three remaining states are classified as lower-middle income countries: Zambia plus the two ministates of Lesotho and Swaziland. Among the

four most prosperous countries, Botswana and Namibia are both desert-dominated, sparsely settled states, with a combined population of less than 4.5 million. Booming Angola has become one of the realm's oil powers, a fast-growing country on the move. South Africa, however, stands apart—the region's giant by every measure.

Since we are working our way through the Subsaharan African realm from north to south, we will also approach the region of Southern Africa from that perspective. We begin with the Northern

© STRINGER/Reuters / Corbis Images

By the time you read this, Angola will probably be Subsaharan Africa's leading oil-producing country. This mid-2015 view of the center of its burgeoning capital city, Luanda, is reminiscent of the Persian Gulf's petrostates during their initial building booms. But oil is a double-edged sword: even though Luanda is the world's most expensive city to live in, the rest of the Angolan population lives in a world of extreme poverty. With 95 percent of the country's foreign income dependent on oil sales, the recent drop in the price of oil has forced government budget cutbacks that only intensify the hardships of life in one of the world's most unequal societies.

Tier of countries, proceed southward into the Middle Tier, and then put preeminent South Africa in the geographic spotlight.

The Northern Tier

In the four countries that extend across the Northern Tier of Southern Africa—Angola, Moçambique, Zambia, and Malawi—change is everywhere. **Angola**, Southern Africa's most resource-dependent country, has been the region's star economic performer in this decade. This country of 21.5 million—which also includes its tiny **exclave 19** (separated outlier) of *Cabinda* located north of the Congo River's mouth—had a thriving (colonial) economy based on a wide range of mineral and agricultural exports at the time of independence from Portugal in 1975. But the Portuguese did not prepare the country for self-rule, and soon Angola fell victim to the Cold War, with its northern peoples opting to follow a communist course and southerners falling under the sway of a rebel movement backed by South Africa and the United States. The results included devastated infrastructure, abandoned farms, looting of diamonds, hundreds of thousands of casualties, and millions of landmines that continue to kill and maim.

Since the mid-2000s, Angola's economic performance has been nothing short of spectacular, and statistically it became the world's fastest-growing national economy. Today, its capital, Luanda, has achieved the dubious distinction of being the most expensive city on Earth (see photo). Almost all of this growth comes from oil, much of which is exported to China, and Angola is on track to surpass Nigeria in 2016 as Africa's top petroleum producer. Yet this bonanza is decidedly a two-edged sword, because the country's reliance on foreign oil income is too great:

this fuel now accounts for half of Angola's GDP, more than 90 percent of its exports, and at least 75 percent of government revenues. Those impressive statistical indicators notwithstanding, they mask both Angola's vulnerability to fluctuations in global oil prices and catastrophic and intensifying domestic inequalities that force a huge majority of the population to survive on no more than two U.S. dollars a day.

On the opposite coast, the other major former Portuguese dependency, **Moçambique** (25.8 million), experienced a comparable trajectory. Upon independence, Moçambique also chose a Marxist course with unfortunate economic and political consequences. Here, too, a rebel movement supported by South Africa caused civil conflict, created famines, and generated a huge stream of refugees toward Malawi.

Moçambique has neither oil nor as much agricultural land as Angola. But it does have considerable bauxite and coal deposits, and its long Indian Ocean coastline provides an advantageous relative location (as well as access to promising offshore natural gas reserves). In recent years, port traffic has been revived, and the country is working with South Africa on building a joint Maputo Development Corridor (Fig. 7-19). But despite the rapid growth of Moçambique's economy over the past decade, here again we have an African country that depends too heavily on a single commodity—bauxite (aluminum ore), which accounts for about one-third of all exports. And, like Angola, this country has performed poorly in distributing the benefits of economic development, barely affecting the severe poverty that still reigns among Moçambique's masses.

Landlocked **Zambia** (14.5 million), another product of British colonialism, shares the mineral riches of the Copperbelt with Katanga Province in its northern neighbor, DRCongo. But the stable commodity prices on which the country depends have fluctuated wildly since independence, and the railways leading to Zambia's outlet ports in Angola and Moçambique were rendered inoperative by Cold War conflicts. More recently, the Chinese have taken an interest in Zambia's minerals, and they have invested in railroad repairs as well as the expansion of mining operations—resulting in an annual economic growth rate averaging about 6 percent over the past decade.

Neighboring **Malawi** (16.4 million), like Zambia, has been able to sustain democracy for more than two decades. Malawi has an almost totally agricultural economic base and is also unremittingly challenged by environmental degradation. This country's dependence on corn as its food staple, its variable climate, and its severely fragmented land-use system have inhibited economic growth—which nevertheless has risen steadily at nearly 6 percent a year since the turn of this century.

The Middle Tier

Five states lie between the countries of the Northern Tier and South Africa. They form Southern Africa's Middle Tier subregion, and all border the Republic of South Africa: Botswana, Namibia, Zimbabwe, and the two ministates of Lesotho and Swaziland

(Fig. 7-19). As the map reveals, four of these five are landlocked. Diamond-exporting (and upper-middle-income) Botswana occupies the heart of the Kalahari Desert and surrounding steppe. Despite its lucrative diamonds, the great majority of its 2.0 million inhabitants are subsistence farmers, and in recent years Botswana has been the most severely AIDS-afflicted country in all of tropical Africa. Lesotho and Swaziland, both traditional kingdoms, depend very heavily on remittances from their workers in South African mines, fields, and factories.

Southern Africa's youngest independent state, Namibia (2.3 million), is the former German colony of South West Africa that was administered by South Africa from 1919 to 1990. The country is named after one of the world's driest deserts (the Namib, which lines its coast). Territorially about as large as Texas and Oklahoma combined, only its far north receives enough moisture to enable subsistence farming, which is why most of the people live close to the Angolan border. Mining and ranching constitute the leading commercial activities, but even though orderly land reform is under way, much of the population still lives in poverty.

The Tragedy of Zimbabwe

Zimbabwe (population: 4.2 million) lies at the heart of Southern Africa, between the Zambezi River in the north and the Limpopo River in the south and between the Great Escarpment to the east and the desert to the west. Landlocked but endowed with good farmlands, cool uplands, and a wide range of mineral resources, Zimbabwe at independence had one of Southern Africa's most vibrant economies and seemed to have a bright future. Its core area straddles the mineral-rich Great Dyke that extends across the heart of the country, which contains copper, gold, asbestos, chromium, and platinum. Its farms are capable of producing tobacco, tea, sugar, and cotton in addition to staples for the local market.

But going back almost four decades, these have not been good times for Zimbabwe. During the colonial period, the tiny minority of whites that controlled what was then called Southern Rhodesia (after Cecil Rhodes, the British capitalist of diamond fame and scholarship honors), took the most productive farmlands and organized and ran the agricultural economy. Following independence, their descendants continued to hold huge estates, only parts of which were being farmed. In the absence of comprehensive land reform, let alone progress toward democratic rule, the persistence of this monstrous inequality impeded broad-based development.

In the wake of their successful joint campaign to end white-minority rule in 1980, the two peoples that constitute most of Zimbabwe's population—the Shona (71 percent) and the Ndebele (16 percent)—engaged in acrimonious ethnic conflict. Proper legal reforms did not take effect, and the rule of newly elected Robert Mugabe turned from ineffectiveness to disaster to extreme paranoia. As he encouraged squatters to invade white farms (where a number of owners were killed and many others fled), the agricultural economy began to collapse. Corruption in the Mugabe government resulted in the transfer of land, not to needy squatters but to cronies of top officials. Foreign investment in other Zimbabwean enterprises, including the mining industry, dried up. By the mid-1990s, Zimbabwe was in economic free fall.

Mugabe also turned against the vital informal sector of the economy, which was all most Zimbabweans had left when jobs on farms and in factories and mines disappeared. He ordered his henchmen to destroy the dwellings of some 700,000 "informal" urban slum dwellers, who had little choice other than to join a growing army of persecuted homeless refugees who soon began to stream out of Zimbabwe, many headed for South Africa.

Heroic Zimbabweans and persistent outsiders have put pressure on 91-year-old Mugabe to end his catastrophic dictatorial rule. But the country remains isolated internationally—except for the embrace of China and South Africa for very different reasons. Beijing is pursuing numerous economic opportunities in Zimbabwe and is almost entirely responsible for the country's period of recent (unsteady) GDP growth. Pretoria, however, remains silent in gratitude for Zimbabwean support of South African revolutionaries during the struggle against their own white-minority-government oppressors in the 1980s. Through it all, Mugabe has remained in power, the oldest and longest serving leader in Africa, and continues to lay the groundwork for his spouse, Grace Mugabe, to succeed him as president. Today more than ever, Zimbabwe stands in the sharpest contrast to what is being achieved in the other countries of Southern Africa.

South Africa

The Republic of South Africa (RSA) is the giant of the region of Southern Africa, with a population of 52.3 million and a modern economy second in size only to Nigeria's within this realm. Its historical geography differs somewhat from much of the rest of Subsaharan Africa. The country's lands were fought over by various African nations before the Europeans arrived and the colonial "scramble for Africa" took place. Indigenous peoples had migrated southward—first the Khoisan-speakers and then the Bantu peoples—into the South African *cul-de-sac*. The Zulu and Xhosa nations were fighting over these lands at about the time the first European settlers arrived.

South Africa is one of the most strategic places on Earth, the gateway from the Atlantic to the Indian Ocean, a key source of provisions on the route to Asia's riches. The Dutch East India Company founded Cape Town as early as 1652, and the Hollanders and their descendants, known as **Boers**, have been a part of the South African cultural mosaic ever since. The British took over about 150 years later, and both colonial powers vied for control. The British came to dominate the Cape, while the Boers trekked into the South African interior and, on the high plateau they called the *highveld*, founded their own republics. A war ensued, but by 1910 the Boers and the British had negotiated a power-sharing arrangement, although the Boers eventually achieved dominance that lasted from 1948 to 1994. Having long since shed their European ties, they called themselves **Afrikaners**, their word for Africans.

For more than 40 years, between about 1950 and the mid-1990s, multicultural South Africa was in the grip of the world's most notorious racist policy, **Apartheid** [20]—the word itself means "apartness," but in practice it promotes strict racial segregation and severe discrimination. Out of the concept of Apartheid grew a notion, promulgated by the white (European) minority then in control of the state, known as ***separate development***. This would apply Apartheid to the entire country, carving it up into racially-based entities whose inhabitants were citizens of those ethnic domains—but not of South Africa as a whole. Predictably, such racist social engineering aroused strong opposition within South Africa and beyond, leading to worldwide condemnation and international sanctions.

By the late 1980s, South Africa seemed headed for a violent revolution, but disaster was averted by a most unexpected turn of events. A leader of the white-minority government that for decades had ruthlessly pursued its Apartheid policies, and a revered leader of the multicultural majority who had for 28 years languished in an island prison not far from Cape Town, struck a momentous accord, ushering in a new South Africa virtually overnight. Nelson Mandela walked out of prison a free man on February 11, 1990, and in the RSA's first democratic election in 1994 he became the country's president—alongside a new, fully representative parliament that was now indeed the country's first "rainbow assembly." At the same time, the country's internal political geography was reorganized: before 1994, the RSA had been divided into four provinces, but this was now replaced with a fairer federal arrangement consisting of nine provinces (see Fig. 7-20, including the inset map).

FIGURE 7-20

Ethnicity and Xenophobia

South Africa has long been the realm's most pluralistic and heterogeneous society. In addition to the various indigenous African nations and the Europeans who settled in South Africa came peoples from Asia. The Dutch brought thousands of Southeast Asians to the Cape to serve as domestics and laborers, as the British did later from their South Asian colonies on the Indian subcontinent. Moreover, a substantial population of mixed ancestry clustered at the Cape, today known as the *Coloured* component of the country's citizenry.

Although heterogeneity marks the urban spatial demography of South Africa, regionalism pervades the national-scale human mosaic. The Zulu nation, 10 million strong, still is largely concentrated in today's Kwazulu-Natal Province (Fig. 7-20). The Xhosa (7.5 million) still cluster in the Eastern Cape below the Great Escarpment, and the Tswana (3.3 million) still occupy ancestral lands along the border with Botswana. Metropolitan Cape Town remains the heartland of the Coloured population; the city of Durban still has the strongest South Asian imprint.

Since the end of the Apartheid era more than two decades ago, international migrations from all over the realm have streamed into South Africa. These influxes have intensified the RSA's multiculturalism and are most visible in major urban areas, especially in their takeover of much of inner-city Johannesburg. But peaceful coexistence does not prevail in this supposedly "rainbow nation": tensions persist at an explosive level, and violent attacks directed at poor foreign immigrants (mainly those with a high profile in street trade) periodically erupt. Such expressions of **xenophobia 21** (extreme dislike and even fear of foreigners) are motivated by the perception that these immigrants are stealing jobs from South Africans—and further fueled by the dominance of low-end commerce by foreign entrepreneurs whose business skills exceed those of natives (see photo). Efforts to defuse xenophobic sentiments continue: South African schoolchildren learning the alphabet are taught that the letter "X" stands for xenophobia.

South Africa's Economic Geography

South Africa's economy, the realm's most modernized, has wide regional influence. On the global scene, the RSA is regarded as an emerging-market economy as well as a gateway to Subsaharan Africa, particularly concerning corporate command and control. With a land area in excess of 1.2 million square kilometers (470,000 sq mi), South Africa contains the bulk of the region's minerals, most of its good farmlands, its largest cities, best harbors, most productive factories, and the best developed transport network. Mineral exports from the Copperbelt and the Great Dyke move through South African ports. Workers from as far away as Malawi and as close as Lesotho work in South Africa's mines, industrial plants, and fields. And entrepreneurs from Somalia, Ethiopia, Bangladesh, and China are increasingly entrenched in the commercial transactions of the country's informal economy.

Ever since diamonds were discovered at Kimberley in the 1860s, South Africa has been synonymous with minerals. Rail lines were laid from the coast to the diamond complex even as fortune seekers, capitalists, and tens of thousands of African workers, many from as far afield as Moçambique, streamed to the site. Subsequently, prospectors discovered what was long to be the world's greatest goldfield on a ridge named the Witwatersrand (Fig. 7-20). There, Johannesburg soon became the gold capital of the world. During the twentieth century, mining was diversified to include coal, iron ore, nickel, and copper, boosting a sizeable domestic metallurgical and manufacturing sector as well as yielding large revenues on world markets. Before the imposition of severe international sanctions in the 1980s to protest the growing outrages of Apartheid, capital flowed into the country, white immigration expanded, farms and ranches were laid out, and overseas markets multiplied.

Along the way, South Africa's cities grew rapidly. Johannesburg became an industrial agglomeration, major financial center, and immigration hub. To its north, Pretoria became the country's administrative capital, and the two cities have now coalesced into the Gauteng conurbation that contains more than 13 million people who produce over one-third of the national economy's output. In Orange Free State (simply Free State since 1994), substantial industrial development matched the expansion of mining. Durban's port served not only the Witwatersrand but a wider regional hinterland as well. And Cape Town has become South Africa's second-largest city and a leading international tourist destination; its port, industries, and productive agricultural hinterland gave it primacy over much of the southernmost RSA.

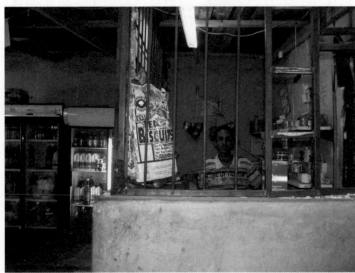

© Jan Nijman/Richard Grant

The newest immigrants from Somalia to the RSA often start out by running *spaza* shops—such as this one in Thokoza Township south of Johannesburg. (*Spaza* is a South African word for a small convenience store in the informal economy.) These Somalis work long hours, live on the premises, and constantly face the hostility of South African *spaza* competitors as well as unemployed youths. Sometimes they are shot at, and a few have even had their shops burned down (as in nearby Soweto Township in early 2015). For many desperate immigrants, working in the *spazas* of dangerous "locations" (their term for the townships) is a rite of passage and the only job that a freshly arrived immigrant can find. Tellingly, the Somali shop-owners who employ them remain in central Johannesburg, choosing to send out only new Somali arrivals to handle the challenges of working in the suburban "locations."

But Apartheid frustrated South Africa's prospects, leaving a deep legacy as well as more than a few major social and economic scars. Not only was the separate development scheme astronomically expensive: its social costs have proven hard to dislodge. Mounting ethnic unrest during the decade preceding the fall of Apartheid created a vast educational gap among young people. And even today, the after-effects of the economic damage caused by international sanctions against the race-obsessed, white-minority regime continue to plague parts of the economy.

Ongoing Challenges

In many respects, South Africa is the most important country in Subsaharan Africa, and the entire realm's fortunes are bound up with it. No African country attracts more foreign investment or foreign workers. Its universities, hospitals, and research facilities are the best on the continent. Few have the free press, effective trade unions, independent courts, or financial institutions to match the RSA's. And with a population greater than 52 million (77 percent black, 9 percent Coloured, 9 percent white, and 2.5 percent Asian), South Africa has a substantial, multiracial middle class.

But South Africa faces daunting political challenges. The African National Congress (ANC) Party—which produced all of the RSA's presidents since 1994—has failed to deliver the jobs and housing it had promised in successive campaigns. Black economic empowerment has mostly benefited the privileged elite; and even though a black middle class has emerged in urban South Africa, it mainly chooses to wall itself off inside gated residential communities well away from the enormous underclass. Most whites who remained in the country after 1994 continue to enjoy the rewards of their economic power despite the national shift in political power. But for tens of millions of South Africans, very little has changed: indeed, they seem poorer than ever compared to the newly affluent black populations and the ever-entrenched whites.

It should therefore come as no surprise that in 2014 South Africa recorded the greatest inequality within any of the world's countries. Predictably, there are massive gaps in education, personal income, and service provision. South Africa now ranks 132nd out of a total of 144 countries in elementary school performance. Urgently needed land reform is stalemated, purportedly by a lack of funding. Although millions of houses have been built and were connected to utilities during the post-Apartheid era, this total falls drastically short of the number of dwelling units actually needed. The government reports an unemployment figure of about 25 percent, but that proportion is easily doubled in the destitute slums of the inner city and the outlying townships. In the rural areas, farm workers and miners organize protests and strikes to demand fairer wages, and many white landowners have been intimidated by homeless invaders. It goes without saying that these developments neither inspire confidence nor help the cause of attracting new investment. Yet South Africa still has Africa's most diversified economy—and China is steadily deepening its ties with the RSA.

South Africa has many of the necessary geographic endowments for a strong economy, but it will take time, stability, and competent management to achieve its full potential. Meanwhile, South Africa's leaders have maneuvered to raise the country's profile in an ever more globalized and competitive world. Its 2011 joining of the exclusive BRICs—the world's four biggest emerging-market economies (Brazil, Russia, India, and China)—raised South Africa's global standing, even though Pretoria has always been the minor power of the group. China was instrumental in lobbying the other BRIC countries to include South Africa in the club. A New Development Bank (NDB) is being established under BRIC auspices and has the potential to be a major player in providing funds for infrastructure and sustainable development projects in the developing world. The RSA will participate, and, as a gesture of its close friendship, Pretoria declared 2015 to be the "Year of China in South Africa"; in response, the BRICs agreed to locate a regional headquarters in Johannesburg.

South Africa's economy, however, is smaller (one-twentieth the size of China's) and much weaker than those of the four BRICs. It also lags well behind the economies of such other emerging-market countries as Turkey, Mexico, and Indonesia. Furthermore, the RSA does not have an oil bonanza and must still depend heavily on commodity sales in fickle world markets. It also appears to be losing its regional advantage in education and skills. Most of all, it lacks capable leadership that can do for the economy what Mandela did for the society.

As this chapter has underscored, we can no longer accept the "conventional wisdom" view of a Subsaharan Africa described in terms of negative stereotypical judgments. All too often, this realm is still portrayed as a single, monolithic, economic basket-case, where dysfunctional government goes hand in hand with economic underperformance, frequent famines, frightening disease outbreaks, and aimless ethnic violence. Now more than ever, this selective characterization misrepresents what is really happening in much of the realm. In the mid-2010s, Subsaharan Africa harbors the world's newest emerging-market economy; in many countries, economic forecasts tell of another decade lying ahead in which growth and development may surpass that of all other geographic realms; and a number of African societies are demonstrating incredible resilience in fighting such monumental challenges as HIV/AIDS and Ebola.

These positives and potentials notwithstanding, we have also seen that Subsaharan Africa still faces a formidable set of challenges today. Democracy is taking hold in an unprecedented number of countries, but elsewhere war and dislocation remain unaffected. Nigeria may have overtaken the RSA to become the realm's largest economy, but the Boko Haram insurgency endangers the very fabric of the state. As for South Africa, its foremost internal threat may well be its appalling, widening inequalities of opportunities, lifestyles, and personal well-being. And as far as the bigger geographic picture is concerned, energy as well as mineral resources are the cornerstones of this realm's progress in the 2010s; but building a prosperous future will require greater economic diversification and tightening cooperation throughout Subsaharan Africa.

- Shipping a new car from China to a Tanzanian port costs less than hauling it inland from Tanzania to neighboring Uganda.

- The realm contains more than four-dozen countries; the number of languages spoken within Subsaharan Africa is estimated to exceed 2000.

- China's impact on Subsaharan Africa is expanding steadily through activities ranging from trade to investment to education. Brazil, India and Russia are also increasing their ties with this realm.

- At the end of the first two decades of this century, Subsaharan Africa is likely to have recorded the fastest economic growth among all world geographic realms.

- South Africa was freed from racist Apartheid over two decades ago, but it now ranks as the most unequal society on Earth.

- A number of low-income Subsaharan African countries have seen major revenue increases from raw material exports. Should they diversify their economies to reduce dependence on commodities?

A Rajasthani infant peering over his mother's shoulder at the market in the north-western Indian town of Pushkar. Across the subcontinent, women tend to marry and bear children at an early age. Yet fertility rates have been declining, and in India the average number of children per woman has dropped from 3.9 in 1990 to 2.4 today—a substantial decrease but still generating a national population growth rate well above the global average.

© Jan Nijman

IN THIS CHAPTER

- A deadly triangle
- Cutting-edge IT, backward agriculture
- The Indian Ocean: A crucial geopolitical arena
- The gender imbalance
- Pakistan: A hard place to govern
- Caste in modern India
- The Maldives going under

CONCEPTS, IDEAS, AND TERMS

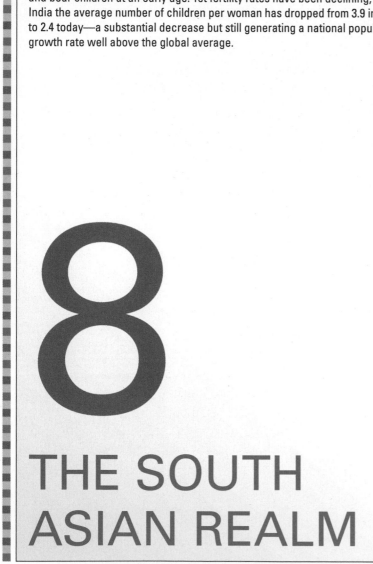

8

THE SOUTH ASIAN REALM

FIGURE 8-1

outh Asia is a realm of almost magical geographic names: Mount Everest, the Ganges River, Kashmir, the Khyber Pass. There was a time when this realm was legendary and prized. Remember that it was "India" and its fabled wealth that the European explorers were after, from Vasco da Gama to Columbus to Magellan. Before them, the fourteenth-century North African geographer Ibn Battuta had traveled overland to South Asia, and his writings about its riches were met with astonishment and even disbelief. From the sixteenth century onward, European trading companies derived enormous profits from commerce in this realm.

By the late nineteenth century, however, South Asia seemed to have become remote from the affairs of the world—hungry, weak, exploited, the prototype realm of the global periphery. Even after independence in 1947, India as well as the other countries of South Asia long remained among the world's poorest. For decades, population growth outstripped economic expansion.

Today, for a number of reasons, South Asia commands the world's attention once again. It became the most populous geographic realm on Earth in 2011. Two of its states, India and Pakistan, often find themselves in conflict and both are nuclear powers. In Afghanistan and the remote mountain hideaways of western Pakistan, a terrorist organization's leaders planned attacks that changed the skyline of New York and presaged the battleground of Iraq. In the ports of India, a growing navy reflects the emergence of the Indian Ocean as a new global geopolitical arena in which China, too, is asserting itself. Meanwhile, outsourcing by U.S. companies to India became a hot topic, and India's spectacular rise in information technology has changed that industry. Our daily lives are increasingly affected by what happens in this fascinating part of the world.

DEFINING THE REALM

THE GEOGRAPHIC PANORAMA

The Eurasian landmass incorporates all or part of 6 of the world's 12 geographic realms, and of these half-dozen very few are more clearly defined by nature than the one we call South Asia. Figure 8-1 shows us why: the huge triangular Indian subcontinent that divides the northern Indian Ocean between the Arabian Sea and the Bay of Bengal is so sharply demarcated by mountain walls and desert wastes that you could take a pen and mark its boundary, from the Naga Hills in the far east through the Great Himalaya and Karakoram in the north to the Hindu Kush and the Iran-bordering wastelands of Balochistan and Afghanistan in the west. Note how short the distances are over which the green of habitable lowlands turns to the dark brown of massive, snowcapped mountain ranges.

MAJOR GEOGRAPHIC QUALITIES

SOUTH ASIA

1. Most of South Asia is clearly defined physiographically, and much of the realm's boundary is marked by mountains, deserts, and the Indian Ocean.

2. South Asia's great rivers, especially the Ganges, have for tens of thousands of years supported enormous population clusters.

3. South Asia, and especially northern India, was the birthplace of major religions that include Hinduism and Buddhism.

4. Because of the realm's natural boundaries, foreign influences in premodern South Asia arrived mainly via a narrow passage in the northwest, the Khyber Pass.

5. Afghanistan is the realm's westernmost country and constitutes a cultural and political transition zone with the North Africa/Southwest Asia (NASWA) realm.

6. The South Asian realm covers less than 4 percent of the Earth's land area but contains one-quarter of the world's human population.

7. South Asia's annual monsoon continues to dominate life for hundreds of millions of subsistence and commercial farmers. Disruption of the monsoon cycle spells economic and social crisis.

8. Certain disputed territories in the realm's remote northern mountain perimeter are potentially a source of dangerous friction between India and both Pakistan and China.

9. South Asia is still predominantly rural and studded with hundreds of thousands of small villages; but it also contains some of the biggest cities in the world.

This realm extends from Afghanistan in the west to Bangladesh in the east. India, the realm's giant, is flanked by six countries (in clockwise order, Pakistan, Nepal, Bhutan, Bangladesh, Sri Lanka, and the Maldives) as well as Kashmir, a remaining disputed territory in the far north. South Asia's kaleidoscope of cultures may be the most diverse in the world, proving that neither formidable mountains nor forbidding deserts could prevent foreign influences from further diversifying an already variegated realm. We will encounter many of these influences in this chapter, but most of South Asia also possessed one unifying force of sorts: the British Empire, which from its late-nineteenth-century heyday through the late 1940s came to hold sway over almost all of it.

Since Islam prevails in Afghanistan and Pakistan, why are they included with South Asia and not with the North African/Southwest Asian realm? The answer lies in historical geography. First, Pakistan's history is closely tied to that of India—going back to the ancient civilization of the Indus Valley and, more recently, to the role of Islam and of British hegemony in the subcontinent. The tight integration of Pakistan with the rest of South Asia will not surprise you after you study the realm's physiography in Figure 8-1: the natural boundary in this part of the realm lies west of the Indus River, not east of it. Pakistan today remains part of a realm that begins to change not in the Punjab, but at the Khyber Pass, the highland gateway to Afghanistan. Second, Afghanistan was a staging ground for Islam in the Indian subcontinent, and the cultural boundary between Afghanistan and Pakistan has remained fluid ever since. In political terms, too, there continue to be intricate connections between the two countries that, in turn, impact relations between Pakistan and India. At the same time, Afghanistan today is caught up in developments across Southwest Asia (as well as Central Asia), which is why that country is considered to be a *transition zone* between South Asia and NASWA.

SOUTH ASIA'S PHYSIOGRAPHY

From snowcapped peaks to tropical forests and from bone-dry deserts to lush farmlands, this part of the world presents a virtually endless array of environments and ecologies, a diversity that is matched by its cultural mosaic. The broad outlines of this realm's physiography are best understood against the backdrop of its fascinating geologic past.

A Tectonic Encounter

As Figure 8-2 shows, the spectacular relief in the north of this realm is the product of a collision between two of the planet's great tectonic plates (see Fig. G-4). About 10 million years ago, after a lengthy geologic journey following the breakup of the supercontinent Pangaea (see Fig. 7-3), the Indian Plate encountered Eurasia. In this huge, slow-motion, accordion-like collision, the Earth's crust surrounding the contact zone was pushed upward, thereby creating the mighty Himalaya mountain range. That process is still

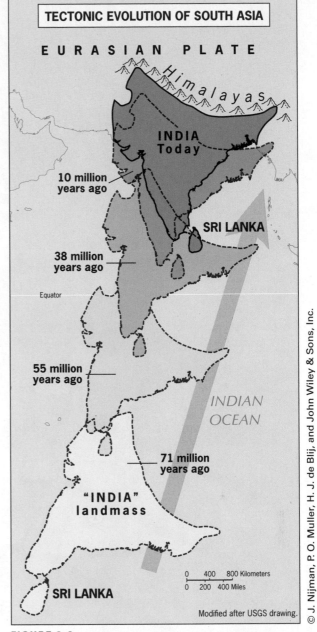

FIGURE 8-2

going on—at the rate of 5 millimeters (0.2 in) per year—and the huge 2015 earthquake in Nepal proved once again that this is one of the most quake-prone areas in the world. One major outcome of this tectonic collision was that the northern margins of the realm were upwardly thrusted to elevations where permanent snow and ice make the landscape appear polar. The march of the seasons melts enough of this snowpack in spring and summer to sustain the great rivers below, providing water for farmlands that support hundreds of millions of people. The Ganges, Indus, and Brahmaputra all have their origins in the Himalaya. Only south of the Ganges Basin does the massive plateau begin that marks the much older geologic core of the Indian Plate as it drifted northeastward toward Eurasia.

© J. Nijman, P. O. Muller, and John Wiley & Sons, Inc.

FIGURE 8-3

The Monsoon

Physical geography, therefore, is crucial here in South Asia—but not just on and below the surface. What happens in the overlying atmosphere is critical as well. The name "South Asia" is almost synonymous with the term **monsoon** [1] because the annual rains that accompany its onset, usually in June, are indispensable to all forms of agriculture in the realm's key country, India.

Figure 8-3 shows how the monsoon works. As the subcontinental landmass heats up during the spring, a huge low-pressure system is formed above it. This low-pressure system begins to draw in vast volumes of air from over the ocean onto the subcontinent. When the inflow of moist oceanic air reaches critical mass in early June, the *wet monsoon* has arrived. It may rain for 60 days or more. The countryside turns green, the paddies fill, and another dry season's dust and dirt are washed away—the region is reborn (see photo pair). The moisture-laden air flowing onshore from the Arabian Sea is forced upward against the Western Ghats, cooling as it rises and condensing large amounts of rainfall. The other branch of the wet monsoon originates in the Bay of Bengal and gets caught up in the convection (rising hot air) over northeastern India and Bangladesh. Seemingly endless rain now inundates a much larger area, including the entire North Indian Plain. The Himalaya mountain wall blocks the onshore-flowing airstream from spreading into the Asian interior and the rain from dissipating. Thus the moist airflow is steered westward, drying out as it advances toward Pakistan. After persisting for weeks, this pattern finally breaks down, and the wet monsoon gives way to periodic rains and, eventually, another dry season. Then the anxious wait begins for the next year's wet monsoon because without it India would face disaster. In much of rural India, life can hang by a meteorological thread.

Physiographic Regions

Figure 8-3 underscores South Asia's overall division into three physiographic zones: northern mountains, southern plateaus, and, in between, a wide crescent of river lowlands.

The **northern mountains** extend from the Hindu Kush and Karakoram ranges in the northwest through the Himalaya in the center (Everest, the world's tallest peak, lies on the crestline that forms the Nepal–China border) to the ranges of Bhutan and the Indian State of Arunachal Pradesh in the east. Dry and barren in Afghanistan and western Pakistan, the ranges become green and tree-studded in Kashmir, forested in the lower-lying sections of Nepal, and even more densely vegetated in Arunachal Pradesh. Transitional foothills, with many deeply eroded valleys cut by rushing meltwater, lead to the river basins below.

The belt of **river lowlands** extends eastward from Pakistan's lower Indus Valley (the area known as Sindh) through India's wide Gangetic Plain and then on across the great double delta of the Ganges and Brahmaputra in Bangladesh (Fig. 8-3). In the east, this physiographic region is often called the North Indian Plain. To the west lies the lowland of the Indus River, which rises in Tibet, crosses Kashmir, and then bends southward to receive its major tributaries from the area known as **Punjab** ("Land of Five Rivers") to the east.

The **southern plateaus** constitute peninsular India, dominated by the **Deccan**, a massive tableland built of lava sheets that poured out when India separated from Africa during the breakup of Pangaea. The Deccan (meaning "South") tilts toward the east, so that its highest areas are in the west and most of the rivers flow into the Bay of Bengal. North of the Deccan lie two other plateaus, the Central Indian Plateau to the west and the Chota Nagpur Plateau to the east (Fig. 8-3). On the map, also note the Eastern and Western Ghats: "ghat" means step, and it connotes the descent from Deccan Plateau elevations to the narrow coastal plains below. The onshore winds of the annual wet monsoon bring ample precipitation to the Western Ghats; as a result, here lies one of India's most productive farming areas and one of southern India's largest population concentrations.

BIRTHPLACE OF CIVILIZATIONS

Indus Valley Civilization

A complex and technologically advanced civilization emerged in the Indus Valley by about 2500 BC, simultaneous with other Bronze Age "urban revolutions" in Egypt and Mesopotamia (see Fig. 6-3 for the map of world culture hearths). The Indus Valley civilization was centered on a pair of major cities, Harappa and Mohenjo-Daro, which may have been capitals during different periods of its history (Fig. 8-4); in addition, there were more than 100 smaller urban settlements. The locals apparently called their state **Sindhu**, and both **Indus** (for the river) and **India** (for the later state) may derive from this name. Although the influence of this civilization extended as far east as present-day Delhi, it did not last because of environmental change and, perhaps, because the political center of gravity shifted southeastward into the Ganges Basin.

Aryans and the Origins of Hinduism

Around 1500 BC, northern India was invaded by the **Aryans** (peoples speaking Indo-European languages based in what is now Iran). As the Iron Age dawned in India, acculturated Aryans began

© Steve McCurry/Magnum Photos, Inc.

© Steve McCurry/Magnum Photos, Inc.

The arrival of the annual rains of the wet monsoon transforms the Indian countryside. By the end of May, the paddies lie parched and brown, dust chokes the air, and it seems that nothing will revive the land. Then the rains begin, and blankets of dust turn into layers of mud. Soon the first patches of green appear on the soil, and by the time the monsoon ends all is green. The photo on the left, taken just before the onset of the wet monsoon in the west-coast State of Goa, shows the paddies before the rains begin; three months later when the rice crop is ready to harvest, this same countryside looks as on the right.

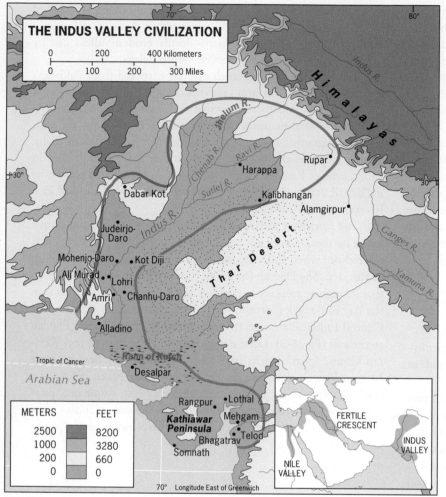

THE INDUS VALLEY CIVILIZATION

FIGURE 8-4

© J. Nijman, P. O. Muller, H. J. de Blij, and John Wiley & Sons, Inc.

ranking higher on the social ladder. In recent years, the caste system appears to be eroding from the combined effects of globalization, economic growth, and urbanization; this is true for India's bigger cities, but much less so for the majority of people who still reside in rural areas.

Although Hinduism spread across South Asia and even reached the Southeast Asian realm (especially Cambodia and Indonesia), Indo-European languages never took hold in the southern portion of the subcontinent. As Figure 8-5 indicates, **Indo-European languages** **3** (several of which are rooted in Sanskrit) predominate in the western and northern parts of South Asia, whereas the southern languages belong to the **Dravidian** **4** family—languages that were indigenous to the realm even before the arrival of the Aryans. But these are not fossil languages: they remain vibrant and have long literary histories. Today, Telugu, Tamil, Kanarese (Kannada), and Malayalam are spoken by more than 200 million people. In India's northern and northeastern fringes, Sino-Tibetan languages predominate, and smaller pockets of Austro-Asiatic speakers can be found in eastern India and neighboring Bangladesh. To the west, Afghanistan consists of a mosaic of Altaic and Indo-European languages, and Pashto (spoken by two-thirds of the Afghan population) spills across the border into neighboring Pakistan.

Buddhism and Other Indigenous Religions

Hinduism is not the only religion that emerged in this realm. Around 500 BC, **Buddhism** arose in the eastern Ganges Basin in what is today the Indian State of Bihar. The famous story of the "enlightenment" of the Prince Siddhartha (the Buddha) took place in the town of Bodh Gaya, and his following soon expanded in every direction. The appeal of Buddhism was (and is) especially strong among lower-caste Hindus, and large numbers of them have converted through the ages. Interestingly, Buddhism evolved inside what is now India, but its most powerful influences were felt beyond the realm in Southeast and East Asia. Today, only 0.8 percent of the population of India adheres to Buddhism (about 80 percent are Hindu), but it is the state religion in Bhutan and the great majority (almost 70 percent) of Sri Lankans are Buddhist as well.

Another (much smaller) indigenous religion that has evolved alongside Hinduism since ancient times is **Jainism**, often described as a more purist, principled, and deeply spiritual form of Hinduism. It is well known for its uncompromising stand on nonviolence and vegetarianism. Jains today constitute about one-half of 1 percent of the population in India. Finally, we should take note of **Sikhism** as another of the realm's indigenous religions, a blend of sorts of Islamic and Hindu beliefs. This religion, practiced by about 2 percent of the population, is of course much younger; it emerged around AD 1500, a few centuries after Islam became a dominant force in much of the South Asian realm.

the process of integrating the Ganges Basin's isolated tribes and villages into a new organized system, and urbanization made a comeback. The Aryans brought their language (Sanskrit, related to Old Persian) and a new social order to the vast riverine flatlands of northern India. Their settlement here was also accompanied by the emergence of a religious belief system, **Vedism**. Out of the texts of Vedism and local creeds there arose a new religion—**Hinduism**—and with it a new way of life.

It is thought that the arrival and accommodation of Aryans in this new society forged a layered system of social stratification that would solidify the powerful position of the Aryans and be legitimized through religion. Starting about 3500 years ago, a combination of regional integration, the organization of villages into controlled networks, and the emergence of numerous small city-states produced a hierarchy of power among the people, a ranking from the especially powerful (Brahmins—highest-order priests) to the weakest. This class-based **caste system** **2** of Hinduism is highly controversial in the West (and among many Indians too) because of its rigidity and the ways in which it justifies structural inequality. Those in the lowest castes, deemed to be there because they deserved it given their past lives, are worst off, without hope of advancement and at the mercy of those

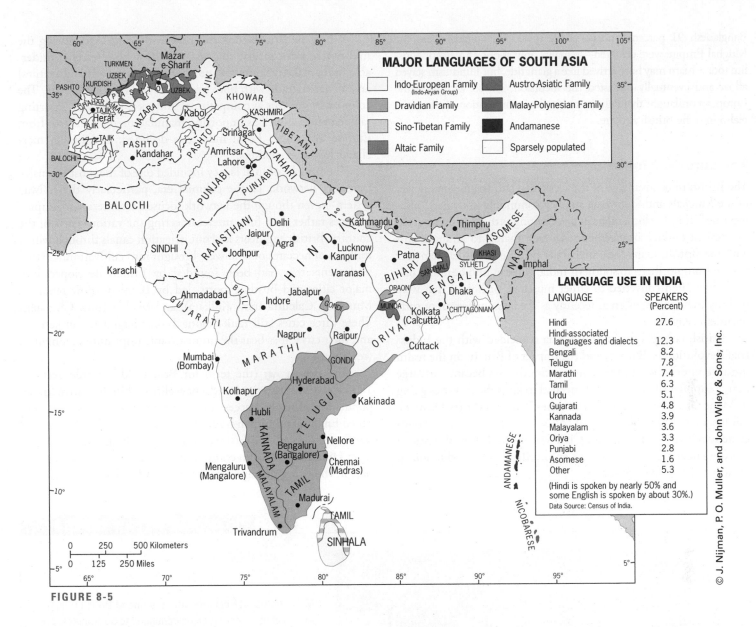

MAJOR LANGUAGES OF SOUTH ASIA

- Indo-European Family (Indo-Aryan Group)
- Dravidian Family
- Sino-Tibetan Family
- Altaic Family
- Austro-Asiatic Family
- Malay-Polynesian Family
- Andamanese
- Sparsely populated

LANGUAGE USE IN INDIA

LANGUAGE	SPEAKERS (Percent)
Hindi	27.6
Hindi-associated languages and dialects	12.3
Bengali	8.2
Telugu	7.8
Marathi	7.4
Tamil	6.3
Urdu	5.1
Gujarati	4.8
Kannada	3.9
Malayalam	3.6
Oriya	3.3
Punjabi	2.8
Asomese	1.6
Other	5.3

(Hindi is spoken by nearly 50% and some English is spoken by about 30%.)
Data Source: Census of India.

© J. Nijman, P. O. Muller, and John Wiley & Sons, Inc.

FIGURE 8-5

FOREIGN INVADERS

The Reach of Islam

In the late tenth century, Islam came rolling like a giant tide across South Asia, spreading across Persia and Afghanistan, through the Khyber Pass into the Indus Basin, and across Punjab into the Ganges Basin—converting virtually everybody in the Indus Valley and foreshadowing the emergence, many centuries later, of the Islamic Republic of Pakistan. By the early thirteenth century, the Muslims had established the long-surviving and powerful **Delhi Sultanate** that controlled much of the northern tier of peninsular India.

In the early 1500s, a descendant of Genghis Khan named Babur conquered Kabol (Kabul) in Afghanistan, and from that stronghold he penetrated the Indus Basin and Punjab and directly challenged the Delhi Sultanate. In the 1520s, his Islamicized Mongol forces ousted the Delhi rulers and established the **Mughal (Mogul) Empire**. By most accounts, Mughal rule was at times remarkably enlightened, especially under the leadership of Babur's grandson Akbar, who forcibly expanded the Empire but adopted tolerant policies toward the Hindus under his sway; Akbar's grandson, Shah Jahan, made his enduring mark on India's cultural landscape by commissioning magnificent architectural creations, particularly the Taj Mahal in the city of Agra.

Nonetheless, by the early eighteenth century, the Mughal Empire was in decline. Maratha, a Hindu state in the west, expanded not only into the peninsular south but also northward toward Delhi, capturing the allegiance of local rulers and weakening Islam's hold. Fractured India now lay open to still another foreign intrusion, this time from Europe.

Despite more than seven centuries of Islamic rule in South Asia, it is remarkable that Islam never achieved proportional dominance over the realm as a whole. Whereas today Pakistan is 97 percent Muslim and

Bangladesh 91 percent, India—where the Delhi Sultanate and the Mughal Empire were centered—remains only about 15 percent Muslim today. Islam may have arrived like a giant tide, but Hinduism stayed afloat and eventually outlasted the incursion. It also withstood the European onslaught that culminated in the incorporation of the entire realm into the British Empire.

The European Intrusion

The British took advantage of the weakened and fragmented power of the Mughals and followed a strategy commonly known as "indirect rule." They left local rulers in place as long as the British extracted the desired trading arrangements. In fact, thanks to arrangements with the British, many local maharajas became wealthier than ever before: from northern to southern India, you can find beautiful palaces (now often converted to either museums or hotels) that were built by these rulers, often as recently as the late nineteenth or early twentieth century.

British colonialism in South Asia coincided with the Industrial Revolution in Europe, and the impact of Britain on the realm must be understood in that context. South Asia became, in large part, a supplier of raw materials needed to keep the factories going in Manchester, Birmingham, and other British industrial centers. For instance, when the supply of cotton from the American South came to a halt during the U.S. Civil War in the 1860s, the British quickly encouraged (and indeed enforced) cotton production in what is today western India.

When the British took full power in South Asia during the mid-nineteenth century, this was a realm with already considerable industrial development (notably in metal goods and textiles) and an active trade with both Southeast and Southwest Asia. The colonialists saw this as competition, and soon India was exporting raw materials as well as importing manufactured products—from Europe, of course. Local industries declined, and Indian merchants lost their markets.

The British colonial legacy included one of the most extensive transport networks of the colonial era, particularly the railway system—even though the network focused on interior-to-seaport linkages rather than fully interconnecting the various parts of the country. British engineers laid out irrigation canals through which millions of hectares of land were brought into cultivation. Coastal settlements that had been founded by Britain developed into major cities and bustling ports, led by Bombay (now renamed Mumbai), Calcutta (now Kolkata), and Madras (now Chennai). These three cities still rank among India's largest urban centers, and their cityscapes bear the unmistakable imprint of colonialism (see photo).

The British *raj* (the term for Britain's direct rule between 1857 and 1947) also produced a new elite within the native Indian population. They had access to education and schools that combined English and Indian traditions, and their Westernization was reinforced through university education in Britain. This elite drew from Hindu and Muslim communities, and it was to play a major part in the rising demands for self-rule and independence. These

@ **FROM THE FIELD NOTES...**

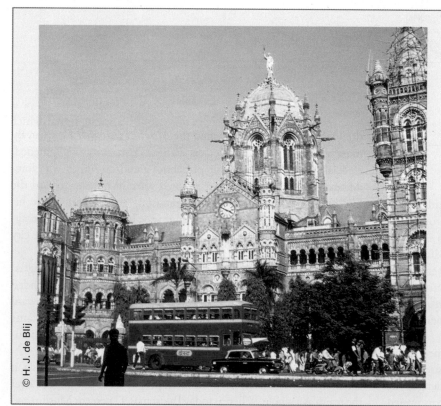

© H. J. de Blij

"More than a half-century after the end of British rule, the centers of India's great cities continued to be dominated by the Victorian-Gothic buildings the colonizers constructed here. This also is evidence of a previous era of globalization, when European imprints transformed urban landscapes. Walking the streets of some parts of Mumbai (the British called it Bombay), you can turn a corner and be forgiven for mistaking the scene for London, double-decker buses and all. One of the British planners' major achievements was the construction of a nationwide rail system, and railway stations were given great prominence in the urban architecture. I had walked up Naoroji Road, having learned to dodge the wild traffic around the circles in the Fort area, and watched the throngs passing through Victoria (now Chhatrapati Shivaji) Station. Inside, the facility is badly worn, but the trains continue to run, bulging with passengers hanging out of doors and windows."

demands started to gather momentum in the early twentieth century and could no longer be denied after World War II came to an end in the mid-1940s.

THE GEOPOLITICS OF MODERN SOUTH ASIA

Partition and Independence

Even before the British government decided to yield to demands for independence, it was clear that British India would not survive the coming of self-rule as a single political entity. As early as the 1930s, Muslim activists were promoting the idea of a separate state. As the colony moved toward independence, a major political crisis developed that eventually resulted in the separation of India and Pakistan. But **partition** **5** was no simple matter. True, Muslims were in the majority in the western and eastern sectors of British India, but smaller Islamic clusters were scattered throughout the realm. Furthermore, the new boundaries between Hindu and Muslim communities had to be drawn right through areas where both sides coexisted—thereby displacing millions (see photo).

The consequences of this migration for the social geography of India were especially far reaching. Comparing the country's 1931 and 1951 distributions of Muslims in Figure 8-6, you can see the impact on the Indian Punjab and in what is today the State of Rajasthan. (Since Kashmir was mapped as three entities before partition and as one afterward, the change there represents an administrative, not a major numerical, alteration.) A Muslim exodus occurred even in the east, as is reflected on the map by the State of West Bengal, adjacent to East Pakistan (now Bangladesh), where the Islamic component of the local population declined substantially.

Flight was one response to the 1947 partition of what had been British India, resulting in the greatest mass population transfer in human history. Here, two trainloads of eastbound Hindu refugees, fleeing then West Pakistan, arrive at the station in Amritsar, the first city inside India.

The world has seen many refugee migrations but none involving so many people in so short a time as the one resulting from British India's partition (which occurred on Independence Day, August 15, 1947). Scholars who study the refugee phenomenon differentiate between "forced" and "voluntary" migrations, but as this case underscores, it is not always possible to distinguish between the two. Many Muslims, believing they had no choice, feared for their future in the new India and joined the stampede. Others had the means and the ability to make a decision to stay or leave, but even these better-off migrants undoubtedly sensed a threat.

The great majority of Hindus who lived on the "wrong" side of the border moved as well. The Hindu component of present-day Pakistan may have been as high as 16 percent in 1947 but is barely more than 1.5 percent today; in Bangladesh, which was named East Pakistan at the time of partition, it declined from 30 percent to around 8 percent today. Partition, therefore, created an entirely new cultural and geopolitical landscape in South Asia.

India–Pakistan

From the moment of their separate creation in 1947, India and Pakistan have had a tenuous relationship. Upon independence, present-day Pakistan was united with present-day Bangladesh, and the two countries were respectively called West Pakistan and East Pakistan. As we have noted, the basis for this scheme was Islam: in both Pakistan and Bangladesh, Islam is the state religion. Between the two Islamic wings of Pakistan lay Hindu India. But there was little else to unify the Muslim easterners and westerners, and their union lasted less than 25 years. In 1971, a costly war of secession, in which India supported East Pakistan, led to the collapse of this unusual arrangement. East Pakistan, upon its "second independence" in 1971, took the name Bangladesh—and since there was no longer any need for a "West" Pakistan, that qualifier was dropped and the name Pakistan remained on the map.

India's encouragement of independence for Bangladesh emphasized the continuing tension between Pakistan and India, which had already led to war in 1965, to further conflict during the 1970s over Jammu and Kashmir, and to periodic flareups over other issues. Armed conflict between the two South Asian countries seemed to be a regional matter—until the early 1990s, when their arms race took on ominous nuclear proportions. Since then, the specter of nuclear war has hung over the conflicts that continue to embroil Pakistan and India, a concern not just for the South Asian realm but for the world as a whole. No longer merely a decolonized, divided, and disadvantaged country trying to survive, Pakistan has taken a crucial place in the political geography of a geographic realm in turbulent transition.

The relationship between India and Pakistan is especially sensitive because so many Muslims still live in India. Massive as the 1947 refugee movement was, it left far more Muslims in India than those who had departed. The number of Muslims in India declined sharply, but it remained a huge minority, one that was growing rapidly to boot. It has now surpassed 180 million, just

© J. Nijman, P. O. Muller, H. J. de Blij, and John Wiley & Sons, Inc.

FIGURE 8-6

under 15 percent of India's total population—not only slightly larger in size than Pakistan's entire population but also the largest cultural minority in the world.

What this means is that a substantial proportion of India's population tends to have more or less "natural" sympathies vis-à-vis Pakistan, even though India's Muslims are generally known to be 'moderate.' Their presence in many instances works as a brake on hawkish Indian policies toward the Islamabad government. At the same time, conflict with Pakistan can have detrimental effects on Hindu-Muslim relations inside India, and over the years has led to communal violence and deadly clashes. Today that issue is further complicated by the alleged involvement of some Indian Muslims in terrorist activities within India that were orchestrated from Pakistan.

Contested Kashmir

When Pakistan became independent following the partition of British India, its capital was Karachi, located on the southern coast near the western end of the Indus Delta. As the map shows, however, the present capital is Islamabad. By moving the capital from the "safe" coast to the embattled interior, and by placing it on the doorstep of the contested territory of Kashmir, Pakistan announced its intent to stake a claim to its northern frontiers. And by naming

the city Islamabad, Pakistan proclaimed its Muslim foundation here in the face of the Hindu challenge. This politico-geographical usage of a national capital can be assertive, and as such Islamabad exemplifies the principle of the **forward capital** 6 .

Kashmir is a territory of high mountains surrounded by Pakistan, India, China, and, along more than 50 kilometers (30 mi) in the far north, Afghanistan (Fig. 8-7). Although known simply as Kashmir, the area actually consists of several political divisions, including the Indian State properly referred to as Jammu and Kashmir, a major bone of contention between India and Pakistan.

When partition took place in 1947, the existing States of British India were asked to decide whether they wanted to be incorporated into India or Pakistan. In most of the States, the local ruler made this decision, but Kashmir was an unusual case. It had about 5 million inhabitants at the time, nearly three-quarters of them Muslims, but the maharajah of Kashmir himself was a Hindu. When he decided not to join Pakistan and instead aimed to retain autonomous status, this was answered with a Muslim uprising supported by Pakistan. The maharajah, in turn, called for help from India. After more than a year's fighting and through the intervention of the United Nations, a cease-fire line left most of Jammu and Kashmir (including nearly four-fifths of the territory's population) in Indian hands. Eventually, this line—now known as the *Line of Control*—began to appear on maps as the final

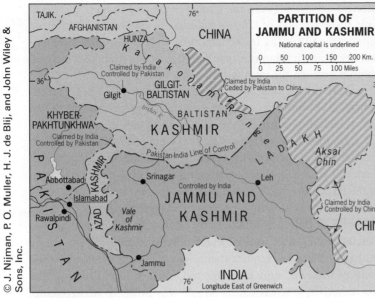

FIGURE 8-7

boundary settlement, and Indian governments have proposed that it be so recognized.

With about two-thirds of the Kashmiris adhering to Islam, Pakistan has for decades demanded a referendum in Jammu and Kashmir in which the people can decide for themselves to remain with India or become part of Pakistan. India has refused, arguing that there is a place for Muslims in secular India but not for Hindus in the Islamic Republic of Pakistan. Given the specter of terrorism in India and the dangerous precedent of a concession in light of India's prodigious ethnic and regional diversity, the Kashmir conflict has been left to smolder and is unlikely to be resolved for some time to come.

The Specter of Terrorism

Comparatively successful as the integration of India's Muslim communities into the fabric of the Indian state has been, the risk of Islamic violence, directed against Indian society in general, has been rising. The highest-profile attack to date was in 2008, when terrorists targeted Mumbai's two most upscale, Westernized hotels. Nearly 200 people perished and hundreds were wounded. Live pictures of smoke billowing from the famous Taj Mahal Hotel in southern Mumbai were broadcast around the world (see photo). The group responsible for these heinous attacks was *Lashkar-e-Taiba* (the Party of the Righteous), a Pakistan-based organization that among other things aims to return Kashmir to Islamic rule.

In the meantime, Pakistan's northwestern frontier is effectively controlled by the Taliban, the Afghan Islamic extremists who also have a history of collaboration with al-Qaeda. This entire border zone with Afghanistan lies well beyond the control of the Pakistani government. U.S. efforts to defeat the Taliban in Afghanistan continued in 2015, but were mostly thwarted by the Taliban's ability to move back and forth across this border—despite the pressure the

United States exerts on Islamabad to confront the Taliban on the Pakistani side of the border in this treacherous mountain refuge.

This remains an especially delicate geopolitical chess match for the Pakistani regime. It must be very careful not to alienate its Islamic base, even though it despises the border-zone extremists, and it also fears an economically stronger India with growing ties to the United States. India is deeply concerned about Pakistan's role in terrorism on Indian soil and, even worse, the possibility of jihadists taking control of Pakistan's nuclear arsenal; at the same time, it must guard against heightening tensions between Hindus and Muslims inside India. India is also frustrated by American reluctance to choose sides in the Kashmir conflict. The United States, in turn, is sympathetic to the world's largest democracy but needs Pakistan as an ally in the global counterterrorism campaign.

Smoke billows from the landmark Taj Mahal Hotel in Mumbai on November 29, 2008. The Taj was one of several sites in southern Mumbai that were simultaneously targeted for attack by Islamic militants. The siege lasted four days and almost 200 people died in the violence. Indian commandos killed all of the terrorists but one, who was captured, tried, sentenced to death, and executed in 2012. It was not the first time that such terrorism had struck India, and several smaller-scale attacks have occurred in this decade.

Each of these parties is walking a tightrope, where the slightest mistake could have deadly consequences.

Chinese Border Claims

Any overview of this realm's geopolitical framework today must also include China's increasingly powerful and sometimes invasive presence. As Figure 8-7 shows, China claims the northeastern extension of India's Jammu and Kashmir State. This issue has been stalemated in recent years, but officially neither China nor India shows any sign of conceding.

China's control over Tibet (called Xizang by the Chinese) and its efforts to influence the lives of Tibetans both within and outside Tibet create additional issues. To the dismay of Beijing, Tibet's exiled Dalai Lama calls India his second home. In recent years, China has been pressuring the government of Nepal, wedged between India and Tibet, to discourage Tibetan immigration and to constrain the activities of Tibetans already in the country. Farther to the east, China claims the bulk of the Indian State of Arunachal Pradesh ("land of the dawn-lit mountains") based on its assertion that the boundary, established in 1914, was never ratified by Beijing—even though it was approved by the then-independent Tibetans themselves. And China even claims rights to the small area of northern India lying between Tibet-adjoining Sikkim and Bhutan, again on the basis that its inhabitants are Tibetans and therefore belong under Chinese jurisdiction.

China's power in South Asia is felt in other ways as well. A few years ago, the Chinese announced plans to construct dams on the Tibetan headwaters of the upper Brahmaputra River, potentially jeopardizing the main water supplies of both northeastern India and Bangladesh. Thus, by no means is Kashmir the only place in the realm's lengthy northern frontier zone where trouble can erupt at any time—and within which India directly borders China for 3500 kilometers (2200 mi).

Indian Ocean Geopolitics

China needs access to markets for its products and supplies of raw materials to sustain its steadily growing industrial production, and a major part of that access involves the Indian Ocean (see Fig. 5-18). In order to protect these interests, China is projecting political and military power by expanding its naval presence in this maritime arena and building bases in Pakistan, Myanmar, and Bangladesh. Meanwhile, increasingly concerned about Chinese intentions around the subcontinent's territorial waters, India is responding by strengthening its own navy and forging new alliances with such Southeast Asian states as Indonesia and Vietnam.

From a broader pan-Asian perspective (South and East), geopolitical developments are increasingly a function of U.S.–China–India relations. These developments are propelled by China's assertions in the Indian Ocean Basin and along its land borders with India; India's economic rise, which gives it an ever higher profile in the global political arena; and the ideological propensity of the United States to side with India. The near-term future will likely see a continued

informal alliance between the United States and India that seeks to limit the expansion of China's boldness in South Asia. American support for admitting India to membership in the United Nations Security Council is but one manifestation of the strengthening relationship between the two countries.

These underlying issues aside, there have also been recent indications of more conciliatory exchanges between the two Asian giants, including mutual state visits of their presidents. In a notable turnaround, China now says it would not oppose Indian membership in the Security Council. More importantly, growing economic interdependence is proving to be advantageous to both countries. China wants to penetrate India's enormous, expanding consumer markets. China–India trade is rapidly rising, now at the level of $70 billion annually (with about two-thirds going from China to India and one-third in the other direction). So far, China has been running a significant surplus, reflecting the fact that the terms of trade are more favorable to China: India mainly supplies raw materials, whereas China largely sells finished, higher-value-added goods.

EMERGING MARKETS AND FRAGMENTED MODERNIZATION

In recent years, optimistic reports in the media have been proclaiming a new era for South Asia, marked by rising growth rates for the realm's national economies, rewards from globalization and modernization, and increasing integration into the world economy. India, obviously the key to the realm, has even been described as "India Shining" during this wave of enthusiasm.

Nevertheless, well over half of India's 1.3 billion people continue to live in poverty-stricken rural areas, their villages and lives virtually untouched by what is happening in the cities (where tens of millions of urban dwellers inhabit some of the world's poorest slums). Afghanistan and Nepal rank among the poorest countries on Earth; fully one-third of Pakistan's population lives in miserable poverty, and—Nobel laureate Malala Yousafzai notwithstanding—female literacy still lingers below 40 percent. Half of the people of Bangladesh, and nearly half of those in the realm as a whole, live on the equivalent of two U.S. dollars a day or less. It is estimated that half the children in South Asia are malnourished and underweight, a majority of them girls—this at a time when the world is able to provide adequate calories for all its inhabitants, if not sufficiently balanced daily meals. It still remains to be seen whether the benefits of newfound economic growth can be spread around widely enough to improve the living conditions of South Asia's masses.

Economic Liberalization

Most countries in this realm have liberalized their economies since the 1980s as part of a worldwide turn toward **neoliberalism 7**. This entails privatization of state-run companies, lowering of international

SOUTH ASIA: GDP PER CAPITA IN CONSTANT US$, 2003–2013

Data Source: World Bank

FIGURE 8-8

incomes. Generally, this progress has continued into this decade, even though variations between countries are quite pronounced (Fig. 8-8). Most of the growth is in manufacturing, services, finance, and, in India, information technology (IT). More open economies have attracted increased foreign investment, and during the past two decades a new, urban middle class has emerged. This steadily expanding social stratum may account for only about 25 percent of the realm's population, but in South Asia that translates into more than 400 million people—a huge new consumer market for an array of products ranging from smartphones to automobiles. At the same time, it leaves well over a billion South Asians unaffected by this upward social mobility. For them almost nothing has changed; they remain overwhelmingly dependent on (stagnant) agriculture and are not likely to log on to the new information economy anytime soon. As in so many other parts of the world, here, too, increased overall growth has mostly been accompanied by ever greater economic and social inequality.

One striking feature of the South Asian realm is that even if the majority of its people live traditional lives in rural villages, there also are a fair number of **megacities** 8 in which social and economic change is the order of the day. The metropolitan areas centered on Mumbai, Delhi–NewDelhi, Kolkata, Dhaka, Chennai, Bengaluru, and Karachi all contain populations that exceed 10 million and have grown rapidly over the past two decades. Their densities are often overwhelming, environmental conditions are poor and worsening,

trade tariffs, reduction of government subsidies, cutting of corporate taxes, and overall deregulation to stimulate business activity.

The results of these reforms can be observed across South Asia in terms of increased economic growth and higher per-capita

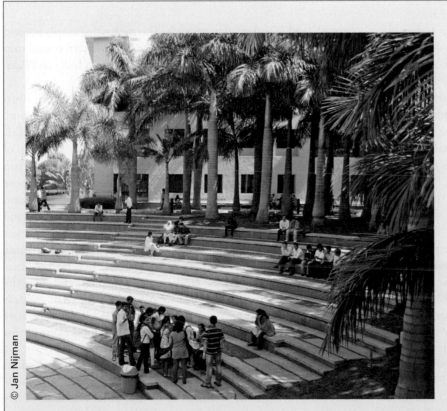

@ FROM THE FIELD NOTES...

"Strolling the grounds of one of India's leading high-tech companies, Wipro, in the southern city of Bengaluru (formerly Bangalore), I was struck by the thought that their use of the word 'campus' was right on target: the great majority of the 33,000 employees at this facility (120,000 in total) are in their mid-20s and almost all have a college degree; a large number of them are involved in research and product development; and the layout of the premises could easily be mistaken for a (well-funded) university. This Silicon Valley-like campus is the heart of the company's booming outsourcing and consulting business. Wipro is active in most major Indian cities and has a sizeable global presence. I was shown around by Rohit, a 26-year-old with a bachelor's degree in engineering from an Indian college and an MBA from Singapore National University. He reveled in the opportunity: 'I joined the company just four months ago in their marketing division and feel extremely fortunate. Everybody wants this job! It's a great professional opportunity and the salary is very good.' And Bengaluru is a pleasant place to live: at 900 meters (3000 ft) above sea level, even the summer weather is tolerable. Bengaluru, however, is growing so rapidly that its infrastructure cannot keep up: thus congestion, traffic jams, and commuting times are all growing as well."

MAP ANALYSIS

FIGURE 8-9 Two Megacities: Metropolitan Dhaka and Los Angeles

South Asia is the most heavily populated geographic realm on Earth. But its overall level of urbanization is quite low at 32 percent because the great majority of people inhabit rural areas. In absolute terms, however, the bigger South Asian cities are growing rapidly, and the realm now contains seven megacities. In comparison, the North American realm has only two: New York and Los Angeles. It is important to realize that significant contrasts mark the world's megacities, as shown in the pair of maps above that differentiate the geographic areas of Dhaka and Los Angeles. Analyze the two maps and provide a concise description of the fundamental differences. Can you think of some reasons that explain these differences? Access your WileyPLUS Learning Space course to interact with a dynamic version of this map and to engage with online map exercises and questions.

and the widening income gap between rich and poor is staggering. But almost always it is better to be poor in the city than in the countryside, simply because cities offer opportunities that do not exist in the villages—which helps to explain the burgeoning flow of rural-to-urban migrants that drives the growth of towns and cities all across South Asia. As Figure 8-9 demonstrates, there is substantial variation in the internal densities and layouts of megacities; those differences also extend to levels of prosperity, and the contrasts between megacities in the Global Core and the Global Periphery are particularly striking.

The Significance of Agriculture

More than half of South Asia's entire workforce is employed in agriculture, ranging from about 43 percent in Pakistan to about 75 percent in Nepal. But overall productivity is low, and the contribution of agriculture to the national economy averages only around 20 percent. Incomes in rural areas are much lower than in major cities, and the same is true for the standard of living. Almost 70 percent of the realm's population is rural, and even those who do not work in agriculture tend to rely on it indirectly.

Millions of lives every year depend on a bountiful harvest. As Figure 8-3 shows, the wet monsoon brings lifegiving rains to the southwestern (Malabar) coast, and a second branch from the Bay of Bengal sweeps across north-central India toward Pakistan, losing

strength (and moisture) as it proceeds. This means that amply watered eastern India and Bangladesh as well as the southwestern coastal strip grow rice as their staple crop; but drier northwestern India and Pakistan raise wheat.

Farmers' fortunes tend to vary geographically, as can be illustrated in the situation on either side of India's Western Ghats upland. The monsoon rains are generally plentiful along the coast-paralleling, western slopes of this linear highland, from the southern tip of the subcontinent as far north as the vicinity of Mumbai. Here you can see the hillside vegetation assume all shades of fresh green come the month of June, a good sign that the harvest will be abundant. But on the eastern (rain shadow) side of the Western Ghats and deeper into the interior of the Deccan Plateau, the rains come less often and do not last as long; farming becomes a gamble with nature, and life becomes precarious. Many of the farmers here are members of the lowest-ranking castes, landless and indebted, and have the hardest time making ends meet. Almost every year, Maharashtra's inland districts report several thousand farmer suicides as desperately poor peasants end their lives because they can no longer provide for their families.

It is clear that the majority of people in this realm depend on agriculture and that governments must aim their economic policies at improving farm productivity to raise rural standards of living. But they have a long way to go. The demands on governments are multiple, and they often seem distracted by economic sectors

@ FROM THE FIELD NOTES...

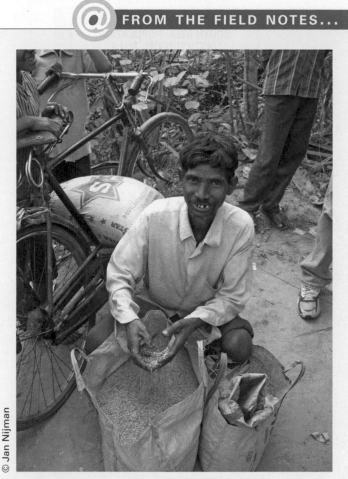

© Jan Nijman

"During my travels in northeastern India in early 2011, I visited some agricultural areas in the Brahmaputra Valley that benefit from plentiful irrigation. Tea and rice are two major crops. Because big companies run the large tea plantations, many of the rice farmers have to make do with small plots of land. Here a farmer in Asom State shows off part of his rice harvest. I encountered him and several others at a small, machine-operated mill shop. Note that the rice is brown—milling it removes the bran layer and makes the rice white. 'We have the best rice in all of India,' he proudly said, as he slowly poured it from his hand back into the bag. Having enjoyed some excellent meals in the area, I could not disagree."

that can make a faster and greater contribution to the tax base, such as manufacturing, financial services, and IT—economic activities that almost always are centered on the bigger cities far from the impoverished countryside.

SOUTH ASIA'S POPULATION GEOGRAPHY

Given its enormous human population, South Asia's areal size is relatively small and totals less than 40 percent of the territory of similarly populous East Asia. Comparing the world's two giants shows that China's territorial extent is almost three times as large as India's. The total population of Subsaharan Africa is less than half of South Asia's, in an area almost five times as large. Adjectives such as "teeming," "overcrowded," and "crammed" are often used to describe the realm's habitable living space, and with good reason. South Asia's intricate cultural mosaic is tightly packed, with only the deserts in the west and the mountain fringe in the north displaying extensive empty spaces. The outlines of the densely populated river basins are clearly visible in the clustering patterns on the population distribution map (Fig. 8-10).

The field of **population geography** 9 focuses on the characteristics, distribution, growth, and other aspects of spatial demography in a country, region, or realm as they relate to soils, climates, land ownership, social conditions, economic development, and other factors. In the South Asian context, it is useful to concentrate on four demographic dimensions: the role of density; the demographic transition; demographic burdens; and the gender bias in birth rates. As we shall see, population issues are often more complicated than they first appear.

Population Density and the Question of Overpopulation

Population density 10 measures the number of people per unit area (such as a square kilometer or square mile) in a country, province, or an entire realm. We distinguish between two types of measures. *Arithmetic density* is simply the number of people per areal unit, usually a country. **Physiologic density** 11 is a more meaningful measure because it takes into account only land that is arable and can be used for food production. In Pakistan, for example, the two measures are quite different because of that country's large deserts and inhospitable mountainous zones: its arithmetic density is 243 per sq km compared to a physiologic density of 838.

Until recently, South Asia's persistent poverty was often related to its enormous, rapidly growing population and its high population densities. The idea was that there were simply "too many mouths to feed"—the realm was "overpopulated." The notion of *overpopulation* can be compelling and seems to make sense at an intuitive level because every country or region can be thought of as having a limited "carrying capacity."

But things are more complex than that. Some countries with high densities, such as the Netherlands or Japan, are highly developed, and not necessarily because they possess an impressive natural resource base (neither does). The point is that high density in itself is not always a problem and that, in certain circumstances, population can be considered as a *human resource*. If productivity is high, there does not appear to be a problem; but if productivity is low, then large populations can be a drain on the economy. Countries with high education levels, institutional effectiveness, and technological know-how are able to use their resources more efficiently. Thus in South Asia, with its substantial number of undereducated and illiterate inhabitants, population functions as an impediment rather than a resource—the problem being that too many people are not sufficiently productive.

SOUTH ASIA: POPULATION
DISTRIBUTION, 2016
One dot represents 75,000 persons

© J. Nijman, P. O. Muller, and John Wiley & Sons, Inc.

FIGURE 8-10

The Demographic Transition

The relevance of population issues to development goes far beyond density, which is really just a snapshot of the population pattern at a moment in time. It gets more interesting—and more complicated—when we relate population change to economic trends. For instance, for a considerable time South Asia's population grew faster than the realm's economy. Clearly, that was a problem because more and more people had to survive on less. Today, fortunately, it is the other way around: in most of the realm, the economy is growing faster than the population.

The term **demographic transition** [12] refers to a structural change in birth and death rates resulting, first, in rapid population increase and, subsequently, in declining growth rates and a stable population (Fig. 8-11). The United States and other highly developed countries had already passed through this transition by the mid-twentieth century, and most of this realm's countries are in the third stage today. Note that Stage 2 and part of Stage 3, with high birth rates and low death rates (due to healthcare advances), entail a sizeable population expansion. In South Asia, this expansion occurred from the 1950s through the 1970s. The

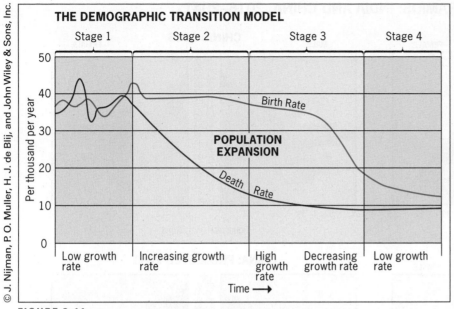

THE DEMOGRAPHIC TRANSITION MODEL

Stage 1 | Stage 2 | Stage 3 | Stage 4

Birth Rate

POPULATION EXPANSION

Death Rate

Per thousand per year

Low growth rate | Increasing growth rate | High growth rate | Decreasing growth rate | Low growth rate

Time →

FIGURE 8-11

key issue, of course, is for birth rates to come down so that overall growth rates will drop and the population will stabilize. This is happening in the twenty-first century, but the process is not yet complete.

Fertility rates 13 (the number of births per woman of child-bearing age) have dropped across the realm over the past quarter-century. Figure 8-12 indicates a continuing decline, with most countries now hovering around 2.5 births. Considering that a fertility rate of about 2.3 reflects a stable population for a realm at this level of development, it is clear that South Asia is still growing. But Afghanistan's extremely high fertility rate is an outlier: recording 4.9 in 2013, double the realm's average.

Demographic Burdens

The immediate significance of demography to economics lies in what is called the **demographic burden 14**. This term refers to the proportion of the population that is either too old or too young to be productive and that must be cared for by the productive population. Typically, the most productive population in developing countries is represented in the age cohorts between 20 and 50 years. A country with low death rates and high birth rates will have a relatively large share of old and young people, and thus a large demographic burden. Obviously, the way to reduce this burden is to lower birth rates.

Let us now examine Figure 8-13 and compare today's **population pyramids 15** (diagrams showing the age–sex structure) for India and China. China has been more successful in curtailing births since 1980, so it now faces a lower demographic burden than India. But, interestingly, what is advantageous today can become a disadvantage down the road. Figure 8-13 also displays the population profiles that are projected to exist 25 years

from now. Assuming that India will be able to further reduce its birth rate in the coming years, its demographic burden will be less than China's one generation from now. In China, today's productive cohort will have moved on to old age, thereby adding to the national demographic burden. This is yet another reason for India boosters to be optimistic about the future. However, other major challenges remain, and certain forms of birth control are as morally reprehensible as they are economically counterproductive.

The Missing Girls

Issues of family planning and birth control shed some fascinating light on the "fragmented modernization" of South Asia. As we have seen, the realm finds itself in an advanced stage of the demographic transition wherein birth rates have begun to decline. But take a good look at India's population pyramid for 2016 and note that among young children boys far outnumber girls. In fact, males outnumber females well into middle age.

Traditionally, boys are valued more than girls because they are thought to be more productive income earners, because they are entitled to land and inheritance, and because they do not require a dowry at the time of marriage. When a couple gets married (often arranged and at a young age), the bride comes into the care of the groom's family, where she also contributes her work in and around the house. For this, the bride's family must provide a dowry, which may impose a major expense for her parents. For

SOUTH ASIA: FERTILITY RATES, 2003–2013

10% — Percent growth, 2003 | 5% — Percent growth, 2013

AFGHANISTAN

PAKISTAN

NEPAL

BHUTAN

INDIA

BANG

SRI LANKA

0 250 500 Kilometers
0 125 250 Miles

Data Source: World Bank

FIGURE 8-12

POPULATION PYRAMIDS: INDIA AND CHINA, 2016–2041

© J. Nijman, P. O. Muller, and John Wiley & Sons, Inc.

Source: U.S. Census Bureau

FIGURE 8-13

these reasons, the birth of a boy is a greater cause for celebration than that of a girl. "Raising a daughter," as one saying goes, "is like watering your neighbor's garden."

One reason for the high fertility rates in the past was that families would continue to have children until there were enough sons to take care of the parents in their old age (the girls, after all, would be taking care of their future husband's parents). Hence, this gender bias is in itself a major factor in South Asia's population growth—but that is not the only problem. When a poor couple repeatedly produces girls and not boys, in some instances the family decides to end the life of the newborn daughter. It is this ***female infanticide*** or *gendercide* that causes the unnatural gender bias in South Asia's population profiles (the same applies to parts of China as well).

But why—as the economic situation has improved, as birth rates have receded, and as modernization has begun to set in—do we still observe this skewed ***sex ratio***? The answer is that with fewer children, the importance of having at least one boy has for many families become even more pressing. And here is where "modernization" throws another curve ball: newly available technologies of ultrasound scanning plus rising incomes (i.e., the growing affordability of a scan) have induced many families to

determine the gender of the unborn child and decide on abortion if the child is female. Thus in recent years the sex ratio has become more, not less, skewed, and the most extreme ratios are now found in some of the most developed parts of South Asia, such as the Indian States of Punjab and Haryana.

In the long run, of course, this leads to a shortage of females, which becomes particularly apparent at marriage age. In some areas, families now face a problem in finding brides for their sons, and this "bachelor angst," in turn, is leading to a change in attitudes. Look at India's population pyramid for 2041 and note that over the next quarter-century the sex ratio is projected to become less skewed.

Discrimination against women and girls in South Asia is expressed in a variety of ways in daily life. One matter that is now receiving more attention is the lack of sanitation for females. In many workplaces, schools, slum neighborhoods, and public spaces, women still have no (or insufficient) access to toilets. For safety and personal reasons women need more privacy, but often that is not available. In India, an estimated 330 million females lack proper access to toilets, with major implications for health and social functioning. Imagine a small rural school with one toilet (if indeed there is one): it will be used by boys at the exclusion of girls. Girls will have no access at all the entire day. This simple fact leads to girls missing school and even dropping out. Recent efforts by development organizations to install toilets for girls have so far only produced significant reductions in female-student dropout rates.

The recent increase of highly publicized incidents of rape in India has triggered debate concerning the country's gender relations. It is an alarming trend, especially where it signals a reaction by men against liberalized behavior among young urban women. But it also reflects the increasingly open disposition of India's mass media as well as a new assertiveness among women (though it is hard to know to what extent this increase is due to a higher rate of reporting and not of rape itself). And, on a relative scale, the incidence of (reported) rape in India is still far below the level of most Western countries, including the United States.

It is difficult to make generalizations about gender relations across this populous realm, in part because of religious and regional diversity as well as rural-urban differences. To be sure, these are in many respects male-dominated societies, especially in the younger age cohorts, and more so in Afghanistan and Pakistan than in India, Sri Lanka, and Bhutan. Nonetheless, keep in mind that Pakistan, India, Sri Lanka, and Bangladesh have all had female prime ministers who held their countries' most powerful political office. That has yet to happen in the United States.

A group of girls in a park in Delhi showing off their henna tattoos. "Mehndi" is a centuries-old festive tradition involving decorative painting on the hands and wrists. These designs usually wear off within a few weeks.

© Jan Nijman

REGIONS OF SOUTH ASIA

The South Asian geographic realm can be subdivided into five regions (Fig. 8-14):

1. **The West** region consists of a pair of Islam-dominated countries. War-ravaged *Afghanistan* currently forms a transition zone between the South Asian and North African/Southwest Asian realms. Beleaguered *Pakistan* revolves around a core area created by the Indus River and its tributaries in Punjab.

2. **India** is the realm's giant and historic centerpiece. Except for its northeastern corner, it is both a region and a state. It is anchored by its core area focused on the broad basin of the Ganges river system.

3. **The East** region is centered on *Bangladesh*. This country sits atop the low-lying double delta formed by the Ganges and Brahmaputra rivers, which creates a unique and challenging geographic environment. Also included is the far northeastern corner of India, whose axis is the lower Brahmaputra Valley northeast of Bangladesh.

4. **The Northern Mountain States** region comprises two noncontiguous Himalayan states: *Nepal* and *Bhutan*. Nepal is plagued by a continuing political struggle following the failure of long-time royal rule; Bhutan remains a peaceful traditional kingdom.

5. **The Southern Islands** region contains two very different countries. Dominantly Buddhist, ethnically divided *Sri Lanka* is

located a short distance southeast of India's southern tip. Farther offshore, on the southwestern side of the subcontinent's tip, lies the island chain known as *The Maldives*, an Islamic ministate endangered by rising sea levels.

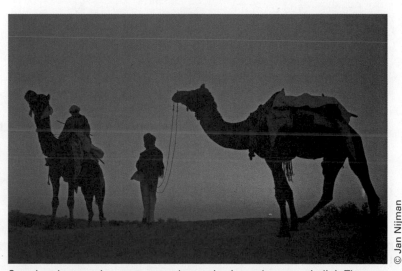

Camel traders gearing up on an early morning in northwestern India's Thar Desert, not far from the border with Pakistan.

© Jan Nijman

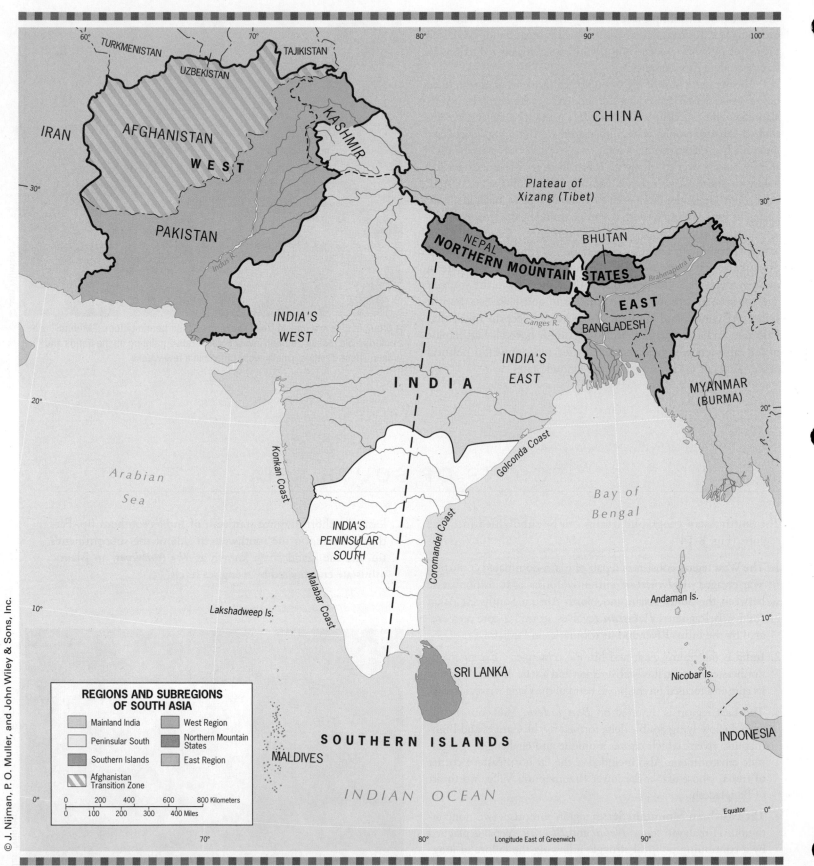

FIGURE 8-14

THE WEST

Afghanistan: Unstable Transition Zone

Geography and history seem to have conspired to divide Afghanistan. As Figure 8-1 shows, the towering Hindu Kush range dominates the center of the country, creating three broad environmental zones: the comparatively well-watered, fertile northern plains and basins; the rugged, earthquake-prone central highlands; and the desert-dominated southern plateaus. Kabol (Kabul), the capital, lies on the southeastern slopes of the Hindu Kush, linked by narrow passes to the northern plains and by the Khyber Pass to Pakistan and the rest of South Asia (Fig. 8-15).

Across this rugged, variegated landscape moved countless peoples: Greeks, Turks, Arabs, Mongols, and others. Some settled here, their descendants today speaking Persian, Turkic, and other languages (see Fig. 8-5). Others left archeological remains or no trace at all. The present-day population of Afghanistan (30.6 million) contains no ethnic majority. This is a country of minorities in which the Pushtuns of the east are the most numerous but make up only 42 percent of the total. The second-largest minority are the Tajiks (27 percent), a world away across the formidable Hindu Kush, concentrated in the far north near Afghanistan's border with Tajikistan. Other groups include Hazaras, Uzbeks, Turkmen, and Baluchis (Fig. 8-5).

Afghanistan has a long history of conflict, complicated by the persistent involvement of outside powers vying for influence in this strategically located territory. The protracted nineteenth-century competition between the Russian and British Empires over control of Afghanistan—with the wider purpose of controlling adjacent Central Asia—went down in history as "the Great Game." In fact, modern Afghanistan owes its existence to this contest: the Russians and British agreed to tolerate it as a geopolitically neutral cushion—or **buffer state 16**—between them. During the second half of the twentieth century, Afghanistan was pulled into the Cold War, suffered a Soviet invasion and ten years of occupation, the emergence of violent Islamic militants (particularly al-Qaeda), a 15-year-long presence of U.S. troops, and most recently a resurgence of the vicious Taliban jihadist forces. And yet, despite this turmoil, Afghanistan remains much as it was before the first inning of "the Great Game"—a feudally balkanized country with a weak and ineffectual government in Kabol. It has also now become a *transition zone*, an arena of spatial change in which the peripheries of adjacent realms join.

The Emergence of the Taliban and al-Qaeda

In 1994, what at first seemed to be just another one of fractious Afghanistan's warring factions appeared on the scene: the **Taliban 17** (seekers of religion), who originated at religious schools (*madrassas*) in Pakistan. Their avowed aim was to end Afghanistan's chronic factionalism and endemic corruption by imposing *Sharia* (strict Islamic) law. Popular support across the war-weary country, especially among the Pushtuns, led to a series of successes, and by 1996 the Taliban had taken Kabol. The harsh Taliban version of *Sharia*, however, was uncompromising: restrictions on the activities of women ended their professional education, employment, and freedom of movement; the lives of children were even more devastatingly impacted; public stonings and amputations rigidly enforced the Taliban's code.

Afghanistan now became a haven for Islamic revolutionary groups whose agendas and goals went far beyond those of the Taliban: they plotted attacks on Western interests throughout southwestern Asia and threatened regimes they deemed to be compliant with Western priorities. Taliban-ruled, cave-riddled, remote and isolated Afghanistan turned out to be an ideal refuge for these outlaws. Already in possession of arms and ammunition (Soviet as well as American) left over from the Cold War, they also took advantage of Afghanistan's huge, thriving, illicit opium trade. Afghanistan has long served as the world's foremost producer of opium, accounting for more than three-quarters of the global heroin supply, and much of this drug revenue was quickly diverted into the coffers of the conspirators.

In 1996, a terrorist organization named **al-Qaeda 18** took root in the country, an expanding global network that would further the aims of the once loosely allied revolutionaries. Afghanistan became al-Qaeda's headquarters, and Osama bin Laden, a notorious Saudi renegade, its director. Al-Qaeda's most fateful operation came with the suicide attacks of September 11, 2001, on New York and Washington, DC, which destroyed the World Trade Center's twin towers, damaged the Pentagon, killed over three thousand people, and inflicted massive psychological damage. Several weeks later, U.S. and British forces, with the acquiescence of Pakistan, attacked both the Taliban regime and al-Qaeda in Afghanistan.

© Aziz Sana/NurPhoto/Corbis Images

Local farm workers harvesting opium sap from a poppy field in southern Afghanistan's Kandahar Province in the spring of 2015. This country is by far the world's largest producer of opium (the raw material of heroin). Cultivating these poppies is considerably more profitable than raising legal crops and is believed to involve various networks of organized crime, including al-Qaeda.

© J. Nijman, P. O. Muller, H. J. de Blij, and John Wiley & Sons, Inc.

FIGURE 8-15

The U.S.-Led Invasion of Afghanistan and Its Aftermath

The American invasion of 2001 initially appeared to be successful. The Taliban were swiftly driven off, a more representative government was taking shape, local warlords were co-opted or sidelined, Pushtun and other refugees were streaming back into Afghanistan from Pakistan and elsewhere, and life in Kabol and other urban centers returned to a semblance of normal. By 2004, Afghanistan had even held an election that produced a representative government. It took nearly a decade of U.S. intelligence work, but in 2011 Bin Laden was finally discovered and eliminated in an American military raid—not in Afghanistan but in the city of Abbottabad in neighboring northern Pakistan. With al-Qaeda markedly weakened, and having

allowed for a transitional period in which to fully train Afghan security forces, the United States at the end of 2014 withdrew the last of its combat troops from Afghanistan (a residual U.S. advisory force departs in 2016).

During that multiyear transition, American advisors found it extremely difficult to foster a workable consensus among various Afghan factions in order to build a more democratic state. One serious problem was the country's economy, which had been all but shattered by decades of unremitting conflict and war, forcing the government to heavily depend on foreign aid. Gauging the new status quo, by mid-2015 Taliban fighting units had regrouped into smaller militias and were again attacking security forces, government facilities, and

Western compounds. Much of this took place in the far north adjacent to Tajikistan (Fig. 8-15), which may well be a testing ground before crossing the Hindu Kush and resuming hostilities in and around Kabol. Although al-Qaeda in Afghanistan had been dealt a stunning blow at the core, the organization has since proven resilient. As discussed in Chapter 6, by 2013 it had branched out beyond Afghanistan to become a major player in unstable countries, such as Pakistan, as well as in the now-failed states of Somalia, Yemen, and Libya.

Pakistan

In the Indus Valley of Pakistan lay South Asia's earliest urban civilizations, whose innovations diffused southeastward into the massive triangular peninsula. Today, this elongated river basin contains South Asia's Muslim frontier, contiguous to the great Islamic realm to the west and irrevocably linked to India's enormous Muslim minority to its east. Pakistan's cultural landscapes bear witness to its transitional location. Teeming, disorderly Karachi is the typical South Asian megacity. Historic, architecturally Islamic Lahore is reminiscent of the scholarly centers of Muslim Southwest Asia. In Pakistan's east, the boundary of the 1947 partition divides a Punjab subregion that continues well into India—a land of villages, wheatfields, and irrigation canals. In the northwest, Pakistan closely resembles the Afghanistan Transition Zone in its huge migrant populations and mountainous borderland.

The Indus River and its principal tributary, the Sutlej, nourish the ribbons of life that form the heart of this populous country (Fig. 8-15). Territorially, Pakistan is not large by Asian standards; its area is about the same as that of Texas plus Louisiana. But Pakistan's population of 182.1 million makes it one of the world's ten most populous states. Among Muslim countries (its official name is the Islamic Republic of Pakistan), only Southeast Asia's Indonesia is larger.

A Hard Place to Govern

At independence (West) Pakistan had a bounded national territory, a capital, a cultural core, and a population—but few centripetal forces to bind state and nation. The disparate subregions of Pakistan shared the Islamic faith but little else. Karachi and the coastal south, the southwestern desert of Balochistan, the city of Lahore and surrounding Punjab, the rugged northwest along Afghanistan's border, and the mountainous far north remain worlds apart. Urdu is the official language, yet English remains the *lingua franca* of the elite. Several other major languages, however, prevail in different areas (see Fig. 8-5), and ways of life vary enormously as well.

Successive Pakistani governments, civilian as well as military, turned to Islam to provide the common bond that history and geography had denied this state. In the process, Pakistan became one of the world's most theocratic countries. But even Islam itself is not unified in restive Pakistan. Roughly 80 percent of the people are Sunni Muslims, and the Shia minority accounts for most of the remaining 20 percent. Sunni fanatics intermittently attack Shi'ites, leading to retaliation and establishing grounds for subsequent counterattacks.

To govern so diverse and fractious a country would challenge any political system, and so far Pakistan has failed the test. Democratically elected governments have repeatedly squandered their opportunities, only to be overthrown by military coups. Pakistan's recent economic boom has not filtered down to the poor; literacy rates are not rising; health conditions are not improving significantly; national institutions are weak (for instance, there are only about 2 million registered taxpayers); and one consequence of the global antiterrorism campaign is that Pakistanis, who used to be overwhelmingly secular in their political choices, are now increasingly joining Islamic parties.

Meanwhile, too little is being done to confront a growing water-supply crisis, an insurgency festers in Balochistan, the army is incapable of establishing control over mountainous Waziristan (where al-Qaeda and Taliban militants reign), the issue of Kashmir costs Pakistan dearly, and relations with neighboring India (which if satisfactory would bring enormous benefits) remain conflicted.

The Four Subregions of Pakistan

Punjab constitutes Pakistan's core area, the Muslim heartland across which the post-independence boundary between Pakistan and India was superimposed (Fig. 8-15). (As a result, India also has a State named Punjab on the other side of the border.) Pakistan's Punjab is home to just over half of the country's population; in the triangle formed by the Indus and Sutlej rivers live more than 100 million people. Punjabi is the language here, and wheat farming is the mainstay.

Three cities anchor this core-area subregion: Lahore, the foremost center of Islamic culture in the realm; Faisalabad; and Multan. Lahore, now home to 8.8 million people, lies close to the border with India. Founded around 2000 years ago, Lahore was situated favorably to become a great Muslim center during the Mughal period when Punjab was the main corridor into India. Punjab's relationship with Pakistan's other three provinces is one of the country's weaknesses. Both the governments and residents of those provinces feel uneasy about the dominance of Punjab, the populous, powerful core of the country from which most of the military is drawn.

The lower Indus River is the key to life in **Sindh**, but Punjab controls the waters upstream, which is one of the issues dividing Pakistan (Fig. 8-15). When the Punjab-dominated regime proposes to build dams across the Indus and its major tributaries, Sindhis (who make up almost one-fifth of the national population) are reminded of their underrepresentation in government and talk of greater autonomy. The ribbon of fertile, irrigated, alluvial land along the lower Indus, where the British laid out irrigation systems, makes Sindh a Pakistani breadbasket for wheat and rice. Commercially, cotton is king here, supplying textile factories in the cities and towns (textiles account for more than half of Pakistan's exports by value).

But the dominant presence in southern Sindh is the chaotic, crime-ridden megacity of Karachi (population: 16.6 million), a place of searing contrasts under a broiling sun. With little effective law enforcement, Karachi has become a hotbed of terrorist activity, but somehow the city still functions as Pakistan's (and Afghanistan's) major maritime outlet and the seat of Sindh's provincial government.

As Figure 8-15 shows, the province of **Khyber Pakhtunkhwa** lies wedged between powerful Punjab to the east and troubled Afghanistan to the west, with the territory long known as the *Tribal Areas* intervening in the south. The name "Pakhtunkhwa" connotes "belonging to the Pushtuns," the Afghan-associated tribes that inhabit this subregion.

The Pakistani government's reach into these parts has always been quite limited. The province's mountainous physical geography reflects its remoteness and the isolation of many of its people. Mountain passes lead to Afghanistan; the Khyber Pass, already noted as the historic route of invaders, is legendary (see photo). Coming from Afghanistan, the Khyber's road leads directly to the provincial capital, Peshawar, which lies in a broad fertile valley where wheat and corn drape the countryside. Khyber Pakhtunkhwa is a conservative, deeply religious, militant province, where Islamic political parties and movements are stronger than in any other part of the country.

The fourth subregion mapped in Figure 8-15 is **Balochistan**. By far the biggest of Pakistan's four provinces, it accounts for nearly half of the national territory (not including Kashmir) but is inhabited by only about 14 million people, barely 5 percent of the country's population. For a sense of its terrain, take a look at Figure 8-1: much of this vast territory is desert, with mostly barren mountains that wrest some moisture from the air only in the northeast. Sheep-raising is the leading livelihood here, and wool the primary export. In its northern extremity, Balochistan abuts the Tribal Areas and Afghanistan. The provincial capital, Quetta, lies in this zone. This province, in fact, could easily be called the "South West Frontier Province."

Balochistan is of considerable economic-geographic importance. Beneath its parched surface lie possibly substantial reserves of oil as well as coal, and already the province produces most of Pakistan's natural gas. The seaport of Gwadar on the southwestern coast not only handles raw materials from Balochistan but is also a transshipment point for oil and gas from Iran and the Caspian Basin, destined for markets in East Asia (China was a leading investor in the construction of this port). But the Baloch people are treated poorly by Pakistan's central government—fully 90 percent of them have no energy supply, and eight out of ten do not even have access to clean water. Short-lived local rebellions have flared up here since Pakistan became independent, and during the 2000s they escalated into a serious insurgency led by the Balochistan Liberation Army (BLA), whose attacks on security forces continue.

Prospects of Afghanistan and Pakistan

This western flank of South Asia is the realm's most critical region, today more so than ever before. Afghanistan is a fractured, partially functional state that has proven impossible to control by outside powers for any prolonged period of time. Pakistan's coherence and stability have now become crucial at a time when the global struggle against terrorism continues to entangle its leaders in Western priorities and foreign policies. The role of Pakistan's government in this struggle is disputed and resented by many of its own citizens, who express their distaste by voting for activist Islamic parties and even voicing support for the Taliban resurgence in neighboring Afghanistan. Moreover, militancy and instability are no longer confined to Pakistan's northwest and the far north: key cities such as Lahore and Karachi are now being targeted repeatedly by terrorists (the government blames the Taliban). Meanwhile, relations between Pakistan and India remain tense and at times perilously close to open conflict, with India's regime accusing the Pakistani government of failing to crack down on the militants in Kashmir and other northern flashpoints. The geopolitical situation involving these three states today has been described as a potentially ***deadly triangle***—in which Afghanistan could become the main source of destabilization.

© Barbara A. Weightman

The pivotal Khyber Pass that links Afghanistan and Pakistan across the Hindu Kush (see Fig. 8-15). One of the most strategic passes in the world, it was no easy passage for invading armies in times past. Nowadays, the road and its tunnels facilitate the movement of refugees, drugs, and weapons, which are used by militant separatists on both sides of the border. Pakistan's Khyber Pakhtunkhwa, especially its provincial capital of Peshawar, is a hotbed of such activities.

INDIA: GIANT OF THE REALM

If you have been reading the press and watching television over the past few years, you have seen the increasing attention being paid to India—not just in North America but around the world. Examples of news coverage include the purportedly growing "strategic partnership" between the United States and India, the outsourcing of American jobs to India, the rapid rise of a new Indian middle class, and the emergence of large Indian companies that are making their presence felt around the globe. Certainly, India has the dimensions to make the world take notice. Not only does it occupy three-quarters of the great land triangle of South Asia: India also is poised to overtake China to become the most populous country on Earth before the middle of this century.

Political Geography

A Federation of States and Peoples

That India has endured as a unified state is a politico-geographical miracle. The country contains a cultural mosaic of immense ethnic, religious, linguistic, and economic diversity and contrast; it is truly a state of many nations. Upon independence in 1947, India adopted a democratic, secular, federal system of government, giving regions and peoples some autonomy and identity, and allowing others to aspire to such status.

The map of India's political geography shows a federation of 29 States, 6 Union Territories (UTs), and 1 National Capital Territory (NCT) (Fig. 8-16). The federal government maintains direct authority over the UTs, all of which are small in both territory and population. The NCT, however, includes most of the Delhi/New Delhi urban region and now contains more than 17 million inhabitants. The country's State boundaries reflect the broad outlines of the country's cultural mosaic: insofar as possible, the system recognizes languages, religions, and cultural traditions. Indians speak 14 major and numerous minor languages, and although Hindi is the official language it is by no means universally used (see Fig. 8-5).

As Figure 8-16 shows, the territorially largest States lie in the heart of the country as well as on the massive southward-pointing peninsula. Uttar Pradesh (200 million, according to the 2011 census) and Bihar (104 million) together cover much of the Ganges Basin and also form the core area of modern India. Maharashtra (112 million), anchored by the great coastal city of Mumbai (called Bombay before 1996), also has a population larger than that of most countries. West Bengal, the State that adjoins Bangladesh, contains more than 91 million residents, 15 million of whom live in its urban focus, Kolkata (known as Calcutta before 2000).

These are staggering numbers, and they do not decline much toward the south. Southern India consists of four States linked by a discrete history and by their distinct Dravidian languages. Facing the Bay of Bengal are Tamil Nadu (72 million) and Andhra Pradesh (50 million), both part of the hinterland of the city of Chennai (formerly Madras) that is located on the coast near their joint border. Facing the Arabian Sea are Karnataka (61 million) and Kerala (33 million).

India's smaller States lie mainly in the northeast, on the far side of Bangladesh, and in the northwest, toward Jammu and Kashmir (Fig. 8-16). The map becomes even more complicated in the remote northeast, beyond the narrow corridor between Bhutan and Bangladesh. The dominant State here is Asom (Assam), home to just over 31 million, famed for its tea plantations, and important because its petroleum and gas production amounts to more than 40 percent of India's total. In the Brahmaputra Valley, Asom resembles the India of the Ganges. But in almost every direction from Asom, things change.

North of Asom, in sparsely populated Arunachal Pradesh (1.4 million), we are in the Himalayan offshoots once again. To the east, in Nagaland (2.0 million), Manipur (2.7 million), and Mizoram (1.1 million), lie the forested and terraced hills that separate India from Myanmar (Burma). This is an area of numerous ethnic groups (more than a dozen in Nagaland alone) and of frequent rebellion against the central government. And to the south, the States of Meghalaya (3.0 million) and Tripura (3.7 million), hilly and still wooded, border the teeming floodplains of Bangladesh. Here in the country's northeast, India confronts one of its strongest regional challenges.

India's Changing Map

Devolutionary pressures have continued throughout India's existence as an independent state. In some of the cases, the federal government and the military come down hard on the insurgents, but in others the government has been more inclined to negotiate. In 2000, for instance, three more new States were recognized: Jharkhand, carved out of southern Bihar State on behalf of 18 poverty-stricken districts there; Chhattisgarh, where tribal peoples had been agitating since the 1930s for separation from the State of Madhya Pradesh; and Uttarakhand (originally named Uttaranchal), which split from India's most populous Ganges Basin, core-area State of Uttar Pradesh on the basis of its highland character and lifeways. Most recently, in 2014, the new State of Telangana (India's 29th) was created out of the northwestern sector of Andhra Pradesh. Telangana is an inland area that long saw itself marginalized by coastal elites. The new State contains more than 35 million people, covers roughly one-third of pre-2014 Andhra Pradesh, and its capital is Hyderabad—one of the country's leading technopoles and its sixth-largest city—which will serve jointly as the capital of the new State of Andhra Pradesh until 2024.

During the past several years, there have been troubling signs of yet another challenge to India's federal government: communist-inspired rebellions that seem to be transforming into a coordinated revolutionary campaign. It is known in India as the ***Naxalite*** movement, named after a village in the State of West Bengal where it was founded in the 1960s. Mainly active in India's poorest and most disaffected States—such as Bihar, Jharkhand, and what is now Telangana—the Naxalites appeal to the poor and other minorities (especially tribal people), whose plight is blamed on India's elites and neoliberal economic policies (Fig. 8-17). They blow up railroad tracks, attack police stations, and intimidate villagers. The Indian government has declared the Naxalites to be the country's greatest internal security threat and has embarked on a major counteroffensive. This appears to have had some success because the death toll associated with Naxalite violence has fallen in the mid-2010s.

Communal Tensions

The term **communal tension** **19** (communal disharmony) refers to the several different categories of conflicts that recur among

© J. Nijman, P. O. Muller, and John Wiley & Sons, Inc.

FIGURE 8-16

India's highly diverse sociocultural groups. Most commonly, these conflicts have a base in (politicized) religion, but they can also be caste-based (discussed below).

The Sikhs

The Sikhs (the word means "disciples") adhere to a religion that was created about five centuries ago to unite warring Hindus and Muslims into a single faith. The Sikhs rejected what they considered to be the negative aspects of Hinduism and Islam, and Sikhism gained millions of followers in Punjab and surrounding areas. During the colonial period, many Sikhs supported the British administration of India, and by doing so they won the respect and trust of the colonialists, who employed tens of thousands of Sikhs as soldiers and police officers. By 1947, Punjab was home to a large Sikh middle class. Today they still exert a strong influence over Indian affairs, far in excess of the approximately 2 percent of the population (about 25 million) they constitute.

For a period after India's independence, Sikh extremists created India's most serious separatist problem, demanding the formation

FIGURE 8-17

of an independent state they wanted to call *Khalistan*. The conflict intensified in the 1970s and culminated in an attack by federal military forces on Sikh rebels barricaded in the Golden Temple in the city of Amritsar, Sikhdom's holiest site (see photo). Shortly after this

1984 attack, then Prime Minister Indira Gandhi was assassinated by her own Sikh bodyguards, unleashing a lengthy period of anti-Sikh violence across much of India. Since these dramatic events of the 1980s, both sides have worked to gain better understanding.

The Muslims

Its Muslim minority may constitute less than 15 percent of India's population, but they number over 180 million, slightly more than the total population of Pakistan. Muslims are in the majority in the States of Jammu and Kashmir, Asom, and West Bengal, and their absolute numbers are substantial in the big States of Uttar Pradesh, Bihar, and Maharashtra. To India's great advantage, the Muslim minority is not geographically concentrated, avoiding what could otherwise lead to a dangerous regional secessionist movement.

The status of Muslims within Indian society is not easy to separate from India's relations with Pakistan, from the confrontation in Kashmir, and from acts of Islamic terrorism that have bedeviled India in recent years. As a result, tensions between Muslims and Hindus have been rising. The most serious threat to the better integration

One of India's most famous landmarks, the Golden Temple in the city of Amritsar in Punjab State, that dates back to the late sixteenth century. Sikh places of worship are called *gurdwaras*, and this is the holiest one of all, known to the faithful as Harmandir Sahib ("The Temple of God"). More than 100,000 Sikhs pray here every day of the year. Adherents of all religions are welcome, but visitors must dress appropriately, remove their footwear, and cover their heads.

© Jan Nijman, Photo by Zach Woodward

"It is a Friday afternoon as we make our way through a main thoroughfare in one of Mumbai's biggest slums. Dharavi 'houses' approximately half a million people on less than 2 square kilometers! About 20 percent are Muslims, and the overwhelming majority of all others are *dalits*. Increasingly, the Muslims live spatially segregated in their own tightly knit neighborhoods. There are several mosques but little space inside them. So the men, taking a break from work nearby, place their mats alongside the road and prepare for prayer outdoors. Having a faith is important to people living and working in Dharavi's filthy and impoverished environment, especially for the Muslim minority."

of Muslims into Indian society arises from their comparatively low level of education and inferior economic standing. Official statistics show that less than 4 percent of the country's Muslims have completed secondary school. And according to one recent government report, Muslims overall have now become as poor as the *dalits* who form India's lowest-ranked Hindu caste.

Hindutva

Hindu extremism or fundamentalism may seem like a contradiction in terms, but a movement has come to the fore in this century that seeks to remake India as a society in which Hindu principles prevail. *Hindutva* **20**, or Hinduness, is variously expressed as Hindu nationalism, Hindu heritage, or Hindu patriotism. It has been the guiding agenda for the Bharatiya Janata Party (BJP), a powerful force in national politics as well as in large States like Maharashtra, Gujarat, and Madhya Pradesh. *Hindutva* fanatics want to impose a Hindu curriculum on schools, change the flexible family law in ways that would make it unacceptable to Muslims, constrain the activities of non-Hindu religious proselytizers, and forge an India in which non-Hindus are essentially outsiders.

It is easy to see how Hindu hardliners and Islamic revivalists can fuel each other's agendas and maintain a vicious cycle of conflict. This naturally worries Muslims as well as other minorities, but it also concerns those who understand that India's secularism—its separation of religion and state—is indispensable to the survival of democracy. Moderate Hindus and non-Hindus in India oppose this strengthening of *Hindutva*, whose notions are as divisive as any India has faced. Furthermore, Indian voters have not rushed to embrace this radicalization of Hinduism, a solid confirmation of the continuing robustness and vitality of India's democracy.

The Persistence of Caste

Hinduism's benign and admirable properties combine with a system of social stratification that is generally derided in the West (and by a large number of Indians as well). The *caste system* has its origins in the early social divisions into priests, warriors, merchants, and farmers, and it is also thought to have had a racial basis (the Sanskrit term for caste, *varna*, means color). More specifically, caste became associated with specific occupations, and, over the centuries, its complexity expanded until India had thousands of castes, some of them containing a few hundred members and others containing millions.

Hindus believe in reincarnation, and a person is believed to be born into a particular caste based on his or her actions in a previous life. Hence, it would not be appropriate to counter such ordained caste assignments by permitting mobility (or even contact) from a lower caste to a higher one. Persons of a particular caste could perform only certain jobs and worship only in prescribed ways at particular places. They and their children could not eat, play, or even walk with people of a higher social status.

The *untouchables* occupying the lowest tier of all were the most debased, wretched members of this rigidly structured social system. Indeed, the term "untouchable" acquired such negative connotations that it was replaced on several occasions. Mahatma Gandhi, a powerful critic of the system, introduced the term *harijans*, meaning children of God. But more recently this label was perceived to have a condescending connotation, and it gave way to the term **dalits** **21** (the oppressed), indicating a greater sense of awareness and assertiveness among them. Today, dalits are estimated to comprise about one-sixth of all Hindus; Brahmins, the highest-ranking caste, account for a similar proportion. The rest of the population are members of a wide range of in-between castes.

Untouchability was officially abolished at independence, but the caste system has proven difficult to dismantle. Many dalits have chosen to convert to other religions, most notably Buddhism. Successive Indian governments have now introduced an elaborate system of affirmative action on behalf of *Scheduled Castes* (the official government label for dalits). This effort has had more effect in urban than in rural areas of India. In any case, dalits now have reserved for them places in the schools, a fixed percentage of State and federal government jobs, and a quota of seats in national and State legislatures. These jobs are often highly desirable because they tend to be white-collar and much better paid than those generally available to dalits.

Just how far India's political pendulum can swing was shown in the 2007 provincial legislative elections in Uttar Pradesh State, where a dalit party won an absolute majority and where a woman named Mayawati Kumari was the first dalit to become a chief minister of one of India's major States. It was a stunning victory that made headlines throughout the country and shook up the political establishment, revealing the growing power of the lowest castes and marking a turning point for India's representative government.

Most lower-caste Indians are still faced with very limited opportunities, widespread discrimination, and the most extreme forms of poverty, particularly in less accessible rural areas. In traditional India, caste provided stability and continuity; in contemporary India, it constitutes an often painful and difficult legacy.

Economic Geography

The most commonly cited, and most clearly evident, regional division of India is between north and south. The north is India's heartland, the south its Dravidian appendage; the north speaks Hindi as its second language, the south prefers English over Hindi; the north is bustling and testy, life in the south seems more measured and less agitated.

East and West

But there is another, as yet less obvious, but potentially more significant divide across India. In Figure 8-18 draw a line from Lucknow, on the Ganges River, south to Chennai, near the northern tip of Tamil Nadu State (see Fig. 8-14). To the west of this line, India is showing signs of economic progress, the kind of productive activity that has brought Pacific Rim countries such as Thailand and Indonesia a new life. To the east, India has more in common with the less-promising countries that also face the Bay of Bengal: Bangladesh and Myanmar (Burma). The manufacturing map may seem to suggest that much of India's industrial strength lies near Kolkata, but the heavy industries built here by the state in the 1950s are now outdated, uncompetitive, and in steady decline. The State of Bihar represents the stagnation that afflicts much of India east of our line: by several measures it ranks among the poorest of the 29 States.

Compare this to western India. The State of Maharashtra, the hinterland of Mumbai, leads India in many categories, and Mumbai leads Maharashtra. Many smaller, private industries have taken root here, manufacturing goods that range from umbrellas to satellite dishes and from toys to textiles. Across the Arabian Sea lie the oil-rich economies of the Persian Gulf countries. Hundreds of thousands of workers from western India have found jobs there, sending remittances back to families from Punjab to Kerala. More importantly, many have used their foreign incomes to establish service industries back home. Outward-looking western India, in sharp contrast to the inward-looking east, has begun to establish additional ties to the outside world. Satellite and fiber-optic cable links have propelled Bengaluru (formerly Bangalore) to become the center of a burgeoning software-producing complex that reaches world markets.

Manufacturing and IT Industries

The industrial geography of India is in part a legacy of colonial times, with coastal Mumbai, Kolkata, and Chennai anchoring major industrial zones, and textiles—the entry-level industry of disadvantaged countries—dominating the manufacturing scene (Fig. 8-18). Other industries, such as steel, machinery, and building materials, have become much stronger, and some of the companies involved have already become major global players. For

INDIA: INDUSTRIALIZATION
- Major industrial region
- Secondary industrial region
- Coalfield
- ✿ Iron and steel manufacturing
- IT Information Technology
- 🎥 Motion Picture Industry
- ▲ Gas and/or oilfield
- Railroad

0 100 200 300 400 500 600 700 Kilometers
0 100 200 300 400 Miles

© J. Nijman, P. O. Muller, and John Wiley & Sons, Inc.

FIGURE 8-18

example, the Tata Steel Group ranked eleventh in size among the world's steel-producing companies in 2014.

India's information-technology (IT) industries, clustered in metropolitan Bengaluru (Bangalore), Hyderabad, Mumbai, and Delhi, draw much international attention. Leading Indian software companies such as Infosys, Wipro, and Cognizant are now household names all across urban India. The growth of software and IT services has been spectacular and now accounts for about 8 percent of GDP and no less than one-quarter of merchandise exports. But this sector employs only about 2 percent of the workforce.

What India needs far more, in terms of employment growth, are manufacturing industries that competitively sell goods in the global marketplace, putting tens of millions to work, and transforming the economy. Clearly, India is very different from China. The two may have comparably-sized populations, but the Chinese economy employs about 10 times as many workers in manufacturing and has a much larger urban middle class. India may have experienced spectacular recent growth, but too much of it is concentrated in the IT sector and benefits only a small, highly educated urban elite.

Life Is in the Harvest

Agriculture provides more jobs in India than any other employment sector, and India's fortunes (and misfortunes) remain strongly tied to farming. More than two-thirds of the population still lives on (and from) the land, spread in and around the country's hundreds of thousands of villages. There, traditional farming methods persist, yields per unit area remain among the world's lowest, and hunger and malnutrition still continue to afflict millions even as grain surpluses accumulate. The relatively few areas of modernization, as in the wheatlands of the Punjab, are islands in a sea of agricultural stagnation.

As everywhere, food production is closely tied to climate and physiography, and farm output varies significantly across the country. In India, the monsoon is absolutely critical, its abundance and timing a reliable predictor of the harvest. Figure 8-19 maps India's agricultural geography and reflects the rainfall patterns and monsoonal cycle depicted in Figure 8-3. Rice dominates all along the Arabian Sea-facing southwestern coast (on the rainy side of the Western Ghats) and in the monsoon-drenched peninsular northeast; where drier conditions develop, wheat and other grains prevail.

Even though the country produces an ample variety of crops, yields usually are disappointingly low as a result of inefficient land ownership and small farmers' lack of access to inputs such as fertilizer, irrigation equipment, and machinery. If there is any to sell, getting produce to market is yet another struggle for millions of farmers. Almost half of India's 600,000-plus villages cannot be accessed by truck, let alone automobile, and in this

era of modern transportation animal-drawn carts still outnumber motor vehicles nationwide.

To meet these challenges, India is now engaged in an unprecedented series of infrastructural improvement projects, ranging from highways to state-of-the-art airports. The most ambitious of these megaprojects is the construction of a 5900-kilometer (3700-mi)-long superhighway known as the Golden Quadrilateral. Completed in 2012, this fifth-longest highway on Earth interconnects the four anchors of the Indian urban system (Delhi, Mumbai, Chennai, and Kolkata) in a gigantic circuit that also passes through 15 other major cities (Fig. 8-16). The impacts of this all-important artery are multiple: it is expanding urban hinterlands; commuters are using it to travel farther to work than ever before; several once-remote rural areas now have a link to markets; and it is accelerating the rural-to-urban migration flow that will continue to transform India well into the future.

Urbanization

India is well known for its enormous and teeming cities, but the country is not yet an urbanized society. Only about 32 percent of the population currently lives in cities and towns—compared to an average of close to 80 percent across the developed world. In absolute terms, however, urbanization in India is a massive phenomenon.

INDIA: AGRICULTURE
DOMINANT CROP ZONE

- Rice
- Wheat
- Cotton
- Groundnut
- Corn
- Millet
- Plantation
- Chickpea
- Shifting Cultivation
- Coconut

0 100 200 300 400 500 600 700 Kilometers
0 100 200 300 400 Miles

FIGURE 8-19

© J. Nijman, P. O. Muller, H. J. de Blij, and John Wiley & Sons, Inc.

© Jan Nijman

"With the help of a real estate agent, I toured some of the new residential developments in India's second-largest city, Mumbai. The new urban middle class wants better and more spacious housing, but given the lack of space in the city it has to be a high-rise far away from the center. From a new construction site I looked out onto the northeastern edge of the city, marked by an already completed condominium complex. These homes are of a much better quality than the average dwelling in Mumbai, containing most of the luxuries taken for granted in the United States. Two-bedroom apartments here sell for about U.S. $75,000 and up—a huge sum of money even for India's upwardly mobile, affluent classes and totally beyond the reach of the city's masses. Half of Mumbai's 22 million residents live in substandard housing, and many in the kind of slums you see in the foreground, nestled up against the coveted middle-class residences. As the new middle class takes off, the contrasts are getting sharper."

People by the hundreds of thousands are arriving in the already over-crowded cities, swelling urban India by about 3 percent annually—twice as fast as the country's overall population growth. Not only do the cities attract as they do everywhere, but many villagers are also driven off the land by desperate conditions in the countryside. As these migrants manage to establish themselves in Mumbai or Kolkata or Chennai, they help their relatives and friends to join them in squatter settlements that often are populated by newcomers from the same locality, bringing their language and customs with them and cushioning the stress of relocation.

Figure 8-16 displays the distribution of major urban centers in India. Except for Delhi–New Delhi, the biggest cities have coastal locations: Kolkata dominates the east, Mumbai the west, and Chennai the south. The overriding influence of these coastal cities is a colonial legacy. But urbanization also has expanded in the interior, notably within the core area (symbolized by Delhi overtaking Mumbai in 2011 to become India's largest city). By 2015, the country had more than 50 metropolitan areas containing populations of at least one million—and an urban population totaling just over 400 million (75 million more than the entire U.S. population).

When you arrive in any Indian city, you are struck by the abundance of small shops everywhere—tiny businesses wedged into every available space in virtually every nonpublic building along every street. Even the upper, walk-up floors are occupied by shops, their advertising signs suspended from windows and balconies. What keeps all these small stores in business? Most of them earn very little and can afford to stay open only because they are part of the **informal sector 22**—they are essentially unregistered, pay no taxes, utilize family labor, have been handed down through generations, and keep going because of the strong social and communal networks that characterize India's societal fabric. Moreover, high densities and the relatively low level of spatial mobility of

© Jan Nijman

"As I set foot into this spacious (and partially air-conditioned) mall from the steamy and crowded Andheri Link Road in Mumbai, I thought of the implications—is this the future in a land of small family-owned shops where 'retail' has a different and historic meaning? Infinity Mall, an upscale shopping center opened in 2005 in suburban Mumbai, caters to the megacity's new middle class. My Indian friend Pankaj thinks this is all for the better. 'On the weekends my wife and I often come here. There is a nice food court and the shopping is great. These are the things you Americans want, and these are the things we Indians want.' Malls have sprung up across India's major cities in the 2010s, transforming both urban landscapes and consumer behavior."

many urbanites combine to create substantial demand for the local availability of daily goods and services.

This bustling retail scene notwithstanding, change is now under way throughout urban India. The country's growing consumerism, driven by a rapidly expanding middle class estimated to number 250 million in 2015, has led to the introduction of a new feature on the landscape of major cities—the shopping mall (see photo). As recently as 2000, India did not have a single shopping center. But only five years later the 100th had opened its doors, and by 2014 nearly 600 were in operation. You will see American fast-food restaurants among the establishments in these malls as well as the brand names of numerous other foreign companies, proving that the tide of globalization has already washed up on India's shores.

Nonetheless, if the 20 percent of the Indian population that makes up the middle class can afford to shop at a mall, the remaining 80 percent (who total no less than one billion!) cannot—and this astonishing inequality widens by the day. Can India's ongoing economic transformation run its course without severe social dislocation?

▌ BANGLADESH

On the map of South Asia, Bangladesh looks like another State of India: the country occupies the area of the **double delta 23** of India's great Ganges and Brahmaputra rivers, and India almost completely surrounds it on its landward side (Fig. 8-20). But Bangladesh is an independent country, born in 1971 after its brief war for independence against Pakistan, with a territory about the size of Wisconsin. Today it remains one of the poorest and least developed countries on Earth, with a population of 157 million that is growing at an annual rate of 1.5 percent.

Bangladesh remains largely a nation of subsistence farmers. Only 26 percent of the population lives in urban settlements, and more than half of the workforce is engaged in agriculture. Dhaka, the megacity capital, is home to about 15 million; the cities of Chittagong, Rangpur, Khulna, and Rajshahi are the only other urban centers of consequence. Moreover, Bangladesh has one of the highest physiologic densities (people per unit area of arable land) in the world: 1848 people per square kilometer/4786 per square mile.

A Vulnerable Territory

Not only is Bangladesh a poor country; it also is highly susceptible to damage from natural hazards. During the twentieth century, two of the deadliest weather disasters in the world struck this small country. The most recent of these occurred in 1991 when a cyclone (as hurricanes are called in this corner of the world) killed more than 150,000 people. Smaller cyclones strike several times every year and almost routinely kill anywhere between dozens and thousands of people.

The reasons for Bangladesh's vulnerability can be deduced from Figures 8-20 and 8-1. Southern Bangladesh consists of the vast deltaic plain of the combined Ganges–Brahmaputra river system, integrating extremely fertile alluvial (river-deposited silt) soils that attract farmers with the low elevations that endanger them when the water rises. Moreover, the shape of the Bay of Bengal forms a funnel that steers cyclones and their storm surges of wind-driven water to barrel into the double delta's coast. Without money to build seawalls, floodgates, elevated shelters in sufficient numbers, or even adequate escape routes, hundreds of thousands of people are at continuous risk—with deadly consequences.

The country's relations with neighboring India have at times been strained over water resources (India controls the Ganges River, which is the lifeline of Bangladesh), cross-border migration (8 percent of the population is Hindu), and transit between parts of India across Bangladesh's north (refer to Fig. 8-18 to see the reason).

Limited Opportunities—Creative Development Strategies

Even though geography seems to offer Bangladesh limited options and the country overall remains very poor, some remarkable successes have been achieved here in recent years. Since 1990, life expectancy in Bangladesh has increased by ten years and now exceeds that of India, despite India's considerably higher incomes. Second, primary school enrollment among girls today is double that of 2010, and the female literacy rate is steadily improving. And third, since 1990 infant (< 1 year) mortality rates have been more than halved, and child (< 5 years) mortality rates as well as maternal mortality rates have dropped by about three-fourths. According to these indicators, Bangladesh outperforms India and also fares much better than Pakistan.

Interestingly, these improvements were achieved in the absence of notable economic growth. Indeed, incomes in Bangladesh are considerably lower than in India, and even Pakistan does better in this regard. Bangladesh is an overwhelmingly Muslim society, but in some ways it has also been a very progressive one, particularly when it comes to the role of women. For example, 30 seats in the national legislature are reserved for women. Much of this success is attributed to the role of **non-governmental organizations (NGOs) 24** that, independently of (but supported by) the national government, have promoted programs to improve the quality of life. An important example concerns family planning programs in which women play the lead role (free contraceptives and education). The fertility rate in Bangladesh fell rapidly from 7.0 in 1970 to 3.4 in 1994 to 2.2 in 2014. Consider this: when Pakistan and Bangladesh split in 1971, each had a population of around 65 million; today, their populations are, respectively, 182 and 157 million.

The NGOs also were the key providers of so-called **micro-credits 25**, small loans at favorable terms to the poor, along with guidance and counseling, allowing them to invest in the means to secure a proper and sustainable livelihood. Often, such loans have gone to small farmers to buy a piece of land or construct a farmhouse, or to starter entrepreneurs to buy machinery or transport equipment. The success of micro-credits has spread from Bangladesh around the less developed world and is now widely considered to be a key element in devising economic development strategies.

BANGLADESH
POPULATION
- 50,000–250,000
- 250,000–1,000,000
- 1,000,000–5,000,000
- Over 5,000,000

National capital is underlined

Railroad — Internal division
Road — Forested

POPULATION DENSITY

per sq km	per sq mi
2,000	5,200
1,000	2,600
800	2,100
600	1,600
300	800

FIGURE 8-20

© J. Nijman, P. O. Muller, H. J. de Blij, and John Wiley & Sons, Inc.

THE NORTHERN MOUNTAIN STATES

As Figures 8-1 and 8-14 show, two landlocked countries lie in the central sector of the mountainous zone that walls India off from China: Nepal and Bhutan. Together they constitute a distinct region of South Asia, but both are relatively small (Nepal is roughly the size of Illinois, and Bhutan only one-quarter of that), Importantly, these countries function geopolitically as buffer states between the two Asian giants.

Nepal

Culturally and (until recently) economically, Nepal lies within India's sphere of influence, but China's presence here is being felt ever more strongly. Maoist-communist groups are supported by China but loathed by India (whose Naxalites espouse a similar ideology), resulting in fractious and contentious governance. China, as it does elsewhere in the world these days, is using its growing economic clout in Nepal as leverage to gain greater political influence.

Nepal, located just to the northeast of India's Hindu core area, has a population of 27.8 million of whom 82 percent are Hindu. Nepal's Hinduism, however, is a unique blend of Hindu and Buddhist ideals. Thousands of temples and pagodas, ranging from the simple to the ornate, grace the cultural landscape, especially in the Vale of Kathmandu, the country's core area. The Nepalese are a people of many origins, including India, Tibet, and interior Asia. Even though well over a dozen languages are spoken, most of the people also speak Nepali, a language related to Indian Hindi.

A closer look at Figure 8-1 shows that Nepal contains three geographic zones: a southern, subtropical, fertile lowland called the Terai; a central belt of Himalayan foothills with swiftly flowing streams and deep canyons; and the spectacular high Himalaya itself (topped by Mount Everest) in the north. The capital, Kathmandu, lies in the east-central part of the country in an open valley of the central hill zone.

The country's soaring Himalayan peaks are a world-renowned tourist attraction, but visitor expenditures are relatively modest and tourism is often disrupted by flareups of political instability. This is a troubled country suffering from severe underdevelopment, recording the lowest per-capita income in the realm, even below Afghanistan's. Many young people go abroad to find work, and Nepal depends heavily on remittances. It also faces strong centrifugal social and political forces. Environmental degradation, overused farmlands, soil erosion, and the blight of deforestation scar the highly corrugated countryside. And in the spring of 2015, Nepal's dysfunction became painfully clear when the government proved to be useless in the effort (taken over by foreign relief agencies) to assist the victims of a massive earthquake that killed more than 8600 people, injured thousands more, and left nearly three million homeless (see photo).

Over the past several decades, Bangladesh has become known as the "textile capital of the world," today accounting for nearly three-quarters of all export earnings. Almost half of all industrial workers are employed in this industry, and most are women. Many of the cheaper items of clothing that are sold worldwide under popular labels from big-name marketers are made in factories in Bangladesh that are unsafe and where the minimum wage is little more than one U.S. dollar a day. Western companies impose strict production quotas and deadlines, and factory managers force their employees to work long hours under often dangerous conditions. There are no unions to protect workers; disastrous fires and building failures often make the international news (worst of all was the 2013 collapse of a factory building in the Dhaka suburb of Savar that killed more than 1300 laborers). Globalization has made Bangladesh's garment industry a valuable contributor to the nation's commercial economy, but those low prices you see advertised in Western stores raise painful questions about what constitutes "fair trade."

© Gavin Gough/NurPhoto / CorbisImages

An earthquake registering a magnitude of 7.8 struck central Nepal on April 25, 2015, the country's most severe quake since 1934. It produced widespread destruction and the loss of over 8600 lives as well as thousands of injuries. More than 70,000 dwellings were destroyed in the initial quake and subsequent aftershocks, leaving 10 percent of Nepal's population homeless and without immediate access to food and clean water. These survivors are searching (by hand) for victims trapped in the rubble of damaged buildings in Bhaktapur, an eastern suburb of the capital, Kathmandu.

Bhutan

Mountainous Bhutan, with a population of less than 800,000, lies wedged, fortress-like, between India and China's Tibet (Fig. 8-1). Bhutan has a long history as a constitutional monarchy, ruled by a king whose absolute power was unquestioned by his subjects. But in 2007, the newly-crowned monarch, who had just succeeded his father—and perhaps with an eye on the recent overthrow of the royal rulers in nearby Nepal—decided to order his subjects to vote for a political party in an abruptly created democracy. Thus Bhutan overnight went from an absolute monarchy to a multiparty democracy, and Thimphu, the capital, became the seat of a newly elected

National Assembly. Throughout the mountainous countryside, the symbols of Buddhism, the state religion, dominate the cultural landscape, and the government's development policies emphasize the importance of spiritual fulfillment alongside the satisfaction of material needs.

Forestry, hydroelectric power, and tourism all have much potential here, and Bhutan also possesses ample mineral resources. The country did not open up to foreign visitors until the 1970s and even today tourism is kept within strict limits. Although isolation and inaccessibility continue to preserve traditional ways of life, Bhutan is not immune to the turbulent social forces that permeate this realm. Tensions between the Buddhist Bhutanese majority and a Nepalese Hindu minority have led many Nepalese Hindus to flee across the sliver of India that separates the two mountain states (Fig. 8-1). To date, the governments of Bhutan and Nepal have been unable to work out a solution to this problem.

THE SOUTHERN ISLANDS

As Figure 8-14 shows, South Asia's subcontinental landmass is flanked by several sets of islands. The largest, Sri Lanka, lies just to the southeast of the southern tip of India. The Maldives form an archipelago farther out in the Indian Ocean to the southwest of southernmost India. And the Andaman and Nicobar Islands (which will not be discussed here), a Union Territory of India, mark the southeastern edge of the Bay of Bengal.

The Maldives

The Maldives consists of some 1200 tiny islands whose combined area is just 300 square kilometers (115 sq mi) and whose highest elevation is less than 2 meters (6 ft) above sea level. The Maldives' low elevation has been singled out repeatedly in assessments of the

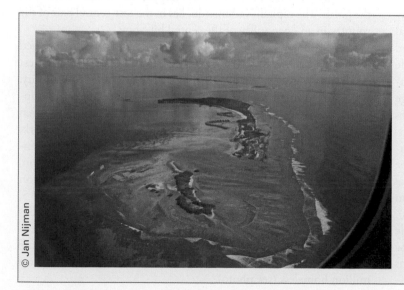

© Jan Nijman

@ FROM THE FIELD NOTES...

"We have just taken off from the Gan airstrip on Addu Atoll, in the southern Maldives. Within seconds, the neighboring island of Villingi comes into full view. It is a resort island, one of many that are vital to the economy of this 'tropical paradise.' Consisting of about 1200 islands with an average elevation of less than one meter (3.3 ft), the Maldives have also been labeled 'ground zero' in debates about global warming and rising sea levels. At the moment it is low tide but you can see how far inland the water reaches at high tide. With stormy weather, flooding is guaranteed. I just spent three days here, and people told me they face growing problems of beach erosion and soil salinity, and that fish are steadily migrating to greater depths farther from the islands. Given the ominous forecasts of global sea-level rise, this may all be gone (or at least depopulated) by the end of the century . . . "

future impact of **rising sea level** 26 caused by global warming (see *Field Note*). The population of 350,000 drawn from Dravidian and Sri Lankan sources is now virtually 100 percent Muslim, with one-quarter of it concentrated on the island of the capital named Maale. This country exhibits the realm's highest GNI per capita (see Fig. 8-8), thanks to the Maldivians having translated their palm-studded, beach-fringed island chain into a tourist mecca that annually attracts tens of thousands of mainly European visitors. The Maldives, however, also faces social and political challenges—its nascent democracy was dealt a blow in 2012 when the first freely elected president was deposed in a *coup* engineered by the country's former dictator and radical Islamists.

Sri Lanka: Paradise Lost and Regained?

Sri Lanka (known as Ceylon before 1972), the compact, pear-shaped island located just 35 kilometers (22 mi) across the Palk Strait from southernmost India (see Figs. 8-1 and 8-14), became independent from Britain in 1948. There were good reasons to create a separate sovereignty for Sri Lanka. This is neither a Hindu nor a Muslim country: nearly 70 percent of its population of 20.5 million are adherents of Buddhism.

The majority of Sri Lanka's population is descended from migrants who came to the island from northwestern India beginning about 2500 years ago. Most of them probably walked across to the island, which was connected to southernmost India by a land bridge (now submerged beneath the Palk Strait) known as *Rama's Bridge*. If you look closely at Figure 8-1, you can see the Sri Lanka-end remains of that land bridge, which was further eroded by a powerful cyclone in the fifteenth century. The migrants brought with them the advanced culture of their source area, building towns and irrigation systems, and introducing Buddhism. Today, their descendants are known as the Sinhalese and speak Sinhala, which belongs to the Indo-European language family of northern India (Fig. 8-5).

The Tamil-speaking Dravidians, who lived on the mainland side of the Palk Strait, arrived much later. During the nineteenth century, the British colonizers brought hundreds of thousands of Tamils to the island to labor on their tea plantations, and they soon became a substantial component of then-Ceylonese society. Not only did the Tamils bring their Dravidian language to the island—they also introduced their Hindu faith. At the time of independence, one year after the India-Pakistan partition, this minority constituted more than 15 percent of the island's population; in the mid-2010s, they comprise just over 11 percent.

When Ceylon became independent, it was one of the great hopes of the postcolonial world. The country had a sound economy plus a democratic government, and it was renowned for its tropical beauty. Its reputation soared when a massive effort succeeded in eradicating malaria, which was followed by a family-planning campaign that reduced the birth rate while the rest of South Asia was experiencing a substantial population explosion. Rivers from the cooler, forested, interior highlands fed the paddies that provided ample rice; commercial crops from the moist southwest paid most

Sinhalese Buddhist monks visit the famous ruins of the twelfth-century Hatadag Temple in the ancient city of Polonnaruwa in north-central Sri Lanka. The temple that once stood here was also known as the "tooth relic temple" because it was said to have treasured an actual tooth of Prince Siddhartha—the Buddha.

© Jan Nijman

of the bills, and the capital, Colombo, grew to reflect the optimism that prevailed.

In the midst of this glowing scenario, the seeds of conflict had already been sown. Sri Lanka's Tamil minority soon began proclaiming its sense of exclusion, demanding better treatment from the Sinhalese majority. Tensions steadily escalated until a full-scale civil war broke out in 1983. Militant activist leaders in the Tamil community now demanded a separate Tamil state encompassing the north and east of the country, coinciding with the Tamil-speaking areas mapped in stripes in Figure 8-5. This violent, devastating conflict endured for 26 years (with a death toll estimated as high as 100,000), finally ending in 2009 when the Sinhalese-dominated government declared victory. In the still bitter aftermath, northernmost Sri Lanka was permitted to have a provincial government dominated by the leading Tamil party—but the central government in Colombo maintains very tight sovereign control.

POINTS TO PONDER ?

- The most skewed sex ratios in the population tend to occur in the most prosperous parts of South Asia.
- According to population growth projections, by 2030 India will be home to the largest national population of Muslims in the world.
- Pakistan, crucially important to the stability of both Southwest and South Asia, is becoming increasingly difficult to govern and faces growing religious fanaticism as well as political turmoil.
- India's IT sector accounts for one-quarter of all exports, but employs barely 2 percent of the national workforce.

China is the world's biggest exporter and home to three of the five busiest container ports in the world: Shanghai, Shenzhen, and Hong Kong (shown here). The other two—Singapore and South Korea's Busan—are also located in the Asian Pacific Rim.

© Jan Nijman

IN THIS CHAPTER

- East Asia: The world's economic powerhouse
- China's breathless urban transformation
- Geopolitics in the seas adjoining China
- China's intensifying regional disparities
- Shenzhen: Wellspring of the Chinese miracle
- The smothering of Tibetan culture
- North Korea's bluster and provocations
- The business of rebooting stalled Japan

CONCEPTS, IDEAS, AND TERMS

Tsunami	1	Economic geography	11
Sinicization	2	Overseas Chinese	12
Hanification	3	Foreign direct investment	13
Asian Tiger	4	Buffer state	14
High-value-added goods	5	Regional complementarity	15
One-Child Policy	6	State capitalism	16
Gender imbalance	7	Conurbation	17
Floating population	8	Demographic burden	18
Hukou system	9	Technopole	19
Special Economic Zone (SEZ)	10		

9
THE EAST ASIAN REALM

FIGURE 9-1

East Asia is a geographic realm like no other. At its heart lies the world's most populous country, the product of what may be the world's oldest continuous civilization. On its Pacific mainland shores an economic transformation has taken shape with no parallel in world history. Its offshore islands witnessed the first use of atomic weapons on civilian populations and the postwar emergence of one of the world's most powerful economies. Few lives in this world were left unaffected, directly or indirectly, by the momentous events that occurred in East Asia over the past two generations. Just look around you. Chinese-assembled smartphones (and so much more), Japanese-made automobiles, South Korean televisions, Taiwanese computers—from toys to textiles and from semiconductors to software—East Asian products fill streets and stores, homes and hotels.

DEFINING THE REALM

It has all happened with astonishing speed. Not that long ago, you were no more likely to find anything useful made in China than you were to buy anything from Russia. But about 50 years ago, things began to change rapidly. Japan led the way, turning its 1945 World War II defeat into postwar economic triumph two decades later. By the mid-1970s, that country's remarkable economic growth—compared to China's seemingly total stagnation—appeared to justify Japan's recognition as a discrete geographic realm, an economic engine for the world and a strong competitor for the top-ranked U.S. economy. On a smaller scale, Hong Kong, South Korea, and autonomous Taiwan soon followed, showing the world what other peoples in East Asia were capable of.

But all this was prologue to the main event—the unprecedented, astonishingly swift rise of China. Less than five years after the 1978 implementation of its game-changing open door and market reform policies, this gigantic country had taken center stage in the realm's economy. Since the mid-1980s, led by burgeoning China, East Asia steadily emerged as the most dynamic realm in the global economy. In the 2010s, it continues to change the world—and in the process is itself undergoing a profound transformation.

THE GEOGRAPHIC PANORAMA

As Figure 9-1 indicates, the East Asian geographic realm forms a roughly triangular wedge between the vast expanses of eastern Russia to the north and the populous countries of South and Southeast Asia to the south, its edges often marked by high mountain

MAJOR GEOGRAPHIC QUALITIES

EAST ASIA

1. East Asia is encircled by snowcapped mountains, vast deserts, and Pacific waters.

2. East Asia was one of the world's earliest culture hearths, and China is one of the world's oldest continuous civilizations.

3. East Asia is the second most populous geographic realm after South Asia; its population remains heavily concentrated in its eastern regions.

4. The People's Republic of China (PRC), the world's largest state demographically, is the current rendition of an empire that has expanded and contracted, fragmented and unified, many times during its long existence.

5. Key components of China's sparsely peopled western regions are now being rapidly developed—not only because they are strategically important to the state, but also because they lie exposed to minority pressures and Islamic influences.

6. An economic transformation launched four decades ago along China's eastern seaboard is today steadily expanding westward, accompanied by a major in-migration of ethnic (Han) Chinese.

7. Widening regional disparities—exacerbated by breakneck, massive urbanization—are straining Chinese society.

8. Japan's economy continues to be ranked among the world's wealthiest, even though it is still mired in a quarter-century-long economic slump, largely the result of the inability of its business culture to adapt to changing regional and global circumstances.

9. Geopolitically, this realm is home to the world's newest superpower as China's economic and political influence is increasingly projected beyond East Asia.

10. The political geography of East Asia contains a growing number of flashpoints capable of generating conflict, including North Korea, Taiwan, and several island groups in the seas adjoining the realm.

ranges or remote deserts. The darker brown on the map designates the highest mountains and plateaus, which create a vast arc north of the Himalayas before bending southward and becoming lower (tan shading) toward Myanmar, Laos, and Vietnam in Southeast Asia. Here in the southwest, where Tibet is located, mountains and plateaus alike are covered by permanent ice and snow, the soaring ranges crumpled up like the folds of an accordion. Three major rivers, their valleys parallel for hundreds of kilometers, disclose the orientation of this high-relief topography. Northward, note how rapidly the mountains give way to broad, flat deserts whose names appear prominently: the Takla Makan in the far west, the Gobi where China meets Mongolia, the Ordos in the embrace of what looks like a huge bend of a river we will learn more about shortly, the Huang He (Yellow River).

In this environment of mountains and deserts, living space is at a premium. And speaking of living space, the green areas on the map, which have the lowest relief and (often) the most fertile soils, are home to the vast majority of this realm's population. Here the great rivers that come from the melting ice and snow in the interior highlands have been depositing their sediment load for eons, and when humans domesticated plants and started to grow crops, this was the place to be. That was thousands of years ago—perhaps as long as 10,000 years—and ever since, this has been the largest human cluster on Earth.

But the East Asian realm is not confined to the mainland of mountains and river basins. You can imagine how its offshore islands were populated: the Korean Peninsula seems to form part of a bridge pointing toward the southernmost large island (Kyushu) of what is today Japan, and from there it seems likely that the early migrants moved farther north until they reached Hokkaido. In warmer times, they may even have ventured beyond, onto the Kurile Islands. And in the south, Taiwan lies even closer to mainland China than Japan does to Korea, whereas tropical Hainan Island, the realm's southernmost extremity, is almost—but not quite—connected to the small peninsula that stretches toward it.

In terms of total land area, though, East Asia is mostly mainland—but the islands and their peoples have played leading roles in forging this realm's regional geography. The waters between mainland and islands (the Taiwan Strait, the South China Sea, the East China Sea, the Yellow Sea, the Korea Strait, among others) also figure prominently in the geographic evolution of this realm. Today, the Japanese and the Chinese are arguing over the ownership of small islands in these waters, specks of land surrounded by possible oil reserves and fishing grounds claimed by both sides. So this realm's map is considerably more complicated than that wedge-shaped triangle in Figure 9-1 initially suggests—even the spellings of many of its contents.

POLITICAL GEOGRAPHY

It is all too easy to refer to China when you mean East Asia because China is East Asia's dominant country, contains more than

85 percent of the realm's population, and has commanded an increasingly prominent role on the global stage. But there are five other political entities on East Asia's map: Japan, South Korea, North Korea, Mongolia, and Taiwan. Note that we refer here to *political entities* rather than *states*. In this realm, the distinction is important. Taiwan refers to itself as the Republic of China (ROC), but it is not recognized as a sovereign state by most members of the international community; the communist administration in Beijing, capital of the People's Republic of China (PRC), regards Taiwan as part of China and as a temporarily wayward province. And North Korea is widely viewed as a rogue state, a brutal and archaic dictatorship that has failed its people terribly. Having compiled one of the world's most dreadful human rights records, North Korea is not even a fully functional member of the United Nations.

Nevertheless, China is the realm's predominant entity—demographically, economically, and politically. It is important to keep in mind, as you read this chapter, that portions of what we map today as regional components of China were not part of the country in the past, and that other areas now lying outside China are regarded by many Chinese as Chinese property (for example, a large sector of the Indian State of Arunachal Pradesh and portions of the Russian Far East). China's imperial past saw the state expand, contract, and then expand again, accumulating unsettled border issues on land as well as at sea. On issues like these, Chinese national sentiments can run quite deeply.

ENVIRONMENT AND POPULATION

To understand the complex physical geography of East Asia mapped in Figure 9-1, it is useful to refer back to Figure G-4 in the introductory chapter. The high snowcapped mountain ranges of the realm's southwestern interior result from the gigantic collision of the Indian and Eurasian tectonic plates (see Fig. 8-2), pushing the Earth's crust upward and creating not only the mountain ranges of which the Himalaya is the most famous, but also popping up the enormous, domelike Qinghai-Xizang (Tibetan) Plateau. In Figure G-4, note the high incidence of earthquakes associated with this collision, converging on a narrow zone of instability that crosses southwestern China and stretches into Southeast Asia. The calamitous 2008 earthquake (magnitude 7.9) in Sichuan Province that killed almost 90,000 people originated in this danger zone but was only one in an endless series that will continue to take its toll (the latest occurring here in April 2013—a 7.0 temblor that killed at least 200 and injured more than 10,000).

From Figure G-4 it is also obvious that the Pacific Ring of Fire, with its lethal combination of volcanism and earthquakes, endangers Japan far more than it does China. This vulnerability became a horrific reality on March 11, 2011, when a monstrous 9.0-magnitude earthquake struck off Japan's northeastern coast near the city of Sendai. Figure 9-2 displays the configuration of the underlying tectonic plates that converge on Japan.

© J. Nijman, P. O. Muller, H. J. de Blij, and John Wiley & Sons, Inc.

JAPAN'S TECTONIC CONFIGURATION

◎ Epicenter of the March 11, 2011 Tohoku earthquake

• Focus of earthquake magnitude 5.0 and higher in the 48 hours before and 12 hours after the Tohoku earthquake

▨ Area of strong shaking

FIGURE 9-2

The Pacific Plate moves westward at an average of 7–10 centimeters (3–4 in) per year, and the only way it can do so is by pressing forward (or subducting) *beneath* the North American Plate (Japan sits atop the western tip of the latter). It was the most powerful Japanese earthquake ever recorded, and one of the five biggest in the world since 1900. Even worse, this massive temblor and its torrent of violent aftershocks triggered a series of devastating **tsunamis** **1** (seismic sea waves) that swept across the narrow, densely populated coastal plains that hug the shoreline of Honshu (Japan's largest island) north of the Tokyo area. And as if the destruction and death left behind was not enough, the leakage of radioactivity from a heavily damaged nuclear-power complex near Fukushima south of Sendai (see photo and Fig. 9-2) may have caused serious future health problems in the local population. Now officially known as the Tohoku earthquake, its final death toll was reckoned to be about 21,000.

Japan's vulnerability to such disasters results from a dangerous combination of circumstances: it is located in a particularly active tectonic-plate collision zone, and many of the country's habitable (and most densely populated) areas are confined, low-lying plains on the islands' eastern coasts open to flooding by Pacific tsunamis. Not surprisingly, the location of most Japanese nuclear power plants along these susceptible shorelines has now become a hotly debated issue.

Looking at the layout of East Asian climates (Fig. 9-3, left map), we should not be surprised that the western and northern sectors of this realm are dominated by conditions that do not

favor substantial population clusters. Permanent snow and ice cover much of the area mapped as *H* (highland climates), including Tibet (Xizang) and Qinghai. Northward, the dry *B* climates (desert and steppe) prevail because this vast expanse lies about as far from maritime influences (and moist air masses) as you can get in Asia. Mongolia is one of the driest countries in the world, but even here—and also in frosty Tibet—there are places where people manage to eke out a living. But the map leaves no doubt as to why most East Asians congregate in the eastern segment of this realm.

When we compare the climates prevailing in the East Asian realm to those familiar to us in North America (Fig. 9-3, right map), it is immediately obvious that the *C* or humid-temperate climates are more extensive in the United States than in East Asia. Note especially the comparative location of the milder *Cfa* climate, which in the United States extends beyond 40° North latitude up to New England, but which in China yields to colder *D* climates at a latitude equivalent to Virginia's. Thus the capital, Beijing, has the warm summers but long, bitterly cold winters characteristic of *D* climates. Take a closer look at Figure 9-3 and it is clear that both the Korean Peninsula and the Japanese islands lie astride this transition from *C* to *D* climates. South Korea is significantly milder and moister than North Korea, and Japan's largest island (Honshu) is temperate in the south but cold in the north. Not unexpectedly, the least densely populated major island of Japan is northernmost Hokkaido, where the climate is quite similar to that of northern Wisconsin.

Comparing just the United States and China, observe that whereas *C* or *D* climates prevail over more than half of the United States, these climates predominate over less than one-third of China—even though the United States has only 316 million people compared to China's 1.4 billion. That is what makes the

DigitalGlobe via Getty Images, Inc.

Satellite image of the heavily damaged Fukushima Daiichi nuclear power complex, three days after the catastrophic Tohoku earthquake and tsunamis struck northeastern Japan in 2011. This strongest earthquake ever recorded in the country triggered the meltdown of three of the facility's six nuclear reactors. Four years later, radiation was still being released into the air and ocean, continuing to threaten distant as well as local areas. Perhaps the longest-lasting impact of this disaster is the opposition to nuclear energy that continues to grow in Japan, Germany, and other key countries around the globe.

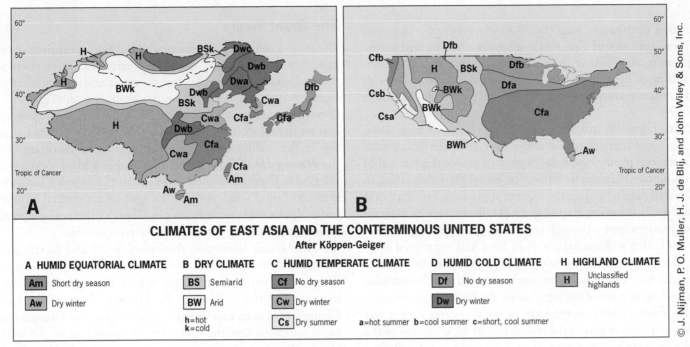

FIGURE 9-3

CLIMATES OF EAST ASIA AND THE CONTERMINOUS UNITED STATES
After Köppen-Geiger

A HUMID EQUATORIAL CLIMATE
| Am | Short dry season |
| Aw | Dry winter |

B DRY CLIMATE
| BS | Semiarid |
| BW | Arid |

h=hot
k=cold

C HUMID TEMPERATE CLIMATE
Cf	No dry season
Cw	Dry winter
Cs	Dry summer

a=hot summer b=cool summer c=short, cool summer

D HUMID COLD CLIMATE
| Df | No dry season |
| Dw | Dry winter |

H HIGHLAND CLIMATE
| H | Unclassified highlands |

EAST ASIA: POPULATION DISTRIBUTION, 2016
One dot represents 75,000 persons

0   600   1200 Kilometers
0   300   600 Miles

Longitude East of Greenwich

FIGURE 9-4

population distribution map (Fig. 9-4) so noteworthy: the overwhelming majority of East Asia's inhabitants are located in the easternmost one-third of the realm's territory, creating the largest and most densely settled population cluster in the world, which is mainly associated with the limited green-colored area in Figure 9-1.

Several times in earlier chapters of this book we have noted the capacity of humans to live under virtually all environmental circumstances; technological developments enable the survival of year-round communities in Antarctica, on oil platforms at sea, in the driest of deserts, and on the highest of plateaus. But the world population distribution map (Fig. G-7) still reminds us how we got our modern start—through crop-raising and herding. The fertile river basins and coastal plains of East Asia supported ever-larger farming populations, whose descendants still live on that same land: for all its ongoing industrialization and urbanization, more than 40 percent of the population of China remains rural to this day. Environment, in the form of terrain, water supply, soil fertility, and climate played the crucial role in the evolution of the population distribution displayed in Figure 9-4—and will continue to do so for centuries to come.

The Great Rivers

China is in some respects the product of four great river systems and their basins, valleys, deltas, and estuaries. These rivers and their tributaries are visible, but do not stand out as clearly in Figure 9-1 as they do on the physiographic map (Fig. 9-5). So important are they as shapers of the realm's regional geography that we should acquaint ourselves with them here. Of the four, the two in the middle are in many ways the most important of all: the *Huang He* (Yellow River) that makes a huge loop around the Ordos Desert and then flows across the North China Plain into the Bohai Gulf, and the *Yangzi* (often spelled Yangtze), probably the most famous river in China's historical geography, known as the *Chang Jiang* (Long River) upstream.

As the map illustrates, the Yellow River and its tributaries form the sources of water essential to the historic core area of China, the North China Plain, where the capital, Beijing, is located. The Yangzi River is the primary artery of the Lower Chang Basin; at its mouth lies China's largest city, Shanghai, and in its middle course the water flow is controlled by the world's biggest dam (see photo). Both the Huang and the Yangzi originate in the snowy highlands of the Qinghai-Xizang (Tibetan)

FIGURE 9-5

Plateau, a reminder that these remote environments are critical to hundreds of millions of people who live thousands of kilometers away.

The other two rivers have much shorter courses, but the **Pearl** River outlet of the one in the south, the **Xi Jiang** (West River), forms an estuary that has become China's (and East Asia's) foremost hub of globalization. This is where you find Hong Kong and, right next to it, the fastest-growing major city in the history of humankind—Shenzhen.

Finally, the northernmost of China's four major rivers, the **Liao** River, originates near the margin of the Gobi Desert and then forms an elbow as it crosses the Northeast China Plain to reach the Bohai Gulf flowing southward. Here the climate is much colder, lowlands scarcer, agriculture lagging, and population smaller than in the more southerly river basins, but mineral resources create opportunities for mining and industry. Each of East Asia's river-based population clusters has its own combination of potentials and problems.

But even these major river systems, together with China's lesser ones, are increasingly pressed to satisfy the country's skyrocketing demands for water. In order to meet these urgent needs, China has embarked on the massive South-North Water Transfer Project to bring new supplies to its thirsty northeastern core area, especially metropolitan Beijing and Tianjin (Fig. 9-5). When completed around mid-century, the entire project will annually divert more than 40 billion cubic meters of water from the Chang River to the north. The Eastern Route (which follows China's Grand Canal) opened in 2013, and the Central Route was completed one year later; but the Western Route, to traverse the rugged highlands of the upper Chang and Huang basins, is barely beyond the planning stage and will not be operational for at least another three decades. This gargantuan project, however, comes with a price tag that is not only monetary: the reservoirs and canals along the Central Route are expected to submerge more than 300 square kilometers of land, forcing well over 300,000 people from their homes in Henan and Hubei provinces.

Along the Coast

East Asia's Pacific margin is a jumble of peninsulas and islands. The Korean Peninsula looks like a near-bridge linking Asia to Japan, and indeed it has served as such in the past. The Liaodong and Shandong peninsulas protrude into the Yellow Sea, which continues to silt up from the sediments of the Huang and Liao rivers. Off the mainland lie the islands that have played such a crucial role in the modern human geography of Asia and, indeed, the world: Japan, Taiwan, and China's Hainan. Japan's environmental range is expressed by cold northern Hokkaido and warm southern Kyushu, but Japan's core area lies on its main island, Honshu. As Figure 9-1 shows, myriad smaller islands flank the mainland and dot the East and South China seas. As we will discover, some of these smaller islands increasingly affect the political geography of this realm.

Three Gorges Dam is emblematic of China's contemporary "era of the megaproject." The dam wall rises 180 meters (600 ft) above the inundated valley floor of the Chang/Yangzi River. It is 2 kilometers (1.2 mi) wide and creates a reservoir—largest of its kind in the world—that extends more than 600 kilometers (400 mi) upstream.

© Wen Zhenxiao/XinHua/XinhuaPress/©Corbis

NATURAL RESOURCES

Given that the East Asian realm contains nearly one-fourth of the world's population, it is not difficult to imagine the magnitude of the demand for natural resources here. The world first received notice of this about 125 hundred years ago when Japan's imperial expansion was in part driven by the need for raw materials to feed its rapidly expanding industrial base. Today, East Asia's demand for natural resources is expressed not in imperialism but in the global marketplace: it is driving a commodity boom around the world, from Russia to Subsaharan Africa to Brazil to Australia.

While China was moribund under its early Mao-led communist administration, and before post–World War II Japan embarked on its headlong rush to become a world economic power, East Asia's requirements remained modest by global standards. But then Japan's economic success, followed by China's swift adoption of market economics, created unimagined and unprecedented needs.

Japan, about the size of Montana but with a population of more than 100 million, showed what lay ahead. With limited domestic resources to support large-scale manufacturing, the Japanese set up global networks through which flowed commodities ranging from oil and natural gas to iron ore and chemicals. Urbanizing and modernizing populations demand ever more consumer goods, and Japanese products poured onto domestic as well as foreign markets. Japanese-owned fleets of freighters now plied the oceans, and for a while Japan even became remote Australia's top-ranked customer for commodities.

When China took off in the 1980s, economic geographers cast a wary eye on the geologic map. Until then, China's biggest resource had been its fertile, river-deposited alluvium (silt): despite the communist regime's best efforts, state-run industries planned to satisfy domestic needs were hardly a match for globalizing Japan. Most Chinese were farmers, and staving off famine was an unrelenting preoccupation. But when China opened its doors to the world, and its industries and cities mushroomed, its needs—for oil, natural gas, metals, food, electricity, water—multiplied. Before long, China had replaced Japan as Australia's leading customer. Moreover, Chinese manufacturers and suppliers were now searching for commodities from Indonesia to Iraq and from Tanzania to Brazil. If China holds an advantage, it is in the so-called *rare earth* elements not commonly known, such as thulium (used in lasers), praseodymium (aircraft parts), lanthanum (electric automobiles), and promethium (X-ray equipment). Nowadays, China's deposits account for roughly 90 percent of the global production of these minerals, which also are increasingly used in missile technology and "green" energy applications.

But China clearly needs the world because East Asia's storehouse of other known resources is not particularly favorable (Fig. 9-6). Widely dispersed coal reserves can satisfy the country's expanding coal-fired energy system—but only at the cost of thousands of miners' lives every year. Oil reserves are also widespread,

but they tend to be rather modest and diminishing. Deposits in northeastern and far western China are the largest, and exploration is proceeding offshore. Yet nothing in China compares with the abundant iron ore deposits and plentiful ferroalloys available to Russia when it industrialized to meet the Nazi challenge; similarly, China cannot begin to match the immense gas and oil reserves of contemporary Russia, let alone the gargantuan deposits of Southwest Asia's Persian Gulf region. Hence China competes with Japan for energy shipments from Russia and with other sources for its industrial raw material imports. So in just a few short years, China has become both the world's biggest consumer as well as its leading exporter.

East Asia's ascent to the globalizing world's principal stage has put it in the forefront of regional development, but at a staggering environmental cost in terms of air and water pollution. The East Asian geographic realm is now in the midst of an economic, social, and political revolution, the outcome of which is far from clear.

EAST ASIA: ENERGY RESOURCES AND SURFACE COMMUNICATIONS

- Gasfield
- Oilfield
- Coalfield
- Nuclear power plant
- Major highway
- Road
- Railroad
- Oil pipeline
- Gas pipeline

Sb	Antimony
Cu	Copper
Fe	Iron
Pb	Lead
Mn	Manganese
Hg	Mercury
Mo	Molybdenum
Sn	Tin
W	Tungsten

National capitals are underlined

FIGURE 9-6

© You-tien Hsing.

Xian preservation site

"Xian in north-central China is well known for its long history and the rich endowments of that heritage. Its unearthed terracotta army of life-sized warriors and horses, an enormous assemblage of terracotta sculptures depicting the military forces of Qin Shi Huang (China's first emperor), is a famous UNESCO World Heritage Site. Xian was the capital city of several imperial dynasties, including those of the prosperous western Han (206 BC–AD 9) and the Tang (AD 618–907). In recent times, Xian has been trying to capitalize on that legacy, asserting its status as the "ancient Rome" of China and investing heavily in historic preservation. But like any major Chinese city today, Xian is also motivated to modernize and expand. In the winter of 2014, I visited the city's Daming Palace complex, a newly designated World Heritage property. The preserved ruins of the palace complex, built sometime during the Tang Dynasty of the 7th–10th centuries AD, lie in a spacious heritage park covering more than 3 square kilometers (1.2 sq mi). Nonetheless, this open space has now become completely encircled by a ring of ultramodern, fortress-like apartment towers, which conveys the impression of a continuous mountain range surrounding a basin. Perceptive visitors to this heritage park soon come to realize the irony of the nostalgia for the past juxtaposed against this present-day expression of hopes for future development."

PEOPLES OF THE EAST ASIAN REALM

Among the many peoples encircling the Chinese are not only the Koreans and the Japanese in the northeast, but also the Mongols and Tatars to the north and northwest; the Kazakhs, Kyrgyz, Tajiks, and Uyghurs to the west; the Tibetans and Nepalese to the southwest; and peoples too numerous to identify individually to the south, including both majorities (Burmans, Thais, Vietnamese) and minorities in states that were still forming in South and Southeast Asia.

China is one of the world's oldest continuous civilizations, and the Chinese imperial state can be traced back to at least 1766 BC when the first ruling dynasty came to power. Most important of all was the formative Han Dynasty (206 BC–AD 220), whose legacy was the molding of the Chinese nation and the consolidation of its territorial base; in fact, to this day the Chinese still define themselves ethnically as the *People of Han*. Like all empires, China expanded and contracted over time. But during its final dynasty, that of the Qing (1644–1911), many of the states and peoples mentioned in the previous paragraph fell under Chinese rule, and the effects of **Sinicization 2**—also known as **Hanification 3**—were considerable even if Chinese domination was resented. This expansion, however, would prove to be the emperors' last hurrah: during the dynasty's final decades, European and Japanese imperialists challenged Qing rule, took control over most of the state's core area, ousted the Chinese from much of the periphery, and left the country in chaos, bringing about the end of nearly 3700 years of dynastic rule in 1911.

By that time, the ethnic geography of East Asia had become highly complicated. Even after losing control over peoples from Korea to Vietnam and from Mongolia to Burma (today called Myanmar), the successor Chinese state of the past century still continues to govern numerous minorities. Thus East Asia remains an extremely complex mosaic of ethnicities and languages.

Figure 9-7 highlights the cultural diversity of a realm in which China dominates numerically but infusions from elsewhere are evident. We already know of the Mongols, the Muslim Uyghurs, and the Buddhist Tibetans; but the most varied and most numerous minority groups inhabit the southeastern corner of this realm, from Hainan Island northward to the mouth of the Yangzi River. For example, the Yue language (middle green on the map—it used to be called Cantonese) is the common language in the pivotal Pearl River Estuary. Many of the other minority tongues shown here have links to Southeast Asian languages.

EAST ASIA'S POST–WORLD WAR II DEVELOPMENT

The fall of the 267-year-old Qing Dynasty was followed by four decades of revolutionary upheaval. Much of this chaotic period was marked by the ultimately futile effort of the Nationalists to gain control over all of China. They successfully expelled the Europeans by 1927 and in 1934 drove their chief domestic enemy—the fighters of the strengthening communist movement—into the remote

© J. Nijman, P. O. Muller, H. J. de Blij, and John Wiley & Sons, Inc.

EAST ASIA: ETHNOLINGUISTIC AREAS

SINO-TIBETAN
- Han Chinese
- Tibetan
- Wu
- North Min
- South Min
- Hakka
- Yue (Cantonese)
- Xiang
- Gan
- Thai
- Miao-Yao

ALTAIC
- Mongolian
- Turkic
- Korean
- Manchu (Tungus)
- Japanese

AUSTRO-ASIATIC
- Mon-Khmer

INDO-EUROPEAN
- Tajik

FIGURE 9-7

western interior. But the Nationalists were unable to resist Japan's expanding imperial thrust into China during the 1930s, and they bore the brunt of the devastating full-scale war that broke out in 1937—soon to be absorbed into the much wider conflict of the Asia-Pacific theater of World War II. Although the Japanese lost that war in 1945, much of China now lay in ruins. The well-armed communists chose that moment to burst out of their far western stronghold, and launched widespread attacks on the exhausted Nationalist forces.

Communist China

With the Japanese finally out of the way, civil war resumed with a vengeance between the Nationalists (commanded by Chiang Kai-shek) and the Red Army (led by the chair of the Communist Party, Mao Zedong). After a four-year losing struggle, Chiang's forces were driven off the mainland, with no other option than to regroup on nearby Taiwan as the newly formed Republic of China. Meanwhile, on October 1, 1949, standing in front of the assembled masses at the Gate of Heavenly Peace in Beijing's Tiananmen Square, Mao proclaimed the birth of the People's Republic of China.

Under communism, Chinese society was completely over-hauled. The communist regime, dictatorial and brutal though it was, attacked China's weaknesses on several fronts, mobilizing virtually every able-bodied citizen in the process. Land was confiscated from the wealthy; farms were collectivized; dams and levees were built with the hands of thousands; the threat of hunger for millions receded; health conditions improved; child labor was reduced; literacy was broadened.

At the same time, the regime committed colossal errors and systematically repressed its citizens at an unheard of scale. The so-called **Great Leap Forward** (the propaganda term for what this aimed to accomplish) was a failed attempt at the end of the

1950s to accelerate the growth of industrial and agricultural productivity; it ended as perhaps the worst human-engineered catastrophe in world history, causing between 30 and 45 million deaths, mostly from starvation.

Another calamitous episode of Mao's rule was the so-called *Great Proletarian Cultural Revolution*, launched during his final decade in power (1966–1976). Fearful that Maoist communism was becoming contaminated by Soviet "deviationism" and concerned about his own stature as its revolutionary architect, Mao unleashed a vicious campaign against what he viewed as emerging elitism in society (most likely, it was a strategy of power consolidation in the name of culture). He mobilized young people living in cities and towns into cadres known as Red Guards and commanded them to attack "bourgeois" and "revisionist" elements throughout China, criticize Party officials, and violently root out "opponents" of the system. As a result, thousands of China's leading intellectuals died; moderate leaders and older revolutionaries, including some of Mao's closest allies, were purged and imprisoned; the economy was staggered, especially agriculture, and by the end at least another 30 million had perished. Although the full economic impact of the decade-long Cultural Revolution is still debated, the repetitive political campaigns, purges, and violence created a culture of interpersonal distrust that would take much time to repair. One of those who survived the political upheaval was a Communist Party leader who had himself been purged and thereafter reinstated—Deng Xiaoping. Deng was destined to lead the country into the new era of economic reform following Mao's death in 1976.

Japan's Defeat and Recovery

After Japan's defeat in 1945, the United States forced it to accept a new constitution as well as minor territorial adjustments. The new constitution stipulated that the country could not spend more than 1 percent of its GDP on the military and it had to accept the permanent stationing of American troops on its soil. Most importantly, this changed postwar condition induced Japan to reset its focus from military might to economic prowess.

EAST ASIA'S ECONOMIC TRANSFORMATION

Japan's Pacesetting Achievements

Japan's accelerated economic recovery during the postwar era that led to its rise as a world economic superpower, was one of the greatest success stories of the second half of the twentieth century. Even though Japan had lost a war, only four decades later it had completely reinvented itself as an industrial behemoth, a technological pacesetter, a fully urbanized society, a political power in the new global arena, and one of the most affluent nations on Earth.

By 1980, the U.S. automobile industry had effectively been competed out of the market by the likes of Toyota and Honda. The very same thing happened in the domain of consumer electronics and other high-technology products. Japan did it better and cheaper than anyone else. Now cities everywhere had reliable Japanese cars on their streets; people across the globe adopted Japanese video and portable audio products; laboratories the world over utilized Japanese precision equipment. Japan today continues to be one of the world's richest countries, but its trading dominance—particularly in the sphere of consumer electronics—has decidedly moderated since 2000. Persistent domestic economic problems have contributed to this leveling off, but an even greater challenge was posed by the steady rise of competitors from within the realm.

The Asian Tigers

Japan's sensational development success did not go unnoticed, particularly in the Asia Pacific's smaller-scale, dynamic, upwardly bound economies that were soon being labeled the four **Asian Tigers** **4**: Hong Kong, South Korea, Taiwan, and Singapore. (The first three belong to the East Asian realm and the last to adjoining Southeast Asia.) In the 1960s and 1970s, all four embarked on similar strategies that resulted in rapid industrialization propelled by the attraction of foreign direct investment and the creation of export processing zones for manufacturing **high-value-added goods** **5**, including computers, mobile phones, kitchen appliances, and a plethora of electronic devices. The highly competitive Tigers (nicknamed "The Four New Japans") quickly became formidable exporters to the most affluent markets of North America and Europe, taking full advantage of their large, ultramodern, fortuitously located ports.

Japan with the Tigers in tow shared an important attribute in their export-driven trajectories of industrial upgrading—the strong intervention of the state. National political leaders worked closely with industrial corporations, targeting such strategic sectors as auto-making, microelectronics, and shipbuilding; nationalized financial institutions were tapped to supply the necessary funding that was soon augmented by the infusion of foreign investment capital. Not unlike newly unified Germany's industrialization in the late nineteenth century, strongly propelled by state capitalism, the economic transformation of the Tigers a century later is embodied in the rise of what social scientists would call the "developmental state" of East Asia.

China's Economic Miracle

After Mao's death in 1976, China began a historic metamorphosis that was to have the widest global impact. The essence of China's transformation—at a much grander scale—was comparable to what the Asian Tigers had achieved earlier: the creation of a favorable environment for foreign investment to support the growth of a state-of-the-art manufacturing sector geared mostly toward exports. Chinese wages, at least at the outset, were kept low, and technology

Forbidden City

© Jan Nijman

"I could not quite believe my eyes when I came upon this Starbucks in the middle of Beijing's Forbidden City: this icon of American consumerism, wrapped in ancient Chinese architecture, in the heart of this revered complex, in a country that calls itself communist. It was back in 2004 that the Chinese authorities had agreed to allow the Starbucks café on the premises in return for a hefty financial contribution to ongoing restoration efforts. But not everybody agreed this was a good idea. Several Chinese groups protested against this "'erosion" of Chinese culture, and by 2007 the government felt compelled to reverse its decision. The deal was rescinded and Starbucks was evicted from the Forbidden City. It was a splendid illustration of the pragmatism that marks the culture of the Chinese and permeates their economic policies. Call it capitalism or authoritarianism, but the Chinese prefer to do whatever works given the circumstances, and they have little time for ideological principles."

transfers in the form of training programs were aimed at constantly upgrading the skills of the local workforce. At the same time, political conditions remained stable because in China, more than anywhere else, the (Party-dominated) government maintains very tight control. This was, and remains, a communist state—but one that proved to be extremely adept at understanding how global capitalism operates and how best to put it to use. These days, pragmatism has overwhelmingly become the hallmark of Chinese policies.

Over the past quarter-century, China has emerged as the most dynamic and fastest-growing component of the global economy. To be sure, Japan and the Tigers had also experienced double-digit growth rates, but it is a far different story when the developmental centerpiece is a country of nearly one and a half billion. In 2010, China surpassed Japan to become the world's second-largest economy. Less than five years later, the International Monetary Fund announced in December of 2014 that China had displaced the United States to become the biggest economy of all.

GEOPOLITICS IN EAST ASIA

Sino-Japanese Relations

Understandably, relations between China and Japan have been complex and plagued by problems during the century following the end of Chinese imperial rule. Japan proudly proclaimed to stand for "pan-Asian" ideals, and that is how it legitimized its invasion (1931) and occupation of China (through 1945). But the Japanese committed unspeakable atrocities during their campaign

in China. Millions of Chinese citizens were shot, burned, drowned, subjected to gruesome chemical and biological experiments, and otherwise wantonly victimized.

Decades later, when China's economic reforms of the 1980s and 1990s led to a renewed Japanese presence in China, the Chinese public and its leaders called for Japan to acknowledge and apologize for these wartime crimes against humanity. The unqualified apology the Chinese desire, in word and deed, has not been forthcoming. Some Japanese history textbooks still avoid acknowledging what happened to the satisfaction of the Chinese, and surveys indicate strong public sentiment on this still-sensitive issue.

But now that China has surpassed Japan in terms of economic prowess; now that China has become not only the biggest exporter but also the largest economy in the world; and now that Japan's economic stagnation has entered its third decade—this time it is China that is on its front foot and brimming with confidence. Although the two countries continue to have close economic ties, their diplomatic relations are strained by historical memory, cultural friction, and clashing interests.

One of the new flashpoints to arise in the 2010s involves the Sino-Japanese-Taiwanese dispute over the Senkaku Islands in the East China Sea (Fig. 9-8). These tiny uninhabited islands were seized by Japan in 1895, yet are claimed not just by China, which calls them the Diaoyu Islands but also by Taiwan. It is not so much the intrinsic value of the islands that is at stake here: it is far more a matter of national pride and entitlement. Moreover, some oil and gas deposits have recently been discovered in this sector, and ownership of the islands grants rights to their surrounding territorial waters and what lies beneath them. At any rate, China's heightened belligerence over these 'rocks' since 2013 has brought

EAST ASIA: GEOPOLITICAL FLASHPOINTS

FIGURE 9-8

tightly closed, unpredictable, and brutally repressive regime: its nuclear missiles may be aimed at South Korea, but (theoretically at least) they can also reach China, Russia, and Japan—and perhaps even North America. The North Korea issue has also deepened divisiveness within the realm. South Korea and Japan are diametrically opposed to the regime based in Pyongyang, whereas China takes a decidedly more neutral position. The Chinese also appear to be using North Korea in their dealings with Japan and the United States, since it is widely thought that China is crucial to containing the North Korean threat.

Ever since the end of World War II, Japan has adhered to a constitution that essentially forbids its rearmament and commits it to a relationship with the United States that includes the stationing of tens of thousands of American armed forces on Japanese soil. But Japan's 2009 election campaign brought to the fore an unusually forceful reappraisal of these issues, and public opinion has been shifting toward a stronger military posture and the departure of U.S. troops. One result of the international community's failure to constrain North Korea's nuclear aspirations may be Japan's military revival. The country has gradually increased its military posture, and vast resources have been invested in a sophisticated missile defense system to protect against potential North Korean (or Chinese) attacks.

The 2011 demise of the North's leader, Kim Jong-il, and the accession of his son, Kim Jong-un, coincided with one of the many critical food shortages that have afflicted the North Korean people since 1945. With the country desperately in need of outside aid, in early 2012 it appeared that an agreement had been reached between the new regime and the United States: the United States would supply emergency food shipments, and in return, the American government extracted a promise from the young Kim to end the testing of nuclear missiles. However, the deal collapsed just a few weeks later, following North Korea's failed launch of a rocket capable of delivering a ballistic missile (when a successful launch occurred months later, all U.S.-North Korean talks were suspended).

Shortly thereafter, North Korea's blustering achieved new heights with repeated threats of "thermonuclear war" against the South and even against the United States. That level of bombast has become part of the standard playbook since 2011 and usually peaks at the time of the annual joint military exercises of South Korean and U.S. forces (Fig. 9-8). Given the unpredictability of the inexperienced Kim Jong-un, these threats continue to raise alarms throughout the realm and well beyond.

Chinese-Japanese relations to their lowest level in decades. It may well be that China has the most compelling claims to these islands because eighteenth- and nineteenth-century maps (emanating from both China and Japan) seem to corroborate that they were in Chinese hands before the Japanese takeover.

The Korea Factor

Throughout history, the Korean Peninsula has repeatedly been divided, partitioned, colonized, and occupied. After Japan—which had annexed Korea as its colony in 1910—was defeated in World War II, the Allied powers divided the peninsula for "administrative" purposes. In 1945, the territory north of the 38th parallel was placed under the control of the forces of the Soviet Union; south of this latitude, the United States was in control. But only five years later, communist armed forces from North Korea invaded the South in a forced-unification drive, unleashing a devastating three-year-long conflict that first swept southward across the peninsula, then back northward across the 38th parallel, and finally drew in the Chinese military who pushed the front southward again. By 1953, a military stalemate halted the hostilities, and the Korean War ended at the cease-fire line not far from where the 38th parallel had marked the original 1945 boundary. Ever since, a heavily armed demilitarized zone (DMZ) has more or less hermetically sealed North from South—with the two Koreas, having grown apart, still in danger of renewed conflict.

But this is not just a bilateral issue. The Korean conflict has long had realmwide and even global implications. A major reason is that North Korea's nuclear capability remains in the hands of a

Taiwan: The Other China

Mention the island of Taiwan in China, and you are likely to be greeted with a frown and a headshake. Taiwan, your host may tell you, is a problem foreigners do not understand. Virtually all of the 23.6 million people of Taiwan are Chinese. Taiwan was part of China during the Qing Dynasty. Taiwan was stolen from China by Japanese imperialists in 1895, when it was known as Formosa. Then, when communists and Nationalists were fighting each other for control of mainland China right after World War II, and the communists were about to win, the Nationalists in 1949 fled by plane and boat to Taiwan, where they overpowered the locals. Even as Mao Zedong was proclaiming the birth of the People's Republic of China (PRC) in Beijing, the loser, Chiang Kai-shek, named his regime in Taiwan the Republic of China (ROC)—and told the world that he headed China's "legitimate" government.

The PRC, of course, ridiculed this assertion, but the ROC had powerful friends, especially the United States. Chiang's regime was soon installed at the United Nations in China's seat. Washington sent massive aid to support the island's economic recovery and weapons to ensure its security. While the PRC languished under communist rule, Taiwan—the name commonly employed for the ROC—advanced economically, and over time its political system matured into a functioning (if turbulent) democracy.

But to the PRC, Taiwan is regarded as a "wayward province" that must be reunited with the motherland. When U.S. President Richard Nixon arrived in Beijing in 1972 for a historic visit, Taiwan was a bargaining chip. Soon, the ROC's United Nations delegation was dismissed, and representatives of the Beijing government were seated in its place. Many countries all over the world that had recognized Taiwan as the legitimate heir to China's leadership now swiftly changed sides. Meanwhile, Beijing's leaders set about trying to isolate the ROC, and to a considerable extent they succeeded.

Geography, however, was to intervene. With billions of U.S. dollars in reserves, fruitful connections to Overseas Chinese in Southeast Asia, and its emergence as an Asian Tiger economic powerhouse, Taiwan had some significant cards to play—and in the late 1970s the Beijing regime discovered it could not afford to deny Taiwanese companies permission to exploit opportunities in the PRC's new development zones. And so, via the "back door" of Hong Kong, Taiwanese entrepreneurs built thousands of factories in mainland China, many of them located directly across the Taiwan Strait. Taiwanese businesspeople now pumped hundreds of millions of dollars into China's development boom, and the economies of Taiwan and the PRC found themselves on a path toward ever tighter integration. In 2015, an estimated 42 percent of Taiwan's exports went to China along with substantial Taiwanese foreign direct investment, and nearly one million Taiwanese were living and working (temporarily) on the mainland.

Today, per capita annual income in Taiwan exceeds U.S. $20,000—more than triple that of China and on a par with South Korea. And even though a sizeable majority of Taiwanese oppose reunification with the PRC, few would want to see their economic well-being imperiled by political adventures.

REGIONS OF EAST ASIA

East Asia presents us with an opportunity to illustrate the changeable nature of regional geography. Our regional classification is based on current circumstances. It is anything but static, and we discuss how this framework may change in the foreseeable future. Today, we can identify seven geographic regions in the East Asian realm (Fig. 9-9), the first three of which comprise the People's Republic of China:

1. ***China's Coastal Core.*** Anchored by the country's eastern coastal provinces, this is not only China's core area but also East Asia's most influential region as well as the most dynamic spatial component of the global economy. The development success of this region since 1980 is the biggest miracle of all—the conversion of a repressed economic backwater of more than a billion people into an ultramodern urban and industrial colossus. In this decade, the seaboard component of the Coastal Core has fully consolidated, and its transformative energy is now being redirected westward; the leading edge of this inland penetration of China's core area follows upgraded transportation corridors that lead to the key interior cities of Chengdu, Lanzhou, and Kunming.

2. ***China's Interior.*** This vast crescent of territory centered on Lanzhou, for centuries the gateway to far western China, is the country's fastest growing region today. Propelled by new government policies to reverse the massive and deepening economic divide between the booming Pacific seaboard and the stagnant, poverty-ridden, still-rural Interior, huge investments are pouring into this region to implement a high-priority "Go West" program to counter this uneven development before it triggers protest movements that could threaten the political stability of the PRC.

3. ***China's Western Periphery.*** Two large Autonomous Regions, remnants of Han and Qing Dynasty imperialism, constitute China's western tier. One is Xizang—better known as Tibet—whose small Buddhist population is dispersed across the enormous, high-altitude Tibetan Plateau that is framed on its south and west by the world's mightiest mountain range. The other is even larger Xinjiang, whose vast desert basins and encircling mountains comprise the homeland of the Muslim Uyghurs, a primary ethnic component in the human geography of the vital frontier borderland where China meets Islamic Central Asia.

4. ***Mongolia.*** The desert state of Mongolia forms East Asia's fourth region. Landlocked Mongolia—like Tibet vast, but sparsely peopled and poverty-ridden—is in the midst of a development boom thanks to rich mineral deposits and massive Chinese investment.

5. ***The Korean Peninsula.*** The Korean people form a single nation, but they have been partitioned by ideology since the end of World War II. Hermetically sealed, China-supported North Korea is a vicious communist dictatorship that relentlessly persecutes its citizens. In the starkest possible contrast, Western-backed South

FIGURE 9-9

Korea has evolved into a vibrant democracy as well as an Asian Tiger economic powerhouse.

6. *Japan.* The miraculous postwar economic transformation of this longtime rival of China is one of modern history's most successful development stories. But an inability to adapt to changing circumstances over the past quarter-century has all but ended the country's growth trajectory. Nonetheless, Japanese society still ranks among the world's most affluent, and the restoration of Japan's late-twentieth-century regional and global position is not impossible if certain social, economic, and political challenges can be surmounted.

7. *Taiwan.* This thriving Asian Tiger continues to maintain its own regional identity within the East Asian realm, even as the PRC views the island as a "wayward province" that must be reunited with the mainland state.

Before embarking on our survey of East Asia's regions, we first need to set the stage with a general overview of China as well as a profile of its political-administrative structure, because this structure is critical to the way in which this huge country is governed.

THE PEOPLE'S REPUBLIC OF CHINA

Citizens of the PRC have no doubt: not only does their country still have the largest population of any state in the world ("still" because India is closing in), but theirs is the oldest continuous civilization on Earth. Yet China's territorial size does not match its demography: even with its 1.4 billion people, the PRC is only slightly larger than the conterminous United States (Fig. 9-10). Distances in China are comparable to those in the Lower-48: it is about the same distance from Beijing to Shanghai as it is from New York to Chicago. And as the map shows, east-west distances are similar as well. Latitudinally, however, China extends farther north as well as south of the conterminous United States, creating

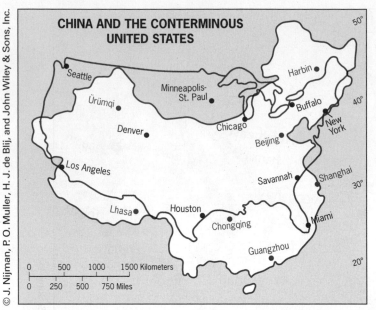

FIGURE 9-10

a greater range of natural environments. In its far northeast, China comes very close to Siberian cold. The extreme south exhibits the Caribbean's tropical warmth—with typhoons (as hurricanes are called in Pacific Asia) to match.

Political-Administrative Divisions

China and the United States may be about the same size, but they have very different governance structures: the United States is a federation, whereas China is an extremely centralized unitary state. For administrative purposes, China is divided into the following units (Fig. 9-11):

4 Central-government-controlled Municipalities (each is known as a *zhi-xia-shi,* or *shi* for short)

22 Provinces

2 Special Administrative Regions (SARs)

5 Autonomous Regions (ARs)

FIGURE 9-11

The four central-government-controlled **Municipalities** (*shi's*) are the capital, Beijing; the nearby port city of Tianjin; China's largest metropolis, Shanghai; and the upper Chang River port of Chongqing in the Sichuan Basin. These megacity *shi's* anchor China's largest internal population clusters (see Fig. 9-4), and direct control over them from Beijing entrenches the PRC government's power.

We should keep in mind that the administrative map of China continues to change—and to pose problems for geographers. The city of Chongqing was made a *shi* in 1996, and its municipal territory was enlarged to incorporate not only the central urban area but an enormous hinterland covering all of eastern Sichuan Province. As a result, the urban population of Chongqing is now officially 30 million, making this the world's "largest metropolis"—but, in truth, the central metropolitan area contains only about 11 million inhabitants. And because Chongqing's population is officially not part of the province that borders it to the west (Fig. 9-11), the official population of Sichuan dropped by 30 million when the Chongqing *shi* was created.

China's 22 **Provinces**, like U.S. States, tend to be smallest in the east and largest toward the west. The territorially smallest are the three easternmost provinces on China's coastal bulge: Zhejiang, Jiangsu, and Fujian. The two largest are Qinghai, flanked by Xizang (Tibet), and Sichuan, China's Midwest. As with all large countries, some provinces are more important than others. Hebei Province nearly surrounds Beijing. Shaanxi Province is centered on the great ancient city of Xian. In the southeast, momentous economic development is occurring in Guangdong Province, whose urban focus is Guangzhou (now China's third-largest city).

In 1997, the British dependency of Hong Kong was taken over by China and became the country's inaugural **Special Administrative Region (SAR)**. In 1999, Portugal similarly transferred Macau to Chinese control, and this former colony, situated opposite Hong Kong on the Pearl River Estuary, became the second SAR under Beijing's administration.

The five **Autonomous Regions (ARs)** were created in order to recognize the non-Han ethnic minorities living there. Some laws that apply to the Han Chinese do not apply to certain minorities. As we saw in the case of the former Soviet Union, however, demographic changes and population movements affect such regions, and the policies of the 1940s may not work in the twenty-first century. Han Chinese immigrants now outnumber several minorities in their own ARs. The label "autonomous" should not be taken literally here, and more than anything else it is a political gesture because Beijing maintains very tight control throughout the PRC. The five Autonomous Regions are: (1) Nei Mongol AR (Inner Mongolia); (2) Ningxia Hui AR (adjacent to Inner Mongolia); (3) Xinjiang Uyghur AR (China's broad northwestern corner); (4) Guangxi Zhuang AR (bordering Vietnam in the far south); and (5) Xizang AR (Tibet, cornerstone of the southwest).

China's "Capitalist" Turn

Although China made considerable progress in developing heavy industry (particularly defense industries) during the 27 years of Mao's regime, this advance cannot begin to be compared with the spectacular achievements of Japan and the trailing Asian Tigers. As the post-Maoist era dawned in the late 1970s, it was clear that the old-fashioned socialist production system had to change because now China was falling ever farther behind, especially in terms of technology.

Taking their cues from the accelerated development of Asia's offshore economic powerhouses, the new leadership saw opportunities that could not even be whispered about under Mao's rule. For example, the Lower Chang Basin—focused on the port of Shanghai (China's largest city) and interconnected by the Yangzi and its tributaries—still retained some of its historical identity and energy. Drab, teeming, and decrepit though it had become, Shanghai had not fully lost its intellectual vigor, artistic individuality, risk-taking entrepreneurs, or even its opponents of communist dogma. And, as always, the city's geography continued to offer immense possibilities. Thus from day one, the new regime in Beijing, led by Deng Xiaoping and his pragmatists, saw in Shanghai what they could not yet foresee in their own backyard. Here was a vibrant hub on the Pacific coast right at the mouth of the realm's greatest river, loaded with talent, and accustomed to taking chances. Moreover, Shanghai's vast hinterland contained much more than a farmscape of flood-prone wheatfields: the Sichuan Basin alone had a population of more than 100 million, growing everything from rice to tea and from fruits to spices.

If China's planners needed any further encouragement, all they had to do was look to southern China's Pearl River Estuary, the main outlet of the Xi River Basin. There, situated at the estuary's mouth, was the burgeoning market economy of Hong Kong, the Asian Tiger still ruled by the British. When Deng took charge, Hong Kong was a not only a successful port city importing raw materials by the shipload, but also disgorged manufactured products that sold on markets all around the world. Hong Kong may have been a British crown colony, but it was Chinese managers and Chinese workers who propelled its beehive economy. If you were a visitor to Hong Kong in the 1970s, you would be taken to a place on a nearby hillside from where you could peer across the fortified border into "Red China"—and observe some villages with duck ponds and rice-producing paddies as well as wooden fishing boats. The contrast was crystal clear.

China's newly formulated policies of the late 1970s, however, did not entail an official departure from communism. This was (and is) still the People's Republic of China: the Communist Party remains firmly in control, the Politburo still calls the shots, Mao continues to be officially revered like a deity, and freedom of speech remains anything but. Yet the Party leadership fundamentally reorganized the way the economy operates in China—beginning with certain vital parts of it. China created highly favorable conditions for foreign investors from around the world with a docile, skilled, hardworking labor force; increasingly efficient facilities; a steadily improving infrastructure; fiscal advantages; and outstanding accessibility vis-à-vis the rest of East Asia, the most rapidly developing regional-scale component in the global economy.

Population Issues

China has long been the world's most populous country, and during the population explosion of the twentieth century it was also one of the fastest growing. Throughout Mao's rule, when China still was a dominantly agricultural society, families were encouraged to produce numerous children, and China grew at a high annual rate of about 3 percent. That was the policy inherited by Deng and his reformers, who immediately grasped that curtailing China's rate of growth was critical to its economic future. Accordingly, the new regime embarked on a severe population-control program that imposed on Han Chinese families (but not minorities) a one-child limit. Enforced by sometimes draconian methods, this policy achieved the desired outcome: by the mid-1980s China's annual population growth had been reduced to 1.2 percent, and by 2014 the rate had dipped to just 0.5 percent (less than half the worldwide average of 1.2 percent).

The **One-Child Policy** 6 had its desired economic impact, but its other results were less salutary. China's is a dominantly patriarchal society, where male children are much preferred; rates of female fetus abortion, infanticide, and abandonment skyrocketed, resulting in a gender imbalance that raises alarms for the future. China's government admits that the PRC has the most serious and prolonged **gender imbalance** 7 on Earth: the latest data for 2014 show that 118 boys were born for every 100 girls. Demographers also calculate that if the one-child policy is not modified, Chinese society will be short some 30 million brides This is already increasing the trafficking of women both within China and from neighboring countries, where widespread resentment of Chinese males pursuing (and sometimes abducting) local females is on the rise.

The One-Child Policy had other outcomes as well. An important one was that China's became an aging population as its proportion of youngsters shrank while the older age cohorts mushroomed. That raised the specter of a population implosion (discussed in Chapters 4 and 5): would there be a sufficient number of workers in the younger age groups to support the ever-expanding older population? For China, this concern was magnified by the prospect that, unlike the situation in European countries and Japan, China might grow old before it grew rich. In the PRC today, public debate over the One-Child Policy is tolerated, reflecting open discussion of it at high levels in the Communist Party. Should China abandon the policy, the fallout will be far-reaching among demographers, especially regarding the projection of world population totals for the rest of the twenty-first century.

China's Urban Transformation

As recently as 1980, barely one-fifth of the Chinese people were urban dwellers; today, more than half live in towns and cities, that milestone surpassed in early 2012. China's urban transformation is both unprecedented and on a scale the world has never seen before. The urban population increased from 18 percent in 1978 to 26 percent in 1990 to 31 percent in 2000. Once into the twenty-first century, however, that proportion skyrocketed to 56 percent in 2015. Much of this growth has been planned and controlled by the government, and that alone is an incredible feat. But it has brought

© Inmaginechina / Corbis

A crowd of Chinese passengers lined up at an entrance to Beijing Railway Station, one of the capital's two main terminals, to return home for the Spring Festival (Chinese New Year). For many migrants, this official week-long holiday is the only one long enough for such a visit; in recent years, this event has become internationally known as the biggest single movement of people on Earth. In 2013, for many this Festival (centered on the lunar-new-year date of February 10th) lasted as long as 40 days, and produced an astounding 3.4 billion trips, an increase of more than 8 percent from the year before. Of that trip total, 3.1 billion were by road, 225 million by rail, and some 35 million by air. In no small way, this army of annual migrants serves to maintain what is left of the ties between China's rapidly developing cities and the far-flung rural hinterlands.

profound changes to Chinese society, uprooting tens of millions, infusing awareness even in remote rural areas (where a quarter of China's villages have been abandoned since 2000), creating inequality both within cities and between city and countryside, and spawning an enormous **floating population** 8 consisting of temporary urban dwellers with restricted residency rights.

The so-called *hukou* **system** 9 is based on residency permits that indicate where individuals are from and where they may exercise such rights as education, healthcare, and housing. The *hukou* tradition dates back to ancient China and is not uncommon in Japan, Vietnam, and other parts of Asia. During Mao's rule, this residency-permit system became far more rigid and was widely used as a tool to control migration, manipulate labor supplies in the urban-industrial as well as the rural-agricultural sector, and minimize government investment in urban services.

As the market reform movement has intensified over the past three decades, so have the liabilities of the *hukou* system. In Shanghai, for instance, millions of migrants continue to pour into China's largest metropolis, their numbers having tripled in the last ten years alone. They now account for over 10 million of the 24 million people who inhabit this mushrooming urban region. They also account for nearly 70 percent of Shanghai's 20–34 age cohort, the critical life-cycle stage during which many couples start a family. Not surprisingly, in 2013 the metropolis contained an estimated 400,000 children below the age of 6 who were officially categorized as migrants lacking a Shanghai *hukou*. With the total of migrant workers now exceeding 300 million, the *hukou* tradition is becoming an ever greater social problem throughout China.

MAP ANALYSIS

FIGURE 9-12   China: Economy and Urban Growth, 2013

This 2013 map shows the location of major cities in China (those with a population greater than 5 million) as well as the size of the economy in each province and autonomous region. China's rate of urbanization in recent years has been so phenomenal that more than 170 cities now contain at least one million people. That is three times as many as in the United States. At the same time, the Chinese economy has steadily expanded to dominate global trade like no other. Generally speaking, what is the geographic relationship between urban growth and economic development in China? How would you explain that relationship? **Access your Wiley PLUS Learning Space course to interact with a dynamic version of this map and to engage with online map exercises and questions.**

These workers maintain that they are major contributors to the expanding economies of the metropolitan centers in which they are employed, yet they are treated as second-class citizens who receive minimal recognition and are denied access to basic local urban services.

China now counts more than 170 cities containing at least one million people, three times as many as the United States. By 2025, there could be dozens more. But the future is uncertain. Will there continue to be enough jobs to employ all these rural-urban migrants? What will be the consequences of China's accelerated program to expand its urban housing stock by millions of new units every year? Can cities that grow at such a frenzied pace be managed in an environmentally sustainable way? Will the Communist Party be able to maintain control over the new concentrations of urbanites with their changing lifestyles, identities, and expectations? For how much longer can the government justify the widening income inequalities within the major cities, and can it meet the steadily increasing demands of industrial labor? Could these enormous economic and social changes at some point force a political transformation—and where might that lead?

CHINA'S COASTAL CORE

China's economic transformation was officially launched in 1980 at a few carefully chosen places. By establishing new economic rules that would initially apply only to certain urban areas on the Pacific coast, most of the rest of the country and the majority of the population would remain comparatively unaffected—at first. Figure 9-13 shows how it was planned to work.

CHINA'S ECONOMIC ZONES
- Special Economic Zones (SEZs)
- Pacific Rim province
- ○ SEZ Capital   ● Open Cities

0 100 200 300 400 500 600 Kilometers
0 100   200   300 Miles

FIGURE 9-13

Special Economic Zones

The government introduced a three-tier system of **Special Economic Zones (SEZs)** 🔟, **Open Cities**, and **Open Coastal Areas**, which would attract technologies and investments from abroad and reshape the economic geography of China's seaboard. In these economic zones, investors were offered numerous incentives that included low taxes, eased import-export regulations, and simplified land leases. The hiring of labor under contract was permitted. Products made in the economic zones could be sold on foreign markets and, under some restrictions, in China as well. Even Taiwanese enterprises could operate here. And profits earned were allowed to be sent back to the investors' home countries.

When the government made the decisions that would reorient China's economic geography, location was a primary consideration. Beijing wanted China to participate in the global market economy, but it also wanted to produce as little impact on interior China as possible—at least in the beginning stages. The obvious answer was to position the Special Economic Zones along the coast. Accordingly, four SEZs were established in 1980, all with particular geographical properties (Fig. 9-13):

1. **Shenzhen**, adjacent to booming, then-British Hong Kong on the Pearl River Estuary in Guangdong Province

2. **Zhuhai**, next to (then still Portuguese) Macau, also on Guangdong's Pearl Estuary

3. **Shantou**, opposite southern Taiwan, a former colonial port, also in Guangdong Province, source of many expatriate Chinese now living in Thailand

4. **Xiamen**, on the Taiwan Strait, another colonial treaty port, in Fujian Province, source of many Chinese now residing in Singapore, Indonesia, and Malaysia

In 1988 and 1990, respectively, two additional SEZs were proclaimed:

5. **Hainan Island**, declared an SEZ in its entirety, its potential success linked to its forward location on the doorstep of Southeast Asia

6. **Pudong**, on the east bank of the Huangpu River directly facing central Shanghai, China's largest city, different from other

© J. Nijman, P. O. Muller, H. J. de Blij, and John Wiley & Sons, Inc.

@ FROM THE FIELD NOTES...

© H.J. de Blij

"Shanghai shows no signs of economic slowdown, even if the environmental consequences are dire. From my vantage point on the top floor of the Jin Mao Tower in Pudong District, buildings beyond the Huangpu River disappeared in a dense smog on a hot spring day that locals described to me as 'pretty clear.' China's prodigious economic growth has made the People's Republic the world's largest polluter of the atmosphere, surpassing the United States. Some projections suggest that, by 2020, China may be emitting twice as much pollution as America, but since China has more than four times the population, the United States would still lead in terms of pollutants per capita. That's just one reason the United States shouldn't invoke the 'China excuse' to weaken its own efforts to reduce pollution from motor vehicles, factories, and other sources. 'Enjoy this view of the Oriental Pearl Television Tower,' Li Sheng told me. 'We're building the Shanghai Tower [China's tallest building as of 2016] right down there where you see the construction site, and it'll be in the way. And it'll be almost twice the height of your Empire State Building.'" [Well, let's hope the view from the top will be clearer.]

Tianjin, the Chinese are building what is planned to be the biggest financial district in the world, *Yujiapu* (see photo).

The grand design of China's economic planners, therefore, was to stimulate economic development in the coastal provinces and to capitalize on the exchange opportunities created by location: (1) the availability of funding; (2) the proximity of foreign investors in Southeast Asia, Taiwan, Japan, and—importantly—still-British Hong Kong; (3) the presence of abundant cheap labor; and (4) the promise of world markets eager for low-cost Chinese products.

To date, the poster child of China's economic planners' grand design is Shenzhen (Fig. 9-14). In the 1970s, the name 'Shenzhen' was hardly known, but within a decade this small fishing village had become the fastest-growing city in human history—from about 30,000 (mainly fishing folk and farmers) in 1980 to 3 million in the early 1990s to more than 10.7 million today. Virtually everyone who lives in Shenzhen comes from somewhere else, and because all are really outsiders they speak to each other in Standard ("Mandarin") Chinese rather than the Cantonese dialect that prevails all around them.

Accompanying this unprecedented growth was the headlong conversion of the surrounding Pearl River Estuary's conurbation into a world-class industrial powerhouse. What caused that to happen? When China opened the Shenzhen SEZ, hundreds of companies moved their factories there from neighboring Hong Kong and offshore Taiwan, capitalizing on lower wages and taxes, looser environmental regulation, and weaker official oversight. At

© Inmaginechina / Corbis

Have you ever seen a construction project of this size? China is studded with these big bets that the country's headlong economic growth will continue apace, but few are more grandiose than the *Yujiapu* financial district (shown here in the fall of 2012). Already nicknamed "China's new Manhattan," this massive complex aspires to be a world-class, highest-order activity hub. It encompasses more than 50 skyscrapers, and is being built within a tight bend of the Hai River on the coastal flats 50 kilometers (30 miles) from central Tianjin. Financed by huge loans from state-owned banks, this immense undertaking is a pet public works project closely associated not only with powerful regional politicians but also Hu Jintao, the PRC's president from 2003 to 2013. Even though Yujiapu is on track to be completed in 2019, it is not at all certain that it will be able to live up to expectations. In fact, recent reports have warned about substantial overcapacity and that many financial firms may well choose to stay in such established hubs as Hong Kong, Shanghai, or nearby Beijing.

SEZs because it was a gigantic state-financed project designed to attract large multinational corporations

And the process continues, with the authorization of the newest SEZ in 2006:

7. **Binhai New Area**, the coastal zone of the northern port city of Tianjin (about 100 kilometers [70 mi] southeast of Beijing), a long-established Open City with considerable foreign investment, now elevated to SEZ status, projected to outperform both Shanghai–Pudong and Shenzhen. On the outskirts of

the same time, the Chinese government built business-friendly infrastructure in the form of state-of-the-art seaport facilities, airports, highways, and railroads. Practically overnight, Shenzhen became the icon of Deng's Open China policy, to be emulated (inasmuch as possible) by the other SEZs.

Think about it: Shenzhen, an urban center less than 40 years old (see photo), now contains 27 percent more residents than New York City, and Guangdong Province has an industrial workforce larger than that of the entire United States! As for the wider Pearl Estuary conurbation, it has become nothing less than the most productive regional economy of its size in the world. Today, it accounts for just about half of all of China's exports, so if you possess a car, television set, laptop computer, smartphone, and/or any similar device, it is certain to contain components that originated in this massive manufacturing complex.

Shenzhen is an ultramodern city of 10.8 million people located in southern China approximately 50 kilometers (30 mi) north of central Hong Kong.

Regional Engines of Growth

None of this accelerated progress would harm Hong Kong's Asian Tiger economy. In fact, it ushered in a new and even more prosperous era for the still-British colony that would rejoin China as its first SAR

in 1997. By the 1990s, Hong Kong had become less of a manufacturing center and far more specialized in international banking, finance, fiscal management, and business services, taking full advantage of the growing skills of its workforce in these spheres. Such occupational transitions are the concern of the field of **economic geography** **11**,

FIGURE 9-14

© You-tien Hsing.

"The Pearl River Estuary of Guangdong Province was the first regional growth engine to emerge under the reforms introduced at the outset of China's post-Mao era. As urban expansion swiftly became a leading development priority, huge tracts of farmland were converted into industrial and residential land uses. Although farmers were compensated for these property losses, their forced relocation often sparked serious social conflict, even violence. In the southern city of Guangzhou, located at the apex of the horseshoe-shaped megalopolis lining the Estuary, I encountered something different. Here I met some well-organized villagers who had bargained shrewdly and successfully with the municipal government, which eventually permitted them to remain on the land in their village homes. As the farmland surrounding their homes was transformed into commercial strips and high-rise buildings, their cluster of houses became a 'village in the city.' Taking advantage of the new premium location of these older homes, the villagers-in-the-city renovated and added several floors to their houses, renting the extra space to migrant workers and vendors. This is one of the largest of these Guangzhou 'villages,' which claims to have a population density more than six times higher than Manhattan's! But look at the inevitable result of cramming together 5- to 8-story buildings to such an extent: with the compressed streets that separate them ranging from only 0.7 to 3 meters (2.3 to 9.8 ft) in width, so little sunlight can enter (even at midday) that this residential environment has acquired the nickname of 'thread-like sky.'"

which focuses on raw-material distributions, historical-cultural factors (such as the aftereffects of colonialism), environmental issues, and particularly the role of spatial economic networks in regional development.

Chinese development strategy was predicated on the understanding that rapid economic growth can be achieved and managed if it is spatially concentrated because local conditions can be manipulated and controlled. Once growth takes off, it can have a stimulating effect on surrounding places, and the benefits can be spread further by a powerful central government. This approach—built on a remarkable blend of capitalist principles and socialist-government control—includes competition among these regional growth poles and cities using tax breaks and other incentives to lure domestic and foreign investors. Thus urban municipalities are allowed important freedoms to stimulate local economic growth. At the same time, the central government encouraged millions of **Overseas Chinese** **12** (most of whom had left China from those very coastal provinces to settle—and often thrive—in other countries) to invest their money in their ancestral homeland, thereby attracting ever more substantial **foreign direct investment** **13** from Japan, the United States, and other key countries.

The Expanding Core

China's economic miracle began with the Special Economic Zones, and the government continues to use this strategy to launch new regional engines of growth through enormous, concentrated investment plans. But economic development has also spilled over into adjoining areas and diffused more widely to other provinces, so an increasing number and variety of localities now carry a "special" designation of one type or another.

We should not get too fixated on the SEZs because some have received far less attention and investment than others. Hainan, for instance, has only experienced minimal increases in manufacturing, mostly in the processing of agricultural products. Furthermore, it does not necessarily require the official status of SEZ to obtain inflows of investment (especially from domestic sources) and to achieve economic growth. Thus the entire tier of seaboard provinces, from Guangdong in the south to Liaoning in the north, now forms the heartland of a highly productive, Pacific Rim-based economy (Fig. 9-13); its relentless growth in recent years has generated the inland expansion of the Coastal Core region as far west as the city of Chengdu in the central Sichuan Basin (Fig. 9-15; see also Fig. 9-9).

Although export industries remain largely clustered along the coast, economic development in this decade is increasingly being steered toward the interior, as shown in the map of GDP growth by province (Fig. 9-16). Much of the growth (resulting from the government's "Go West" program referred to earlier) is occurring around such thriving inland manufacturing centers as Wuhan, Chongqing, Zhengzhou, and Xian, all now engulfed by the expanding Core (Fig. 9-9). As for the leading edge of this inland-moving region, by 2015 three extended lobes had reached the major gateway cities to the western half of China: *Lanzhou* on the Huang River in the north, the jumping-off point for accessing the crucial westernmost ARs of Tibet and Xinjiang; Sichuan's capital, *Chengdu*, in the center; and *Kunming* in the south, routeway hub for travel into mainland Southeast Asia.

We have already noted that the Chinese government is engaged in a race against time to spread the benefits of economic growth from east to west. However, the challenges of mitigating the inequalities produced by this uneven regional development are formidable: an enormous chasm not only persists in the level of prosperity between the high-flying cities and the deprivation-studded rural areas, but also at the national scale between the enlarged Coastal Core and most of the remainder of the country (Fig. 9-15).

CHINA'S INTERIOR

China's Interior region offers a stark contrast to the affluent environs of Shenzhen, Shanghai, or Beijing. Situated between the Coastal Core and Western Periphery, it stretches in a vast crescent from the Russian border in the far northeast to the city of Yumen (midpoint of the Hexi Corridor leading to Xinjiang) in the northwest to the Vietnamese border in the farthest reaches of the south (Fig. 9-9). The Interior can be divided into three subregions: the Northeast, Central China, and the South.

The Northeast

Most of this subregion, formerly known as Manchuria, consists of Liaoning, Jilin, and Heilongjiang provinces as well as the northern prong of the Nei Mongol (Inner Mongolia) AR (Fig. 9-11). It is

© J. Nijman, P. O. Muller, H. J. de Blij, and John Wiley & Sons, Inc.

CHINA: GDP BY PROVINCE, 2014, IN U.S. $BILLION

- More than 400
- 200–399
- 120–199
- 50–119
- Less than 50

Data source: Deutsche Bank

FIGURE 9-15

relatively well endowed with coal and metallic mineral reserves (Fig. 9-17), and Mao's regime made the industrial redevelopment of the war-ravaged Northeast a priority. Large state-owned, heavy industrial enterprises were established here, particulary steel mills. By the late 1950s, it produced more than one-quarter of China's manufacturing output, but its state-run factories never achieved real efficiency. Most of the Northeast's industrial plant was abandoned soon after the reformers came to power, and the 1980s and 1990s were marked by layoffs, factory closures, worker protests, and out-migration.

This decline notwithstanding, in recent years the Northeast has again become a planning priority, and parts of this former 'rustbelt' have changed for the better (Fig. 9-18). Liaoning Province, in particular, has undergone significant development and is now fully integrated into the Coastal Core region—linking Core and Northeast as never before. Indeed, since 2010 the Northeast has grown faster than any other regional economy in China (with the exception of the rest of Inner Mongolia, part of Central China to be discussed next), much of it due to accelerating investment in electronics and the auto parts industry from nearby South Korea and Japan. Undoubtedly, proximity to Beijing and superior coastal access were major factors in this transformation. Liaoning's leading city, Shenyang, today showcases a diversified economy that, besides heavy industry, includes aerospace, electronics, and banking. The Northeast's primary outlet port—Dalian, located where the Bohai Gulf meets the Yellow Sea—has become a leading export focus for trade with Japan and the Koreas.

Central China

The Central subregion of the Interior borders Mongolia in the north, Xinjiang and Tibet in the west, and the Chengdu protrusion of the Coastal Core in the south (Fig. 9-9). The sparsely populated western margins of this subregion are dominated by Tibetan pastoralists. Lanzhou, focal point of the upper basin of the Huang He, is the subregion's leading urban center. This key crossroads city for overland travel to the Western Periphery is also known for its oil refineries and petrochemical industries. Much of the central Interior is arid country that includes parts of the Gobi and Ordos deserts and, farther west, the salt lakes of the Qaidam Basin (Fig. 9-5). But this subregion also includes the middle course of the Huang with its agriculturally productive Loess Plateau (loess consists of fertile, windblown deposits of pulverized rock). Add the water of this great river plus an adequate growing season, and a sizeable population can be supported.

FIGURE 9-16

© J. Nijman, P. O. Muller, H. J. de Blij, and John Wiley & Sons, Inc.

FIGURE 9-17

To the south, Central China includes the northern half of the Sichuan Basin, another highly fertile area, crossed by China's other great waterway, the mighty Chang Jiang. Despite its vulnerability to serious earthquakes (most recently in 2013), this huge natural amphitheater has supported human communities for millennia, and you can readily discern its population cluster in Figure 9-4. The Sichuan Basin, encircled as it is by mountains, is also one of the world's most sharply delineated physiographic regions, and the settlement pattern of its roughly 125 million inhabitants reflects that configuration.

Although there are ethnic minorities in the Interior, the great majority of the people are Han Chinese. Serious regional disparities have long existed between the central interior and coastal China, and the economic gap has widened during the era of market reform. China's Pacific Rim transformation drew tens of millions to the coastal factories and myriad construction sites in what turned out to be the largest internal national migration in human history. Those who stayed behind were too remote from the opportunities introduced by the reform, and now they faced major new

problems in the form of water shortages and the negative effects of rapid farmland conversion to nonagricultural uses. Not surprisingly, protests and social conflicts have steadily risen throughout Central China.

In conjunction with its "Go West" crusade, Beijing's leadership is responding to the development imbalance by extending the privileges enjoyed by the coastal SEZs to the urban centers of the Interior. As workers' wages steadily rise in the seaboard provinces, a growing number of companies are finding it profitable to move their production facilities into this sector of the Interior region, abetted by the new economic zones that operate in cities such as Lanzhou and Kunming. A further incentive is the large-scale construction of public works in Central China, especially the ongoing diversion of waters from the Chang to the increasingly thirsty northern cities of the Core (see Fig. 9-5). Yet another lure involves the rapidly improving linkages to the Coastal Core in the form of thousands of kilometers of new expressways, a steadily expanding high-speed rail network (Fig. 9-17, inset map), and dozens of new or expanded airports.

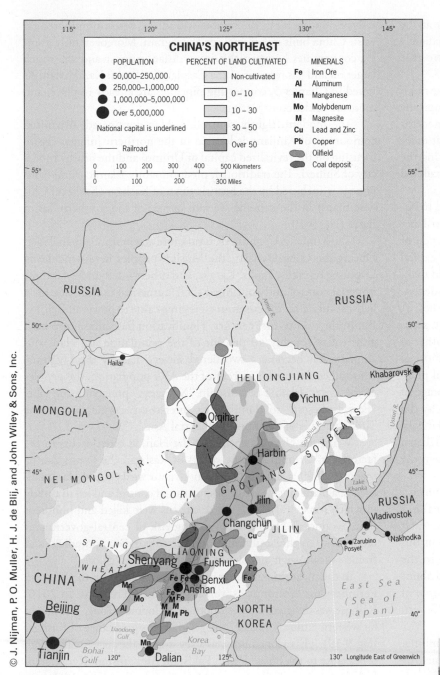

CHINA'S NORTHEAST

POPULATION

- 50,000–250,000
- 250,000–1,000,000
- 1,000,000–5,000,000
- Over 5,000,000

National capital is underlined

— Railroad

PERCENT OF LAND CULTIVATED

- Non-cultivated
- 0 – 10
- 10 – 30
- 30 – 50
- Over 50

MINERALS

- **Fe** Iron Ore
- **Al** Aluminum
- **Mn** Manganese
- **Mo** Molybdenum
- **M** Magnesite
- **Cu** Lead and Zinc
- **Pb** Copper
- Oilfield
- Coal deposit

FIGURE 9-18

The South

The South subregion of the Interior, like its two counterparts, is dominated by Han Chinese, but also includes sizeable concentrations of minorities, especially near the borders with Myanmar, Laos, and Vietnam (Fig. 9-7). Indeed, adjoining Vietnam in the southeast we find the Guangxi Zhuang AR, which was specially created with these minorities in mind (Fig. 9-11).

The South is largely covered by the Yunnan Plateau, source of the streams that feed the Xi Basin and its Pearl Estuary outlet. Much of southeastern China is quite hilly and in places even mountainous. Among other things, this challenging terrain has inhibited overland contact between China and Southeast Asia—an obstacle soon to be

overcome by new routes extending outward from Kunming (see Chapter 10). Yunnan Province itself, in addition to its geopolitical significance, possesses important mineral deposits as well.

CHINA'S WESTERN PERIPHERY

Historically, empires have expanded and contracted, acquiring and losing territory as well as subjects, and leaving their imprints where they once ruled. Russians in Central Asia, British in East Africa, French in Indochina, Japanese in Taiwan all came and went, leaving behind languages, religions, infrastructures, and traditions.

And so it was with China. The Qing Dynasty had forged a sprawling empire covering eastern Asia but lost most of its periphery after 1850. Nonetheless, Figure 9-9 reveals that China is still an empire today. The PRC's beyond-the-Han domain includes a pair of large Autonomous Regions (Xizang and Xinjiang) as well as a third one lining the Mongolian border (Nei Mongol) that is undergoing rapid integration into Han China itself. Even though they are called autonomous regions, the governance structures of China's far western minority entities are fully integrated into the nationwide Party and government hierarchies. Ethnically, the bureaucracies of these ARs may be a joint force of minority and Han administrators, but almost always under the top leadership of a Han.

Xizang (Tibet)

Tibet (called *Xizang* by the Han) is the icebound heartland of Tibetan-Buddhist culture that extends northeastward into Qinghai Province and, most importantly, into a corner of far northeastern India in the Indian State of

The city of Chengdu, capital of Sichuan Province, has grown and modernized at a dizzying pace since 2000, part of China's unbelievably rapid urbanization. Not long ago, this was part of the Interior; now the city has not only been drawn into China's westward-advancing Core region but also forms its spearhead. Traditionally known as one of the country's most laid-back and liveable cities, Chengdu has quickly acquired a new reputation as a pacesetting economic center that specializes in information technology, finance, and biopharmaceutical industries.

© Jan Nijman

Arunachal Pradesh. Tibet shook off Chinese domination at the end of the nineteenth century, but in 1950, almost immediately after the communist regime took control in Beijing, it ordered the Red Army into Xizang to recapture the territory.

Tibetan society had been organized around the fortress-like monasteries of Buddhist monks who paid allegiance to their supreme leader, the Dalai Lama. The Chinese government wanted to modernize this feudal system, but the resistant Tibetans clung to their traditions, and in 1959 the army had to crush an uprising. During the Cultural Revolution, the Chinese destroyed much of Tibet's cultural heritage, looting its religious treasures and works of art.

Under Deng's rule, part of this heritage was returned to Xizang, but Beijing also encouraged Han Chinese to move to Tibet and built "the world's highest railroad" (reaching an altitude of 5072 meters/16,640 feet) across the Tibetan Plateau to the capital of Lhasa (Fig. 9-17). While the Dalai Lama travels the world making the 3-million-plus Tibetans' case and asking not for independence but for genuine autonomy, Hanification continues. A complex set of cultural policies has come to prevail in Tibet: separate school systems, one taught in Chinese Mandarin, the other in Tibetan; dual healthcare systems, one specializing in traditional Tibetan therapeutics, the other in Han (Western) medical practices; and a quota system mandating that a certain percentage of government employees be Tibetan. Nevertheless, driven by the twinned processes of Hanification and modernization, Tibetan culture is existentially threatened. Keep in mind that most Tibetan citizens were born after the PRC seized control, and for them the Chinese presence has become a fact of life.

Tibet today is fiercely controlled by the Chinese government, not only for geopolitical reasons (e.g., the border with India) but also because so many critical parts of China depend on the water provided by the two great eastward-flowing rivers that rise in this soaring highland. In fact, the Chinese government refers to the Tibetan Plateau as "China's water tower." Its resources are so abundant that planners are even considering the diversion of some of these waters to serve the expanding populated areas of the Western Periphery's arid northern interior.

Economic growth in Tibet is strong and incomes are climbing. But popular resistance continues, intermittently accompanied by violence and drama. Since 2008, almost 150 self-immolation suicides have occurred, mainly by Buddhist monks, to protest the Chinese occupation, the marginalization of Tibetan culture, and the callous destruction of the Tibetan environment through mining. Starkly underscoring these grievances is the 2013 landslide disaster at a major copper and gold mining complex near Lhasa that claimed at least 80 lives.

Xinjiang

Xinjiang constitutes the westernmost margin of China's modern empire and is even larger than Xizang (Fig. 9-19). With just under 23 million inhabitants, about 40 percent of them Han Chinese, the Xinjiang-Uyghur AR is even more important than its Buddhist, Tibetan neighbor. Here China meets the peoples of Central Asia and the faith of Islam; here China has highly significant energy

reserves; and here, in the remote, cloudless, desert-dominated far west, China built its original space program. Moreover, only a single country lies between this inner Asian outpost and the much larger oil and gas reserves of the Caspian Sea Basin: Kazakhstan. As noted in Chapter 5, connecting pipelines are already in place, and more are under construction.

The modern, tightly controlled, Han-dominated, urbanized component of Xinjiang is located in the northern Junggar Basin, focusing on the revitalized capital of Ürümqi and the nearby model city of Shihezi. The traditional, predominantly Islamic, rural component is anchored by the historic town of Kashgar (Kashi on Chinese maps) in the extreme western corner of the southern Tarim Basin (Fig. 9-19).

The Xinjiang-Uyghur AR is no more autonomous than Tibet. During the Qing Dynasty, the Muslim peoples here—including Uyghurs, Kazakhs, Tajiks, Kyrgyz, and others—fell under Chinese imperial control and continued their numerical dominance for a half-century after the communist regime came to power in Beijing. But in the twenty-first century, Hanification has shifted into high gear: today roughly 40 percent of the population are Han, and Xinjiang has become a two-tiered society. Led by the Uyghurs, who still constitute 45 percent of the AR's population, the peoples of the lower, Islamic tier disdain "Xinjiang," preferring the label "East Turkestan" to signify their inter-realm connection to the Turkic nations of neighboring Central Asia.

When the Uyghurs encounter the Han, as often occurs in the factories of Ürümqi and the streets of Kashgar, the results can be tragic. In 2009, a skirmish between Uyghur workers and Han managers at a plant outside Ürümqi spun out of control and led to rioting in which some 200 Han Chinese citizens were killed and hundreds more wounded. Despite the growing risk of central-government retaliation, Uyghur militants continue to pursue a sporadic—and fruitless—campaign of violence against Han domination.

In the meantime, Beijing is tightening its grip on the Western Periphery for economic as well as geopolitical reasons. In 2010, Kashgar was designated an "economic development zone," the first in far western China. Consider the geographic significance of Kashgar (Fig. 9-19): it has a long history as a key stop on the ancient Silk Road; it lies close to the borders with Pakistan, Afghanistan, Kyrgyzstan, Uzbekistan, and Tajikistan; and it is already Central Asia's largest marketplace. Thus Kashgar is by far the most effective link between China and this portion of Central Asia, and plans call for its development into a major trading and logistical bridge.

Both Xizang and Xinjiang exhibit the properties of peripheries, where indigenous cultures and traditional economies are increasingly intertwined with expanding national interests and global systems. As a result, the widening disparities between the Western Periphery and the Coastal Core at the national scale are being reproduced at the autonomous-region level, propelled by a deepening, core-periphery-type schism between the successful Han and the struggling peoples who constitute the local social mosaic.

Even though China has made impressive progress and now holds a position of considerable power on the global stage, it still faces some formidable internal challenges. First, rising (regional) inequality will

FROM THE FIELD NOTES...

© Jan Nijman

"China's urban transformation began about 30 years ago in its eastern coastal provinces. Today, just one generation later, it has not only spread to the surrounding provinces of the interior but also to the remote autonomous regions of the far west. In early 2015, I took this photo of the large-scale construction of apartment buildings on the outskirts of Tibet's capital, Lhasa. China now controls Tibet for its crucial water supplies as well as for obvious geopolitical reasons. To these we can also add Tibet's economic contributions: the PRC's government is steadily increasing its investments in and around Lhasa, and the growth rate of the local economy now matches those of China's fastest-developing urban regions. The official Chinese presence in Tibet is symbolized by the red flags near the right margin of the photo. The developers and construction companies are all said to be Chinese, and so will the future owners of these apartments. Racial/ethnic segregation has been prevalent in Lhasa since China's takeover in 1959, and the ongoing developments are very likely to intensify the longstanding social divide. The Han Chinese influx into this urban area is expected to continue well beyond 2020, and within a decade Lhasa's Tibetan population is likely fall below the 50-percent level."

be very difficult to rationalize by a regime that still pays homage—at least on paper—to socialist principles. "Let some people get rich first," were the famous words of the great reformer Deng Xiaoping nearly four decades ago. Yet now it seems that getting rich has become an obsession for many but is attainable for relatively few. That trend is evident in measures of income disparity: China's global rank is comparatively high—behind South Africa and Brazil but well ahead of the United States. Second, there has been a major shift in the constituencies of the Communist Party: peasants and workers, the traditional pillars of support for the regime, have grown increasingly critical as they feel ever more left behind. With the Party today most strongly supported by the urban middle class and the *nouveau riche*, how much longer can it maintain the ideological foundation of its policies? And third, the proliferation of the Internet and social media is providing new channels of information flow and opinion expression. Although most online forums are nonpolitical and the "Great Firewalls" were imposed to limit and police Internet communication, this powerful information-transmission technology is bound to become a new platform of political contestation in China.

MONGOLIA

We now turn to the four East Asian regions beyond China and begin by looking to the north, to Mongolia, where the effects of the realm's economic transformation have only recently arrived. This immense, landlocked, isolated country wedged between China and Russia, with an area larger than Alaska but a sparse

population of just 2.8 million, suggests a steppe-and-desert-dominated vacuum between two of the world's most powerful countries (Fig. 9-9).

Mongolia used to be the domain of a powerful people who, many centuries ago, swept westward to challenge the Russians and southeastward to rule China. But in more modern times, Mongolia became a weak and vulnerable country whose 800,000-plus herders and their millions of sheep today follow nomadic tracks along the fenceless fringes of the immense Gobi Desert, where Siberian cold periodically causes severe human and livestock losses. During the Soviet era, the location of the capital of Ulaanbaatar symbolized the country's security against Chinese encroachment, and Mongolia functioned as the typical **buffer state 14** that separated two political adversaries.

In the 2010s, Chinese involvement and investment are soaring, and much of this revolves around Mongolia's enormous storehouse of raw materials. Major deposits of gold, copper, and coal have been found, and exploration by large foreign companies is under way. Mongolia was the fastest-growing national economy in the world in 2011 (nearly 18 percent), has continued to surge since then, and the International Monetary Fund predicts double-digit annual growth for years to come. But the country has now become overly dependent on a small number of gigantic foreign mining companies, and negotiations concerning taxes and regulations have proven to be a cause of economic volatility. The main consumer of Mongolia's commodities is China, and Chinese investments continue to pour in, largely for infrastructural projects. It is not clear whether ordinary Mongolians, among the poorest people in Asia, will benefit from this bonanza anytime soon; that uncertainty has

FIGURE 9-19

not restrained some economists, who speculate that Mongolia is evolving into another Qatar—where a small population grew rich from its enormous natural endowments.

THE KOREAN PENINSULA

Take another look at Figure 9-9 and you will see one prominent (bicolored) peninsula that seems to reach out from the East Asian mainland toward the islands of Japan. As Koreans will tell you, this isn't just another peninsula. This is a place where human geography

took momentous turns as a corridor of migration, an incubator of culture, a cauldron of warfare, and a cradle of economic miracles. Today, this Idaho-sized tongue of land has a population of 75 million (Idaho: 1.7 million) that is divided between two states steeped in the same linguistic and cultural traditions.

The Korean Peninsula has been a turbulent stage. Thousands of years ago, even before there was a China, lower sea levels made the crossing from Korea to Japan possible and migrants from the mainland began to challenge the even earlier inhabitants of the islands. Later, long-stable states, always influenced by rising China, forged a durable cultural geography with distinctly Korean attributes including ethnicity, language, and customs. Then, more than a century originated as a brutal capitalist dictatorship but transformed itself into both a free-wheeling democracy and an Asian Tiger. The partition of this region remains at the heart of what many still call the world's saddest politico-geographical tragedy.

North Korea

As the map shows, North Korea is territorially larger than South Korea, but its population of 24.9 million is half the size of South Korea's. Seven decades of the most barbaric variety of communist rule have turned North Korea into one of the poorest, hungriest, and most regimented countries on Earth, but with one of the largest standing armies as well as advanced nuclear weapons and missile technologies. A regime that incarcerates its citizens for the slightest offenses operates a gulag of prison camps reputedly worse than that of the Stalinist-era Soviets, starves its people as punishment, and isolates its subjects from the rest of the world—yet invests heavily in weaponry with which to blackmail its neighbors and infect the world. Refugees' and escapees' stories tell of the most extreme forms of poverty and misery. But China, North Korea's ideological ally, supports the regime based in the capital of Pyongyang in the interest of regional 'stability' rather than pressuring it to ameliorate its policies.

North Korean dictators rule with powers unimagined even by the likes of Fidel Castro, Saddam Hussein, and Bashar al-Assad, and the North Korean people have become a society of informers—on each other. Any sign of disaffection risks imprisonment or worse, and mass displays of regimented solidarity reminiscent of Soviet public travesties attend dynastic transfers of power. So it was when the ostensibly "affable" Kim Jong-un, son of the previous dictator Kim Jong-il, took control of the state in 2011. Shortly thereafter he threatened to resume nuclear testing, he and his ruling elite kept comfortably in power by China's provision of food and energy.

South Korea

The tragedy of the Korean Peninsula is especially painful when put in geographic perspective. As environmental maps reveal, North Korea and South Korea actually need each other. The North has raw materials demanded by South Korean industries. The South produces food the North could urgently use. This condition is known as **regional complementarity 15**, but in Korea it is precluded by political segregation. This makes the narrow demilitarized zone (DMZ) that separates the two Koreas an ultimate expression of core-periphery demarcation because even as the North stagnated, the South became one of the legendary Asian Tigers.

As noted above, South Korea (50.2 million) emerged from the Korean War in 1953 as an unstable, dictatorial, politically and economically corrupt nation where, as in the North, dissent could cost you your life. But South Korea's rulers also encouraged industrial development, driven by powerful industrial conglomerates. This kind of **state capitalism 16** triggered rapid growth, and the politicians knew that workers would demand better treatment

The Chinese government's grip on Xinjiang was markedly tightened on December 26, 2014 with the opening of the Lanxin (Lanzhou-Xinjiang) high-speed rail line. This 1776-km (1100-mi) line, China's longest and the first high-speed railway built across a high-altitude zone, operates at 250 kph (155 mph) and reduces the travel time between Lanzhou and Ürümqi from about 24 hours to just under 12. The top photo shows the new bullet train at Hami station near Ürümqi a few days before its maiden eastbound trip. This significant improvement in the connectivity of the Western Periphery is part of a massive effort to modernize China's most remote region. However, the lower photo, which was taken about a year before the inaugural Lanxin run, is a reminder that the Han-Uyghur ethnic conflict continues to percolate and overshadow Xinjiang's progress. This 2013 assemblage of armed military police in central Ürümqi was a deliberate show of force in response to a surge of Uyghur unrest (at least partially inflamed by Muslim extremists) in western Xinjiang. But so far the Chinese authorities have been careful to exercise restraint because any major anti-Islamic campaign could exacerbate Uyghur and other budding separatist movements.

ago, disaster struck as Japan embarked on a campaign of colonial conquest that eventually overran not only Korea, but also large parts of neighboring China. Subsequently, in the aftermath of Japan's World War II defeat, the Korean Peninsula became the scene of a terrible great-power conflict between China-supported communist armies and American-led anticommunist forces. For more than three years (1950–1953) this devastating conflict swept back and forth across the peninsula, claiming more than 3 million lives and ending in an armistice whose cease-fire line can be seen in Figure 9-20. Ever since, that line has separated—almost hermetically sealed—a vicious communist dictatorship in North Korea from a South Korea that

The map of South Korea (Fig. 9-20) reveals several of these achievements, most notably the metropolitan agglomeration formed by Seoul (the capital) and neighboring Incheon, which today constitutes one of Asia's greatest urban complexes and contains about one-quarter of the country's population. Even before democracy took hold, South Korea's rulers decided to further enhance the country's infrastructure by introducing a high-speed, bullet-train network to connect Seoul to the traditionally separated subregions of Geyongsang and Jeolla. For a long time, workers in parts of the country not favored by politicians and manufacturers had trouble finding employment in fast-growing industrial areas, an economic-geographic pattern that has still not completely been erased. As regional integration progressed, the southeastern subregion of Geyongsang, centered on the city of Busan, was always ahead of the southwest, where Jeolla was historically marginalized by the powers in Seoul. Today Geyongsang remains in the lead: here 300,000-ton tankers are built along with smaller freighters and warships; here are the famous Hyundai and Kia automobile plants; and here is the home base of POSCO, headquartered in Pohang, one of the world's largest steelmaking corporations.

Despite all this success, surveys indicate that the South Koreans are not especially happy as a society. Undoubtedly, a major factor is the stress level associated with the ever-present danger of

FIGURE 9-20

Songdo New City, South Korea's answer to China's Pacific Rim hub cities. Songdo is the centerpiece of the ultramodern urban complex springing up around the country's second-largest port of Incheon at the southwestern corner of the Seoul-centered conurbation. Modeled after Singapore, Songdo is not only designed to be "the world's gateway to Northeast Asia" (more than 25 percent of the world's population lives within four hour's flying time) but also to be a prototype "Smart City". The gateway function is based on the global accessibility enabled by adjacent, state-of-the-art Incheon International Airport—making Songdo an international magnet or "aerotropolis" serving globalized businesses as well as offering a Las Vegas-style "pleasure carnival" featuring mega-casino/resorts, theme parks, and a Jack Nicklaus-built golf course. At the same time, still-building Songdo's smart-city technology is regarded as a cutting-edge example of sustainable urbanization, one that so highly impressed the UN's Green Climate Fund that it located its headquarters here in 2012.

and that the economic and political landscape would have to change. And so it was: dictatorial rule ended, representative government took hold, corruption was confronted, and the economy, once described by early postwar observers as undevelopable, took off like a rocket. South Korea in short order became a boisterous democracy, an economic giant, a nation in universal pursuit of education, the world's leading shipbuilder, a manufacturer of iron and steel as well as chemicals and electronics plus countless other industrial products in addition to producing ample food supplies. South Korean social institutions became ever stronger, and the media were now as free as any on Earth. Freedom of religion prevails, Korean art and culture find a global market, and Korean athletes excel in the Olympic Games and other international sports events. As a key member of the international community, South Korea participates in peacekeeping and security operations worldwide.

@ FROM THE FIELD NOTES...

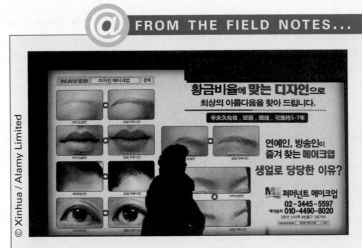

© Xinhua / Alamy Limited

"When I visited Seoul recently, I encountered a large concentration of cosmetic surgery clinics in the trendy, newly developed neighborhood of Gangnam. Significantly, these clinics represent an emerging cultural phenomenon in rapidly changing East Asia. South Korea, formerly an uncool country known mostly for such heavy industrial exports as cargo ships and motor vehicles, has of late become the hottest arbiter and trendsetter of fashion and pop culture in East Asia. Following the Asian financial crisis of 1998, the South Korean government's effort to diversify the economy has scored major successes in both the IT sector and the entertainment industry, especially developing content for TV dramas, films, pop music, and video games. As this new Korean cultural wave rolled across the realm, it triggered the rise of a Korean esthetic. Celebrity K-pop singers and actors have inspired millions of their fans to wear Korean-designed clothing, to use Korean brands of cosmetic products, and even to undergo Korean cosmetic surgeries to look just like their idols! This form of plastic-surgery-dominated medical tourism now attracts at least 200,000 visitors a year to South Korea, with the largest contingent originating in China. Thus the South Korean development model has expanded from the export of goods to that of tastes. Once again, the state was an indispensable driver in this process."

an attack from bellicose North Korea. But there is another dimension that goes beyond the North-South contrasts. In states that achieve great economic success, women tend to participate in the action even if they are not paid as well as men. Yet South Korea's triumphant story is often recounted as being dominated by males, despite the remarkable contribution of female workers to the economic miracle, especially their labor activism that resulted in much improved working conditions throughout the country. Nonetheless, the proportion of women gainfully employed in South Korea today is still not representative of a developed economy, and the number of women holding senior positions remains disturbingly low. Hopes that this serious gender gap may finally be addressed have risen since 2013 when Park Geun-hye, the daughter of a postwar South Korean dictator, took office as the country's first female president.

JAPAN

Japan consists of four main islands—Honshu, the largest; Hokkaido to the north; and Shikoku and Kyushu to the south—in addition to numerous small islands and islets (Fig. 9-21). Most of the country is mountainous and steep-sloped, geologically young, earthquake-prone, and studded with volcanoes. A mere 18 percent of the country is considered to be habitable. Japan's high-relief topography has compressed its economic development. Except for the ancient capital of Kyoto, all of Japan's major cities are perched along the coast, and virtually all lie partly on artificial land reclaimed from the sea.

Sailing into Kobe harbor near Osaka (the country's second-largest city), one passes artificial islands designed for high-volume shipping and connected to the mainland by automatic, space-age trains. Enter Tokyo Bay, and the refineries and factories to the east and west stand on huge expanses of landfill that have pushed the bay's shoreline outward.

With just over 127 million people, seven-eighths of whom reside in towns and cities, Japan uses its habitable living space very intensively—and expands it wherever possible. As Figure 9-21 indicates, farmland in Japan is both limited and regionally fragmented. Urban sprawl has invaded much of the cultivable land. In the hinterland of Tokyo, around Osaka, and surrounding Nagoya, major farming zones are under relentless urban pressure. All three of these lowlands lie within Japan's fragmented but well-defined core area (delimited by the red line on the map), the heart of the country's prodigious manufacturing complex.

Coastal Development

Figure 9-21 also tells us a great deal about the nature of Japan's external orientation and its dependence on foreign trade. All of the country's primary and secondary regions are found on the coast. Dominant among them is the *Kanto Plain*, the heart of Japan's core area focused on metropolitan Tokyo, which contains about one-third of the country's population (see satellite image). Among its advantages are an unusually extensive area of low relief, a fine natural harbor at Yokohama, a relatively mild and moist climate, and centrality with respect to the country as a whole. (Its principal disadvantage lies in its vulnerability to major earthquakes [see Fig. 9-21].)

The second-ranking primary economic region within Japan is named the *Kansai District* (Fig. 9-21); it contains the Osaka-Kobe-Kyoto triangle and is located at the eastern end of the Seto Inland Sea between Honshu and Shikoku. Osaka and Kobe are leading industrial centers and busy ports, but the Kansai District also yields large harvests of rice, Japan's staple food. Between the Kanto Plain and the Kansai District lies the *Nobi Plain* (Fig. 9-21), centered on the city of Nagoya. And, as the map shows, the Japanese core area is anchored in the west by the conurbation **17** focused on *Kitakyushu*, situated not on Honshu Island but in the northwestern corner of Kyushu, Japan's southernmost major island. This still-expanding edge of the core area is particularly favored by its location relative to South Korea and China.

A Trading Nation

Japan is a remarkable country that despite its poor natural endowments (except for its advantageous location) has done extremely well. It became the dominant power of East and Southeast Asia through military force prior to World War II, and its postwar economic miracle astounded the world. In the 1970s and 1980s, people referred to East and Southeast Asia as the "Yen Bloc," one that was dominated by the Japanese economy.

The country defeated the odds by calling on organizational efficiency, massive productivity, dedication to quality, and tight adherence to common goals. Keep in mind that this has always been an exceptionally homogeneous, collectively oriented, and consensus-minded nation with a strong work ethic and a deep-seated conviction that common interests override those of the individual. It is hard to overestimate this cultural context in Japan's outstanding achievements, even if it can become a liability, as when collective rigidity impedes adaptation to new circumstances.

More concretely, Japan dealt with its own limited geographic opportunities by engaging the alternatives. Look at Figure 9-6 as well as Figure 9-21 and note Japan's substantial reliance on nuclear energy—a strategy not without major risks in this earthquake-prone region, as became so painfully clear in the aftermath of the catastrophic Tohoku quake/tsunami in 2011 (see Fig. 9-2). Most important of all, Japan had no choice but to engage other countries, near and far, in order to compensate for what it does not possess itself. Japan, therefore, became wealthy through trade, imported raw materials, the accomplished production and processing of high-value-added goods, and the ability to turn these goods into lucrative exports.

The Challenges of Two Lost Decades

At the end of the 1980s, it seemed that Japan's meteoric rise had been so prodigious that it would only be a matter of time before it would surpass even the United States' economy. But everything came to a rather sudden halt during the early 1990s, and Japan has unexpectedly been stuck in a rut ever since. This now more than 20-year-long economic slowdown had many causes, ranging from government mismanagement and inefficiency to intensifying international competition (such as from South Korean cars and Taiwanese electronic products).

One cause of the slowdown was Japan's deficiency in adaptability: its culturally embedded economic system could not cope with rapidly changing conditions. With bankruptcy a near-taboo due to the tight internal networks of firms and their tradition of social responsibility, weak companies were often kept alive. This kind of economic sclerosis greatly inhibits the creation of new businesses (the United States in any given year has three times the number of start-ups as Japan and twice the number of bankruptcies).

Japan's conservative business culture had now become a liability. Whereas women constitute an ever greater share of Japan's highly educated workforce, only 7 percent of all senior management executives are female (compared to 20 percent in the United States and 31 percent in China). For the first time in decades,

Japan suffered a crisis of confidence and, more tangibly, something unfamiliar to the burgeoning Japan of the late twentieth century—rising unemployment. Faced with much leaner economic times, men especially seemed to seek refuge in the past as growing numbers preferred to stay with the same stagnant big companies instead of launching their own businesses.

Japan is a rapidly aging society as well, projected to decline from 127 million today to barely 97 million in 2050 and less than 60 million by the end of this century. It faces an enormous and increasing **demographic burden 18**: by 2025, it is projected there will be only two workers for every retiree. Ethnically uniform Japan has historically resisted immigration, so when the government tried to recruit ethnic Japanese living in Brazil (and elsewhere) to return home, the experiment was not very successful. Therefore, a shrinking base of qualified workers threatens the government's capacity to sustain social programs for all.

Still another set of problems has to do with Japan's international relations. Japan never signed a peace treaty with the (then) Soviet Union after World War II because the Russians had occupied and refused to return four small island groups in

Planet Observer/Universal Images Group via Getty Images, Inc.

This large Pacific inlet is central Honshu's Tokyo Bay, and Japan's capital city is located just to the west of the bay's northernmost corner. But even this Landsat-7 satellite image—measuring roughly 80 by 115 kilometres (50 by 70 mi)—cannot capture the entirety of the massive, Tokyo-centered conurbation. Colored light gray, the far-flung urban landscape covers more than half of the land area in this view, and is home to at least 40 million of the total Japanese population of 127 million. Tokyo Bay also ranks among the world's finest natural harbors, giving rise to Yokohama (located halfway to the ocean from Tokyo on the bay's western shore), widely considered to be the most productive port anywhere.

JAPAN: MANUFACTURING, LAND, AND LIVELIHOODS

POPULATION
- 50,000–250,000
- 250,000–1,000,000
- 1,000,000–5,000,000
- Over 5,000,000

National capital is underlined

— Railroad
— Road
⛏ Coal
Zn Zinc
Cu Copper
▬ Core area

▨ Primary region
▨ Secondary region
▨ Agriculture

◎ Epicenter of March 11, 2011 Tohoku earthquake
⚓ Nuclear reactor
⚓ Leaking nuclear reactor, 2011
— Coastline affected by 2011 tsunami

0   100   200 Kilometers
0   50   100 Miles

130° Longitude East of Greenwich

© J. Nijman, P. O. Muller, H. J. de Blij, and John Wiley & Sons, Inc.

FIGURE 9-21

the Kurile chain northeast of Hokkaido (Fig. 9-21). Failed negotiations for the return of these "Northern Territories," as the Japanese call them, have cost Japan the opportunity to play a key role in the economic development of the Russian Far East, where crucial energy as well as mineral resources abound. Furthermore, relations with both South and North Korea are troubled—with the South over ownership of a small island group in the East Sea (Sea of Japan) and over memories of Japanese misconduct during World War II; and with the North over the kidnapping of Japanese citizens by North Korean agents and, above all else, the threat posed by North Korea's nuclear weapons development. And add to all this a number of lingering issues with China over Japanese actions during the colonial and wartime periods as well as the Senkaku Islands dispute (see Fig. 9-8).

TAIWAN

When we see where Taiwan is located, it would seem to be inextricably bound up with China because it lies a mere 200 kilometers (125 mi) from the mainland. Not surprisingly, Taiwan has a history of immigration from coastal Chinese provinces. As noted earlier, China's rulers view Taiwan as a "wayward province" that must

one day return to the People's Republic. Taiwan lost its seat in the United Nations in 1971 and was replaced by the PRC. The United States at the time also transferred its official recognition of China from Taiwan to the PRC in order to improve relations with the mainland government.

Taiwan's Island Geography

Taiwan, as Figure 9-22 shows, is not a large island. It is smaller than Switzerland but has a much larger population (23.6 million), most of it concentrated in a crescent lining the western and northern coasts. The Chungyang Mountains, an area of high elevations (some more than 3000 meters [10,000 ft]), steep slopes, and dense forests, dominate the eastern half of the island. Westward, these mountains yield to a zone of hilly topography and, facing the Taiwan Strait, a wide, substantial coastal plain. Streams from the mountains irrigate the paddyfields, and farm production has more than doubled since 1950, even though hundreds of thousands of farmers have left the land for employment in Taiwan's booming industrial sector.

Today, the lowland urban-industrial corridor lining western Taiwan is anchored by the capital, Taipei, at the island's northern end and rapidly growing Kaohsiung in the far south. The Japanese developed Chilung, Taipei's outport, to export nearby coal, but now the raw materials flow the other way. Taiwan imports raw cotton for its

FIGURE 9-22

Taiwan's high-tech manufacturers constitute a force to be reckoned with in the global science industries. Domestically, one of the leading contestants for the title of "Taiwanese Silicon Valley" is the burgeoning technopole centered on the city of Hsinchu—a metropolis of nearly 600,000 located on the Taiwan Strait 45 minutes southwest of the capital, Taipei. R&D predominates here, and the top photo shows the Synchrotron Radiation Research Center, focused on its new-generation nuclear particle accelerator. Hsinchu is not without rivals, however, and an hour to its south lies the smaller city of Taichung whose high-tech centerpiece is its thriving Science Park. The lower photo shows one of this complex's state-of-the-art factories, whose owner is one of the world's foremost manufacturers of flat-screen panels. Demand for the latter has surged in recent years, prompting the AU Optronics Corporation to expand production onto the nearby Chinese mainland.

textile industry, bauxite (aluminum ore) from Indonesia, oil from Brunei, and iron ore from Africa. Taiwan has a growing steel industry, nuclear power plants, shipyards, a substantial chemical industry, and modern transport networks. Most of all, it has become a leading producer and exporter of high-technology products, especially personal computers, telecommunications equipment, and precision electronic instruments. Taiwan offers a wide range of brainpower resources,

and many foreign firms join in the research and development effort. Assisted by the government, a world-class **technopole** **19**, focusing on microelectronics and personal-computer development, has been established at Hsinchu on the northwestern coast (see photo pair). Farther south, the budding "science city" of Tainan specializes in microsystems and information-technology innovations.

The creation of China's Special Economic Zones went a long way toward mitigating longstanding problems between Beijing and Taipei. China needed investment capital and Taiwan had it, and in the almost-anything-goes environment of the SEZs, Taiwanese entrepreneurs could buy or build factories just as "real" foreigners were able to do. In fact, the Xiamen SEZ was laid out directly across from Taiwan for this purpose (people there will tell you that almost all of the locals have relatives across the Taiwan Strait). Consequently, thousands of Taiwanese-owned factories began operating in the PRC, their owners and investors having a strong interest in avoiding violent confrontation over political issues.

Taiwan's Future

Taiwan's unresolved status entails risks, but there are signs that both parties want to find a solution (the Taiwanese reelected President Ma Ying-jeou in 2012, effectively voting for another four years of rapprochement with the PRC). Given that the PRC will never allow Taiwan to attain independence, and the growing prospect that the United States, Taiwan's chief guarantor, could not secure Taiwan against Chinese military intervention, many Taiwanese as well as their allies and adversaries are seeking a long-range, negotiated solution. Among the options is one that appears to be working in Hong Kong, where the principle of ***One Nation, Two Systems*** has functioned more successfully than many observers anticipated—despite the sporadic protests led by the political opposition that mark the ongoing debate over the policy's effectiveness.

POINTS TO PONDER ?

- The fortunes of China and Japan have always seemed to be inversely related: when one was rising, the other was in decline.
- In the past decade, China's defense spending has increased by about 12 percent annually, much of it intended to match and eventually surpass the military strength of the United States in the islands and waters off the Asian mainland.
- Increasing social inequalities resulting from China's rapid economic growth are becoming ever more difficult to justify by China's self-proclaimed communist government.
- After 70 years of dictatorship, malnutrition, and famine, North Koreans at the age of 21 are an average 6 centimeters (2.4 in) shorter than their South Korean counterparts.

Cambodia's Angkor Wat, one of the world's most famous historic religious sites, is a blend of Hindu and Buddhist influences. This treasure-filled architectural triumph embodies the cultural crossroads that is Southeast Asia.

10

THE SOUTHEAST ASIAN REALM

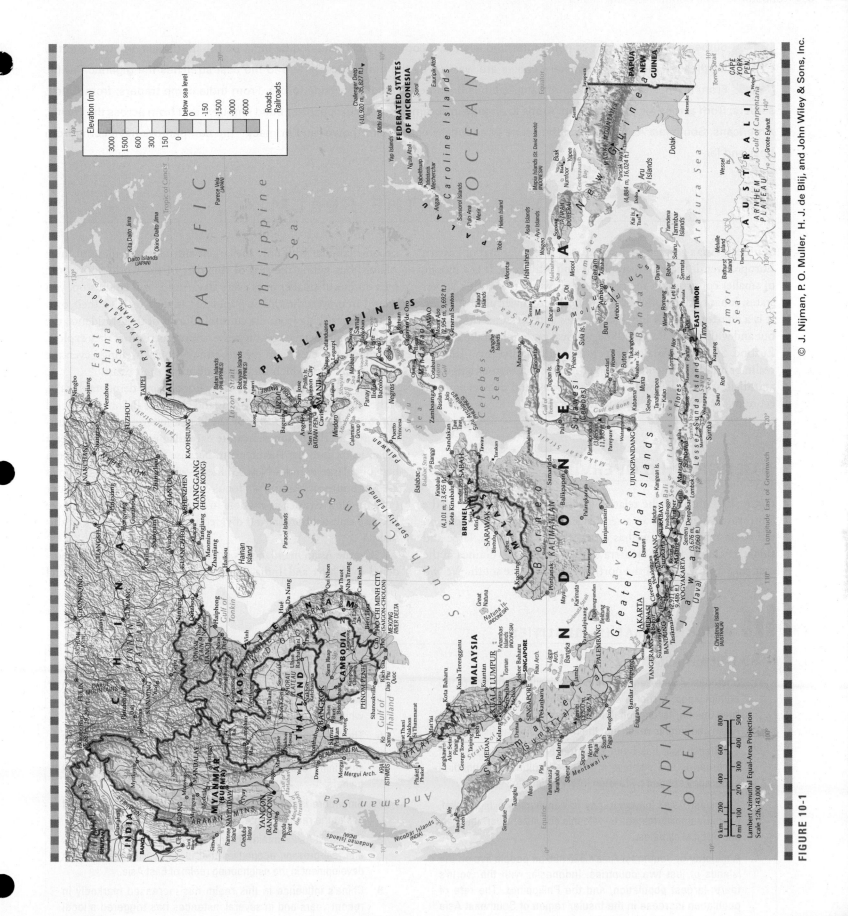

FIGURE 10-1

© J. Nijman, P. O. Muller, H. J. de Blij, and John Wiley & Sons, Inc.

Southeast Asia is a realm of peninsulas and islands, a corner of Asia bounded by India on the northwest and China to the northeast (Fig. 10-1). Its western coasts are washed by the Indian Ocean, and to the east stretches the gigantic Pacific. From all these directions, Southeast Asia has been penetrated by outside forces. From India came traders; from China, settlers; from across the Indian Ocean, Arabs to engage in commerce and Europeans to build empires; and from across the Pacific, Americans. Southeast Asia has been the scene of countless contests for power and primacy—the competitors have come from near and far.

DEFINING THE REALM

Southeast Asia's geography in some ways resembles that of Eastern Europe, even if the physiography is quite different. It is a mosaic of smaller countries on the periphery of two of the world's largest states. It has been a **buffer zone** **1** between powerful adversaries. It is a **shatter belt** **2** in which stresses and pressures from without and within have fractured the political geography. Like Eastern Europe, Southeast Asia exhibits great cultural diversity. This is a realm of hundreds of cultures and ethnicities, numerous languages and dialects, global as well as local religions, and diverse national economies ranging from high to low income.

A GEOGRAPHIC OVERVIEW

Figure 10-1 displays the relative location and dimensions of the Southeast Asian realm, an assemblage of 11 countries situated on the Asian mainland as well as on thousands of islands, large and small, extending from Sumatera in the west to New Guinea in the east, and from Luzon in the north to Timor in the south. We will become familiar with only the largest and most populous of these islands, but sailing the seas of this part of the world you would pass dozens of smaller ones every day. And if you could stop, you would find even the tiniest populated islands to have their own character resulting from the cultural sources of their residents, the environmental challenges they face, the opportunities they exploit, their modes of dress, and the structure of their dwellings, even the vivid colors with which they often decorate their boats.

The giant of this realm, both territorially and demographically, is the far-flung **archipelago** **3** (island chain) of Indonesia, labeled appropriately in Figure 10-1 by the largest lettering. In its easternmost sector, the Indonesian state extends beyond the Southeast Asian realm into the Pacific realm because it controls the western half of a large island—New Guinea—whose indigenous peoples are not Southeast Asian. We focus on this unusual situation later (and on New Guinea as a whole in Chapter 12), but note it here because unusual borders and divided islands are a hallmark of this realm.

On the Asian mainland, as noted above, Southeast Asia is bordered by China and India, both sources of immigrants, cultural

MAJOR GEOGRAPHIC QUALITIES

SOUTHEAST ASIA

1. Southeast Asia extends from the peninsular mainland to the archipelagos offshore. Because Indonesia controls part of New Guinea, its functional region reaches from the Southeast Asian geographic realm into the neighboring Pacific realm.

2. Southeast Asia has been a shatter belt between powerful adversaries and displays a fractured cultural and political geography shaped by foreign intervention.

3. Southeast Asia's physiography is dominated by high relief, crustal instability marked by volcanic activity and earthquakes, and tropical climates.

4. A majority of Southeast Asia's 620 million people live on the islands of just two countries: Indonesia, with the world's fourth-largest population, and the Philippines. The rate of population increase in the Insular region of Southeast Asia exceeds that of the Mainland region.

5. Although the great majority of Southeast Asians have the same ancestry, cultural divisions and local traditions abound, which the realm's divisive physiography nurtures.

6. The Mekong River, Southeast Asia's Danube, has its source in China and borders or traverses five Southeast Asian countries, sustaining tens of millions of farmers, fishing people, and boat owners.

7. Singapore is the leading world-city in Southeast Asia and lies at the realm's center of trade and business relations.

8. Southeast Asia contains a number of rapidly emerging markets and fast-growing economies that in certain respects imitate development in the neighboring realm of East Asia.

9. China's influence in this realm has increased markedly in recent years and in several instances has triggered a local backlash.

infusions, economic initiatives, and other relationships evident in the realm's cultural landscapes. Also originating there are rivers that play a pivotal role in the lives of many millions of people south of the realm border. As we will find, the core areas of many of Southeast Asia's most populous countries are located in the basins of major rivers whose sources lie inside China.

Before we get started, it is helpful to become acquainted with the key states of this realm, some of which will already be familiar. North of Indonesia lies the Philippines, well known to Americans because this was once an American colony and is still a source of many immigrants to the United States. Also north of Indonesia is Malaysia, easy to find on the map because its core area lies on the long Malay Peninsula that almost connects mainland Asia with Indonesia. At the tip of that peninsula lies another famous geographic locality: Singapore, the spectacular economic success story of Southeast Asia and a world-city by every measure.

In the realm's mainland component, the name Vietnam still resonates in America, which fought a bitter and costly war there in the 1960s and 1970s. As the map shows, Vietnam looks like an elongated strip of land extending from its border with China to the delta of the greatest of all Southeast Asian rivers, the Mekong. Looking westward, neighboring Laos and Cambodia may not be all that familiar, but next comes centrally positioned Thailand, the dream (and reality) of millions of tourists and one of the world's most fascinating countries. And then, on the realm's western margin, we come to Myanmar (formerly Burma): here human potential, natural endowment, and opportunity were thwarted by a half-century of extremely harsh military rule that plunged the country into devastating poverty—but that quite unexpectedly relaxed its vise-like grip in 2012 and reopened Myanmar to the world.

SOUTHEAST ASIA'S PHYSICAL GEOGRAPHY

Physiographically, Southeast Asia is in some ways reminiscent of Middle America, a fractured realm of islands and peninsulas flanking a populous mainland studded with high mountains and deep valleys. It is useful to look back at Figure G-4 to see why: both Southeast Asia and Middle America are dangerous places, where the Earth's crust is unstable as tectonic plates collide, earthquakes are a constant threat, volcanic eruptions take their toll, tropical cyclones lash sea and land, and floods, landslides, and other natural hazards make life riskier than in most other parts of the world.

In terms of its economic geography Southeast Asia may be part of the Pacific Rim, but in physiographic terms it forms part of the Pacific Ring of Fire—and all the geologic hazards this designation carries with it (Fig. G-4). In 2004, an undersea earthquake off westernmost Indonesia produced a **tsunami** 4 (seismic sea wave) in the Indian Ocean that killed more than 230,000 people along coastlines from Sumatera to Somalia. This was the latest in a string of geologic disasters originating in Southeast Asia whose effects were felt far beyond the realm's borders. In 1883, the Krakatau volcano lying between Sumatera and Jawa exploded, resulting in a death toll estimated at more than 36,000. In 1815, the Tambora volcano in the chain of islands east of Jawa known as the Lesser Sunda Islands blew up, darkening skies throughout the world and affecting climates around the planet (the year that followed is still known as the "year without a summer" when crops failed, economies faltered, and people went hungry as far away as Egypt, New England, and France). Research on what may well have been the most calamitous of all such eruptions suggests that, about 73,000 years ago, the Toba volcano on Sumatera exploded with such force that its ash and soot not only darkened skies and affected weather, but changed global climate for perhaps as long as 20 years and threatened the very survival of the human population, which then was still small in number and widely scattered. In fact, a number of scientists postulate that the widespread impact of this eruption gave rise to so many casualties that human genetic diversity was significantly diminished.

Thus high relief dominates Southeast Asia, from the Arakan Mountains in western Myanmar to the glaciers (yes, glaciers!) of New Guinea. Take a close look at Figure 10-1 and you can see how

@ **FROM THE FIELD NOTES...**

"After a three-hour drive by bus southward from Hanoi, we reached the Van Long Nature Reserve in northeastern Vietnam's Red River Delta. It lies in a limestone terrain known as karst, part of the same geology that continues to the north through Halong Bay and into southern China beyond. Karst landscapes are formed through the slow but steady dissolution of limestone by (naturally) acidic water; the harder rock formations are more resistant to erosion and remain as towers, resulting in unique and sometimes whimsical landforms (left photo). This nature reserve is a biologically rich ecotourism site that was established in 2001 and involves local communities. Women from nearby villages guide visitors around in their basket boats, and they welcome the extra income in an area barely touched by Vietnam's emerging economy (right photo)."

© Jan Nijman

many elevations approach or exceed 3000 meters (10,000 ft), and, using the elevation guide in the map legend, how mountainous and hilly much of the realm's topography is. Lengthy ranges form the backbones not only of islands such as Sulawesi and Sumatera, but also of the Malay Peninsula, most of Vietnam, and the border zone between Thailand and Myanmar. In Figure G-4 it is possible to trace these volcano-studded mountain ranges by their earthquake epicenters and volcanic records.

Exceptional Borneo

But among the islands there is one significant exception. The bulky island named Borneo in Figure 10-1 has high elevations (Mount Kinabalu in the north reaches 4101 meters [13,455 ft]), yet has no volcanoes and negligible Earth tremors. This island has been called a stable "minicontinent" amid a cauldron of volcanic activity, a slab of ancient crust that long ago was uplifted high above sea level by tectonic forces and then subsequently eroded into its present landscapes. Borneo's soils are not nearly as fertile as those of the volcanic islands, so that an equatorial rainforest developed here that long survived the human onslaught, giving sanctuary to countless plant and animal species including Southeast Asia's great ape, the orangutan. That era is clearly ending as human encroachment on Borneo's tropical habitat is accelerating, logging is destroying vast swaths of the remaining forest, and roads and farms are penetrating its interior. Borneo, along with other parts of Indonesia and Malaysia, has also experienced the rapid expansion of palm oil plantations at the expense of tropical woodlands.

Deforestation and Palm Oil Plantations

Palm oil is used worldwide in processed foods and myriad other products. This form of commercial agriculture is quite lucrative because the yield of oil palms per hectare (equal to 2.5 acres) is much higher than the yield of such competing edible oil crops as soybeans and sunflowers. Moreover, the price of this commodity on global markets has risen steadily over the past several years, triggering successive rounds of production increases. Malaysia and Indonesia have benefited the most, together accounting for 90 percent of the world's palm oil. But because these countries take advantage of economies of scale and produce palm oil on huge, ever-expanding plantations, this crop has become the single biggest threat to tropical forest preservation (Fig. 10-2; see photo). Most at risk for survival are such key wildlife species as orangutans, pygmy elephants, Sumatran rhinos, and tigers.

Borneo's tropical forests are a remnant of a much larger stand of equatorial rainforest that once covered most of this realm. Besides Borneo, eastern Sumatra still contains limited expanses of rainforest, including a few orangutan sanctuaries. So does less populous and more remote New Guinea, which was never reached by these great apes. As we noted in the introductory chapter, equatorial rainforests can still be found today in three low-latitude zones: the Amazon Basin of South America, west-equatorial Africa, and here in Southeast Asia. The climatic conditions that

sustain these forests—persistently warm temperatures and year-round rainfall—produce the biologically richest and ecologically most complex vegetation regions on Earth. Enormous numbers and varieties of trees and other plants grow in very close proximity, vying for space and sunlight both horizontally and vertically. Yet despite all this luxuriant growth, the soils beneath rainforests are nutrient poor. Most of the surface nutrients are derived from the decaying vegetation on the forest floor, which nurture the growth of the next generation of plants. Long-term indigenous inhabitants have developed methods of shifting cultivation to use these soils, plants, and animals to eke out a living—but when migrant farmers from elsewhere remove the trees to utilize the soil, mistakenly assuming it will be able to support their crops, failure swiftly results. This reality notwithstanding, intensifying population pressure throughout the tropics continues to shrink what remains of the world's rainforests—an issue also discussed in the chapters on Middle and South America (Chapters 2 and 3).

FIGURE 10-2

© J. Nijman, P. O. Muller, H. J. de Blij, and John Wiley & Sons, Inc.

A section of a vast palm oil plantation in Sarawak State in the southern part of Malaysian Borneo. These continually enlarging plantations have proliferated in Malaysia and Indonesia (see Fig. 10-2), earning sizeable export revenues but at ever-increasing environmental cost.

© G. Bowater/Corbis Images

Relative Location and Biodiversity

Look again at Figure 10-1, but in a more general way. The Malay Peninsula, adjacent Sumatera, Jawa, and the Lesser Sunda Islands east of Jawa seem to form a series of stepping stones toward New Guinea, and in the extreme lower-right corner of the map you can see Australia's Cape York Peninsula also appearing to reach toward New Guinea. Throughout Southeast Asia's geologic history, sea levels have repeatedly dropped, periodically drying up those narrow bodies of water lying between the islands as well as certain peninsulas. During such prehistoric intervals, the orangutans whose descendants survive in Indonesia today were able to migrate from the mainland into even warmer equatorial latitudes because islands were temporarily connected by land bridges. Later, early human arrivals here also received assistance from nature: Australia's Aborigines probably managed to cross what remained of the deeper trenches between islands by building rafts, but their epic journey was facilitated by land bridges as well.

Thus time and again, Southeast Asia was a receptacle for migrating species. Combine this with the realm's tropical environments, and it is no surprise that it is known for its biodiversity **5**. Scientists calculate that fully 10 percent of the Earth's plant and animal species are found in this comparatively small realm. Biogeographers can trace the progress of many of these species from the Asian mainland across the Indonesian archipelago toward Australia, but some were halted by deeper trenches that remained filled with water even when sea levels dropped. An especially deep trench lies in the Lombok Strait between the islands of Bali (next to Jawa) and Lombok (Fig. 10-1), a key biogeographical boundary (discussed in Chapter 11) first recognized by the naturalist Alfred Russel Wallace, a contemporary of Charles Darwin.

As we shall presently observe, Southeast Asia's biodiversity had a fateful impact on its historical geography. Among the realm's specialized plants, it was the spices that attracted outsiders from India, China, and Europe, with consequences still visible on the map today.

Four Major Rivers

Water is the essence of life, and Southeast Asia is comparatively well endowed with moisture (Fig. G-6). Ample, occasionally even excessive, rainfall fills the rice-growing paddies of Indonesia and the Philippines. On the realm's mainland, where annual rainfall averages are somewhat lower and the precipitation is more seasonal, major rivers and their tributaries fill irrigation channels and form fertile deltas. The map of the mainland's population distribution is dominated by this spatial relationship between rivers and people, expressed most clearly in distinct coastal clustering atop the biggest deltas (see Fig. 10-5).

As Figure 10-1 reminds us, rivers that are crucial to life in one realm may sometimes flow from sources in neighboring realms, a geographic issue that can lead to serious regional discord. What right does an upstream state in one realm (or even the same realm) have to dam or otherwise siphon water away from a river whose flow is vital to a state (or states) downstream? In Southeast Asia's case, three of its four principal rivers originate in China: (1) the **Mekong**, which crosses the mainland from north to south; (2) the **Red** River, which reaches the sea via its course across northern Vietnam; and (3) the **Irrawaddy**, which forms the lifeline of Myanmar. The fourth is an intra-realm river, the **Chao Phraya**, which functions as Thailand's leading artery.

The Mighty, Endangered Mekong

From its headwaters high in the frozen uplands of China's Qinghai-Xizang (Tibetan) Plateau, the Mekong River rushes and flows some 4200 kilometers (2600 mi) to its delta in southernmost Vietnam. This "Danube of Southeast Asia" traverses or borders five of the realm's countries, supporting rice farmers and fishing people, forming a vital transport route where roads are few, and providing electricity from dams upstream. Tens of millions of people depend on the waters of the Mekong, from subsistence farmers in Cambodia to apartment dwellers in China. The Mekong Delta in southern Vietnam is one of the realm's most densely populated areas and produces enormous harvests of rice. The Mekong Basin as a whole (mapped in green in Fig. 10-3) is reputed to be the world's largest and most productive area for freshwater fishing.

FIGURE 10-3

But the Mekong is increasingly under threat, and so are the livelihoods of millions in its Basin, as demand for hydroelectric power accelerates all across mainland Southeast Asia. To the north, China has already built seven dams for generating hydropower, and many more are either under construction or planned (Fig. 10-3). And in the Mekong's middle basin, impoverished Laos considers hydropower its most strategic commodity and promising route to development, and is making big plans to sell much of that energy to China, Thailand, and Vietnam. Favored by its hilly terrain, Laos is planning seven new dams, and even its lowland downstream neighbor, Cambodia, wants to build another three.

Among these projects, it is the ongoing construction of the Xayaburi Dam in Laos that has by far drawn the most attention and criticism from environmental activists because it sits squarely in the Mekong's main stream (Fig. 10-3; see photo). As such, it will substantially reduce the river's downstream transportation of silt, the alluvial sediment that is highly crucial to the formation of fertile soils on the floodplains of central Cambodia and particularly the massive delta in southern Vietnam. And the sheer number of dams planned for the Mekong Basin is certain to damage ecosystems all along the Southeast Asian courses of the river and its tributaries (starting with the disruption of the migratory routes of freshwater fish). In the long run, therefore, prioritizing energy over environment and even food production may well prove to be a disastrous choice.

STR/EPA / Newscom

Construction of the controversial Xayaburi Dam in northern Laos, located on the Mekong River upstream from the capital, Vientiane. For obvious reasons, this site is zealously guarded, and current photos are hard to find (this one dates from late 2012). Opposition groups say this dam will displace and adversely affect countless communities along the Mekong as well as produce significant ecological harm. Meanwhile, the Laotian government is primarily focused on developing the dam's hydropower potential, which is planned to be a main driver of national economic growth through the rising sale of electricity to China as well as neighboring countries. Xayaburi Dam is expected to be fully up and running soon after its expected completion in 2019.

POLITICAL GEOGRAPHY: TERRITORIAL CONFIGURATIONS

Political boundaries define and delimit states. They also create the mosaic of interlocking territories that give individual countries their shape or **state territorial configuration** **6**. A country's shape can affect its condition, and in certain cases even its survival. Vietnam's extreme elongation has influenced its existence since time immemorial. And, as noted later in this chapter, Indonesia has tried to redress its fragmented layout—consisting of thousands of far-flung islands—by promoting unity through the "transmigration" of residents from its most populous, core-area island (Jawa) to many of the others.

Political geographers have identified five major categories of state territorial configuration, all of which we have encountered in our world regional survey but have not classified until now. All but one of these prototypical shapes can be seen on the political map of Southeast Asia,* and Figure 10-4 displays the terminology and examples:

- **Compact states** **7** have territories shaped somewhere between round and rectangular, without major indentations. This encloses a maximum amount of territory within a minimum length of boundary. Southeast Asian example: Cambodia.

- **Protruded states** **8** (sometimes called *extended*) have a substantial, usually compact territory from which extends a peninsular or other corridor that may be landlocked or coastal. Southeast Asian examples: Thailand and Myanmar.

- **Elongated states** **9** (also known as *attenuated*) have territorial dimensions in which the length is at least six times the average width, creating a state that lies astride environmental and/or cultural transitions. Southeast Asian example: Vietnam.

- **Fragmented states** **10** consist of two or more territorial units separated by foreign territory or a substantial body of water. Subtypes include mainland-mainland, mainland-island, and island-island. Southeast Asian examples: Malaysia, Indonesia, the Philippines, and East Timor.

For so comparatively small a realm, the 11 countries of Southeast Asia display a considerable variety of shapes. But one note of caution: states' territorial configurations do not determine their viability, cohesion, unity, or lack thereof; they can, however, influence these characteristics. Cambodia's compactness, for instance, has not ameliorated its divisive political geography.

---

*The missing category is the ***perforated state***, which is completely surrounded by the territory of another state (the latter thereby having a "hole" in it). The best example on the world map is South Africa, which is perforated by the ministate of Lesotho (see Fig. 7-20).

STATE TERRITORIAL CONFIGURATIONS

CAMBODIA

Phnom Penh

COMPACT

THAILAND

Bangkok

PROTRUDED

VIETNAM

Hanoi

ELONGATED

Manila

FRAGMENTED

PHILIPPINES

Ho Chi Minh City
(Saigon-Cholon)

FIGURE 10-4

POPULATION GEOGRAPHY

Examine the map of Southeast Asia's population distribution (Fig. 10-5), and you are immediately struck by the huge concentration of people on a relatively small island in Indonesia—a cluster larger than any other in the realm, the four mainland river deltas included. This is Jawa, and its population of close to 150 million not only accounts for well over half of Indonesia's national total but also exceeds that of every other country in the realm.

This population concentration is particularly noteworthy because Indonesia is not yet a highly urbanized country. Today half of all Indonesians still live in rural areas, and although Jawa is the country's most urbanized island, more than 60 million of its inhabitants still live off the land. What makes all this possible is a combination of highly fertile volcanic soil, abundant water, and extremely warm temperatures that enable Jawa's farmers to raise three crops of rice in a single paddy during a single year, helping feed a national population still growing faster than the global average. But we should also be aware that the thickest red clusters on Jawa denote fast-growing urban areas, cornerstones in the building of a new economy with increasingly global linkages.

Within Indonesia, the contrasts between Jawa and the four other major islands—Sumatera, Borneo (Kalimantan), Sulawesi, and most especially Indonesian New Guinea—reflect the core-periphery relationship between these two sectors of the country. As the map suggests, similar contrasts, less sharply defined, also mark other Southeast Asian countries, and the primate cities of this realm (such as Bangkok, Manila, Jakarta, and Kuala Lumpur) are particularly dominant. In fact, Vietnam even contains two such anchors, which respectively represent its historic northern (Hanoi) and southern (Ho Chi Minh City) core areas (Fig. 10-4).

The states of Southeast Asia are not, by world standards, especially populous. Indonesia today is the world's fourth-ranking country in population size (just over 250 million), but no other Southeast Asian country contains even half that total. Three mainland countries—Vietnam, Thailand, and Myanmar—contain between 50 and 100 million people. But take Laos, quite a large country territorially (about the size of the United Kingdom) and note that its population is less than 7 million; similarly, Cambodia, half the size of Germany, is home to only slightly more than 15.5 million. In part, such modest numbers on the mainland reflect natural conditions less favorable to farming than those prevailing on volcanic soils or in fertile river basins; but

FIGURE 10-5

more generally, this realm did not grow as explosively as neighboring realms did during the past century. Indeed, four countries in Southeast Asia (the three big mainland states just mentioned plus Singapore) are now growing at a rate below the global average of 1.2 percent annually.

The Ethnic Mosaic

Southeast Asia's peoples come from a common stock just as (Caucasian) Europeans do, but this has not prevented the emergence of regionally or locally discrete ethnic or cultural groups. Figure 10-6 displays the broad distribution of ethnolinguistic groups in the realm, but be aware that this is a generalization. At the scale of this map, myriad smaller groups cannot be represented (for example, Myanmar alone has some 135 different ethnic groups).

Figure 10-6 shows the rough spatial coincidence on the mainland between major ethnic group and contemporary political state. The Burman dominate in the country formerly called Burma (now Myanmar); the Thai inhabit the state once known as Siam (now Thailand); the Khmer form the nation of Cambodia and extend northward into Laos; and the Vietnamese occupy the long strip of territory facing the South China Sea.

Territorially, by far the largest population mapped in Figure 10-6 is classified as Indonesian, the inhabitants of the great island chain that extends from Sumatera west of the Malay Peninsula to the Malukus (Moluccas) in the east and from the Lesser Sunda Islands in the south to the Philippines in the north. Collectively, all of these peoples shown on the map—the Filipinos, Malays, and Indonesians—are known as Indonesians, but they have been divided by history and politics. And note as well that the Indonesians in Indonesia itself include Javanese, Madurese, Sundanese, Balinese, and other major groups; moreover, hundreds of smaller groups again cannot be mapped at this scale. In the Philippines, too, island isolation and contrasting ways of life are reflected in the cultural mosaic. Also part of this Indonesian ethnic-cultural complex are the Malays, whose heartland lies on the Malay Peninsula but who form minorities in other areas as well. Like most Indonesians, the Malays are Muslims, although Islam is a more powerful force within Malay society than it generally is in Indonesian culture.

FIGURE 10-6

In the northern margins of mainland Southeast Asia, numerous minorities inhabit remote corners of the countries in which the Burman (Burmese), Thai, and Vietnamese dominate. Those minorities tend to occupy such peripheral areas because the terrain is mountainous and the forest is dense, thereby hindering the governments of their national states from exerting complete control. This remoteness and sense of detachment give rise to aspirations of secession, or at least resistance to government efforts to establish full authority, often resulting in bitter ethnic conflict.

Immigrant Communities

Figure 10-6 reminds us that Southeast Asia also is home to major ethnic minorities from outside the realm. On the Malay Peninsula, note the South Asian (Hindustani) cluster. Such Hindu communities with Indian ancestries exist in many parts of the peninsula, but in the southwest they form the majority in a small area. In Singapore, too, South Asians form a significant minority. These communities emerged during the European colonial period. However, South Asians had arrived in this realm many centuries earlier, propagating Buddhism and leaving their architectural and cultural imprints on landscapes as far removed as Jawa and Bali. Most South Asian communities blended in reasonably well, but in the far northwestern corner of the realm one glaring exception has come to the fore in this decade: the Muslim Rohingyas of Myanmar's Bay of Bengal coastland. Originally from what is now adjacent Bangladesh, this long-downtrodden ethnic group has now been singled out for intensified persecution—ironically at the hands of Myanmar's dominant Buddhist majority.

The Rohingya Refugee Crisis

Since 2012, the relaxation of harsh military dictatorship in Myanmar, accompanied by tentative democratic openings, has been widely celebrated. But this transition has also unleashed suppressed ethnic tensions: in 2013, the world was stunned by the wave of savage violence perpetrated by Buddhists against a small Islamic minority residing in Myanmar's northwestern Rakhine State. Known as the Rohingyas (see Fig. 10-6), they have lived in Myanmar for generations and total about 1.3 million today. These Muslims, who have never been accepted into Myanmar's society, not only have been denied citizenship but also have long been subject to discrimination and intermittent acts of violence. The Buddhist majority refers to them as "Bengalis," a reference to their origins in neighboring Bangladesh that borders Rakhine on the north (Fig. 10-6).

Since this upsurge in ethnic violence began in 2013, some 140,000 Rohingyas have been interned "under protection" in (or fled to) refugee camps operated by the government. Many more fled the country to neighboring Thailand. But when Myanmar's leaders (pandering to inflamed anti-Muslim sentiment) closed down these overland escape routes in 2015, Rohingyas by the

thousands desperately took to the Indian Ocean. By following coast-paralleling routes southward through the Andaman Sea, they hoped to reach the presumably friendlier shores of Thailand, Malaysia, or even distant Indonesia (Fig. 10-1; see left photo). Unfortunately for most, this exodus became yet another horrifying experience: many of their flimsy, poorly equipped, overcrowded boats proved unseaworthy, and hundreds drowned; those who survived the arduous journey beyond Myanmar's waters were often initially denied landing by other countries; and most who finally found safe harbor were forced to move around from one place to another, receiving minimal sustenance and enduring inhumane living conditions. By mid-2015, this refugee crisis had finally attracted sufficient international attention to exert pressure for a regional response led by the three countries that the Rohingya boat people had targeted.

The Chinese

By far, the largest immigrant minority in Southeast Asia is of Chinese origin. The Chinese began arriving here during the Ming and early Qing (Manchu) dynasties, with the largest exodus occurring in the late colonial period (1870–1940), when as many as 20 million immigrated. The European powers at first encouraged this influx, employing the Chinese in administration and trade. These **Overseas Chinese 11** soon began to congregate in the realm's major cities, where they established Chinatowns and gained control over much of the local commerce. By the time the Europeans tried to limit Chinese immigration, World War II was about to break out and the colonial era would end soon thereafter.

Figure 10-7 shows the migration routes and current concentrations of Chinese in Southeast Asia. Most migrants originated in

The ethnic conflict between the Buddhist majority and the Rohingya Muslim minority in Myanmar's Rakhine State has now become one of Southeast Asia's major news stories of the decade. It first came to light in 2012 and then prominently resurfaced in the spring of 2015 when the accelerating flight of Rohingyas created an international refugee crisis. In the left photo, dated May 20, 2015, Rohingya migrants await rescue from an overloaded boat in the Andaman Sea just offshore from the northernmost section of Sumatera, one of Indonesia's main islands (Fig. 10-1). The right photo, taken just one week later in Myanmar's largest city (Yangon), captures the protest march of several hundred irate Buddhist monks to denounce foreign criticism of the country's mistreatment of its Rohingya population.

FIGURE 10-7

CHINESE IN SOUTHEAST ASIA
- → Migration route
- Major Concentration
- Secondary Concentration
- • City

National capitals are underlined

© J. Nijman, P. O. Muller, H. J. de Blij, and John Wiley & Sons, Inc.

themselves of Chinese descent—even though they are third-, fourth-, or fifth-generation, and have to varying degrees blended in with surrounding Malay culture (Singapore and Malaya were both part of the same British colonial entity for more than 130 years before independence in the 1960s). However, that shared Sinic ethnicity has not prevented Singaporeans from becoming ever more critical of the ongoing influx of 'newly rich' mainland Chinese. Resentment continues to rise concerning the role of purportedly arrogant, wealthy, immigrants in driving up real-estate prices, commandeering the best jobs, and disrespecting the city-state's strict codes of public behavior. Overall, this clash increasingly erodes social relations because 'old' Chinese and 'new' Chinese today refer to distinctly different identities within Singapore's multicultural mosaic.

RELIGIONS OF SOUTHEAST ASIA

Southeast Asia, more than any other realm, is a historic crossroads of religions (see Fig. G-9). With the migrants from the Indian subcontinent came their faiths: first Hinduism and Buddhism, later Islam. The Muslim religion, promoted by the increasing number of Arab traders who appeared on the scene, soon became the predominant religion in Indonesia (where 88 percent of the population adheres to Islam today). But in Myanmar, Thailand, and Cambodia, Buddhism remained supreme, and in all three of these mainland countries at least 90 percent of the people are currently adherents. In culturally diverse Malaysia, the Malays are Muslims (to be a Malay *is* to be a Muslim), and almost all Chinese are Buddhists; but most Malaysians of Indian ancestry remain Hindus.

Although Southeast Asia has generated its own local cultural expressions, most of what remains in tangible form has resulted from the infusion of foreign elements. For instance, take Angkor Wat, the enormous complex of religious structures built in Cambodia during the twelfth century AD and today one of the world's most famous monuments (see chapter-opening photo). It was originally constructed as a Hindu temple, dedicated to the god Vishnu. The carvings on the walls of Angkor Wat tell the stories of the Hindu epics, and the temple's designs are closely associated with Hindu cosmology. But during the fourteenth century, Buddhism took over this part of the realm, and the temple complex became a major place of worship for Buddhists.

Vietnam represents what is surely the ultimate fusion of diverse religious influences in Southeast Asia. In this country, for centuries there has been an almost casual blending of early Hinduism with Buddhism, Daoism, and Confucianism—all mixing comfortably with age-old traditions of ancestor worship.

southern China's Fujian and Guangdong provinces, and a large number invested much of their wealth back in China when the country opened up to foreign businesses in the 1980s. Clearly, the Overseas Chinese of Southeast Asia have played a significant role in shaping the PRC's economic miracle.

Today, Southeast Asia is home to about 27 million Overseas Chinese, more than half of this Sinic diaspora's worldwide total. Their lives here have often been challenging. The Japanese relentlessly persecuted Chinese living in Malaya during World War II. Later, in the 1960s, Chinese in Indonesia were accused of communist sympathies, and hundreds of thousands were killed. In the late 1990s, Indonesian mobs again attacked Chinese and their property because they resented their relative wealth—and because many Chinese had converted to Christianity during the colonial era and were now targeted by Islamic throngs. Such hostility continues to smolder in various parts of Southeast Asia, and episodic flare-ups against the Chinese can be expected to persist.

During this decade, Singapore is experiencing its own unique challenges with the latest wave of Overseas Chinese. Unique, because the overwhelming majority of native Singaporeans are

© H.J. de Blij

@ FROM THE FIELD NOTES...

"Like most major Southeast Asian cities, metropolitan Bangkok includes a large and prosperous Chinese sector. Roughly 14 percent of Thailand's population of 67 million is of direct Chinese ancestry, and most reside in the country's urban areas. However, this large non-Thai population is well integrated into local society, and intermixing is so widespread that Chinese ethnicity marks as much as half of Thailand's population. Still, Bangkok's 'Chinatown' remains a distinct element in the great city's landscape, and there is no mistaking its limits: Thai commercial signs shift to Chinese, goods offered for sale also change (Chinatown contains many shops selling gold, for example), and the urban atmosphere, from street markets to bookshops, becomes predominantly Chinese. This is a boisterous, noisy, energetic part of multicultural Bangkok, a vivid reminder of the Chinese commercial success in Southeast Asia."

THE IMPRINT OF COLONIALISM

When the European colonizers arrived in Southeast Asia, they encountered a patchwork of kingdoms, sultanates, principalities, and other traditional political entities whose leaders they tried to co-opt, overpower, or otherwise fold into their imperial schemes. These colonizers forged empires here, often by playing one state off against another; the Europeans divided and ruled. Out of this foreign intervention emerged the modern map of Southeast Asia, and only Thailand (formerly Siam) survived the colonial era as an independent entity—because it conveniently served as a buffer between the French sphere to the east and the British to the west.

Figure 10-8 shows the colonial framework in the late nineteenth century, (before the United States assumed control over the Philippines in 1898). The colonial powers divided their possessions into administrative units as they did in Africa and elsewhere. Some of these political entities became independent states when the colonial powers withdrew or were ousted by force. The French named their empire Indochina **12**. The *Indo* part of Indochina refers to cultural imprints received from South Asia: the Hindu presence; the importance of Buddhism, which came to Southeast Asia via Ceylon (Sri Lanka) and its seafaring merchants; the influences of Indian architecture and art (especially sculpture), writing, and literature as well as social structures. The *China* in the name Indochina signifies the role of the Chinese here: its emperors coveted Southeast Asian lands, and China's power penetrated deep into this realm. French Indochina consisted of present-day Vietnam, Cambodia, and Laos.

The British ruled a pair of major entities in Southeast Asia—Burma and Malaya—in addition to a large part of northern Borneo and many small islands in the South China Sea. Burma (now Myanmar) became independent in 1948 and the Federation of Malaysia was created in 1963 by the political unification of recently independent mainland Malaya, Singapore, and the former British dependencies on the largely Indonesian island of Borneo. Singapore, however, left this Federation in 1965 to become a sovereign city-state, and the remaining components were later restructured into peninsular Malaysia and, on Borneo, Sarawak and Sabah. Thus the term *Malaya* properly refers to the geographic territory of the Malay Peninsula, including Singapore and other nearby islands; the term *Malaysia* identifies the politico-geographical entity of which Kuala Lumpur is the capital.

The Dutch had come to Southeast Asia in the late sixteenth century, primarily attracted by the so-called spice trade, and they ended up controlling most of what is now Indonesia. Among the plants domesticated by the local people on these islands of the "East Indies" were black pepper, cloves, cinnamon, nutmeg, ginger, turmeric, and other condiments essential to flavorful meals. In Figure 10-1 you can detect, in eastern Indonesia between Sulawesi and New Guinea, a group of small islands called the Maluku Islands (formerly the Moluccas). The Dutch colonizers called these the ***Spice Islands*** because of the lucrative commerce in spices long carried on by Arab, Indian, and Chinese traders. Java (Jawa), the most populous and productive island, became the focus of Dutch administration; from its capital at Batavia (now Jakarta), Dutch colonialism threw a girdle around Indonesia's more than 17,000 islands, paving the way for the creation of the realm's largest and most populous state that is now home to a quarter-billion people.

Finally, the Philippines were controlled by Spain since 1571. But these colonizers were confronted with a major uprising here

© H. J. deBlij

@ FROM THE FIELD NOTES...

"Walking along Tunku Abdul Rahman Street in Kuala Lumpur, Malaysia, I had just passed the ultramodern Sultan Abdul Samad skyscraper when this remarkable view appeared: the old and the new in a country that seems to have few postcolonial hang-ups and in which Islam and democracy coexist. The British colonists designed and effected the construction of the Moorish-Victorian buildings in the foreground (now the City Hall and Supreme Court); behind them rises the Bank of Commerce, one of many banks here in the Malaysian capital. Look left, and you see the Bank of Islam, not a contradiction here in this economically diversified country. And just a few hundred yards away stands St. Mary's Cathedral, across the street from yet another bank. Several members of the congregation told me that the church was thriving and that there was no sense of insecurity here. 'This is Malaysia, sir,' I was told. 'We're Muslims, Buddhists, Christians. We're Malays, Chinese, Indians. We have to live together. By the way, don't miss the action at the Hard Rock Café on Sultan Ismail Street.' Now there, I thought, was a contradiction as remarkable as this scene."

just as the Spanish-American War broke out over the Caribbean island of Cuba in 1898 (there, as well, the locals revolted against Spanish domination). As part of the settlement of that brief war, the victorious United States replaced Spain in Manila as colonial proprietor and eventually guided the Philippines to independence in 1946. Today all 11 of Southeast Asia's states are independent, but centuries of imperial rule have left deeply embedded cultural imprints. In its urban landscapes, education systems, languages, and many other spheres, Southeast Asia will long carry the legacies of its colonial past.

SOUTHEAST ASIA'S EMERGING MARKETS

If Japan and the Asian Tigers set the early example in the 1960s and 1970s, with post-Mao China following suit since the mid-1980s, several Southeast Asian countries have since the turn of the century joined the ranks of the world's rapidly **emerging markets 13**. Vietnam, Indonesia, and Malaysia, particularly, have in recent years attracted substantial foreign investment and exhibited decidedly robust economic growth rates.

But this realm is marked by decidedly extreme economic disparities. The tiny oil-rich state of Brunei has an income higher than that of the United States (although this figure is not indicative of the earnings of its relatively large foreign workforce). However, it is comprehensively developed Singapore that records Southeast Asia's highest income—at a level about halfway between those of wealthy

Luxembourg and Switzerland. Malaysia comes in a distant third and is followed by Thailand. Populous Indonesia and the Philippines exhibit per-capita incomes less than one-twelfth of Singapore's. Finally, Vietnam, Laos, Cambodia, Myanmar, and East Timor, in that order, rank at the bottom of the list. Indeed, economically this realm is so internally dissimilar that its countries fall into all four World Bank income categories (see Fig. G-10).

Singapore's Leadership

With its soaring per-capita income, there can be little doubt that Singapore is the economic heart of Southeast Asia. With a mere 5.4 million citizens and only 619 square kilometers (240 sq mi) of territory, we are obviously not referring to size—this is all about the connections and centrality that enable Singapore to function as the leading **node 14** in a realmwide economic network. At the same time, geographers also rank Singapore as a top-tier world-city because it has major international linkages and exerts global influence. Singapore's container port is not only the largest in the realm but also the second-biggest in the world, underscoring its key role within and far beyond Southeast Asia (see photo). Moreover, ultramodern Changi Airport Singapore (SIN) is consistently ranked as one of the world's best international airports.

Singapore's commanding regional position is the result of its exceptional relative location. In the seventeenth century, it was Malacca farther up the Malay Peninsula that was the leading hub for trade and shipping in Southeast Asia. The Strait of Malacca (Fig. 10-8), named after that town, was already the most important

FIGURE 10-8

© J. Nijman, P. O. Muller, H. J. de Blij, and John Wiley & Sons, Inc.

sea route, providing access to the realm's waters for ships coming from the west. When the British sought to displace Dutch dominance here during the eighteenth century, they discovered an even better local base of operations: Singapore Island. Because it possessed a larger and deeper natural harbor than Malacca to accommodate the larger steamships of the time, Singapore swiftly rose to prominence as the British consolidated their power over this key corner of the world.

Most importantly, Singapore today is a symbol of modernity, a model for Southeast Asia's future. Throughout the realm, those who can afford it go there to shop, to connect to international flights, to transact business, to invest in real estate, or to send their children to one of the city-state's highly ranked universities. Imagine other Southeast Asian localities following in Singapore's footsteps over the next several years and consider how that might impact the realm and the world!

Singapore, Southeast Asia's unmatched world-city, has a state-of-the-art container port whose annual cargo-handling volume is second only to Shanghai's. This thriving city-state simultaneously functions as the nucleus of the realm's spatial economic system as well as its vitally important gateway to the global economy.

Prospects of Realmwide Integration: ASEAN

The overall development of Southeast Asia still has a long way to go. Political stability and increased regional integration will facilitate the process, and that is the long-term goal of **ASEAN 15**, the Association of Southeast Asian Nations. Founded during the late 1960s, this supranational organization has primarily been concerned with security. But that has been a constantly challenging effort because a wide range of conditions mark its ten member-states. With the lone exception of minuscule East Timor, ASEAN encompasses all of the realm's countries. These include one influential city-state (Singapore); an Islamic oil state (Brunei); two impoverished communist regimes (Vietnam and Laos); and a reforming military dictatorship whose population is rising from the ranks of the world's most deprived (Myanmar).

Another problem that ASEAN has failed to resolve, one that literally affects health across much of this realm, is the recurrent air pollution caused by Indonesia's massive, human-ignited forest fires. Depending upon weather conditions and prevailing winds, thick plumes of smoke emanating from Indonesia's Sumatera (where most of the burning occurs, often to enlarge oil palm plantations) stream out toward Jawa, Singapore, Malaysia, and countries farther afield. Repeatedly, Singapore's government has had to advise residents to stay indoors (a new record-high pollution level was set in 2013). Indonesia has been especially slow to address this environmental crisis and even refused to ratify the 2002 ASEAN Agreement on Transboundary Haze Pollution.

In 1992, 25 years after its founding, ASEAN was able to expand into the economic domain through **AFTA 16**, the ASEAN Free Trade Agreement, and here the payoff has been more substantial. AFTA has both engendered the lowering of tariffs and encouraged the expansion of trade within Southeast Asia. With lower wages than China, certain foreign-investment flows (e.g., in the garment industry) have shifted to Southeast Asia. And intra-realm trade has indeed surged in recent years—a critical forward step to avoid being completely overshadowed by China.

CHINA IN SOUTHEAST ASIA TODAY

Since the turn of this century, Southeast Asia has increasingly been drawn into the economic (and political) orbit of China. In 2010, ASEAN and China concluded a landmark free-trade agreement, with the Chinese most interested in acquiring raw materials from mainland Southeast Asia as well as accessing the realm's growing export markets. As Figure 10-9 shows, trade shifts resulting from this agreement have been significant.

East Asia and Southeast Asia are now steadily growing closer economically. This heightened interaction is not only about trade but also about investment flows, industrial development, divisions of labor, and infrastructural integration. With wages in Southeast Asia currently about one-third of China's, this differential is fueling the relocation of much industrial activity from China to Southeast Asian countries. Those manufactures are then exported directly from—for example, Vietnam—to China and/or other parts of the world. Increasingly, it is low-skill jobs that move to Southeast Asia, whereas the better-paying higher-skilled jobs remain in China.

In order to facilitate the increased production and transportation of goods within as well as to and from China, the Chinese established the Asian Infrastructure Investment Bank (AIIB) in 2015. The AIIB's members include most of the realm's states, and several European countries plus New Zealand have also joined (the United States, for geopolitical reasons, has not). With the Chinese now investing heavily to exploit Southeast Asia's raw materials, infrastructural improvements to upgrade connections to China are a leading priority. Most important of all is the high-speed rail network originating in Kunming, interior southern China's routeway hub for continental Southeast Asia and capital of Yunnan Province. The main rail corridor leads south from Kunming across northern Laos to Vientiane, the Laotian capital on the Mekong River that also forms the border with Thailand. From there, state-of-the-art rail lines will fan out in stages to Thailand, Cambodia, Vietnam, Malaysia, and eventually Singapore. Another high-priority project is the recently completed pair of oil and gas pipelines from Myanmar's Indian Ocean coast to Kunming (discussed later in the section on Myanmar).

As a result of this Chinese regional economic stimulation, Southeast Asia is growing faster than any other developing realm. And since the building of this inter-realm relationship has been fast-tracked, East and Southeast Asia are already more economically integrated than either North America with Middle America or Europe with Africa. Yet even though China's boost to the realm's economic development has been quite successful so far, there is another side to this story. Because the Chinese have not been reticent in asserting themselves, their continuing drive for dominance is creating both uneasiness and growing concern in a number of the realm's quarters. Some of this anxiety stems from

MAP ANALYSIS

FIGURE 10-9

Figure 10-9 allows you to study major trade shifts from 2000 to 2012. The map on the left shows the largest import origins for Southeast Asian countries in 2000 and in 2012; the map on the right shows the largest export destinations for the same years. For how many Southeast Asian countries was China the biggest export destination in 2012?

For how many countries was China the leading provider of imports? Which two powers outside the realm lost substantial ground (to China) in Southeast Asian trade since 2000? Access your WileyPLUS Learning Space course to interact with a dynamic version of this map and to engage with online map exercises and questions.

China's sheer economic prowess, which produces rather asymmetrical relationships; for instance, many infrastructure projects are formidable joint ventures that require comparatively larger investments from these much smaller Southeast Asian states. And, as is usually the case, the benefits are more favorable for China than for the host nation. Moreover, this issue is not entirely about economics: for countries such as Vietnam or Japan, historic sensitivities and/or longstanding rivalries inevitably come into play. Still another issue is a perceived lack of Chinese diplomacy as well as consideration for regional and national interests. Consider this remark made by China's foreign minister not long ago: "China is a big country and other countries are small countries, and that is just a fact."

GEOPOLITICS IN THE SOUTH CHINA SEA

If China's economic role in Southeast Asia is frequently viewed with mixed feelings, its recent geopolitical and military forays are increasingly met with indignation and opposition. Much of this revolves around China's maritime ambitions and its claims to territorial waters in the South China Sea. Since 2009, the PRC government

has circulated its so-called **nine-dashed-line map** **17** displaying Chinese claims in these waters—a map said to date back to 1949, the year the PRC was founded. The 'nine dashes' refer to the delimitation of Chinese maritime claims that effectively cover most of the South China Sea, and are mapped in red in Figure 10-10. In the words of one high-ranking Chinese official in 2012, "China does not want all of the South China Sea, it just wants 80 percent." But that 80 percent includes some 40 islands (which China says are illegally occupied by other countries) as well as international shipping lanes and seafloor zones potentially rich in deposits of petroleum and natural gas.

The disputes and conflicts over territorial waters in the South China Sea are numerous and complicated. The most important concern: (1) the *Paracel Islands*, claimed by China, Taiwan, and Vietnam; (2) the *Spratly Islands*, claimed by China, Vietnam, Malaysia, and Brunei; and (3) the *Scarborough Shoal*, claimed by China, Taiwan, and the Philippines (Fig. 10-10). ASEAN has been rather hopelessly divided on these geopolitical controversies, with Vietnam, Malaysia, and the Philippines strongly countering Chinese claims; Cambodia and Laos (tacitly) supporting China; and Singapore and Indonesia trying to steer a more neutral, diplomatic course.

FIGURE 10-10

When satellite imagery in mid-2015 confirmed that China was dredging enormous quantities of sand to reclaim and enlarge several major reefs (especially Mischief Reef) in the southern part of the Sea to construct runways and naval bases, ASEAN sounded a more alarmist tone—but words were not followed by actions. That did not prevent Vietnam, Thailand, and the Philippines from individually pursuing closer ties to the United States, clearly in search of an appropriate counterweight to Chinese leverage. In fact, all three quickly concluded military agreements with the U.S., and joint military exercises in the area have been initiated.

Whereas the United States has declared its neutrality on the matter of who owns which islands and shoals, it has emphasized the paramount importance of free and open international access to the waters of the South China Sea. Why? Because this is one of the most essential maritime trade corridors in the world. Think not only of the major shipping routes within the Southeast Asian realm that interconnect its chief ports of Singapore, Manila, Ho Chi Minh City, Bangkok, and Jakarta, but also of all the intercontinental trade coming from the west—from Europe and Africa as well as South and Southwest Asia—that passes through the Strait of Malacca, around Singapore, and then across the South China Sea on the way to Taiwan, South Korea, Japan, and China itself (not to mention all the seaborne cargoes that are constantly transported along this route in the opposite direction). According to some estimates, nearly half of all international oceanic trade passes through the South China Sea.

REGIONS OF SOUTHEAST ASIA

Southeast Asia's first-order regionalization recognizes the mainland–island dichotomy that is obvious from any map of this realm. But things are quite a bit more complicated than that because the southern part of the Malay Peninsula, occupied by the states of Malaysia and Singapore, exhibits physiographic, historical, and cultural characteristics to justify its inclusion in the *Insular* (island) rather than the *Mainland* region. Using the political framework as our grid, we map the broadest regional division of Southeast Asia in Figure 10-11:

- **Mainland region:** Vietnam, Cambodia, Laos, Thailand, and Myanmar (Burma)
- **Insular region:** Malaysia, Brunei, Singapore, Indonesia, East Timor (Timor-Leste), and the Philippines

Special note should also be taken of the discordance between the political matrix and the cultural-geographic reality along the eastern periphery of the Insular region. Here, even though Indonesia controls the western half of the island of New Guinea, all of New Guinea belongs to the neighboring Pacific realm.

MAINLAND SOUTHEAST ASIA

Five countries form the Mainland region of Southeast Asia (Fig. 10-11): the three remnants of Indochina—Vietnam, Cambodia, and Laos—in the east; Thailand in the center; and Myanmar (formerly Burma) in the west. Although a single religion, Buddhism, dominates cultural landscapes, this is a multicultural, multiethnic region (see Fig. 10-6). This may be one of the less urbanized regions of the world, yet the Mainland contains several major cities and the pace of urbanization is quickening today. Moreover, as Figure 10-11 shows, two countries (Vietnam and Myanmar) possess dual core areas.

We approach this region from the east, beginning our survey in Indochina where the United States fought and lost a disastrous war that ended in 1975, but whose impact on America continues to be felt today. After the Indochina War formally began in 1964 (U.S. involvement in Vietnam actually began earlier), some scholars warned that the conflict might spill over from Vietnam into Laos and Cambodia, and from there into Thailand, Malaysia, and even then-Burma. This view was based on the **domino theory** **18**, which holds that destabilization and conflict from any cause in one country can result in the collapse of order in one or more neighboring countries, triggering a falling-domino chain of events that can affect a series of contiguous states in a region. That did not actually happen, and in hindsight, at least, it is clear that some of the rhetoric of the time was infused with alarmist views from the West.

Vietnam: The Scars of War

Vietnam (population: 89.7 million) still carries the scars of the Indochina War (1964–1975), even though the overwhelming majority of the Vietnamese have no personal memory of that terrible conflict. Japan had invaded Vietnam during World War II, after which the French attempted to regain control, fighting Vietnamese nationalists in a war that lasted nine years before they were decisively ousted in 1954. But newly independent Vietnam did not become a unified state. Separate regimes took control: a communist one in the north based in Hanoi and a noncommunist counterpart in the south headquartered at Saigon. Vietnam's pronounced elongation had made things difficult for the French, and now it played a similar role during the postcolonial era. Note, in Figure 10-12, that Vietnam is widest in the north and south, with a slender "waist" in its middle zone—ensuring that North and South Vietnam were worlds apart in a number of ways. For more than a decade, the United States tried to prop up the Saigon regime that controlled the south, but the communists eventually prevailed; like China, Vietnam is still ruled by a communist government today. After hostilities terminated in 1975, as many as 2 million Vietnamese refugees set out in often-flimsy watercraft onto the South China Sea; of those "boat people" who survived, a majority settled in the United States.

Vietnam in Transition

Vietnam officially became a unified state in 1976, and since then the contrasts between north and south have slowly diminished. The northern capital, Hanoi, has long lagged behind bustling Saigon (now renamed Ho Chi Minh City to honor North Vietnam's revolutionary leader), but today its growing skyline reflects modernization and the overland links to its outport of Haiphong have been upgraded. With 3.6 million residents, Hanoi anchors the northern (Tonkin Plain) core area of Vietnam, the lower basin of the Red River (its agricultural hinterland).

A leading development focus in Vietnam since 2000 has been the expansion of agricultural production. One result is that the country now ranks among the world's five leading coffee producers; however, because Vietnam's mass-produced bulk coffee is raised plantation-style in the sun, the diffusion of this crop into the interior highlands has sparked substantial deforestation, worsening the annual monsoon-related floods that plague this now-unstable countryside. Other major Vietnamese crops, led by rice, include rubber, tea, sugar, and spices—all with increasingly favorable export opportunities.

The south has experienced more significant change and faster economic growth (see photo pair) as the country has opened up to the global economy. Today, Ho Chi Minh City is a burgeoning metropolis of 7.3 million, propelled by the rapid development of

SOUTHEAST ASIA: REGIONS AND STATES

POPULATION
- · Under 250,000
- · 250,000–1,000,000
- ● 1,000,000–5,000,000
- ● 5,000,000–10,000,000
- ● Over 10,000,000

National capitals are underlined

— Railroad
▨ Core area
▨ Transition Zone

0 100 200 300 400 600 800 Kilometers
0 100 200 300 400 Miles

FIGURE 10-11

FIGURE 10-12

the east bank of the Saigon River, which includes a Chinese-style Special Economic Zone as well as the booming New Saigon business/residential district. In fact, the Saigon area now accounts for more than a quarter of Vietnam's industrial output and about a third of its tax revenues.

Somewhat ironically, Vietnam these days has become one of the most pro-American countries in Southeast Asia despite its persistence as a communist state. Undoubtedly, these good feelings toward their former enemy in the "American War" at least partially reflect the growing perception of dominance by China, against whom the United States can provide a counterbalance. As noted above, American-Vietnamese relations have warmed considerably during the 2010s, with Vietnam even opening its strategically located port at Cam Ranh Bay to U.S. naval forces—the original

@ **FROM THE FIELD NOTES...**

"On a walking tour of Ho Chi Minh City (which many here still call Saigon), I encountered two contrasting images of present-day Vietnam, just minutes apart: a statue of the still-revered former communist leader Ho Chi Minh in front of City Hall; and, just three blocks away, the country's first Louis Vuitton store. The latter caters to the expanding base of higher-income Vietnamese consumers—not only a definite sign of national change but also that upscale designer fashion is truly global in scope."

builder of these facilities during the Indochina War a half-century earlier (Fig. 10-12).

Cambodia

Compact Cambodia is heir to the ancient Khmer Empire whose capital was Angkor and whose legacy is a vast landscape of imposing monuments including the majestic Buddhist-inspired temple complex, Angkor Wat (see the chapter-opening photo). Today, more than 85 percent of Cambodia's 15.1 million inhabitants are ethnic Khmers, with most of the remainder divided between Vietnamese and Chinese. The present capital, Phnom Penh, lies on the Mekong River, which crosses Cambodia before it enters and forms its massive delta in Vietnam (Fig. 10-12).

From 1970 to 1976, Cambodia suffered one of the most brutal and murderous regimes of the twentieth century. The so-called *Khmer Rouge* (*Red Khmer*, the name being a reference to its self-declared communist convictions and to the glorious past of the Khmer people) embarked on a malevolent course of terror and destruction in order to reconstruct *Kampuchea*—as leader Pol Pot referred to Cambodia—as a rural society. They drove townspeople into the countryside where they had no place to live or work, emptied hospitals to send the sick and dying into the streets, and outlawed both religion and family. By the time they were ousted from power by neighboring Vietnamese forces, as many as 2 million Cambodians (out of a population of 8 million at the time) had lost their lives in what forever will be known as the Cambodian Genocide.

Even four decades later, the country's postwar trauma still lingers. Many young children of the 1970s, now in their forties or fifties, became orphans and faced an uphill struggle in life.

Almost every family can tell of parents and/or close relatives who were murdered by the regime. The long-lasting effects of social and economic dislocation, especially in the countryside, have still not been overcome. Today, despite Cambodia's continuing status as one of Southeast Asia's poorest countries, it has finally begun to achieve meaningful economic growth. Income from oil exports began in 2012 (from newly discovered reserves off the coast) and could approach U.S. $2 billion annually by the end of this decade. Investments from abroad are multiplying, mostly from producers seeking to relocate their operations out of China (where wages are steadily rising) in order to capitalize on the much lower cost of labor in Cambodia and other disadvantaged Southeast Asian countries. Cambodians also are optimistic that tourism, largely focused on the Angkor temple complex in the northwest, is growing and may soon become a mainstay of the economy.

Laos

Landlocked Laos has no fewer than five neighbors, one of which is East Asia's giant, China (Fig. 10-12). The Mekong River forms a long stretch of its western boundary, and the important sensitive border with Vietnam to the east lies in mountainous terrain. With only 6.8 million people—more than half of them ethnic Lao, related to the Thai of Thailand—the country has a woefully inadequate infrastructure, hardly any industry, and less than 5 percent of its land is suitable for agriculture. Laos is only one-third urbanized, but the capital and largest city, Vientiane, does have a fast-growing population of right around one million.

Communist Laos has been very slow to follow the lead of its neighbors in opening up its economy, but momentum is now beginning to build. The country's first stock market began operating in 2011. Around the same time, a major new hydropower project on the central Laotian Nam Theun River, a tributary of the Mekong, began producing electricity for much of central Laos as well as selling surplus power to Thailand. And we have already discussed the much bigger Xayaburi Dam project on the Mekong (which despite drawing much regional criticism is on track to open at the end of the decade) in addition to the soon-to-be-completed high-speed rail line from Kunming to Vientiane that is certain to pull Laos even more tightly into China's orbit.

Thailand

In virtually every way, Thailand is the leading Mainland state. In contrast to its awakening neighbors, Thailand has been a strong participant in the realm's economic development for decades. This country of 67 million also exhibits some impressive demographic indicators: its annual growth rate is the lowest in the realm (even below Singapore's); it is the only Southeast Asian country whose projected 2050 population is lower than today's total (by some 8 percent); and Thailand leads its region in literacy and life expectancy. Its capital, Bangkok, is one of the world's most prominent **primate cities** 19 and the largest urban agglomeration in the Mainland region—on the verge of becoming a megacity with 9.3 million residents in 2015.

At least some of Thailand's internal political instability of recent decades can be attributed to the overreaching dominance of Bangkok and its elites as well as this primate city's long distance from the sizeable electorates of the more peripheral rural areas. Marked by frequent government turnovers and episodes of military intervention, Thailand has yet to face up to the challenge of legitimizing national politics across this sprawling, unevenly developed country. And given Bangkok's location, particularly in the context of Thailand's protruded territorial configuration, the task of national governance is not made any easier by the presence of a large Malaysian (and overwhelmingly Islamic) minority in the Malay Peninsula, Thailand's most remote corner that abuts the Malaysian border in the far south (Fig. 10-13).

As Figure 10-11 indicates, Thailand occupies the heart of the Mainland region. Even though it has no Red, Mekong, or Irrawaddy Delta, this country's central lowland is watered by a network of streams that flow off the northern highlands and the Khorat Plateau to the east. One of these waterways, the Chao Phraya, is the Mississippi of Thailand and forms the axis of the country's core area (Fig. 10-13). From the head of the Gulf of Thailand to Nakhon Sawan, this river is a highway for boat traffic. Barge trains loaded with rice head toward the coast, ferries move upstream, and freighters transport tin and tungsten (of which Thailand is among the world's leading producers). Bangkok sprawls along both sides of the lower Chao Phraya floodplain, here flanked by skyscrapers, pagodas, modern factories, boatsheds, ferry landings, luxury hotels, and myriad modest dwellings, all built in crowded confusion on the swampy terrain where the central lowland meets the Gulf of Thailand.

The Restive Peninsular South

As noted above, Thailand is a classical protruded state, with its southward-pointing land corridor extending away from the more compactly shaped main body of the state. Thailand's protrusion extends for hundreds of kilometers into the upper Malay Peninsula, narrowing to less than 32 kilometers (20 mi) in the vicinity of the Kra Isthmus (Fig. 10-13). A recheck of Figure 10-6 shows the dominance of ethnic Malays in the southernmost periphery of the protrusion—a population that is about 90 percent Muslim (the figure for Thailand as a whole is less than 6 percent). Moreover, the international border between Thailand and Malaysia is quite porous and can be crossed practically at will.

For more than a century, these southernmost provinces of Thailand have had closer ties with Malaysia across the border than with the Thai capital 1000 kilometers (600 mi) to the north. Today, rising Islamic militancy and violence here pose a mounting challenge to the Bangkok government (more than 5000 have been killed and thousands more maimed by local terrorist attacks since the mid-2000s). These events also have economic consequences: Phuket and other coastal resorts facing the Andaman Sea in this area—key components of Thailand's important tourist industry—are located just northwest of the troubled Malaysian border zone and are regarded by many as increasingly vulnerable to a spillover of continuing hostilities (see Fig. 10-13, inset map).

Myanmar (Burma)

Thailand's western neighbor, Myanmar (often still referred to as Burma, its former name), is one of Asia's poorest countries. Long oppressed by one of the world's most corrupt and brutal military dictatorships, its government turned a corner in 2011 and soon surprised the world by taking bold steps forward. Grinding poverty, crippling isolation along with sanctions (imposed by much of the world because of human rights violations), and growing dependence on China must finally have played a part in pushing Myanmar's leadership in a new direction. When the West embargoed and worked to isolate Myanmar's military regime in years past, China took the opportunity to invest an estimated U.S. $27 billion here; nonetheless, the Chinese are likely to retain and even increase their influence in Myanmar as the country opens up to the rest of the world.

Myanmar's Geographic Challenges

Myanmar's challenges are in some ways comparable to those of Thailand: both are protruded states, and in both cases the primary urban region is quite distant from the outermost rural peripheries (Fig. 10-14). Yangon (previously named Rangoon) is the country's main port, economic center, and primate city of nearly 5 million. In 2006, the military junta capriciously relocated the capital from Yangon 300 kilometers (200 mi) northward to a newly constructed headquarters named Naypyidaw, with the government asserting that its more central location would yield greater administrative

FIGURE 10-13

efficiency as well as better protection against (unspecified) enemies. After a slow start, the new capital city now houses just over one million people, not far behind the population size of the second city, the precolonial hub of Mandalay in central Myanmar (Fig. 10-14).

As Figure 10-14 indicates, the peripheral peoples of Myanmar occupy significant parts of the country that are designated as seven different States. The majority Burman-dominated areas (68 percent of a national population of 53.3 million) are designated as

Divisions. Among the leading minorities, the Shan of the northeast and far north, who are related to the neighboring Thai, account for 9 percent of the population, or 4.8 million. The Kayin (Karen), who constitute 7 percent (3.7 million), live in the neck of Myanmar's protrusion and have proclaimed their desire to create an autonomous territory within a federal Myanmar. The Rohingyas (2.4 percent; 1.3 million) of Rakhine State on the northwestern coast comprise the besieged Muslim minority discussed earlier in this chapter, whose plight has now escalated into a refugee crisis of

@ FROM THE FIELD NOTES...

© H.J. de Blij

"Thailand exhibits one of the world's most distinctive cultural landscapes, in both its urban and rural areas. Graceful pagodas, stupas, and spirit houses of Buddhism adorn city and countryside alike. Gold-layered spires rise above and beautify cities and towns that would otherwise be drab and featureless. The architecture of Buddhism has diffused into public architecture, so that many secular buildings, from businesses to skyscrapers, are embellished by something approaching a national style. The grandest expression of the form undoubtedly is a magnificent assemblage of structures within the walls of the Grand Palace in the capital, Bangkok, of which a small sample is shown here. Climb onto the roof of any tall building in the city's center, however, and the urban scene will display hundreds of such graceful structures, interspersed with the modern and the traditional townscape."

lengthy era of repression was the resurfacing of ugly, long-simmering ethnic tensions. Besides the appalling ethnic cleansing of the Rohingya community in Rakhine State, there have been violent clashes in the northernmost State of Kachin between government forces and the Kachin Independence Army. This strife also has potential international ramifications because the Kachin people—800,000 strong, constituting 1.5 percent of Myanmar's population—are part of a larger ethnic community that spills across both the Chinese and Indian borders.

Yet another challenge from the periphery is running its course in the mountainous eastern margins of Shan State, which has long been part of the infamous opium-producing region known as the **Golden Triangle** [20]. This name refers to the huge profitability of opium (the raw material of heroin, harvested from the sap of poppies) on the global market together with the triangular shape of this area that encompasses northwestern Laos and northernmost Thailand in addition to eastern Shan State (Fig. 10-14). Although poppy cultivation has almost been wiped out in Thailand, it is as strong as ever in Laos and especially Myanmar. Today, with only Afghanistan producing more opium than Myanmar, it is hard to discern if the Naypyidaw government is truly interested in clamping down on the opium trade: this illicit narcotic remains the source of enormous foreign revenues that, through elusive business connections, tend to find their way into the formal economy–including real estate transactions and infrastructural works in Myanmar's cities.

Myanmar's Regional Opportunities

Myanmar's opening is loaded with new possibilities for itself, its neighbors, and the world beyond—reinforced in no small part by the country's considerable geographic assets. To demonstrate the range of Myanmar's opportunities, we focus here on three major infrastructural projects that are presently underway, involving the neighboring countries of India, Thailand, and China.

The first is a joint venture with India to renovate and expand port facilities in Sittwe, the strategically located coastal city facing the northern end of the Bay of Bengal (Fig. 10-14). India's overriding interest here is to increase the accessibility of its landlocked easternmost States, a deeply impoverished highland periphery difficult to reach via overland routes. A fully developed port at Sittwe would allow the opening of a shorter, easier southern routeway: ships sailing around Bangladeshi waters from nearby Kolkata can offload their goods onto smaller vessels at Sittwe; the smaller vessels could then travel up the Kaladan River to the Indian border, from where trucks using upgraded roads would cover the final leg (Fig. 10-14).

The second project, located in the far south, is a far more significant enterprise from Myanmar's point of view—the so-called Dawei Development Project, a joint initiative with neighboring Thailand to convert the small southern city of Dawei on the Andaman Sea into a world-class port facility. A key initial step is the construction of a 160-kilometer (100-mi) highway eastward from Dawei to Thailand's Kanchanaburi (a town with first-rate surface

realmwide proportions. The Mon (2 percent; 1.1 million) were in what is today Myanmar long before the Burmans and even introduced Buddhism to the Irrawaddy Basin; today they still seek the return of ancestral lands from which they were ousted.

When Myanmar finally shook off the excesses of dictatorship in the early 2010s and embraced long-awaited reforms, it was hoped this would usher in a period of national progress and a more democratic society. To some extent, advances were made—political prisoners were freed, press censorship was largely ended, and long overdue elections took place. But a byproduct of the ending of that

FIGURE 10-14

connections to the lower Chao Phraya Basin), located halfway between Dawei and Bangkok (Fig. 10-14). Plans further call for Dawei to become Myanmar's chief industrial node, a newly built complex to be modeled after China's Special Economic Zones. For Myanmar, this could spark a noteworthy entry into the global economy, while far more advanced Thailand would gain valuable direct access for its burgeoning core area to the nearby Andaman Sea and India beyond it.

The third infrastructural project has already been briefly noted: China's completion of a dual set of oil and gas pipelines from a new port terminal at Kyaukphyu, south of Sittwe, 1150 kilometers (700 mi) northeastward to the key Chinese city of Kunming (Fig. 10-14).

These linkages are of huge importance to the PRC because they directly provide southern China with much needed energy, thereby circumventing the far longer maritime route through the Singapore Straits and the South China Sea. And Myanmar benefits as well: the pipelines pass through the center of the country and allow for a series of delivery points in such major cities as Mandalay.

INSULAR SOUTHEAST ASIA

On the peninsulas and islands of Southeast Asia's southern and eastern periphery lie 6 of the realm's 11 states (Fig. 10-11). Few regions in the world contain so diverse a set of countries. Malaysia, the former British colony, consists of two major territories separated by hundreds of kilometers of South China Sea. The realm's southernmost state, Indonesia, sprawls across thousands of islands from Sumatera in the west to New Guinea in the east. North of the Indonesian archipelago lies the Philippines, another island-chain country that once was a U.S. colony. These are three of the most acutely fragmented states on Earth, and each has faced the challenges that such politico-geographical division presents. This Insular region of Southeast Asia also contains three small but significant sovereign entities: a city-state, a sultanate, and a ministate. The city-state is Singapore, once a part of Malaysia (and one instance in which internal centrifugal forces were too powerful to be overcome). The sultanate is Brunei, an oil-rich Muslim territory on the island of Borneo that seems to be transplanted from the Persian Gulf. The ministate is severely impoverished East Timor, which only achieved independence (from Indonesia) in 2002. Few parts of the world are more varied or interesting geographically.

Malaysia

One portion of Malaysia's national territory lies on a continent and the other on one or more islands. The country is a colonial political artifice that combines a pair of decidedly disparate components into a single state: the southern end of the Malay Peninsula and the northern sector of the island of Borneo. These are referred to, respectively, as **West Malaysia** and **East Malaysia** (Fig. 10-11). The appellation "Malaysia" came into use in 1963, when the original Federation of Malaya, on the Malay Peninsula, was expanded to incorporate the territories of Sarawak and Sabah in Borneo. When the name *Malaya* is used, it refers to the peninsular part of the Federation, whereas *Malaysia* refers to the total political entity.

The Malays of the peninsula constitute 50 percent of the country's population of 29.7 million. They possess a compelling cultural identity expressed in adherence to the Muslim faith, a common language, and a strong sense of territoriality. The Chinese came to the Malay Peninsula as well as to northern Borneo in substantial numbers during the colonial period, and now they account for roughly 23 percent of Malaysia's population. Hindu South Asians were in this area long before the Europeans, and for that matter before the Arabs and Islam arrived on these shores; today they still form a substantial minority of almost 7 percent of the population, clustered, like the Chinese, on the western side of the peninsula (see Figs. 10-6 and 10-15).

FIGURE 10-15

Malaysia's ethnic and racial groups have long coexisted in relative harmony, but relations today are marked by rising social tensions. As Malays increasingly assert claims based on 'majority rights,' the Chinese find themselves in much the same minority position that characterizes the Overseas Chinese across the realm. The Indian population is much smaller and feels ever more crowded out. Add religious tensions to the mix (that occasionally include Muslim demands for a separate *Sharia* legal system), and the notion of a common Malaysian identity now looks more fragile than ever before. The ruling ethnic Malay government extended its hold on power in the 2013 elections, but did so by the narrowest of margins—a sign of the country's deepening political polarization heightened by growing expressions of dissent among its minorities.

The populous peninsular component of Malaysia remains the country's dominant sector, containing 11 of its 13 States and approximately 80 percent of its population. Here, the Malay-dominated government has very tightly controlled economic and social policies as it pursues nationwide modernization. The Strait of Malacca (Melaka) to the west continues to be one of the world's busiest and most strategic waterways, and forms yet another crucial *choke point* in the flow of resources and goods between major geographic realms. A longstanding problem in the Strait, which has sharply resurfaced in the mid-2010s, is piracy. The spatial configuration of this narrow, 1000-kilometer (625-mi)-long waterway makes it highly susceptible to such attacks. Moreover, the Strait is always a particularly inviting target because the 50,000-plus vessels that annually transit through it constitute fully one-third of the world's shipping and carry one-quarter of its oil.

Malaysia, despite the loss of Singapore and notwithstanding its recurrent ethnic troubles, has become a leading player in contemporary Southeast Asia. Its new **Multimedia Supercorridor** not only anchors the national core area but also showcases the country's commitment to twenty-first-century technology—symbolized by Putrajaya, the recently built high-tech administrative capital. Elsewhere, the strong skills and modest wages of the local workforce have attracted many companies, and the government has capitalized on its opportunities, such as encouraging the creation of an ultramodern manufacturing complex on the far northwestern island of Pinang, where Chinese outnumber Malays by two to one.

East Malaysia

The decision to combine the 11 Sultanates of Malaya with the States of Sabah and Sarawak on Borneo (Fig. 10-11), creating the country we now call Malaysia, had far-reaching consequences. Those two offshore States may constitute 60 percent of Malaysia's territory, but they are home to barely 20 percent of the population. They endowed Malaysia with abundant energy resources, huge stands of timber, and vast tracts of land suitable for palm oil plantations (the country is now the world's leading exporter, with Sarawak and Sabah producing 55 percent of the national crop). But they also complicated Malaysia's ethnic complexion because each State contains more than two dozen indigenous groups (for

example, immigrant Chinese form Sarawak's largest ethnic group). These locals complain that the federal government in Kuala Lumpur treats Malaysian Borneo as a colony, and politics here are contentious and fractious. It is therefore quite likely that Malaysia will eventually confront devolutionary forces in East Malaysia.

Brunei

Also located on Borneo—along the coast near where Sarawak and Sabah meet—is Brunei (Fig. 10-17 inset map), a rich, oil-exporting Islamic sultanate far from the Persian Gulf. Brunei, which became independent from the United Kingdom in 1984, is only slightly larger than Delaware and has a population of just 450,000. Thus the sultanate is a mere ministate—except for the discovery of oil and natural gas, which made it Southeast Asia's richest country. And there are indications that further discoveries will be made in the offshore zone owned by Brunei. The Sultan of Brunei rules as an absolute monarch; his palace in the capital, Bandar Seri Begawan, is reputed to be the world's largest. He will have no difficulty finding customers for his oil in energy-poor eastern Asia.

Brunei's growing population—in which immigration plays a larger role than internal natural increase—is ethnically two-thirds Malay and about 12 percent Chinese. Most inhabitants live and work near the offshore oilfields in the northern district of Brunei Muara that includes the capital city. Evidence of profligate spending is everywhere, ranging from sumptuous palaces to magnificent mosques to the most luxurious of resort hotels. In this respect Brunei offers a stark contrast to surrounding East Malaysia, but even within Brunei there are significant differences. The country's interior, where a small minority of indigenous groups survive, remains an area of subsistence agriculture and rural isolation, its villages a world apart from the modern splendor of Bandar Seri Begawan.

Singapore

In 1965, a fateful event occurred in Southeast Asia. Singapore, the crown jewel of British colonialism in this realm, seceded from the recently independent (1963) Malaysian Federation and became a sovereign state, albeit a tiny city-state (Fig. 10-16). With its unparalleled relative location, its multiethnic and well-educated population, and its no-nonsense government, Singapore then overcame the severe limitations of space and the absence of raw materials to become one of the four original Asian Tigers.

With a mere 619 square kilometers (239 sq mi) of territory and 5.4 million people, space is at a premium in Singapore, and this is a constant challenge for the government. The highest priority is to develop cutting-edge but space-conserving manufacturing and service industries. Benefiting from its relative location, the old port of Singapore had become one of the world's busiest even before independence. It thrived as an **entrepôt** 21 between the Malay Peninsula, the rest of Southeast Asia, Japan, and other emerging economic powers on the Asian Pacific Rim. Crude oil from Southeast Asia still

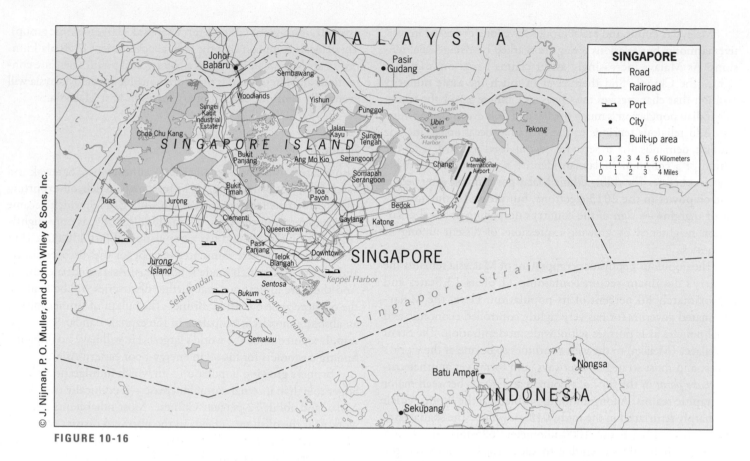

FIGURE 10-16

© J. Nijman, P. O. Muller, and John Wiley & Sons, Inc.

is unloaded and refined in Singapore, then shipped to East Asian destinations. Raw rubber from the adjacent Malay Peninsula and from Indonesia's nearby island of Sumatera is shipped to China, Japan, the United States, and many other countries. Timber from Malaysia, rice, spices, and other foodstuffs are processed and forwarded through Singapore. In return, automobiles, machinery, and equipment are imported into Southeast Asia and distributed almost exclusively via Singapore.

Singapore's current development strategy stresses two primary objectives. The first is to focus its industries more tightly on leading-edge information technology, automation, and biotechnology. The second is to forge an ever stronger **Growth Triangle 22** involving Singapore's developing neighbors, Malaysia and Indonesia. Accordingly, Malaysia and Indonesia would supply the necessary raw materials and relatively inexpensive labor, and Singapore would supply the capital and technological expertise.

The ethnic composition of Singapore's population today is just below 75 percent Chinese, 13 percent Malay, and 9 percent South Asian. The government is Chinese-dominated, and its policies have served to sustain ethnic-Chinese control. Indeed, the government of Singapore has encouraged immigration from the PRC to stabilize the ethnic mosaic of the city-state, where natural-increase rates, especially among its native Chinese, have for some time been well below the replacement level. But, as noted earlier, this may not be a long-term solution because tensions are mounting between the resident and newly arrived immigrant Chinese communities.

Indonesia

The fourth-largest country in the world in terms of human numbers is also the globe's most expansive archipelago. Spread across a chain of more than 17,000 mostly volcanic islands, Indonesia's 251 million people live both separated and clustered—separated by water and clustered on islands large and small. With more than 300 discrete ethnic clusters, over 250 languages, and just about every religion practiced on Earth, Indonesia's survival as a unified state is remarkable. Indeed, Indonesia's national motto is *bhinneka tunggal ika*—diversity in unity. Nevertheless, Indonesia nominally is the largest Muslim country in the world: 88 percent of the people adhere to Islam, and in the cities the silver domes of neighborhood mosques rise densely above the townscape. Although until recently Indonesian Islam has been relatively moderate, as noted earlier more overt Islamization has been on the rise, with new laws now in effect to ban public displays of affection and the kinds of clothing women may wear in public.

The Major Islands

With almost 150 million inhabitants, **Jawa** is one of the world's most densely peopled places (see Figs. G-7 and 10-5) as well as one of the most agriculturally productive, with its terraced rice-growing paddies rising up the highly fertile flanks of myriad active volcanoes. Jawa also is the most highly urbanized part of a country in which half the people still live off the land. The primate city of

INDONESIA

Jakarta National capital
Jambi Provincial capital

POPULATION
• Under 250,000
• 250,000–1,000,000
● 1,000,000–5,000,000
● 5,000,000–10,000,000
● Over 10,000,000

▲ Mountain over 1800 m (5900 ft.)

— Railroad — Road

Core area
Forest
Park; reserve
Main area of human activity
Swamp; mangrove

BRUNEI
— Oilfield
— Gasfield

0   100   200   300   400 Kilometers
0   100   200   300 Miles

Longitude East of Greenwich

FIGURE 10-17

© J. Nijman, P. O. Muller, and John Wiley & Sons, Inc.

381

Jakarta has today become the heart of the much larger Greater Jakarta conurbation that now exceeds 30 million, or 25 percent of Indonesia's urban population. Thousands of factories, their owners taking advantage of low prevailing wages, have been built in this sprawling conurbation (which goes by the acronym *Jabodetabek*, after its constituent cities), badly straining its infrastructure and overburdening the port of Jakarta.

Sumatera, spelled 'Sumatra' in the colonial era, is Indonesia's westernmost island and forms the western shore of the critical Strait of Malacca; Singapore lies across the Strait from approximately the middle of this island. Although much larger than Jawa, Sumatera has only about one-third as many people (roughly 50 million). In colonial times, the island became a base for rubber and palm oil production; its high relief makes possible the cultivation of a wide variety of crops, and neighboring Bangka and Belitung yield petroleum and natural gas. Palembang is the key urban center in the south. In recent years, the northeastern coastlands of Sumatera, along with parts of mainland Malaysia and southern Kalimantan, have been blanketed by huge and expanding palm oil plantations to drive up export revenues—but at enormous cost to the regional environment, as discussed earlier in this chapter (also see satellite image). In Sumatera's far north lies Aceh, the land area closest to the epicenter of the 2004 undersea earthquake that triggered a disastrous tsunami that claimed about 230,000 lives in the Indian Ocean Basin. The province of Aceh was designated a "Special Territory" within Indonesia because of the separatist movement of its indigenous Acehnese population; but that crisis has become defused thanks to the massive post-tsunami relief effort and follow-up of the central government—and relations between the province and Jakarta today are much improved.

Kalimantan is the Indonesian sector that comprises 73 percent of the island of Borneo, a slab of the Earth's crystalline crust whose mountainous backbone is of erosional, not volcanic, origin. Larger than Texas, Borneo has a deep, densely forested interior that is a last refuge for some 35,000 orangutans as well as dwindling numbers of Asian elephants, rhinoceroses, and tigers. Along with a number of indigenous peoples, these wildlife species survive even as loggers, plantation builders, and small farmers relentlessly penetrate their shrinking habitat. Borneo is home to a comparatively small human population (roughly 19 million on the Indonesian side, less than 8 percent of the country's total) on poor tropical soils. Aboriginal peoples, principally the Dayak clans, have traditionally had less impact on the natural environment than the Indonesian and Chinese immigrants as well as the multinational corporations that log the rainforests and clear space for new farms. As Figure 10-17 shows, the only towns of any size in Kalimantan lie on or near the coast; access routes into the interior are still minimal.

Sulawesi consists of a set of intersecting, volcanic mountain ranges rising above sea level; the 800-kilometer (500-mi) Minahasa Peninsula, propelled by volcanic activity, continues to build itself into the Celebes Sea. This northern peninsula, a favorite of the Dutch colonizers, remains the most developed sector of an otherwise rugged and remote island, with Manado its relatively prosperous urban focus. Seven major ethnic groups inhabit the valleys and basins scattered among the mountains, but the population of about 18 million also includes many immigrants from Jawa, especially in and around the southern center of Ujungpandang. Subsistence farming is the leading mode of life, although logging, scattered mining, and fishing augment the economy. Clashes between Muslims and Christians occur intermittently in remote areas.

Papua, the Indonesian name for the western half of the island of New Guinea, has become an issue in national politics. Papua is bordered on the east by a superimposed, mostly straight-line boundary that separates it from the state of Papua New Guinea on the eastern half of the island (Fig. 10-17). This territory was forcibly taken over by Indonesia from the Dutch in 1969, a remnant of the Netherlands East Indies colony that was held back when Indonesia became independent in 1945. Papua contains about 22 percent of Indonesia's territory, but its (fast-growing) population is only 3.5 million—just under 1.5 percent of the country's total. The indigenous inhabitants of this territory—which is essentially an Indonesian colony—are Papuan, with most living in the remote reaches of this mountainous, densely forested island. Papua is economically important to Indonesia because it contains what is reputed to be the world's richest gold mine as well as its third-largest open-pit copper mine. But political consciousness has now reached the Papuans: the Free Papua Movement remains active, holding small demonstrations in the capital (Jayapura), displaying a Papuan flag, and demanding recognition.

Jeff Schmaltz, LANCE/EOSDIS Rapid Response / NASA Earth Observatory

In June of 2013, human-ignited forest fires on Sumatera generated enormous plumes of thick smoke that drifted eastward to engulf Singapore and surrounding areas of Malaysia and Indonesia. The Singaporean authorities warned the public to stay indoors, and the city-state's government loudly insisted to its Indonesian counterpart that a greater effort must be made to prevent plantation owners and smaller-plot farmers from deliberately burning vast stands of trees to clear land for additional palm oil production.

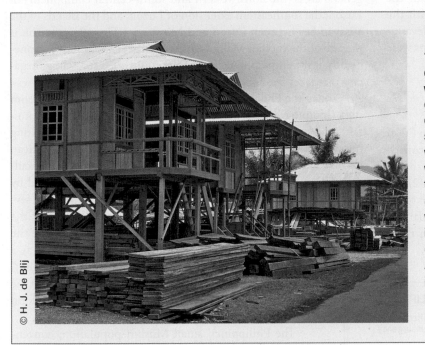

© H. J. de Blij

"I drove from Manado on the Minahasa Peninsula in northeastern Sulawesi to see the ecological crisis at Lake Tondano, where a fast-growing water hyacinth is clogging the water and endangering the local fishing industry. On the way, in the town of Tomolon, I noticed this side street lined with prefabricated stilt houses in various stages of completion. These, I was told, were not primarily for local sale. They were assembled from wood taken from the forests of Sulawesi's northern peninsula, then taken apart again and shipped from Manado to Japan. 'It's a very profitable business for us,' the foreman told me. 'The wood is nearby, the labor is cheap, and the market in Japan is insatiable. We sell as many as we can build, and we haven't even begun to try marketing these houses in Taiwan or China.' At least, I thought, this wood was being converted into a finished product, unlike the mounds of logs and planks I had seen piled up in the ports of Borneo awaiting shipment to East Asia."

Indonesia's Challenges and Prospects

Indonesia's development challenges are complicated by spatial fragmentation and an especially uneven population distribution. From 1974 to 2001, Indonesia's central government instituted a policy known as **transmigration** **23**, inducing millions from the densely populated inner islands (especially Jawa, Bali, and Madura) to relocate to such sparsely inhabited, peripheral islands as Kalimantan and Sulawesi. As many as 8 million Jawanese were relocated to outer islands, but half of them experienced a decline in their standard of living; many were reduced to bare subsistence to scratch out a living in the tropical woodlands confiscated from indigenous inhabitants—which turned out to be unsuitable for the farming methods used by the settlers. In the end, cultural conflict, ecological havoc, and rampant deforestation finally led the Indonesian government to admit failure and cancel this grandiose social program in 2001.

Today, Indonesia's stature within Southeast Asia is steadily rising. This is not just the biggest country in the realm: it is also striving to achieve a commensurate level of economic development and prosperity. The economy has expanded vigorously during the past decade, and incomes here now exceed those of Laos, Cambodia, and Myanmar. Yet they are still ranked well below those of Thailand or Malaysia. At the same time, centrifugal forces are intensifying (growing Islamic militancy in Jawa is of particular concern), and ongoing decentralization and devolution throughout this immense archipelago will continue to challenge the Indonesian government.

East Timor

As Figure 10-17 shows, Timor is the easternmost of the sizeable Lesser Sunda Islands, and throughout the colonial Dutch East Indies period the Portuguese maintained their own colony on the eastern half of it. In 1975, Indonesian forces overran that colony and formally annexed it the next year. Indonesian rule, however, was even less benign than that of the Portuguese, and soon a bitter struggle for independence was under way. Eventually, in 1999, under UN supervision, the people of East Timor voted overwhelmingly for independence, but Indonesia still refused to let go, and unleashed a brutal military occupation. It took another three years of armed conflict and foreign intervention (led by Australia, which previously had supported Indonesia) until independence was finally realized in 2002. By then, the violence had devastated most of this ministate and its infrastructure. East Timor's leaders proclaimed the official name of their country to be *Timor-Leste*—Portuguese for "Timor East." Not surprisingly, more than a decade later nation-building in this Connecticut-sized country remains a formidable proposition.

As Figure 10-17 shows, East Timor is a fragmented country, with a dominant east (where the coastal capital, Dili, is located) and a tiny **exclave** **24** on the northern coast of Indonesian West Timor named Ocussi. Even though East Timor has always been overwhelmingly agricultural, farming was badly neglected during the turmoil unleashed under the Indonesian occupation. To make matters worse, the country's population of 1.2 million continues to grow explosively, exhibiting a yearly natural-increase rate of 2.7 percent (35 percent higher than that of Laos, the realm's second-fastest-growing country); East Timor's fertility rate of 5.7 children per woman of childbearing age in 2013 ranks among the world's highest. Recently, the economic growth rate has been more encouraging, and there is the prospect of billions of dollars to be earned from future oil revenues (the country shares the rights to offshore oil reserves in the Timor Sea with Australia). But the nation-building project here has barely begun to deal with the country's extreme poverty as well as the lingering social devastation of a prolonged conflict in which some 100,000 civilians were killed by the Indonesian military.

The Philippines

North of Indonesia lies a lengthy archipelago of more than 7000 islands inhabited by 98.4 million people. This island chain constitutes the Philippines, and it can be subdivided into three groups: (1) Luzon, largest island of all, and Mindoro in the north; (2) the Visayan group in the center; and (3) Mindanao, second-largest, located in the south (Fig. 10-18). The Philippines' population, concentrated where productive farmlands lie, is densest in three general areas: the northwestern and south-central components of Luzon; the southeastern extension of Luzon; and the islands of the Visayan Sea between Luzon and Mindanao (Fig. 10-5). Luzon is the site of the capital, Manila-Quezon City (12.9 million, or one-eighth of the entire population of the country), a sprawling megacity that faces the South China Sea.

Today the Philippines, adjacent to the world's largest Muslim state (Indonesia), is 81 percent Roman Catholic, 9 percent Protestant, and only 5 percent Muslim. Out of the Philippines melting pot, where Malay, Arab, Chinese, Japanese, Spanish, and American elements met and mixed, has emerged the distinctive Filipino culture. It is not a homogeneous or unified culture, as is reflected by the dozens of Malay languages in use in the islands, but it is in several ways unique. At independence in 1946, the largest of the Malay languages, Tagalog (also known as Pilipino), became the country's official language. But English is widely learned as a second language, and a Tagalog-English hybrid, "Taglish," is increasingly heard today. The Chinese component of the population is small (1.6 percent) but dominates the local business scene.

The Philippines' small Muslim population, concentrated in the southernmost flank of the archipelago, and especially on densely forested Basilan Island (Fig. 10-18), has long decried its marginalization in this predominantly Christian country. Over the past generation, a half-dozen Islamic organizations have promoted the Muslim cause through tactics that have ranged from peaceful negotiation with the government to violent insurgency.

Agriculture continues to dominate the Philippines' economy. Alluvial as well as volcanic soils, together with ample moisture in this largely humid-tropical environment, enable self-sufficiency in rice and other staples and make the Philippines a net exporter of farm products despite its high population growth rate of 1.8 percent. Unemployment is rampant, trade linkages are insufficient, and additional land reform is badly needed, as is social restructuring to reduce the controlling influence of a comparatively small group of families over national affairs.

Perhaps more than any other people, Filipinos take jobs in foreign countries in massive numbers, proving their capacity to succeed in jobs they cannot find at home. More than 10 million Filipinos, over one-quarter of the total workforce, are now employed abroad. The global merchant marine would not exist without Filipino sailors, and Filipina nurses and domestic workers can be found from Dubai to Dublin to Dubuque. Funds sent home to family members by the emigrants make the Philippines a world leader in remittances, a monetary inflow that is the basis for at least 10 percent of the country's GDP in any given year.

Overall, the Philippines now sustains a lower-middle-income economy, and given a longer period of political stability and success in mitigating the problems outlined above, it should be able to rise to the next level and finally take its place among the Asia-Pacific's growth poles.

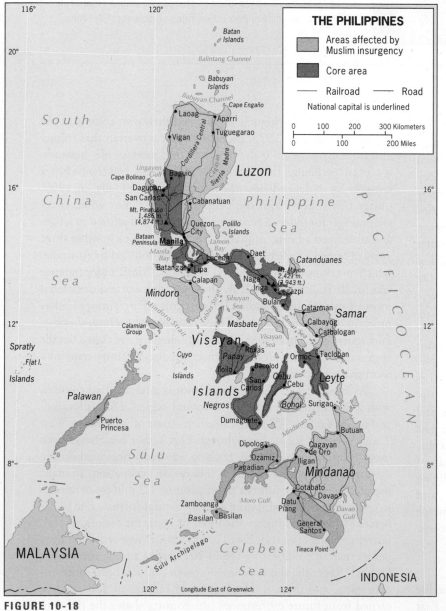

FIGURE 10-18

© J. Nijman, P. O. Muller, and John Wiley & Sons, Inc.

Currently, the Philippines is establishing itself in a pair of niches in the global economy: international call centers and the outsourcing of digital services. The latter operate through such Internet websites as *oDesk*—one of several that serves as a global marketplace for digital services by freelancers who perform data entry and other routine back-office work. Although such outsourcing is highly unpredictable as a growth industry and compensation is low by worldwide standards, the **business process outsourcing 25** sector today employs about one million full-time workers and is the second-largest contributor (after remittances) to the Philippines' foreign exchange earnings.

In its foreign relations, this country is changing course. A quarter-century ago, it terminated the lease for the huge U.S. Subic Bay Naval Base—then the largest of all American foreign military installations. But now the Philippines government is once again seeking closer relations with the United States. As noted earlier, this is related to China's growing maritime-territorial assertiveness in Southeast Asia—not only regarding the South China Sea in general but also longstanding specific disputes between China and the Philippines concerning the Spratly Islands and Scarborough Shoal.

POINTS TO PONDER

- The two smallest countries of the realm are also by far the wealthiest.

- Is Southeast Asia becoming a Chinese periphery, or is China going to face a backlash against its assertive strategies in this realm?

- In your daily life, how many products do you use that contain palm oil? How can you find out? As a consumer, do you contribute to deforestation?

- Under Malaysia's constitution, a person of Malay ancestry (unlike someone with Chinese or South Asian roots) is automatically registered as a Muslim.

- If all current plans come to fruition, China will have built 27 dams in the upper basin of the Mekong River before the waterway enters the Southeast Asian realm in Laos.

- With the deepening involvement of India, China, and Thailand in Myanmar over the past few years, it is hard to imagine that this country could ever return to its isolationist postcolonial past.

© Andrew Bertuleit Photography / Corbis Images

Ayers Rock, located in the heart of the immense, interior Outback near the geographic center of Australia, is the most famous landform in the Austral realm. This miniature tableland is all that is left of a much larger surrounding plateau that has been completely eroded away to form the flat plain that frames this iconic formation of red sandstone rock. Geomorphologists use the term *inselberg* (German for "island mountain") to describe this kind of geologic remnant, which consists of rocks that are far more resistant to the local forces of erosion than those that once surrounded it. Ayers Rock—called *Uluru* by the Aboriginal people who first settled Australia—stands about 350 meters (1150 ft) above the plain and has a circumference of nearly 10 kilometers (5.8 mi). Despite its extreme remoteness from where most Australians live, about 300,000 visitors annually make the long trek out here, and the tourist infrastructure has kept pace.

11

THE AUSTRAL REALM

IN THIS CHAPTER

- The realm's Asian turn
- Australia's exceptional biogeography
- China covets Australian commodities
- Aboriginal claims to land and resources
- New Zealand's Maori question

CONCEPTS, IDEAS, AND TERMS

Austral	1	Outback	10
Southern Ocean	2	Primary sector	11
Subtropical Convergence	3	Environmental degradation	12
West Wind Drift	4	*El Niño*	13
Biogeography	5	Desalinization	14
Wallace's Line	6	Aboriginal land issue	15
Aboriginal population	7	APEC	16
Federation	8	Peripheral development	17
Unitary state	9		

FIGURE 11-1

The Austral realm is the only one that lies entirely in the Southern Hemisphere. It is also unique in that it has no land link of any kind to a neighboring realm and is thus completely surrounded by oceans and seas. It is second only to the Pacific as the world's least populous geographic realm. Appropriately, its name refers to its location (the word **austral** 1 derives from the Latin for "south")—a location far from the sources of its dominant cultural heritage, but relatively close to its newfound economic partners on the Asian Pacific Rim.

DEFINING THE REALM

Two countries constitute the Austral realm: Australia, in every way the dominant national entity, and New Zealand, physiographically more varied but demographically much smaller than its giant partner (Fig. 11-2). Between them lies the South Pacific Ocean's Tasman Sea. To the realm's west lies the Indian Ocean, to its east the Pacific, and to the south the frigid Southern Ocean.

Today, this southern realm is at a crossroads. On the doorstep of populous eastern Asia, the realm's Anglo-European legacies have been diversified accordingly by the infusion of many Asian cultural strains. Markets throughout the Asia Pacific are buying ever larger quantities of raw materials. Chinese and other Asian tourists fill hotels and resorts. The streets of Sydney and Melbourne display a multicultural panorama unimagined just two generations ago. Indigenous minorities—the Polynesian Maori in New Zealand and Aboriginal communities in Australia—are demanding greater rights and wider acknowledgment of their cultural heritage. All these changes have stirred political debate. Issues ranging from immigration quotas to indigenous land rights dominate, exposing social fault lines (city versus Outback in Australia; North Island versus South Island in New Zealand). Aborigines and Maori first settled this realm, then the Europeans arrived, and now Asians are an increasingly significant economic and cultural element.

LAND AND ENVIRONMENT

Physiographic contrasts between massive, compact Australia and elongated, fragmented New Zealand are related to their locations with respect to the Earth's tectonic plates (see Fig. G-4). Australia, carrying some of the geologically most ancient rocks on the planet, lies at the center of its own Australian Plate. New Zealand, younger and less stable, lies at the convulsive convergence of the Australian and Pacific plates. Earthquakes are rare in Australia, and volcanic eruptions are unknown; New Zealand has plenty of both. This locational contrast is also reflected by differences in relief (Fig. 11-1). Australia's highest relief occurs in what Australians call the Great Dividing Range, the mountains that line the east coast from the Cape York Peninsula in the north to southern Victoria State, with an outlier in Tasmania. The highest point along these old, now eroding mountains is Mount Kosciusko, 2230 meters (7316 ft) tall. In New Zealand, entire ranges are higher—Mount Cook in the Southern Alps, for example, reaches 3754 meters (12,316 ft). West of Australia's Great Dividing Range, the physical landscape mostly exhibits low relief, with some local exceptions such as the Macdonnell Ranges near the continental center; plateaus and plains dominate (Fig. 11-1; Fig. 11-3, top map). The Great Artesian Basin is a key physiographic region, providing underground

▮▮▮ MAJOR GEOGRAPHIC QUALITIES

AUSTRAL REALM

1. Australia and New Zealand constitute a geographic realm by virtue of relative location, territorial dimensions, and dominant cultural landscape.

2. Despite their inclusion in a single geographic realm, Australia and New Zealand differ physiographically. Australia has a vast, dry, low-relief interior; New Zealand is mountainous and has a temperate climate.

3. Australia and New Zealand are marked by peripheral development—Australia because of its aridity, New Zealand because of its topography.

4. The populations of Australia and New Zealand are not only peripherally distributed but also highly clustered in urban centers.

5. The economic geography of Australia and New Zealand is dominated by the exporting of livestock products and specialty goods such as wine. Australia also is a significant producer of wheat and possesses a rich, diverse base of mineral resources.

6. Australia and New Zealand are now integrated into the economic framework of the Asian Pacific Rim, principally as suppliers of raw materials. Mineral-rich Australia's newly discovered and exploited gas reserves are adding to its export base.

7. As immigration from the western Pacific Rim increases, Australia has, in a cultural sense, "returned" to Asia.

© J. Nijman, P. O. Muller, H. J. de Blij, and John Wiley & Sons, Inc.

AUSTRAL REALM: POPULATION DISTRIBUTION, 2016

One dot represents 50,000 persons

| 0 | 400 | 800 | 1200 | 1600 Kilometers |
| 0 | 250 | 500 | 750 | 1000 Miles |

FIGURE 11-2

water sources in what is otherwise desert country. To its south lies the continent's predominant Murray-Darling river system. The area mapped as *Western Plateau and Margins* in the lower map of Figure 11-3 contains much of Australia's mineral wealth.

Climates

Figure G-6 reveals the effects of latitudinal position and interior isolation on Australia's climatology. In this respect, Australia is far more varied than New Zealand, its climates ranging from tropical in the far north, where rainforests flourish, to Mediterranean in two corners of the south. The interior is dominated by desert and steppe conditions, the semiarid steppes providing the grasslands that sustain tens of millions of livestock. Only in the extreme east does Australia possess a zone of humid temperate climate, and here lies most of the country's core area. New Zealand, by contrast, is totally under the influence of the Southern and Pacific oceans, creating moderate, moist conditions, temperate in the north and progressively colder in the south.

The Southern Ocean

Twice now we have referred to the **Southern Ocean** 2 , but try to find this ocean on maps and globes produced by such well-known cartographic organizations as the National Geographic Society. From their maps you would conclude that the Atlantic, Pacific, and Indian oceans reach all the way to the shores of Antarctica. Australians and New Zealanders know better. They experience the frigid waters and persistent winds of this great weather-maker on a daily basis.

For us geographers, it is a good exercise to turn the globe upside down now and then. After all, the usual orientation is quite arbitrary.

Modern mapmaking started in the Northern Hemisphere, and the cartographers put their hemisphere on top and the other at the bottom. That is now the norm, and it can distort our view of the world. In souvenir shops in the Southern Hemisphere, you see upside-down maps showing Australia and Argentina at the top, and Europe and Canada at the bottom. But this matter has a serious side. An inverted view of the globe shows us how vast the ocean encircling Antarctica is (see Fig. 11-4). The Southern Ocean may be remote, but its existence is real.

Where do the northward limits of the Southern Ocean lie? This ocean is bounded not by land but by a marine transition called the **Subtropical Convergence** 3 . Here the very cold, extremely dense waters of the Southern Ocean meet the warmer waters of the Atlantic, Pacific, and Indian oceans. It is quite sharply defined by changes in temperature, chemistry, salinity, and marine fauna. Flying over it, you can actually observe it in the changing colors of the water: the Antarctic side is a deep gray, the northern side a greenish blue. Even though the Subtropical Convergence shifts seasonally, its position does not vary far from latitude 40° South, which also is the approximate northern limit of Antarctic icebergs. Defined this way, the great Southern Ocean is a huge body of water that circulates clockwise (from west to east) around Antarctica, which is why we also call it the **West Wind Drift** 4 .

©Theo Allofs/Latitude/Corbis

The breathtakingly beautiful Great Barrier Reef, off the coast of the State of Queensland in northeastern Australia (Fig. 11-3, top map). This is the largest coral reef on the planet, sometimes described as the only living thing on Earth that is visible from space. It was designated a World Heritage Site in 1981, and most of it now lies within Great Barrier Reef Marine Park. Not surprisingly, it ranks as one of the world's most prominent ecotourism destinations.

AUSTRALIA: PHYSIOGRAPHY

Desert and semi-desert

Tropical and sub-tropical grassland and forest

Mediterranean scrub

0   200   400   600   800 Kilometers
0   100   200   300   400 Miles

PHYSICAL LANDSCAPES

Eastern Uplands   Central Lowlands   Western Plateau & Margins

0   400   800   1200   1600 Kilometers
0   500   1000 Miles

FIGURE 11-3

© J. Nijman, P. O. Muller, H. J. de Blij, and John Wiley & Sons, Inc.

Biogeography

One of this realm's defining characteristics is its wildlife. Australia is the land of kangaroos and koalas, wallabies and wombats, possums and platypuses. These and numerous other *marsupials* (animals whose young are born very early in their development and then are carried in an abdominal pouch) owe their survival to Australia's early isolation during the breakup of Gondwana (see Fig. 7-3). Before more advanced mammals could enter Australia and replace the marsupials, as happened in every other part of the world, this landmass was separated from Antarctica and India, and today it contains the world's largest assemblage of marsupial fauna. Australia's vegetation has distinctive qualities as well, notably the hundreds of species of eucalyptus trees native to this geographic realm. The study of fauna and flora in spatial perspective integrates the disciplines of biology and geography in a field known as **biogeography** **5**, and Australia serves as a gigantic research laboratory for its practitioners.

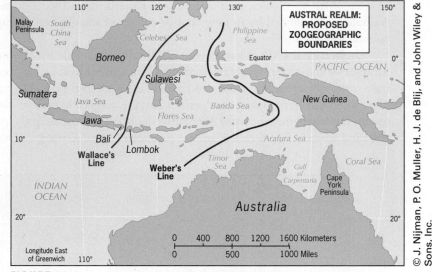

FIGURE 11-4

Biogeographers are especially interested in the distribution of plant and animal species, as well as the relationships between plant and animal communities and their natural environments. (The spatial analysis of plant life is called *phytogeography*, and that of animal life is known as *zoogeography*.) In 1876, one of the founders of biogeography, the great British naturalist Alfred Russel Wallace, posited that the zoogeographic boundary of Australia's fauna was located beyond Australia in the Sunda island chain to the northwest, between Borneo and Sulawesi, and just east of Bali (Fig. 11-4).

Wallace's Line **6** soon was challenged by other researchers, who found species Wallace had missed and who visited islands Wallace had not. There was no question that Australia's zoogeographic realm terminated somewhere in Indonesia's Sunda archipelago, but where? Western Indonesia was the habitat of nonmarsupial animals such as tigers, rhinoceroses, and elephants in addition to primates; New Guinea clearly was part of the realm of the marsupials. How

far had the more advanced mammals progressed eastward along the island stepping stones toward New Guinea? The zoogeographer Max Weber discovered evidence that led him to postulate his own ***Weber's Line***, which, as Figure 11-4 shows, lay very close to New Guinea.

In Australia, the arrival of the **Aboriginal population** **7** about 50,000 years ago appears to have triggered an ecosystem collapse. Widespread burning of the existing forest, shrub, and grassland vegetation all across Australia probably led to the spread of desert scrub and to the rapid extinction of most of the continent's large mammals soon after the human invasion occurred. Those species that survived faced a second crisis tens of thousands of years later when the European colonizers introduced their livestock, leading to the further destruction of remaining wildlife habitats. Survivors include marsupials such as the koala bear and the wombat, but the list of extinctions is much longer.

REGIONS OF THE REALM

Australia is the predominant component of the Austral realm, a continent-scale country in a size category that also includes China, Canada, the United States, and Brazil. For two reasons, however, Australia has fewer regional divisions than these other countries: its comparatively uncomplicated physiography and its small human population.

▌ AUSTRALIA

Positioned on the Pacific Rim, almost as large as the 48 contiguous U.S. States, well endowed with farmlands and vast pastures, rivers, groundwater supplies, mineral deposits, and energy resources, served

by good natural harbors, and populated by just under 24 million mostly well-educated people, Australia is one of the most geographically fortunate countries in the world.

Historical and Political Geographies

From the late eighteenth century onward, the Europeanization of Australia doomed the continent's Aboriginal societies. The first to suffer were those situated in the path of British settlement on the coasts, where penal colonies and free towns were founded. Distance protected the Aboriginal communities of the northern interior longer than elsewhere; but in Tasmania, the indigenous Australians died off in just decades after having lived there for perhaps 45,000 years. Eventually, the major coastal settlements became the centers of seven different colonies, each with its own hinterland; by 1861, Australia was delimited by its now-familiar pattern of straight-line boundaries (Fig. 11-5).

On January 1, 1901, following years of difficult negotiations, the Australia we know today finally emerged: the Commonwealth of Australia, consisting of six States and two Federal Territories (Fig. 11-5). The two Federal Territories are the *Northern Territory*, assigned to protect the interests of the substantial Aboriginal population concentrated there and agitating for Statehood, and the *Australian Capital Territory*,

carved from southern New South Wales to accommodate the federal capital of Canberra that was completed in 1927.

Australia's six States are New South Wales (capital Sydney), at 7.6 million the most populous and politically powerful; Victoria (Melbourne), small but populous by Australian standards with 5.9 million residents; Queensland (Brisbane), at 4.8 million, with the Great Barrier Reef offshore and tropical rainforests in its north; South Australia (Adelaide), with 1.7 million, where the Murray-Darling river system reaches the sea; Western Australia (Perth) with barely more than

**AUSTRALIA:
POLITICAL DIVISIONS, CAPITALS,
AND COMMUNICATIONS**

POPULATION

- Under 50,000
- 50,000–250,000
- 250,000–1,000,000
- 1,000,000–5,000,000

Canberra National capital
Sydney State capital
—— Railroad
—— Main Road
Australian core area

0 200 400 600 800 1000 Kilometers
0 250 500 Miles

© J. Nijman, P. O. Muller, H. J. de Blij, and John Wiley & Sons, Inc.

FIGURE 11-5

2.6 million people in an area of more than 2.5 million square kilometers (nearly 1 million sq mi); and Tasmania (Hobart), containing 515,00, the island across the Bass Strait from the mainland's southeastern corner that lies in the path of Southern Ocean storms.

In earlier chapters, we referred to the concept of federalism, which is a form of politico-territorial organization. The word "federal" comes from the Latin *foederis*, implying alliance and coexistence, a union of consensus and common interest—a **federation 8**. It stands in contrast to the idea of a centralized or **unitary state 9**. For this, too, the ancient Romans devised a term: *unitas*, meaning "unity." Most of Europe's states are unitary, including the United Kingdom. Although most Australians came from that tradition, they managed to overcome their differences and forge a Commonwealth that was, in effect, a federation of States with different viewpoints, economies, and objectives, separated by enormous distances along the rim of a particularly remote island continent.

The Role of Distance

Australians often talk about distance. One of their leading historians, Geoffrey Blainey, labeled it a tyranny—an imposed remoteness from without and a divisive part of life within. Even today, Australia is far from nearly everywhere on Earth. A trans-Pacific jet flight from Los Angeles to Sydney takes about 14 hours nonstop and is correspondingly expensive. You may have had the fortune of traveling abroad several times, yet never made it to Australia.

Freighters carrying products to European markets take ten days to two weeks to get there. Within Australia, distances are of continental proportions as well, and Australians pay the price—literally. Until some upstart private airlines started a price war, Australians paid more per mile for their domestic flights than air passengers anywhere else in the world.

But distance also was an ally, permitting Australians to ignore the obvious. Australia was a British progeny, a European outpost. Once you had arrived as an immigrant from Britain or Ireland, there was a wide range of environments to choose from, magnificent scenery, vast open spaces, and seemingly limitless opportunities. When the Japanese Empire expanded in the 1930s and 1940s, Australia's remoteness saved the day. When boat people by the hundreds of thousands fled Vietnam in the mid-1970s aftermath of the Indochina War, almost none reached the shores of Australia. However, it should be noted that from the beginning in 1901, the "White Australia Policy" operated to prevent immigration from becoming an issue: its self-perceived comforts of isolation led Australia to adopt an all-white admission policy, the last vestiges of which were not officially terminated until 1978.

Core and Periphery

As Figure 11-2 shows, Australia is a large landmass, but its population is almost entirely concentrated in a (discontinuous) core area that lies in the east and southeast, most of which faces the Pacific

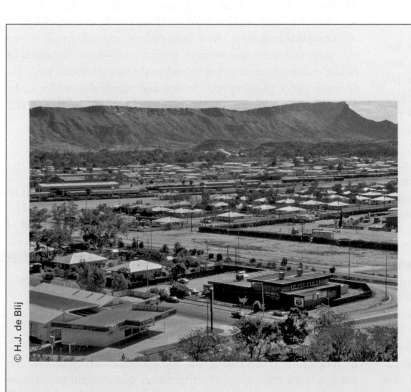

© H.J. de Blij

@ FROM THE FIELD NOTES...

"My most vivid memory from my first visit to Alice Springs in the heart of the Outback is spotting vineyards and a winery in this parched, desert environment as the plane approached the airport. I asked a taxi driver to take me there and got a lesson in economic geography. Drip irrigation from an underground water supply made viticulture possible; the tourist industry made it profitable. None of this, however, is evident from the view seen here: a spur of the Macdonnell Ranges overlooks a town of bare essentials under the hot sun of the Australian desert. What Alice Springs has is centrality: it is the largest settlement in a vast area Australians often call "the centre." Not far from the midpoint on the nearly 3200-kilometer (2000-mi) Stuart Highway from Darwin on the Northern Territory's north coast to Adelaide on the Southern Ocean, Alice Springs also was the northern terminus of "The Ghan," the Central Australia Railway line (before it was extended north to Darwin in 2004), seen in the middle distance. The shipping of cattle and minerals is a major industry here. You need a sense of humor to live here, and the locals have it: the town actually lies on a river, the ephemeral Todd River. An annual boat race is held, and in the absence of water the racers carry their boats along the dry river bed. No exploration of Alice Springs would be complete without a visit to the base of the Royal Flying Doctor Service, which brings medical help to the farthest-flung villages and homesteads."

Ocean's Tasman Sea. Figure 11-5 shows that this crescent-shaped Australian heartland extends from north of the city of Brisbane to the vicinity of Adelaide and includes the largest city, Sydney; the capital, Canberra; and the second-largest city, Melbourne. A secondary core area has developed in the far southwest, centered on Perth. Beyond the core areas lies the enormous periphery that Australians call the **Outback 10** (see *Field Note*).

To better understand the evolution of this spatial arrangement, it helps to refer again to the map of world climates (Fig. G-6). Environmentally, Australia's most favored strips face the Pacific and Southern oceans as well as a sector of the Indian Ocean, and they are not large. We can describe the country as a coastal rimland with cities, towns, farms, and forested slopes giving way to the immense, arid, interior Outback. On the western flanks of the east coast's Great Dividing Range lie the extensive grassland pastures that catapulted Australia into its first commercial age—and on which still graze one of the largest sheep herds in the world (about 72 million sheep in 2014, producing a major share of all the apparel wool sold in the world). Where it is moister, to the north and east, cattle by the millions graze on ranchlands. This is frontier Australia, over which livestock have ranged for nearly two centuries.

See an urban scene like this, and you realize why Australians refer to their land as 'the lucky country.' This is Sydney Harbour on a sunny, breezy day when sailboats by the hundreds emerge from coves and inlets along the estuary that connects the Pacific Ocean with the booming center of Sydney (whose skyline has grown both vertically and horizontally since this photo was taken in 1987). Captains of ocean liners plying the world often remark that Australia's biggest city is not only their favorite port but also the grandest harbor entrance of them all.

An Urban Culture

Despite the vast open spaces and romantic notions of frontier and Outback, Australia is an urban country, with 89 percent of all Australians living in cities and towns. The large cities lie along the coast, the centers of manufacturing complexes as well as the foci of agricultural areas. For all its vastness and youth, Australia has developed a remarkable cultural identity, a sameness of urban and rural landscapes that persists from one end of the continent to the other.

Sydney, often called the New York of Australia, lies on a spectacular estuarine site, its compact, high-rise central business district overlooking a port bustling with ferry and freighter traffic. Sydney today is a widely spread out metropolis of 4.5 million, with multiple outlying centers dominating its burgeoning suburban ring; brash modernity and reserved British ways somehow blend here. Melbourne (4.2 million), sometimes regarded as the Boston of Australia, prides itself on its more interesting architecture and more cultured ways. Brisbane (2.2 million), the capital of Queensland, which also anchors Australia's Gold Coast and adjoins the Great Barrier Reef, is the Miami of Australia; unlike Miami, however, its residents can find nearby relief from the summer heat in the mountains of its immediate hinterland (as well as at its beaches). Perth (1.9 million) is separated from its eastern Australian neighbors by three-quarters of a continent, and from Southeast Asia and Africa by thousands of kilometers of ocean—but due to the ever-expanding mining activities of Western Australia it is increasingly drawn into the global economy.

And yet, each of these cities—as well as the capitals of South Australia (Adelaide), Tasmania (Hobart), and, to a lesser extent, the Northern Territory (Darwin)—exhibits an Australian character of unmistakable quality. Life is both orderly and unhurried; streets are clean; slums are uncommon; graffiti rarely seen. Standards of public transport, urban schools, and healthcare provision are high. Spacious parks, pleasing waterfronts, and plentiful sunshine make Australia's urban life more agreeable than just about anywhere else on Earth.

Economic Geography

Australia today ranks twelfth among the world's countries in GDP, and for the vast majority of Australians life is quite comfortable. In terms of key development indicators, Australia is far ahead of all its western Pacific Rim competitors except Japan and Singapore. The country's initial prosperity was achieved in the mines and on the farms, not in the cities. Australia has material assets of which other Asia Pacific countries can only dream.

Agricultural Abundance

In agriculture, sheep-raising was the earliest commercial venture, but it was the technology of refrigeration that brought world markets within the reach of Australian beef producers. Wool, meat, and wheat have long been the country's big three income earners; Figure 11-6 displays the immense pastures in the east, north, and

AUSTRALIA:
AGRICULTURE AND MINERAL RESOURCES

Pastoralism: sheep
Pastoralism: cattle
Dairy farming
Commercial grain
Sugarcane
Mediterranean (livestock, grain, fruit)
Specialized horticulture
Hunting and gathering
Fishing grounds
No agriculture or pastoralism

▲	Natural gas	**Ab** Asbestos
⚒	Crude oil	**Ag** Silver
●	Iron ore	**Au** Gold
⛏	Coal	**Bx** Bauxite

Cu	Copper	**Pt**	Platinum
Mn	Manganese	**Sn**	Tin
Ni	Nickel	**U**	Uranium
Pb	Lead	**W**	Tungsten
		Zn	Zinc

0   200   400   600   800   1000 Kilometers
0   125   250   375   500 Miles

—— Rivers
- - - Intermittent rivers

Adapted (in part) and updated with permission from *The Jacaranda Atlas*, 4 rev. ed., p. 29. Jacaranda Press (Milton, Qld., Aust.), 1992.

© J. Nijman, P. O. Muller, H. J. de Blij, and John Wiley & Sons, Inc.

FIGURE 11-6

west that constitute the ranges of Australia's huge herds. The zone of commercial grain farming forms a broad crescent extending from northeastern New South Wales through Victoria into South Australia, with a major outlier covering much of the hinterland of Perth.

And keep in mind the scale of this map: Australia is only slightly smaller than the continental United States. Commercial grain farming in Australia is big business. As the climate map (Fig. G-6) suggests, sugarcane grows along most of the humid, subtropical,

coastal-lowland strip of Queensland, and Mediterranean crops (including grapes for Australia's highly prosperous wine industry) cluster in the hinterlands of Adelaide and Perth. Mixed horticulture concentrates in the Murray River Basin, including rice, grapes, and citrus fruits, all under irrigation. In addition, as elsewhere across the world, dairying has developed close to the large metropolitan areas. With its considerable range of environments, Australia yields a great variety of crops.

Mineral Wealth but Limited Manufacturing

Australia's mineral resources, as Figure 11-6 shows, also are diverse. Major gold discoveries in Victoria and New South Wales produced a ten-year gold rush starting in 1851 and ushered in a new economic era. By the middle years of that decade, Australia was producing 40 percent of the world's gold. Subsequently, the search for more gold led to the discoveries of additional minerals including nickel, copper, iron ore, uranium, tungsten, bauxite, and more. Since the 1990s, the enormous demand for raw materials from China as well as emerging India has marked a long-running commodity boom in Australia.

New finds are still being made today, and even oil and natural gas have been discovered both on land and offshore (Fig. 11-6). Most promising are the undersea deposits off the northwest coast, which has sparked an energy boom that is transforming Darwin. This northernmost city of Australia directly faces the fossil-fuel riches of the Timor Sea and is becoming a major processor of natural gas, liquefying it for export to East and Southeast Asia. Another liquid natural gas (LNG) boomtown is the east-coast, central Queensland port of Gladstone: here, an ultramodern processing-plant complex produces LNG from piped-in coal-seam gas, which is extracted by new hydraulic fracturing technologies from coal deposits in interior Queensland's Bowen and Surat Basins (Fig. 11-6).

Australian manufacturing caters mainly to the domestic market and is quite diversified, producing a wide array of industrial goods ranging from machinery made of locally produced steel to chemicals to textiles to paper products. These industries cluster in and near the major urban areas where consumers are located. The domestic market in Australia is not large, but it remains relatively affluent. This makes it attractive to foreign producers, and Australia's retail outlets overflow with high-priced goods from China, Japan, South Korea,

About 320 kilometers (200 mi) south-southeast of Port Hedland in the Pilbara region of Western Australia, Mount Whaleback near the town of Newman is one of the world's leading sources of high-grade iron ore. China's insatiable demand for this ore has generated the planet's largest "open-cut" iron mine, a huge and growing gash in the natural landscape—just one of many such negative impacts resulting from Australia's role as raw-material supplier to industrializing economies.

© Photoshot Holdings Ltd/Alamy

the United States, and Europe. Overall, the continuing prominence of the **primary sector** **11** (mining and agriculture) and heavy reliance on raw-material exports indicates that the manufacturing sector is only partially developed, in spite of a long history of protectionist government policies. Whereas exports continue to be dominated by ores, fuels, and other commodities, the leading imports include cars, trucks, and medical equipment and supplies. In 2014, Australia's top five partners in two-way trade were China, Japan, the United States, South Korea, and Singapore.

Australia's Environmental Problems

Environmental degradation **12**, unfortunately, is practically synonymous with Australia. First the Aborigines, then the Europeans and their livestock, inflicted catastrophic damage on Australia's natural environments and ecologies. Great stands of magnificent forest were ravaged. In Western Australia, centuries-old trees were simply ringed and left to die so that the sun could penetrate through their leafless crowns to nurture the grass below for pasture for introduced livestock. In island Tasmania, where Australia's native eucalyptus tree reaches its greatest concentrations (comparable to California's redwood stands), tens of thousands of hectares of this irreplaceable treasure have been lost to chain saws and pulp mills. Many of Australia's unique marsupial species have been driven to extinction, and many more are endangered or threatened. "Never have so few people wreaked so much havoc on the ecology of so large an area in so short a time," observed one prominent Australian geographer not long ago.

More recently, the intensification of mining activities as well as oil and gas exploration have raised ecological stress to the next level. The energy-related development of Gladstone referred to above offers a telling example. Environmental activists are sounding alarms concerning the new hydraulic fracturing techniques for coal-seam gas production because they threaten Queensland's agricultural resources. Moreover, these same groups are vehemently protesting the expansion of port facilities because the coastal zone surrounding Gladstone lies within the supposedly protected confines of the Great Barrier Reef.

Another environmental problem involves Australia's wide and long-term climatic variability. In a predominantly arid continent, droughts in the moister fringes are the worst enemy, and Australia's history is replete with devastating dry spells. Australia is vulnerable to **El Niño** **13** events (see Chapter 3), but global warming may also be playing a role in the process. Figure 11-7 shows what are believed to be the effects of global warming on Australia between 1997 and 2011: whereas the northern (tropical) zones have experienced enhanced rainfall, the populated southwestern and southeastern coastal regions received less precipitation and have been more vulnerable to drought. The most serious drought (see photo pair), in fact the worst ever recorded, lasted throughout the 2000s and imperiled the entire Murray-Darling river basin, Australia's breadbasket). Part of this was a water deficit stemming from natural causes—but it was greatly exacerbated by rising demands for water in Australia's burgeoning urban areas in addition to

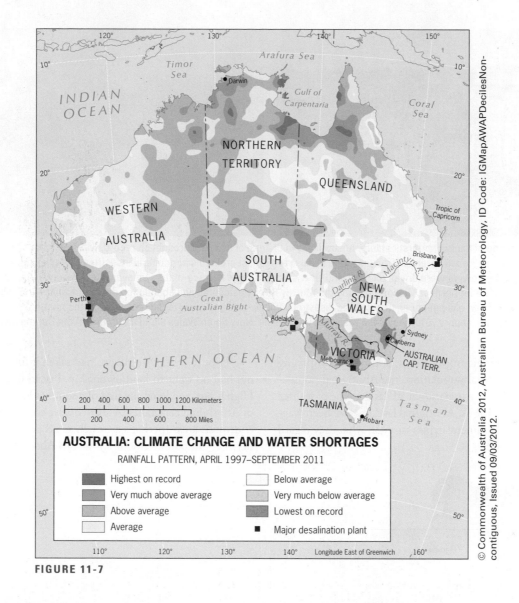

© Commonwealth of Australia 2012, Australian Bureau of Meteorology, ID Code: IGMapAWAPDecilesNon-contiguous, Issued 09/03/2012.

AUSTRALIA: CLIMATE CHANGE AND WATER SHORTAGES

RAINFALL PATTERN, APRIL 1997–SEPTEMBER 2011

- Highest on record
- Very much above average
- Above average
- Average
- Below average
- Very much below average
- Lowest on record
- ■ Major desalination plant

FIGURE 11-7

excessive damming, well-drilling, and water diversion in the valleys of upstream tributaries of these two vital rivers. The Australian government responded by initiating the first fully coordinated drainage-basin-control program in the country, but all stakeholders recognize that it will take many years (and delays caused by future dry spells) as well as agricultural adjustments for the full benefits to materialize.

Western Australia, too, faces colossal challenges in the provision of water for its urban, agricultural, and industrial land uses. Here, much like in the western United States, water supplies in this dry environment are increasingly stressed by population growth—and especially all those massive new mining projects. The largest population cluster here is centered on west-coast Perth, and some scientists have warned that it could become the first "ghost metropolis," abandoned because of acute water shortages. To stave off that scenario, Perth's prospects have improved through the use of innovative **desalinization** 14 technologies that convert ocean water into a viable source of supply for households and businesses. Today, two desalination plants south of Perth provide about half of this

metropolitan area's water needs. Moreover, this technology is now beginning to be applied to Australia's southeastern coast, and a number of plants have already come on line (Fig. 11-7).

The Aboriginal Challenge

Despite Australia's generally positive fortunes, not everyone shares sufficiently in the national wealth. The indigenous (Aboriginal) population, though a small minority today of 680,000, remains disproportionately disadvantaged in almost every way—shorter life expectancies, elevated unemployment, lower high school graduation rates, and much higher imprisonment ratios.

For the past several decades, the Aboriginal challenge focused on two questions: (1) official acknowledgment, by the government as well as the Australian majority, of mistreatment of the Aboriginal minority accompanied by appropriate apologies and reparations; and (2) land ownership. The first question was resolved in 2008 when Prime Minister Kevin Rudd offered the long-awaited formal apology for the historic mistreatment of the Aborigines. The second

During the first decade of the twenty-first century, Australia lay in the grip of its worst drought on record, with calamitous consequences ranging from deadly wildfires to parched farmlands. One of Australia's most profitable agricultural industries, winemaking, suffered severely as many winegrowers were driven off their land. The left photo shows a once-mature vineyard near Yarrawonga, on the New South Wales-Victoria border north of Melbourne along the Murray River, desiccated and abandoned during the summer of 2009. When the rains finally came (with a vengeance), they caused disastrous floods, washing away topsoil and eroding the heat-baked countryside. The right photo, taken in January 2010, shows the results at Coonamble, New South Wales along a tributary of the Darling River.

question has major geographic implications. Although comprising less than 3 percent of the total population, the Aboriginal population has been gaining influence in national affairs over the past three decades: as far back as the 1980s, Aboriginal leaders have been waging a campaign to obstruct mineral exploration on what they regard as ancestral and sacred lands.

Prior to 1992, Australians had taken it for granted that Aborigines had no right to land ownership, but in that year the Australian High Court rendered the first of a series of landmark rulings in favor of Aboriginal claimants. A subsequent court decision implied that enormous areas (perhaps as much as 78 percent of the entire continent) could potentially be subject to Aboriginal claims. Today, the

MAP ANALYSIS

FIGURE 11-8

The Aboriginal population totals less than 700,000 today and remains marginalized within Australian society. Not unlike the situation with Native Americans in the United States, land ownership and entitlement are important and contentious issues. Figure 11-8 shows the spatial distribution of both the Aboriginal (bars) and overall Australian (dots) populations across the different States. Where are the Aborigines most heavily concentrated, and how does that compare with the distribution of the total national population? And what would you say are some of the implications of the clustering of Aborigines in relation to the distribution of Australia's largest cities (Fig. 11-5) and its geography of natural resources (Fig. 11-6)? Access your WileyPLUS Learning Space course to interact with a dynamic version of this map and to engage with online map exercises and questions.

Aboriginal land issue **15** remains mostly (though not exclusively) an Outback issue, but it has the potential to overwhelm Australia's court system and to constrain economic growth.

Immigration and the Changing Cultural Mosaic

Fifty years ago, when Australia had less than half the population it has now, 95 percent of its citizens were of European ancestry, and more than three-quarters of them came from Britain and Ireland. As we know, the White Australia immigration policy maintained this situation until the late 1970s. Today, the picture is dramatically different: of 23 million Australians, only about one-third claim British or Irish origin, and Asian immigrants outnumber those of European ancestry as well as reproducing at a faster annual rate. During the early 1990s, about 150,000 legal immigrants arrived in Australia each year, mainly from Hong Kong, Vietnam, China, the Philippines, India, and Sri Lanka. Annual immigration quotas were subsequently reduced but then were allowed to rise again—reaching 212,000 by 2014—with Asian immigrants continuing to outnumber those from Western sources. Sydney, the leading recipient of the Asian influx, has become a mosaic of ethnic neighborhoods, some of which have gone through periods of gang violence and drug dealing, but have stabilized and even prospered over time. Still, as the economy has grown, particularly in the mining sector, the country will have to rely on immigration to meet growing skilled-labor demands. Multiculturalism is sure to remain a long-term challenge for Australia.

Australia's Place in the World

Since World War II, as its historic ties to the United Kingdom have faded, Australia has become ever more embedded in the Asia

Cabramatta, one of Sydney's western suburbs, is seen as the "multicultural capital of Australia," a tourist attraction and proof of Australia's capacity to accommodate non-Europeans and especially Asians. This "Freedom Gate" in the Vietnamese community is flanked by a Ming horse and a replica of a Forbidden City lion—all reflecting Australia's growing ties to the Asia Pacific.

Pacific—economically, culturally, and politically. The country's strengthening orientation toward its Asian neighbors is reflected in economic policies and patterns of trade and investment. Australia is now a key member of the Asia Pacific Economic Cooperation forum, APEC **16**; China and Japan, by far, have become its most important trading partners. Australia itself, as the result of changing immigration patterns, is becoming culturally more heterogeneous and increasingly Asian. And, politically, Australia has become more involved with the geographic realms to its north, especially the countries of Indonesia, East Timor, and Papua New Guinea.

Geostrategically, Australia finds itself in a relatively secure location. However, security issues in the South China Sea and the Indian Ocean have become a growing concern precisely because Australia's interdependence with Asia is intensifying. Also of much interest is that Australia has been seeking closer relations with the United States as a counterweight to China's rapidly increasing influence across these realms, especially Southeast Asia. To nobody's surprise, the United States is pleased to reciprocate: since 2012, several thousand American troops have been stationed at Australian military bases, and the U.S. Air Force has been given access to Australian airfields in the Northern Territory that are within easy flying distance of the South China Sea.

Territorial dimensions, relative location, and raw-material wealth have helped determine Australia's place in the world and, more specifically, on the western Pacific Rim. Australia's population may still be less than 25 million, but the country's importance in the international community greatly exceeds its human numbers.

NEW ZEALAND

Twenty-four hundred kilometers (1500 mi) east-southeast of Australia, in the Pacific Ocean across the Tasman Sea, lies New Zealand, also known as *Aotearoa* in Maori (meaning "land of the long white cloud"). In an earlier age, New Zealand would have been part of the Pacific geographic realm because its population was entirely Maori, a people with Polynesian roots (see Chapter 12). But New Zealand, like Australia, was invaded and occupied by Europeans. Today, its population of 4.4 million is about two-thirds European, and the Maori comprise a substantial minority in excess of 600,000, with many of mixed Euro-Polynesian ancestry (including Pacific Islanders).

New Zealand consists of two large mountainous islands and many scattered smaller islands (Fig. 11-9). The two main islands, with the South Island somewhat larger than the North Island, look diminutive in the great Pacific Basin, but together they are bigger than Britain. In contrast to generally low-relief Australia, the more rugged terrain of the two large islands contains several peaks rising far higher than any on the Australian landmass. Along its western margin, the South Island has a spectacular snowcapped range appropriately called the Southern Alps, with a number of

© Robert Francis/Alamy

© J. Nijman, P. O. Muller, H. J. de Blij, and John Wiley & Sons, Inc.

FIGURE 11-9

summits reaching beyond 3300 meters (10,000 ft). The smaller North Island has proportionately more land under low relief, but it also has an area of central highlands along whose lower slopes lie the pastures of New Zealand's chief dairying district. The convergence of the Australian and Pacific tectonic plates underlies much of New Zealand, and it makes the country vulnerable to volcanic eruptions and earthquakes.

Human Spatial Organization

The most promising areas for habitation, therefore, are the lower-lying slopes and lowland fringes on both islands. On the North Island, the largest urban area, Auckland, occupies a comparatively low-lying peninsula. On the South Island, the largest lowland is the agriculture-dominated Canterbury Plain, focused on the city of Christchurch (now fully recovered from a pair of devastating earthquakes, only months apart, in 2010–2011). What makes these lower-elevation zones so attractive, apart from their availability as cropland, is their magnificent pastures. The range of soils and pasture plants allows both summer and winter grazing. Moreover, the Canterbury Plain, the chief farming region, also produces a wide variety of vegetables, cereals, and fruits. About half of all New Zealand is pasture land, and much of the farming provides fodder for the pastoral industry. About 30 million sheep, 6

@ FROM THE FIELD NOTES...

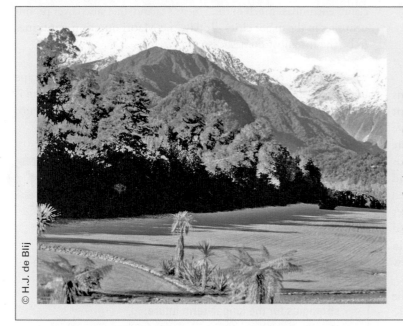

"The drive from Christchurch to Arthur's Pass on the South Island of New Zealand was a lesson in physiography and biogeography. Here, on the eastern side of the Southern Alps, you leave the Canterbury Plain and its agriculture and climb into the rugged terrain of the glacier-cut, snowcapped mountains. A last pasture lies on a patch of flatland in the foreground; in the background is the unmistakable wall of a U-shaped valley sculpted by moving ice. Natural vegetation ranges from pines to ferns, becoming even more luxuriant as you approach the moister western side of the island."

© H.J. de Blij

million dairy cattle, and 4 million beef cattle dominate these livestock-raising activities, with wool, milk products, and meat providing about 75 percent of the islands' export revenues.

Despite their contrasts in size, shape, physiography, and history, New Zealand and Australia share a number of characteristics. Apart from their joint British heritage, they share a sizeable pastoral economy with growth in specialty goods such as wines, a small local market, the problem of great distances to world markets, and a desire to stimulate (through protectionist policies) domestic manufacturing. The very high degree of urbanization in New Zealand (86 percent of the total population) once again resembles Australia: substantial

employment in city-based industries—mostly the processing and packing of livestock and farm products—as well as government jobs.

Spatially, New Zealand further shares with Australia its pattern of **peripheral development 17** and population distribution (Fig. 11-2), imposed not by deserts but by high rugged mountains and the fragmented layout of the country. The major cities—Auckland and the capital city, Wellington, on the North Island; Christchurch and Dunedin on the South Island—are all situated on the coast, and the rail and highway networks are therefore dominantly peripheral in their configuration. Furthermore, the two main islands are separated by Cook Strait, a windswept waterway that can only be crossed by

@ FROM THE FIELD NOTES...

"It was Sunday morning in Christchurch on New Zealand's South Island, and the city center was quiet. As I walked along Linwood Street I heard a familiar sound, but on an unfamiliar instrument: the Bach sonata for unaccompanied violin in G minor—played magnificently on a guitar. I followed the sound to the artist, a Maori musician of such technical and interpretive capacity that there was something new in every phrase, every line, every tempo. I was his only listener; there were a few coins in his open guitar case. Shouldn't he be playing before thousands, in schools, maybe abroad? No, he said, he was happy here, he did all right. A world-class talent, a street musician playing Bach on a Christchurch side street, where tourists from around the world were his main source of income. Talk about globalization."

© H.J. de Blij

ferry or plane (Fig. 11-9). On the South Island, the Southern Alps are the country's most formidable barrier to surface communications.

The Maori Factor and New Zealand's Future

Like Australia, New Zealand has had a history of difficult relations with its indigenous population. The Maori, who today account for 13.6 percent of the country's total population, appear to have reached the islands during the tenth century AD. By the time the European colonists arrived, the Maori had also had a tremendous impact on the islands' ecosystems, especially on the North Island where most of them lived. In 1840, the Maori and the British signed a treaty at Waitangi that granted the colonizers sovereignty over New Zealand but guaranteed the Maori rights covering established tribal lands. Although the British abrogated parts of the treaty in 1862, the Maori still had reason to believe that vast reaches of New Zealand, as well as offshore waters, were theirs in perpetuity.

As in Australia, judicial rulings during the 1990s supported the Maori position, which led to expanded claims and growing demands. Culturally, the declaration of Maori as an official language in New Zealand and its teaching throughout the school system are seen as significant progress toward an acceptance of the Maori cultural heritage. But the most persistent Maori complaint concerns their slow pace of integration into modern New Zealand society. Although Maori land claims encompass much of rural New Zealand, they also cover prominent sites in the major cities. Today, the Maori question continues to be the foremost domestic issue.

Dominant cultural heritage and prevailing cultural landscape form two criteria on which the delimitation of the Austral realm is based. But in both Australia and New Zealand, the convergence with neighboring geographic realms is well underway.

POINTS TO PONDER ?

- In this era of globalization, Australia is becoming more strongly tied to neighboring geographic realms as its links to faraway Europe steadily erode.

- Australia's advancing desalinization technologies provide relief for the country's immediate water needs, but are delaying the much needed formulation of conservation strategies for the longer-term future.

- New Zealand experiences more than 10,000 earthquakes each year; as many as 150 can be felt, and about 20 are strong enough to cause damage on the human landscape.

The wide world of the South Pacific Ocean, as seen from Chile's Easter Island, located about 3500 kilometers (2200 mi) west of that coastal South American country. The rocky outcrops below are a bit more than a kilometer (1 mi) offshore. The closest inhabited land from here—2075 kilometers (1289 mi) due west—is the Pitcairn group of islands, the last remaining British Overseas Territory in the Pacific Basin (see Fig. 12-1).

© Jan Nijman

12

PACIFIC REALM AND POLAR FUTURES

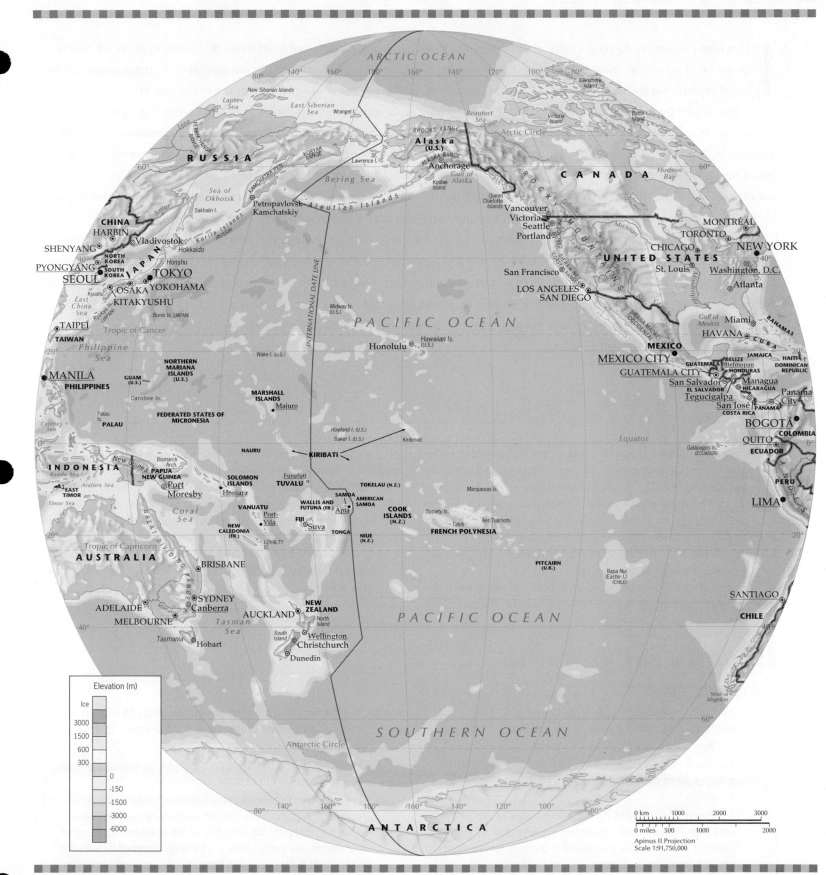

FIGURE 12-1

© J. Nijman, P. O. Muller, H. J. de Blij, and John Wiley & Sons, Inc.

n this final chapter we focus on three of the world's gigantic expanses, so different from those discussed in previous chapters that we need an entirely new perspective. Water and ice, not land and soil, dominate the physiography of these areas. Of the utmost importance is that environmental change occurring in all three has consequences for the entire planet.

The largest by far of these expanses, the ***Pacific realm***, covers almost half the globe from the Bering Sea to the Southern Ocean and from Central America to Indonesia (Fig. 12-1). The Pacific Ocean, larger than all the world's landmasses combined, is studded with tens of thousands of islands, of which New Guinea is by far the largest. The two polar zones to which it is linked could hardly be more different. The ***Arctic*** consists entirely of water, frozen at the surface, but you can reach the North Pole in a submarine. The ***Antarctic*** is mainly land, weighed down by the planet's largest accumulation of permanent ice; the surface journey to the South Pole, first accomplished more than a century ago, has cost many an explorer's life.

Climate change is altering geography everywhere, as is the enhanced technology of mineral (especially oil and gas) exploration and extraction. Rising sea levels would imperil thousands of populated low-elevation islands. Areas of ocean floor once beyond the scope of human intervention are now coming within reach. Melting ice may be clearing new maritime shipping routes. At the same time, as we will see, international rules to govern who owns what in these remote frontiers are subject to dispute. When land boundaries jut outward into the sea, who owns what depends on who draws the maps. We should be aware of the potential for competition, even conflict.

DEFINING THE REALM

Our survey begins with the Pacific geographic realm, which covers just about an entire hemisphere, the one commonly called the Sea Hemisphere (Fig. 12-1). This Sea Hemisphere meets the Russian and North American realms in the far north and merges into the Southern Ocean in the south. Despite the preponderance of water, this fragmented, culturally complex realm does possess regional identities. It includes the Hawaiian Islands, Tahiti, Tonga, Fiji, and Samoa—fabled names in a world apart.

In terms of modern cultural and political geography, Indonesia and the Philippines are not part of the Pacific realm, though Indonesia's political system reaches into it; nor do Australia and New Zealand belong to it. Before the European invasion and colonization, Australia would have been included (because of its Aboriginal population) as well as New Zealand (its Maori population has Polynesian affinities). But the Europeanization of their countries has engulfed indigenous Australians and Maori New Zealanders, and the regional geography of the Austral realm today is decidedly not Pacific. In New Guinea, by contrast, Pacific peoples, numerically and culturally, remain the dominant element.

MAJOR GEOGRAPHIC QUALITIES

PACIFIC REALM

1. The Pacific realm's total area is the largest of all geographic realms. Its land area, however, is the smallest, as is its population.

2. The island of New Guinea, with 10.8 million people, alone contains more than three-quarters of the Pacific realm's population.

3. The Pacific realm, with its huge expanses of water and numerous islands, has been strongly affected by United Nations Law of the Sea provisions regarding states' rights over economic assets in their adjacent waters.

4. The highly fragmented Pacific realm consists of three island regions: Melanesia (including New Guinea), Micronesia, and Polynesia.

5. The Pacific realm's islands and cultures may be divided into volcanic high-island cultures and coral-based low-island cultures.

6. In Micronesia, U.S. influence has been particularly strong and continues to affect local societies.

7. In Polynesia, indigenous culture exhibits remarkable consistency and uniformity throughout the region, its enormous dimensions and dispersal notwithstanding. Yet at the same time, local cultures nearly everywhere are severely stressed by external influences. In Hawai'i, as in New Zealand, indigenous culture has been largely submerged by Westernization.

Aerial view of a circular atoll in the Fijian archipelago, which marks the eastern edge of Melanesia. This atoll is almost completely submerged except for the tiny sliver of sand visible at the extreme left. From the plane, atolls can often first be detected first by the clouds clustered above them: the shallow water atop an atoll (for that matter, any low island) heats up more than the deeper ocean around it—causing the air to rise, condense into clouds, and occasionally trigger a local rainstorm.

© Jan Nijman

In examining the Pacific realm, it is important to keep in mind the dimensional contrasts this part of the world presents. The realm may be enormous, but its total land area is a mere 975,000 square kilometers (377,000 sq mi), about the size of Texas plus New Mexico, with over 90 percent constituted by the island of New Guinea. Indeed, the realm's inhabitants are so widely and thinly scattered that a population distribution map would be impractical; moreover, the population totals only around 14 million (about as many people as there are in Kolkata, India).

Physiographically, the main landforms of the Pacific can be distinguished in high islands, low islands, and atolls. **High islands 1** are seafloor-based volcanic peaks and ridges that protrude above the ocean surface and can reach thousands of meters in elevation. **Low islands 2** are far more common, mainly consist of coral, and sometimes barely rise above sea level. Livelihoods tend to vary among these different environments: farming prevails on high islands (thanks to fertile soils and orographic rainfall) whereas fishing dominates the economy of low islands (where freshwater supplies are limited). The third landform category is the **atoll 3**, a ring-shaped coral reef that encircles a lagoon, lying more or less at sea level (see photo). The circular shape results from the reef growing on the rim of an underwater volcano (which may long have eroded away); typically, atolls are uninhabitable.

COLONIZATION AND INDEPENDENCE

The Pacific islands were colonized by the French, British, and Americans—an indigenous Polynesian kingdom in the Hawaiian Islands was annexed by the United States in 1898, and since 1959 has functioned as the fiftieth State. Because political geography

evolves slowly within the Pacific Basin, today's map is still an assemblage of colonial territories and independent microstates (Fig. 12-1). France controls New Caledonia and French Polynesia. The United States administers Guam as well as American Samoa, the Line Islands, Wake Island, the Midway Islands, and several smaller island groups; the United States also has special relationships with other territories, former dependencies that are now nominally independent. The British, through New Zealand, have responsibility for the Pitcairn group of islands, and New Zealand administers and supports the Cook, Tokelau, and Niue Islands. Easter Island, that storied speck of land in the southeastern Pacific, is part of Chile (see chapter-opening photo). And, as discussed in Chapter 10, Indonesia rules Papua, the western half of the island of New Guinea.

Other island groups have now become independent states. The largest are Fiji, once a British dependency; the Solomon Islands, also formerly British; and Vanuatu, until 1980 ruled jointly by France and Britain. Also on the current map are such impoverished microstates as Tuvalu, Kiribati, Nauru, and Palau, which must depend on foreign aid for their survival. Tuvalu, for instance, has a total area of 26 square kilometers (10 sq mi), a population of barely 10,000, and a per-capita GNI of about U.S. $6000, derived from fishing, sales of copra (coconut meat used to make coconut oil), and some tourism. But what really keeps Tuvalu going is an international trust fund set up by Australia, New Zealand, the United Kingdom, Japan, and South Korea. Annual grants from that fund, as well as remittances sent back to families by workers who have left for New Zealand and elsewhere, allow Tuvalu to survive.

THE PACIFIC REALM AND ITS MARINE GEOGRAPHY

Certain land areas may not be part of the Pacific realm (we mentioned the Philippines and New Zealand), but the Pacific Ocean extends from the shores of North and South America to mainland East and Southeast Asia and from the Bering Sea off Alaska southward to the Subtropical Convergence around latitude 40°S. This means that several seas, including the South China Sea, the East China Sea, the Sea of Japan (East Sea), the Java Sea, and the Tasman Sea, are all part of the Pacific Ocean. As we are about to discover, this relationship matters. Pacific coastal countries, large and small, mainland as well as island, compete for jurisdiction over the waters that bound them.

The ocean-focused Pacific realm, therefore, provides an ideal setting for introducing some basic principles of **marine geography 4**. This field encompasses a variety of approaches to the study of oceans and seas; some marine geographers concentrate on the biogeography of coral formations, others on the geomorphology of beaches, and still others on the circulation of ocean currents. A most fascinating specialty within marine geography has to do with the definition and delimitation of political boundaries at sea. Here geography intersects both political science and maritime law.

The State at Sea

Littoral (coastal) states do not end where atlas maps suggest they do. States have claimed various forms of jurisdiction over coastal waters for centuries, closing off bays and estuaries and ordering foreign fishing fleets to stay away from nearby fishing grounds. Thereby arose the notion of the **territorial sea** 5, where all the rights of a coastal state would prevail. Beyond lay the **high seas** 6, free, open, and unfettered by national interests—so-called international waters.

It was in the interest of colonizing and/or trading states to keep territorial seas narrow and high seas wide, thereby interfering as little as possible with their commercial fleets. In the seventeenth and eighteenth centuries, the territorial sea was 3, 4, or at most 6 nautical miles wide, and the colonizing powers claimed the identical widths for their colonies (1 nautical mile = 1.85 kilometers or 1.15 statute miles).*

During the twentieth century, these constraints weakened. States without trading fleets saw no reason to limit their territorial seas. States with nearby fishing grounds traditionally exploited by their own fleets wanted to keep the increasing number of foreign trawlers away. States with shallow **continental shelves** 7 (offshore continuations of coastal plains) wished to control their resources on and below the seafloor, made more accessible by improved technology. States also disagreed on the methods by which offshore boundaries, whatever their width, should be defined. Early efforts by the League of Nations (the precursor to the United Nations) during the 1920s to resolve these issues met with only partial success, mainly in the technical sphere of boundary delimitation.

Scramble for the Oceans

In 1945, the United States helped precipitate what has become known as "the scramble for the oceans." President Truman issued a proclamation that claimed U.S. jurisdiction and control over all the resources "in and on" the continental shelf down to its margin, around 100 fathoms (183 meters/600 ft) deep. In some areas, the shallow continental shelf of the United States extends more than 300 kilometers (200 mi) offshore, and Washington did not want foreign countries drilling for oil just beyond the 3-mile territorial sea.

Few observers foresaw the impact the Truman Proclamation would have, not only on U.S. waters but on the oceans everywhere, including the Pacific. It set off a rush of additional claims. In 1952, a group of South American countries, some with little continental shelf to claim, issued the Declaration of Santiago, claiming exclusive fishing rights up to a distance of 200 nautical miles off their coasts. Meanwhile, as part of the Cold War competition, the Soviet Union claimed a 12-mile territorial sea and urged its allies to do likewise.

---

* Here and in the rest of this section all distances involving maritime boundaries are given in nautical miles only. Throughout the modern history of maritime law, this unit of measurement has been the only one used for such boundary-making. Any distance stated in nautical miles can be converted to its metric equivalent by multiplying it by 1.85.

UNCLOS Intervention

At this point the United Nations intervened, and a series of UNCLOS (United Nations Conference on the Law of the Sea) meetings began. These meetings addressed issues ranging from the closure of bays to the width and delimitation of the territorial sea, and after three decades of negotiations they achieved a convention that changed the political and economic geography of the world's oceans forever. Among its key provisions were the authorization of a 12-mile territorial sea for all countries as well as the establishment of a 200-mile (230-statute-mi/370-km)-wide **Exclusive Economic Zone (EEZ)** 8 over which a coastal state would have total economic rights. Resources in and under this EEZ (fish, minerals, oil, and gas) belong to the coastal state, which could either exploit them or lease, sell, or share them as it saw fit. These zones were noted in the discussion of the new competition for the Arctic Ocean in Chapter 5, but as we shall see, EEZs take on even greater importance in the Pacific realm.

These new provisions had a far-reaching impact on the world's oceans and seas (Fig. 12-2), especially the Pacific. Unlike the Atlantic Ocean, the Pacific is studded with myriad islands, so that a microstate consisting of a single small island suddenly acquired an EEZ covering 166,000 square nautical miles. European colonial powers that still controlled minor Pacific possessions (most notably France) saw their maritime jurisdictions vastly expanded. Small, low-income archipelagos could now bargain with large, rich fishing nations over fishing rights in their now huge EEZs. And for all the UNCLOS Convention's provisions for the "right of innocent passage" of shipping through EEZs and via narrow straits, the geographic extent of the world's high seas has been diminished accordingly.

Maritime Boundaries

The extension of the territorial sea to 12 nautical miles and the EEZ to an additional 188 nautical miles created new **maritime boundary** 9 problems. Waters less than 24 nautical miles wide separate many countries all over the world, so that **median lines** 10, equidistant from opposite shores, have been delimited to establish their territorial seas. And even more countries lie closer than 400 nautical miles apart, requiring further maritime-boundary delimitation to determine their EEZs. In such maritime arenas as the North Sea, the Caribbean Sea, and the East and South China seas (and more recently the Arctic Ocean), a maze of maritime boundaries emerged, some of them subject to dispute. Furthermore, political changes on land can also lead to significant modifications at sea.

EEZ Implications

The UNCLOS provisions created new opportunities for some states to expand their spheres of influence. Wider territorial-sea and EEZ allocations raised the stakes: claiming an island now entailed potential control over a huge maritime area. In Chapter 5, we mentioned the impending need to further carve up the Arctic

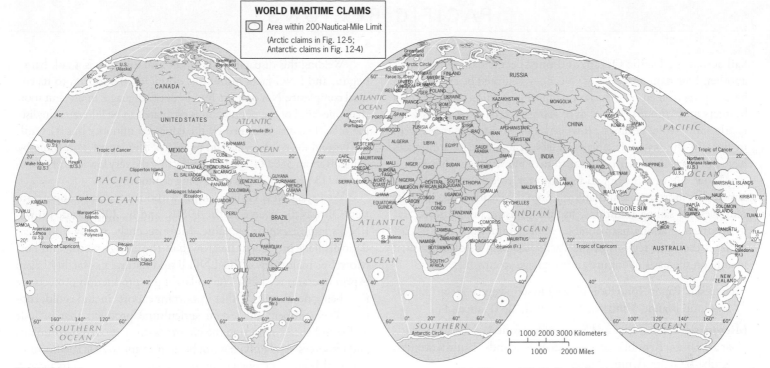

WORLD MARITIME CLAIMS

◯ Area within 200-Nautical-Mile Limit

(Arctic claims in Fig. 12-5;
Antarctic claims in Fig. 12-4)

FIGURE 12-2

© J. Nijman, P.O. Muller, H.J. de Blij, and John Wiley & Sons, Inc.

Basin as the ice melts and untapped resources beckon. In Chapters 9 and 10, we referred to several island disputes off mainland East and Southeast Asia, which involve Japan and Russia, Japan and South Korea, Japan and China, China and Vietnam, and China and the Philippines. Ownership of several offshore Asian islands is uncertain, and small specks of island territory have become large stakes in the scramble for the oceans. For instance, in the case of the Spratly Islands, several countries claim ownership, including both Taiwan and China (see Fig. 10-10). China's island claims in the South China Sea (within its alleged "Nine-Dash-Line" maritime boundary) support Beijing's contention that this body of water is part and parcel of the Chinese state—a position that increasingly troubles other states with coasts facing it.

Figure 12-2 indicates what EEZ regulations have meant to Pacific realm countries such as Tuvalu, Kiribati, and Fiji, with nearly circular EEZs now surrounding the clusters of islands in this vast oceanic space. Japan, Taiwan, and other nations with fisheries have purchased fishing rights in these EEZs from the island governments. Nonetheless, violations of EEZ rights do occur; not long ago, Vanuatu and the Philippines were at odds over unauthorized Filipino fishing in Vanuatu's EEZ. The process of boundary delimitation continues to evolve, and the Pacific (and world) map of maritime boundaries remains very much a work in progress.

That point is underscored by a development that occurred at the end of the 1990s, when the implications of provisions in Article 76 of the 1982 UNCLOS Convention were reexamined. Whenever a continental shelf extends beyond the 200-mile limit of the EEZ, the Convention apparently permits a coastal state to claim that extension as a natural prolongation of its landmass. Although there is as yet no regulation to also allow the state to extend its EEZ to the edge of the continental shelf, it is not difficult to foresee such an amendment to the Convention. As it stands, states are now delimiting their proposed natural prolongations under a current deadline, which puts poorer states at a disadvantage since the required marine surveys are quite expensive. Today, this issue has become quite heated, especially in the Arctic Ocean as countries scramble to claim prolongations (most notably Russia, as we shall see at the end of this chapter). The big winners are likely to include Russia, the United States, Canada, Australia, New Zealand, and India—although a total of 60 coastal countries could ultimately benefit to some extent from the provision.

Another complicating factor is the role of new technologies that are accelerating deep-sea exploration and the exploitation of seabed resources. Today, the quest to discover additional mineral deposits is increasingly an underwater endeavor, led by the search for so-called massive sulfides—enormous concentrations of gold, silver, copper, and other metallic minerals formed below the ocean floor in sulfide-rich volcanic environments. The technological breakthroughs entail sensors, advanced robotic devices, and other state-of-the-art gear (such as 'sea gliders') developed by the off-shore oil and gas industry. The growing use of sea gliders is of special interest because these small, slow-moving, unmanned vessels can track enormous distances as they collect detailed information on everything below—from marine life to seabed resources to the movements of submarines. And so the scramble for the oceans marches ever onward, and with it the constriction of the open waters of the world's high seas.

PACIFIC REGIONS

Sail across the Pacific Ocean, and one spectacular vista follows another. Dormant and extinct volcanoes, sculpted by erosion into basalt spires draped by luxuriant tropical vegetation and encircled by reefs and lagoons, tower over azure waters. Low atolls with nearly snow-white beaches, crowned by stands of palm trees, seem to float in the water. Pacific islanders, where foreign influences have not overtaken them, appear to take life with enviable ease.

More intensive investigations, however, reveal that such Pacific sameness is more apparent than real. Even the Pacific realm, with its centuries-old sailing traditions, its still-diffusing populations, and its historic migrations, has a durable regional framework. Figure 12-3 outlines the three regions of the Pacific that can be distinguished mainly on the basis of culture and language groups:

Melanesia: Papua (province of Indonesia), Papua New Guinea, Solomon Islands, Vanuatu, New Caledonia (France), Fiji

Micronesia: Palau, Federated States of Micronesia, Northern Mariana Islands, Republic of the Marshall Islands, Nauru, western Kiribati, Guam (United States)

Polynesia: Hawaiian Islands (U.S.), Samoa, American Samoa, Tuvalu, Tonga, eastern Kiribati, Cook and other New Zealand-administered islands, French Polynesia, Easter Island (Chile)

MELANESIA

The large island of New Guinea lies at the western end of a Pacific region that extends eastward to Fiji and includes the Solomon Islands, Vanuatu, and New Caledonia (Fig. 12-3). The human mosaic here is complex, both culturally and ethnically. Most of the 10.8 million people of New Guinea (divided between the Indonesian province of Papua and the independent state of Papua New Guinea) are Papuans, and a large minority is Melanesian. Altogether there are more than 800 communities speaking different languages; the Papuans are most numerous in the densely forested highland interior and in the lowland south, whereas the Melanesians inhabit the north and east. Minus Papua, Melanesia contains around 9 million inhabitants, making this the most populous Pacific region by far.

With 7.3 million people today, **Papua New Guinea (PNG)** became a sovereign state in 1975 after nearly a century of British and Australian administration. Almost all of PNG's limited development is taking place along its coasts, whereas most of the interior remains hardly touched by the changes that transformed neighboring Australia. Perhaps four-fifths of the population is part of the self-sufficient subsistence economy, cultivating root crops, raising pigs, hunting wildlife, and gathering forest products. Age-old traditions of the kind lost in Australia continue on here, protected by remoteness and the rugged terrain.

Welding this disparate population into a nation is a task hardly begun, and PNG faces numerous obstacles in addition to its cultural complexity. Not only are hundreds of languages in use, but more than one-third of the population is estimated to be illiterate. English, the official language, is used by the educated minority but is of little use beyond the coastal zone and its towns. The capital, Port Moresby, contains about 350,000 residents, reflecting the very low level of urbanization (13 percent) in this developing economy.

Yet Papua New Guinea is not without economic opportunities. Oil was discovered in the 1980s, and by 2000 crude oil was PNG's largest export by value. More recently, newly found gas deposits have propelled the growth of the energy sector, and Port Moresby's major new PNG-LNG liquid natural gas facility began export operations in 2014 (even larger gas production projects are now being developed). Other important exports include gold, copper, silver, timber, and several agricultural products that include coffee and cocoa, reflecting the country's diversity of environments and resources. And PNG has not been immune to Asia Pacific influences: although the biggest share of exports still go to its nearest neighbor, Australia, Japan ranks second, and fast-rising China is close behind.

Turning eastward, it is a measure of Melanesia's cultural fragmentation that as many as 120 languages are spoken in the approximately 1000 islands that make up the **Solomon Islands** (about 80 of them support almost all the inhabitants, who number almost 600,000). Inter-island, historic animosities among the islanders were worsened by the events of World War II, when U.S. forces moved thousands of Malaitans to Guadalcanal. This started a postwar cycle of violence that finally led Australia to intervene in 2003.

New Caledonia, still under French rule, is in a very different situation. Only around 42 percent of its population of roughly 265,000 are Melanesian; not that far behind are the 28 percent of European ancestry (known as the Caldoches), many of them descended from the inhabitants of the penal colony France established here during the nineteenth century. Nickel mines, based on reserves that rank among the world's largest, dominate New Caledonia's export economy (see *Field Note*). The mining industry attracted additional French settlers, and social problems soon arose. Most of the French population lives in or near the capital city of Nouméa, steeped in French cultural landscapes, located in the southeastern quadrant of the island. Melanesian demands for an end to colonial rule have led to violence, and the two communities remain engaged in a long-running process of coming to terms.

On its eastern margins, Melanesia includes one of the Pacific realm's most interesting countries, **Fiji**. On two larger and over 100 smaller islands live about 900,000 Fijians, of whom 54 percent are Melanesians and 38 percent South Asians, the latter brought to Fiji from India during the British colonial occupation to work on the sugar plantations. When Fiji achieved independence in 1970, the indigenous Fijians owned most of the land and held political

FIGURE 12-3

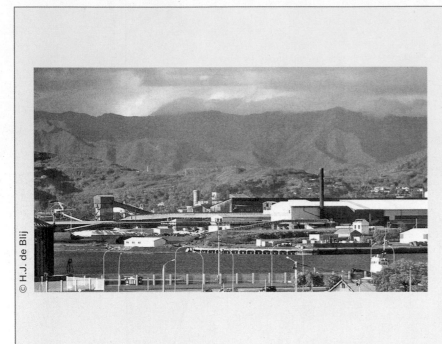

© H.J. de Blij

"Arriving in the capital of New Caledonia, Nouméa, is an experience reminiscent of French Africa in the postwar era. The French tricolor was much in evidence, as were uniformed French soldiers. European French residents [the Caldoches] occupied hillside villas overlooking palm-lined beaches, giving the place a Mediterranean cultural landscape. And New Caledonia, like Africa, is a source of valuable minerals. It is one of the world's major producers of nickel, and from this vantage point you could see the huge treatment plants, complete with concentrate ready to be shipped (left, under conveyor). What you cannot see here is how southern New Caledonia has been ravaged by the mining operations, which have denuded whole mountainsides. Working in the mines and in this facility are the local Kanaks, Melanesians who make up between 40 and 45 percent of the population of roughly 265,000. Conflict between Kanaks and French Caldoches has obstructed government efforts to change New Caledonia's political status in such a way as to accommodate pressures for independence as well as continued French administration."

control, whereas the Indians were concentrated in the towns (chiefly Suva, the capital) and dominated commercial life. It was a recipe for trouble, which surfaced as soon as the British had departed,

Nearly a half-century of ethnic conflict and political malfunctioning has cast a long shadow across the country. In 2009, after some 40 years of strife and violations of democratic rule, Fiji was suspended from the British Commonwealth. The sugar industry was damaged by the nonrenewal of Indian-held leases by Fijian landowners, which also resulted in a substantial movement of Indians from the countryside to the towns, where unemployment is already high. Only after democratic elections were finally held in 2014, was Commonwealth status reinstated; more importantly, at the same time both Australia and New Zealand restored diplomatic relations. It is also worth pointing out that while Fiji was shunned by these other countries, China stepped in to fill the void. Chinese foreign aid to Fiji has increased sevenfold since the last military coup in 2006, and Chinese investments in infrastructural projects have increased concomitantly. This, of course, is yet another sign that China's influence is now being projected globally—and not always in a manner reassuring to the international community.

◼ MICRONESIA

North of Melanesia and to the east of the Philippines lie the islands that constitute the second Pacific region, Micronesia (Fig. 12-3). The name refers to the size of the islands: the 2000-plus islands of Micronesia are not only tiny (many of them no larger than one square kilometer [half a square mile]), but they are also much lower-lying, on average, than those of Melanesia. Some are volcanic high islands, but they are outnumbered by low coral islands and atolls that barely reach above sea level (see photo). Guam, with 550 square kilometers (210 sq mi), is Micronesia's largest island, and no elevation anywhere in Micronesia reaches 1000 meters (3300 ft).

Take, for example, the country of **Kiribati**, consisting of three main island groups that lie across Micronesia and Polynesia (Fig. 12-3). The average elevation of Kiribati's 33 islands is 2 meters (6.5 ft), and, like all other low islands in this realm, they are particularly vulnerable to rising sea level driven by global warming.

© H.J. de Blij

A classic example of a low island in the Kiribati archipelago, which lies near the point where the equator and the International Date Line intersect. As Figure 12-3 indicates, the realm's three regions also meet in this immediate vicinity.

The high-island/low-island dichotomy, introduced in the earlier discussion of Pacific physiography, is particularly applicable in Micronesia. The high islands wrest substantial orographic rainfall from the moist maritime air, tend to be well watered, and are usually endowed with productive volcanic soils. As a result, agriculture can thrive, life is reasonably secure, and populations tend to be larger than those of low islands. On those low islands, drought is the rule, and fishing and the coconut palm are the mainstays of life; most of the islands can only support small communities, and over time many have died out. Thus the major precolonial, indigenous migrations, which sent fleets to populate islands from Hawai'i to New Zealand, mainly originated in the high islands.

From 1945 until the mid-1980s, Micronesia was largely a United States Trust Territory (the last of the post–World War II trusteeships supervised by the United Nations). But, as Figure 12-3 shows, Micronesia is now divided into countries bearing the names of independent states. These days the Marshall Islands, where the United States tested nuclear weapons (giving prominence to the name *Bikini*), is a republic in "free association" with the United States, having the same status as Palau and the Federated States of Micronesia. The Northern Mariana Islands comprise a commonwealth "in political union" with the United States, which in effect provides billions of dollars in assistance to these countries, in return for which they commit themselves to avoid foreign policy actions that are contrary to American interests.

The Republic of Palau is another fascinating example of how some of these small Pacific nations make a living. This group of islands some 800 kilometers (500 mi) east of the Philippines was granted independece in 1994 and has a total population of roughly 21,000. Palau has no military but entered into an agreement with the United States to provide security in return for a 50-year lease on a military base. Though highly dependent on American foreign assistance, Palau also receives funds and investments from Taiwan—a reward for Palau's official recognition of Taiwan's status as the Republic of China as well as for withholding recognition of the PRC.

Also part of Micronesia is the U.S. territory of Guam, where independence is not in sight and where U.S. military installations and tourism provide most of its income. And then there is the remarkable Republic of Nauru, with a population of barely 10,000 and just 20 square kilometers (8 sq mi) of land. Nauru got rich by selling its phosphate deposits to Australia and New Zealand, where they are used as fertilizer. At one point, per-capita income had risen above U.S. $12,000, making Nauru one of the realm's high-income societies; but the main surface deposits of phosphate have now been depleted, and great uncertainty prevails as the mining of a secondary layer proceeds.

In this region of tiny islands, almost all inhabitants subsist on farming or fishing, and virtually every country needs infusions of foreign aid to survive. The natural economic complementarity between the high-island farming cultures and the low-island fishing communities is all too often negated by distance, spatial as well as cultural. Life here may seem idyllic to the casual visitor, but for too many Micronesians it is a difficult daily struggle.

POLYNESIA

To the east of Micronesia and Melanesia lies the heart of the Pacific, enclosed by a great triangle stretching from the Hawaiian Islands to New Zealand to Chile's Easter Island (see *Field Note*). This third Pacific region is Polynesia (Fig. 12-3), which consists of numerous islands (*poly* means many) ranging from volcanic mountains rising far above the Pacific's waters (Mauna Kea on Hawai'i reaches over 4200 meters [nearly 13,800 ft]), clothed by luxuriant tropical forests and drenched by well over 250 centimeters (100 in) of rainfall each year, to low coral atolls where, if there is just enough elevation, a few palm trees form the only vegetation and drought is a persistent problem.

Despite its vast extent (about twice the areal size of Melanesia and Micronesia combined), Polynesia clearly constitutes a distinct Pacific region. Polynesian culture, although spatially fragmented, exhibits a remarkable consistency and uniformity from one island to the next, from one end of this widely dispersed region to the other. This consistency is particularly expressed in vocabularies, technologies, housing, and art forms. The Polynesians are uniquely adapted to their marine environment, and long before European sailing ships began to arrive in their waters, Polynesian seafarers had learned to navigate their enormous expanses of ocean in huge double canoes as long as 45 meters (150 ft). They traveled hundreds of kilometers to favorite fishing zones and engaged in inter-island barter trade, using maps constructed from bamboo sticks and cowrie shells and navigating by the stars. However, modern descriptions of a Pacific Polynesian paradise of emerald seas, lush landscapes, and gentle people distort harsh realities. Polynesian society was forced to accept much loss of life at sea when storms claimed their boats; families were ripped apart by accident as well as migration; hunger and starvation afflicted the inhabitants of smaller islands; and the island communities were often embroiled in violent conflicts and cruel retributions.

The political geography of Polynesia is complex. In 1959, the Hawaiian Islands became the fiftieth State to join the United States. The State's population is now 1.5 million, with more than 70 percent living on the island of Oahu. There, the superimposition of cultures is symbolized by the panorama of Honolulu's seaside skyscrapers framed against the famous extinct volcano at nearby Diamond Head. The Kingdom of Tonga became an independent country in 1970 after seven decades as a British protectorate; the British-administered Ellice Islands were renamed Tuvalu and received independence from the UK in 1978. Other islands continue under French control (including the Marquesas Islands and Tahiti), under New Zealand's administration (Cook Islands), and under British, U.S., and Chilean flags.

In the process of politico-geographical fragmentation, Polynesian culture has suffered severe blows. Land developers, hotel builders, and tourist dollars have set Tahiti on a course along which Hawai'i has already traveled far. The Americanization of eastern Samoa has created a new society different from the old. Today, Polynesia has lost much of its ancient cultural consistency, and the region has become a patchwork of new and old—the

© Jan Nijman

@ FROM THE FIELD NOTES...

"Easter Island, known as Rapa Nui in the native language, is a typical high island that is constituted by three extinct volcanoes protruding above the ocean surface. In the foreground is the spectacular crater and lake named Rano Kau, its rim reaching about 330 meters (1000 ft) above sea level. We have just taken off from the airport runway aligned left to right behind the crater where the slopes of Rano Kau end. Our pilot executed a 180° right turn to let us have this splendid view of the island. Just beyond the left end of the runway is the town of about 5000 people. Most high islands of the Pacific carry abundant vegetation, and Easter Island was indeed once blanketed by palm trees. However, during the seventeenth century the local population effectively cut down all the trees (for reasons that are still not understood), eliminating an essential resource and triggering massive soil erosion. Today, some scholars consider Easter Island to be a symbol of environmental destruction and societal collapse. The trees now found on Easter Island were all replanted. Thus, most food and supplies for the islanders—who total about 6000—have to be brought in from Chile, 3500 kilometers (2200 mi) away."

new often bleak and barren, with the old under ever mounting pressure.

The countries and cultures of the Pacific realm lie in an enormous ocean basin along whose rim a great drama of economic and political transformation will play itself out during the twenty-first century. Already, the realm's own former margins—in Hawai'i in the north and in New Zealand in the south—have been so thoroughly reshaped by foreign intervention that very little remains of the kingdoms and cultures that once prevailed. In the 2010s, the world's largest country and next superpower (China) faces the richest and most powerful (the United States) across the Pacific. How will the microstates of the Pacific realm fare?

POLAR FUTURES

PARTITIONING THE ANTARCTIC

South of the Pacific realm lies Antarctica and its encircling Southern Ocean (discussed in Chapter 11). The combined area of these two gargantuan geographic expanses constitutes 40 percent of the entire planet—two-fifths of the Earth's surface containing a mere one one-thousandth of the world's population.

Is Antarctica a geographic realm? In exclusively physiographic terms, yes—but not on the basis of the criteria we use in this book. Antarctica is a continent, practically twice the size of Australia, but virtually all of it is covered by a dome-shaped ice sheet nearly 3.2 kilometers (2 mi) thick near its center. The continent frequently is referred to as "the white desert" because, despite all of its ice and snow, annual precipitation is extremely low—averaging less than 15 centimeters (6 in) per year. Temperatures are frigid, with winds so powerful that Antarctica also is called the home of the blizzard. For all its size, no functional regions have developed here, nor have any towns or transport networks except the supply lines of research stations. Antarctica still is a frontier, even a scientific frontier only grudgingly giving up its secrets. Underneath all that ice lie some 70 lakes of which Lake Vostok is the largest at over 14,000 square kilometers (5400 sq mi), and it may be as much as 600 meters (2000 ft) deep.

Like virtually all frontiers, Antarctica has long attracted explorers and pioneers. Whale and seal hunters destroyed huge populations of Southern Ocean fauna during the eighteenth and nineteenth centuries, and explorers planted the flags of their countries on Antarctic shores. By the first decade of the twentieth century, the quest for the South Pole had become an international obsession; Roald Amundsen, the Norwegian, reached it first in 1911. All this activity soon led to national claims in Antarctica during the interwar era (1918–1939).

The geographic effect was the partitioning of Antarctica into pie-shaped sectors centered on the South Pole (Fig. 12-4). In the

© H.J. de Blij

"The Antarctic Peninsula is geologically an extension of South America's Andes Mountains, and this vantage point in the Gerlach Strait leaves you in no doubt: the mountains rise straight out of the frigid waters of the Southern Ocean. We have been passing large, flat-topped icebergs, but these do not come from the peninsula's shores; rather, they form on the leading edges of ice shelves or on the margins of the main-land where continental glaciers slide into the sea. Here along the peninsula, the high-relief topography tends to produce jagged, irregular icebergs. The mountain range that forms the Antarctic Peninsula continues across Antarctica under thou-sands of feet of ice and is known as the Transantarctic Moun-tains. Looking at this place, you are reminded that beneath all this ice lies an entire continent, still little known, with fossils, minerals, even lakes yet to be discovered and studied."

least frigid area of the continent, the Antarctic Peninsula, British, Argentinian, and Chilean claims overlapped—and they still do. One sector, Marie Byrd Land (shown in neutral beige on the main map), was never formally claimed by any country.

Why should states be interested in territorial claims in so remote and difficult an area? Because these remote areas near the poles contain substantial resources and are attractive for possible future use, despite the formidable challenges of access. Antarctica is no different as both land and sea contain raw materials that may someday become crucial: proteins in the waters, and fossil-fuel as well as mineral deposits beneath the ice sheet. As noted above, Antarctica at 14.2 million square kilometers (5.5 million sq mi) is nearly twice the size of Australia, and the Southern Ocean is almost as large as the North and South Atlantic combined. How-ever distant actual future exploitation may be, countries want to hang onto their stakes here.

But the states claiming Antarctic territory recognize the need for cooperation. During the late 1950s, they participated in the International Geophysical Year (IGY) that launched major scien-tific programs and established a number of permanent research stations across the continent. This spirit of cooperation led to the 1961 signing of the **Antarctic Treaty 11**, which ensures continued scientific collaboration, prohibits military activities, safeguards the environment, and holds national claims in abeyance. In 1991, when the treaty was extended under the terms of the Wellington Agreement, concerns were raised that it does not do enough to control future resource exploitation.

Notice, in Figure 12-4, that the map shows no maritime claims off the pie-shaped sectors that blanket most of Antarctica. In fact, some of the claimant states did draw maps that extended their claims into the Southern Ocean, but the 1991 treaty extension terminated that initiative and restricted the existing claims to the landmass. It is, of course, a possibility that those claims could be reinstated should

the Wellington Agreement disintegrate: national self-interest has abrogated many an international treaty in the past. But delimiting maritime claims off Antarctica is particularly challenging for practical reasons. As Figure 12-4 indicates, Antarctica in a number of places is flanked by *ice shelves* attached to its coast, permanent slabs of float-ing ice (such as the Ronne and the Ross). These ice shelves are replen-ished on the landward side and break off (calve) on the seaward side, where huge icebergs float northward into the Southern Ocean to become part of the wide belt of *pack ice* that encircles the continent. From where would any territorial sea or EEZ be measured? From the inner edge of the ice shelf? That would put the territorial "sea" bound-ary on the ice shelf! From the ever-changing outer edge of the ice shelf? Measuring a territorial sea or EEZ from such an unstable perimeter would not work either. No UNCLOS regulations would be applicable in situations like this, so no matter what the maps showed, such claims are legally unsupportable. That is just as well: the last thing the world needs is a scramble for Antarctica.

In an age of growing national self-interest and increasing raw material consumption, the possibility exists nonetheless that Antarc-tica and its offshore waters may yet become an arena for international rivalry. Until now, its remoteness and its forbidding environments have saved it from that fate. The entire world benefits from this because evidence is mounting that Antarctica plays a critical role in the global environmental system, so that human modifications are likely to have worldwide (and unpredictable) consequences.

GEOPOLITICS IN THE ARCTIC BASIN

To observe how different the physiographic as well as the political situation is in the Arctic, consider the implications of Figure 12-5. Not only does the North Pole lie on the floor of a relatively small body of water (grandiosely labeled the Arctic Ocean), but the

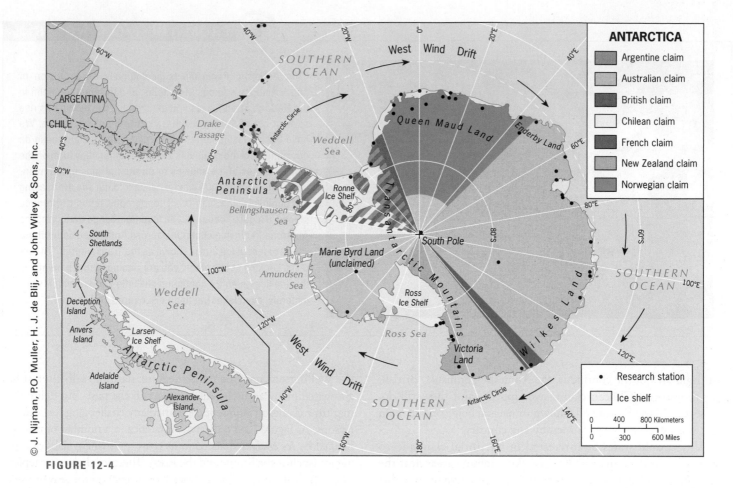

© J. Nijman, P.O. Muller, H. J. de Blij, and John Wiley & Sons, Inc.

FIGURE 12-4

entire Arctic is ringed by countries whose EEZs, delimited under UNCLOS rules, would allocate much of that ocean floor (the *subsoil*, to use its legal designation) to those states. Moreover, again under UNCLOS regulations, states can expand their rights to the seafloor even farther than the 200-mile EEZ if they are able to prove their continental shelves continue beyond that limit—in fact, up to 350 nautical miles offshore.

As the main map shows, the Siberian Continental Shelf is by far the largest in the Arctic; indeed, it is the largest in the world. It extends from Russia's north coast beneath the waters of the Arctic Ocean and under the floating ice cap at the ocean's center. And because several island groups off the mainland (such as Franz Josef Land, North Land, and the New Siberian Islands) belong to Russia, the Russians can claim virtually all of Eurasia's northern continental shelf under existing regulations. But even that is not

enough: the Russians want to extend their claim all the way to the North Pole itself, where in 2007 they sent a submarine to plant a Russian flag on the seafloor nearly 4000 meters (13,000 ft) below the Pole (see photo).

To bolster their claim, the Russians assert that the Lomonosov Ridge, a submerged mountain range, is a "natural extension" of their Siberian Continental Shelf. Few if any neutral observers might agree that this ridge, or several others rising from the deep floor of the Arctic Ocean, is a continental-shelf landform. But in 2014, the Danes made a new claim on the same grounds: following "ten years of scientific research," their conclusion was that the Lomonosov Ridge is actually connected to Greenland's shelf and therefore lies in Danish territory.

Take another look at Figure 12-5 and observe how far the Danes have extended their claims, which now overlap with those of

MAP ANALYSIS

Over the past few decades, the Arctic has become an increasingly contested geopolitical arena. The main map in Figure 12-5 shows the Exclusive Economic Zones (EEZs) of the affected countries plus additional claims that extend to latitude 90°N—the North Pole itself. Which two countries claim territory beyond their EEZs? Next, compare this map to Figure 12-4: why wouldn't an Antarctic-type, pie-shaped, sectoral partitioning work in the Arctic? Now look closely at the two inset maps in

Figure 12-5 and examine how much, and where, the polar ice cap has been melting since 1980. Finally, is the Northeast Passage or the Northwest Passage becoming the more important new oceanic route, and why? Access your WileyPLUS Learning Space course to interact with a dynamic version of this map and to engage with online map exercises and questions.

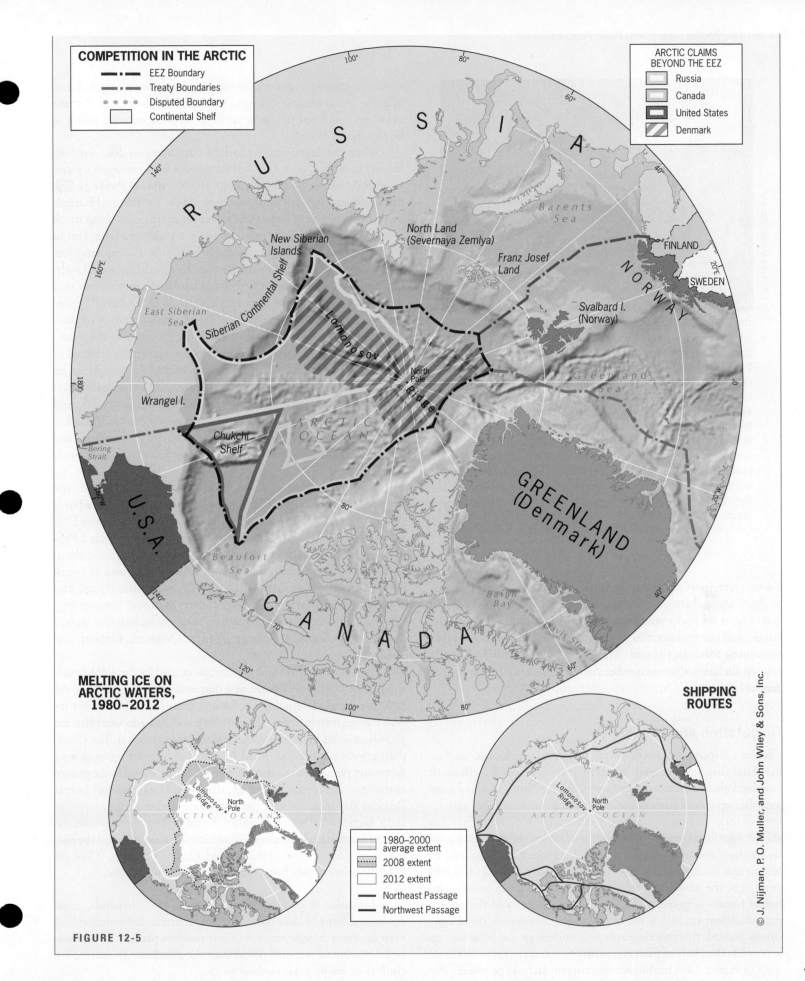

COMPETITION IN THE ARCTIC

- — ·· — EEZ Boundary
- — · — Treaty Boundaries
- ·· ·· ·· Disputed Boundary
- ☐ Continental Shelf

ARCTIC CLAIMS BEYOND THE EEZ

- Russia
- Canada
- United States
- Denmark

R U S S I A

R U S S I A

Barents Sea

North Land (Severnaya Zemlya)

Franz Josef Land

FINLAND

SWEDEN

N O R W A Y

Svalbard I. (Norway)

New Siberian Islands

East Siberian Sea

Siberian Continental Shelf

Lomonosov Ridge

North Pole

Greenland Sea

0°

Wrangel I.

Chukchi Shelf

ARCTIC OCEAN

Bering Strait

U.S.A.

GREENLAND (Denmark)

Beaufort Sea

Baffin Bay

C A N A D A

Davis Strait

MELTING ICE ON ARCTIC WATERS, 1980–2012

Lomonosov Ridge

North Pole

ARCTIC OCEAN

- 1980–2000 average extent
- 2008 extent
- 2012 extent
- Northeast Passage
- Northwest Passage

SHIPPING ROUTES

Lomonosov Ridge

North Pole

ARCTIC OCEAN

© J. Nijman, P. O. Muller, and John Wiley & Sons, Inc.

FIGURE 12-5

On August 3, 2007, tens of millions of Russian television viewers witnessed live the scene pictured here: the operator of a Russian mini-submarine planting a metal Russian flag at the North Pole, on the seafloor of the Arctic Ocean beneath the polar ice-cap. As Figure 12-5 shows, the North Pole lies far beyond Russia's continental shelf. Yet in future international negotiations to settle national claims in the Arctic Basin, the Russians will claim that the Lomonosov Ridge actually constitutes an extension of their continental shelf, thereby entitling Russia to draw international boundaries up to, and even beyond, the North Pole. Other countries with actual and pending Arctic claims are making less dramatic but equally assertive moves in the run-up to negotiations that have the potential to transform the map of the Arctic region. With new sea routes as well as newly exploitable oil and gas reserves in prospect, diplomacy will be difficult but crucial.

Canada and, more significantly, Russia—the North Pole is included in the Danish claim as well. The stakes are high: estimates of the quantity of oil and natural gas reserves to be found beneath the Arctic seafloor and ice run as high as 25 percent of the world's remaining total. In the end, whatever method is used to finally resolve the Lomonosov and other claims, Russia will be the biggest winner.

Disputation and Navigation

Even as a dispute over Arctic maritime boundaries looms, another key issue that has emerged in recent years involves the effects of sustained global warming. The Greenland Ice Sheet, in many ways a smaller version of Antarctica's, is now in a melting phase, and the zone of seasonally expanding and contracting Arctic sea ice has also shown significant attrition in both surface extent and thickness. Researchers with the Greenland Ice Core Project have reported in *Science* that during a previous warming episode, more than 450,000 years ago, the southernmost part of Greenland was covered by boreal forest—suggesting that both the ice sheet and the sea ice may disappear entirely if the ongoing global warming trend continues. Indeed, this process seems well underway: the polar ice cap is now less than half of what it was in 1979, and the lower-left inset map in Figure 12-5 highlights the current shrinkage phase. The

ecological consequences are far-reaching, threatening the habitats of polar bears, whales, walruses, seals, and other species. But the economic- and politico-geographical implications are going to be significant as well.

When the Arctic's sea ice melts in summer and does not fully recover in winter, once-blocked waterways begin to open up and accessibilities can be transformed. The **Northeast Passage** 🄬, that legendary high-latitude route between the Far East and Europe, has been the object of hope and despair for centuries, causing much loss of human life in the quest to forge a viable sea lane. But in 2007, for the first time in recorded history, it was completely ice-free for a brief period during the late-summer peak of seasonal melting. Since then, the Northeast Passage has been ice-free every year during that August/September window; by 2013, a total of 71 specially equipped vessels—carrying well over a million tons of cargo—were able to successfully navigate the route (mapped in the lower-right inset map of Fig. 12-5). That map also shows the presence of a similar route around the northern perimeter of North America, the **Northwest Passage** 🄭. This other icon of (unsuccessful) pre-twentieth-century polar exploration would connect the Pacific and Atlantic oceans, following an eastward watercourse along Alaska's northern coast, through Canada's massive Arctic Archipelago, and then a final southward traverse via the Davis Strait between Canada and Greenland.

If either or both of these waterways can be activated, the impact on global shipping will be momentous. Consider what it would mean if ships bound for Asia's Pacific Rim from the U.S. eastern seaboard, and too large to fit through the upgraded Panama Canal, could shorten their voyages by several days and thousands of kilometers by rounding North America instead of South America! No wonder countries like China, India, Japan, and even Singapore have applied for 'observer status' at the meetings of the Arctic Council, whose regular members include the United States, Russia, Canada, Norway, Sweden, Denmark, Finland, and Iceland.

But who really owns the waterways vacated by the ice? Canada regards the Northwest Passage as a domestic waterway, giving it the right to control all shipping. The United States holds that the Northwest Passage is an international waterway, and American and Russian ships have sailed through it without Canada's permission. The Canadian government has affirmed its position by starting work on a deepwater port at Nanisivik on Nunavut's Baffin Island at the eastern entrance to the Northwest Passage, and by building a military base at Resolute Bay to the west; it has also announced the construction of eight Arctic patrol vessels to police the route. In the meantime, the United States is opening a new Coast Guard station to patrol the narrow Bering Strait between Alaska and Russia. The lines, if not yet battle lines, are already being drawn in the warming Arctic.

Humankind's growing numbers, escalating demands, environmental impacts, and technological capacities are transforming even the most remote recesses of our resilient planet. No international resolution is more critical than the avoidance of destructive conflict in fragile polar environments.

- Governments of low-lying island-states in both the Pacific and Indian oceans are demanding international assistance as they are confronted by the threat of the world ocean's rising sea level.

- Fiji receives support from Beijing for officially recognizing the People's Republic of China (PRC) but not Taiwan; Palau gets support from the Taiwanese for officially recognizing Taiwan but not the PRC.

- Singapore, located almost directly on the equator, has requested and been granted 'observer status' to monitor the meetings of the Arctic Council because of the impact the possible opening of ice-free, far northern passages will have on global shipping.

GLOSSARY*

Aboriginal land issue The legal campaign in which Australia's **indigenous peoples** have claimed title to traditional land in several parts of that country. The courts have upheld certain claims, fueling Aboriginal activism that has raised broader issues of indigenous rights.

Aboriginal population Native or *aboriginal* peoples; often used to designate the inhabitants of areas that were conquered and subsequently colonized by the **imperial** powers of Europe.

Absolute location The position or place of a certain item on the surface of the Earth as expressed in degrees, minutes, and seconds of **latitude** and **longitude**.

Acculturation Cultural modification resulting from intercultural borrowing. In **cultural geography**, the term refers to the change that occurs in the **culture** of **indigenous peoples** when contact is made with a society that is technologically superior.

AFTA The ASEAN Free Trade Agreement that since 1992, through lowered tariffs and other incentives, has fostered increased trade within Southeast Asia. This is the economic centerpiece of the **Association of Southeast Asian Nations (ASEAN)**, a **supranational** organization whose members include 10 of that realm's 11 states (only East Timor does not participate).

Agribusiness The agricultural operations of large, often multinational, corporation.

al-Qaeda The terrorist organization that evolved into an expanding global network under the directorship of Osama bin Laden between the mid-1990s and his elimination by the U.S. in 2011. It sought to coordinate the efforts of once loosely allied Muslim revolutionary movements, and unleash a *jihad* aimed at what it perceived to be Islam's enemies in the West.

Altiplano High-elevation plateau, basin, or valley between even higher mountain ranges, especially in the Andes of South America.

Altitudinal zonation Vertical regions defined by physical-environmental zones at various elevations (see Fig. 2-4), particularly in the highlands of South and Middle America.

American Manufacturing Belt North America's near-rectangular Core Region, whose corners are Boston, Milwaukee, St. Louis, and Baltimore.

Antarctic Treaty International cooperative agreement on the use of Antarctic territory.

Apartheid Literally, *apartness*. The Afrikaans term for South Africa's pre-1994 policies of racial separation, a system that produced highly segregated socio-geographical patterns.

APEC The Asia-Pacific Economic Cooperation, a forum for 21 Pacific Rim member-economies that promotes free trade throughout the Asia-Pacific region

Archipelago A set of islands grouped closely together, usually elongated into a *chain*.

ASEAN The **Association of Southeast Asian Nations (ASEAN)**, a **supranational** organization whose members include 10 of that realm's 11 states: Brunei, Cambodia, Indonesia, Laos, Malaysia, Myanmar (Burma), the Philippines, Singapore, Thailand, and Vietnam (only minuscule East Timor does not participate).

Asian Tigers The Asia-Pacific's four smaller-scale, dynamic, upwardly bound economies: Hong Kong, South Korea, Taiwan, and Singapore (the first three are located in East Asia; the last in Southeast Asia). They emerged during the 1970s in the wake of Japan's much larger economic success. All followed a similar development strategy of modernization, rapid industrialization, luring foreign direct investment, and building economies focused on the exporting of high-quality manufactures.

Atoll A ring-like coral reef surrounding an empty lagoon that probably formed around the rim of a now-completely-eroded volcanic cone standing on the seafloor. They are common in certain tropical areas of the Pacific Ocean where they are classified among that realm's **low islands**.

Austral South.

Balkanization The fragmentation of a **region** into smaller, often hostile political units. Named after the historically contentious Balkan Peninsula of southeastern Europe.

Barrio Term meaning "neighborhood" in Spanish. Usually refers to an urban community in a Middle or South American city.

Bauxite Aluminum ore; usually deposited at shallow depths in the wet tropics (past or present).

Biodiversity hot spot A much higher than usual, world-class geographic concentration of natural plant and/or animal species. Tropical rainforest environments have dominated, but their recent ravaging by **deforestation** has had catastrophic results.

Biodiversity Shorthand for *biological diversity*; the total variety of plant and animal species that exists in a given area.

Biogeography The study of *flora* (plant life) and *fauna* (animal life) in spatial perspective.

Boko Haram The violent, jihadist, terrorist organization that is based in—and controls parts of—northeastern Nigeria. It engages in extreme terrorist attacks, increasingly beyond its territory as it seeks to expand into other parts of the country.

Borderland A formal or functional region that straddles a political boundary.

Boreal forest The subarctic, mostly **coniferous** snow-forest that blankets Canada south of the **tundra** that lines the Arctic shore; known as the **taiga** in Russia.

Break-of-bulk A location along a transport route where goods must be transferred from one carrier to another. In a port, the cargoes of oceangoing ships are unloaded and put on trains, trucks, or perhaps smaller river boats for inland distribution. An *entrepôt*.

BRICs Acronym for the four biggest emerging national markets in the world today—**B**razil, **R**ussia, **I**ndia, and **C**hina.

Buffer state or Buffer zone A country or set of countries separating ideological or political adversaries. For example, in southern Asia, Afghanistan, Nepal, and Bhutan were parts of a buffer zone set up between British and Russian-Chinese imperial spheres. In Southeast Asia, Thailand was a buffer state between British and French colonial domains.

Business process outsourcing Often abbreviated as BPO, refers to overseas outsourcing by companies of their business processing activities to low-cost countries, such as the Philippines or India. Examples of such activities include processing of insurance claims and data file management.

Caliphate An imperial-scale Islamic government led by a caliph, considered a direct successor to the Prophet Muhammad, who rules and exerts moral authority over Muslims worldwide.

Caste system The strict **social stratification** of people—specifically in India's Hindu society—on the basis of ancestry and occupation.

Central business district (CBD) The downtown heart of a central city; marked by high land values, a concentration of business and commerce, and the clustering of the tallest buildings.

Centrality The strength of an urban center in its capacity to attract producers and consumers to its facilities; a city's "reach" into the surrounding region.

Centrifugal forces A term employed to designate forces that tend to divide a country—such as internal religious, linguistic, ethnic, or ideological differences.

Centripetal forces Forces that unite and bind a country together—such as a strong national culture, shared ideological objectives, and a common faith.

Cerrado Regional term referring to the fertile savannas of Brazil's interior Central-West that make it one of the world's most promising agricultural frontiers. Soybeans are the leading crop, and other grains and cotton are expanding. Inadequate transport links to the outside world remain a problem.

Choke point A narrowing of an international waterway causing marine-traffic congestion, requiring reduced speeds and/or sharp turns, and increasing the risk of collision as well as vulnerability to attack. When the waterway narrows to a distance of less than 38 kilometers (24 mi), this necessitates the drawing of a **median line (maritime) boundary**.

*Words in **boldface type** within an entry are defined elsewhere in this Glossary

City-state An independent political entity consisting of a single city with (and sometimes without) an immediate **hinterland**.

Climate The long-term conditions (over at least 30 years) of aggregate **weather** over a region, summarized by averages and measures of variability; a synthesis of the succession of weather events we have learned to expect at any given location.

Command economy The tightly controlled economic system of the former Soviet Union, whereby central planners in Moscow assigned the production of particular goods to particular places, often guided more by socialist ideology than the principles of **economic geography**.

Commercial agriculture For-profit **agriculture**.

Communal tension Persistent stress among a country's sociocultural groups that can often erupt into communal violence.

Compact state A politico-geographical term to describe a state that possesses a roughly circular, oval, or rectangular territory in which the distance from the geometric center to any point on the boundary exhibits little variance.

Complementarity Exists when two regions, through an exchange of raw materials and/or finished products, can specifically satisfy each other's demands.

Connectivity The degree of direct linkage between a particular location and other locations within a regional, national, or global transportation **network**.

Continental drift The slow movement of continents controlled by the processes associated with **plate tectonics**.

Continental shelf Beyond the coastlines of many landmasses, the ocean floor declines very gently until the depth of about 660 feet (200 m). Beyond the 660-foot line the sea bottom usually drops off sharply, along the *continental slope*, toward the much deeper mid-oceanic basin. The submerged continental margin is called the continental shelf, and it extends from the shoreline to the upper edge of the continental slope.

Continentality The variation of the continental effect on air temperatures in the interior portions of the world's landmasses. The greater the distance from the moderating influence of an ocean, the greater the extreme in summer and winter temperatures. Continental interiors also tend to be dry when the distance from oceanic moisture sources becomes considerable.

Conurbation General term used to identify a large multimetropolitan complex formed by the coalescence of two or more major **urban areas**.

Core area In geography, a term with several connotations. *Core* refers to the center, heart, or focus. The core area of a **nation-state** is constituted by the national heartland, the largest population cluster, the most productive region, and the part of the country with the greatest **centrality** and **accessibility**—probably containing the capital city as well.

Core-periphery relationship The contrasting spatial characteristics of, and linkages between, the *have* (core) and *have-not* (periphery) components of a national, regional, or the global **system**.

Cultural diffusion The **process** of spreading and adopting a cultural element, from its place of origin across a wider area.

Cultural geography The wide-ranging and comprehensive field of geography that studies spatial aspects of human **cultures**.

Cultural landscape The forms and artifacts sequentially placed on the **natural landscape** by the activities of various human occupants. By this progressive imprinting of the human presence, the physical (natural) landscape is modified into the cultural landscape, forming an interacting unity between the two.

Cultural pluralism A society in which two or more population groups, each practicing its own **culture**, live adjacent to one another without mixing inside a single **state**.

Culture hearth Heartland, source area, or innovation center; place of origin of a major **culture**.

Dalit The term now used for members of the lowest-ranking social layer in the **caste system** of Hindu India.

Deindustrialization Process by which companies relocate manufacturing jobs to other regions or countries with cheaper labor, leaving the newly-deindustrialized region to convert to a service economy while struggling with the accompanying effects of increased unemployment and meeting the retraining needs of its workforce.

Demographic burden The proportion of a national population that is either too old or too young to be productive and that must be cared for by the productive population.

Demographic transition Multi-stage **model**, based on Western Europe's experience, of changes in population growth exhibited by countries undergoing industrialization. High **birth rates** and **death rates** are followed by plunging death rates, producing a huge net population gain; birth and death rates then converge at a low overall level.

***Dependencia* theory** Originating in South America during the 1960s, it was a new way of thinking about economic development and underdevelopment that explained the persistent poverty of certain countries in terms of their unequal relations with other (i.e., rich) countries.

Desalinization The process of removing dissolved salts from water, thereby producing fresh (drinking) water from seawater or brackish water

Development The economic, social, and institutional growth of national **states**.

Devolution The **process** whereby regions within a **state** demand and gain political strength and growing autonomy at the expense of the central government.

Distance decay The various degenerative effects of distance on human spatial structures and interactions.

Domino effect The belief that political destabilization in one **state** can result in the collapse of order in a neighboring state, triggering a chain of events that, in turn, can affect a series of **contiguous** states.

Domino theory Theory that holds that destabilization and conflict from any cause in one country can result in the collapse of order in one or more neighboring countries, triggering a falling-domino chain of events that can affect a series of contiguous states.

Double delta South Asia's combined **delta** formed by the Ganges and Brahmaputra rivers. All of Bangladesh lies on this enormous deltaic plain, which also encompasses surrounding parts of eastern India. Well over 200 million people live here, attracted by the fertility of its soils that are constantly replenished by the **alluvium** transported and deposited by these two of Asia's largest river systems. Natural hazards abound here as well, ranging from the flooding caused by excessive **monsoonal** rains to the intermittent storm surges of powerful cyclones (**hurricanes**) that come from the Bay of Bengal to the south.

Dravidian languages The language family, indigenous to the South Asian realm, that dominates southern India today; as opposed to the Indo-European languages, whose tongues dominate northern India.

Economic geography The field of geography that focuses on the diverse ways in which people earn a living and on how the goods and services they produce are expressed and organized spatially.

Economic integration The benefits of forging new international partnerships.

Economies of scale The savings that accrue from large-scale production wherein the unit cost of manufacturing decreases as the level of operation enlarges. Supermarkets operate on this principle and are able to charge lower prices than small grocery stores.

Ejidos Mexican farmlands redistributed to peasant communities after the Revolution of 1910–1917. The government holds title to the land, but user rights are parceled out to village communities and then to individuals for cultivation.

El Niño A periodic, abnormal warming of the sea surface in the eastern equatorial Pacific Ocean. Disturbs weather patterns across much of the world, especially in northwestern and northeastern South America.

Elongated state A **state** whose territory is decidedly long and narrow in that its length is at least six times greater than its average width.

Elongation A long and narrow political territory whose length is at least six times greater than its average width.

Emerging market The world's fastest growing national market economies as measured by economic growth rates, attraction of **foreign direct investment**, and other key indicators. Led by the **BRICs** (Brazil, Russia, India, and China), but this club is now expanding to include many other countries.

Endemic Refers to a disease in a host population that affects many people in a kind of equilibrium without causing rapid and widespread deaths.

Entrepôt A place, usually a port city, where goods are imported, stored, and transshipped; a **break-of-bulk point**.

Environmental degradation The accumulated human abuse of a region's **natural landscape** that, among other things, can involve air and water pollution, threats to plant and animal **ecosystems**, misuse of **natural resources**, and generally upsetting the balance between people and their habitat.

Environmental determinism A geographic school of thought, popular in the first quarter of the twentieth century, that maintains the physical environment determines human behaviour and/or social outcomes.

Epidemic A local or regional outbreak of a disease.

Estuary The widening mouth of a river as it reaches the sea; land subsidence or a rise in sea level has overcome the tendency to form a **delta**.

Eurasian Customs Union A **supranational** economic alliance formed in 2010 by Russia, Belarus, and Kazakhstan, and joined in 2015 by Armenia and Kyrgyzstan.

European state model A **state** consisting of a legally defined territory inhabited by a population governed from a capital city by a representative government.

Exclave A bounded (non-island) piece of territory that is part of a particular **state** but lies separated from it by the territory of another state.

Exclusive Economic Zone (EEZ) An oceanic zone extending up to 200 **nautical miles** (370 km) from a shoreline, within which the coastal **state** can control fishing, mineral exploitation, and additional activities by all other countries.

Expansion diffusion The spreading of an innovation or an idea through a fixed population in such a way that the number of those adopting it grows continuously larger, resulting in an expanding area of dissemination.

Failed state A country whose institutions have collapsed and in which anarchy prevails.

Favela Shantytown on the outskirts or even well within an urban area in Brazil.

Federal system A political framework wherein a central government represents the various subnational entities within a country where they have common interests—defense, foreign affairs, and the like—yet allows these various entities to retain their own identities and to have their own laws, policies, and customs in certain spheres.

Federation A country adhering to a political framework wherein a central government represents the various subnational entities within a **nation-state** where they have common interests—defense, foreign affairs, and the like—yet allows these various entities to retain their own identities and to have their own laws, policies, and customs in certain spheres.

Fertility rate The average number of children born to women of childbearing age in a given population.

First Nations Name given Canada's **indigenous peoples** of American descent, whose U.S. counterparts are called Native Americans.

Floating population China's huge mass of mobile workers who respond to shifting employment needs within the country. Most are temporary urban dwellers with restricted residency rights, whose movements are controlled by the *hukou* system.

Foreign direct investment A key indicator of the success of an **emerging market** economy, whose growth is accelerated by the infusion of foreign funds to supplement domestic sources of investment capital.

Formal economy The part of a national economy that is registered with government agencies and complies with laws and regulations, especially taxation.

Formal region A type of **region** marked by a certain degree of homogeneity in one or more phenomena; also called *uniform region* or *homogeneous region*.

Forward capital Capital city positioned in actually or potentially contested territory, usually near an international border; it confirms the **state's** determination to maintain its presence in the region in contention.

Fossil fuel The energy resources of **coal**, natural gas, and petroleum (oil), so named collectively because they were formed by the geologic compression and transformation of tiny plant and animal organisms.

Four Motors of Europe *Rhône-Alpes* (France), *Baden-Württemberg* (Germany), *Catalonia* (Spain), and *Lombardy* (Italy). Each is a high-technology-driven region marked by exceptional industrial vitality and economic success not only within Europe but on the global scene as well.

Fragmented modernization A checkerboard-like spatial pattern of **modernization** in an **emerging-market** economy wherein a few localized regions of a country experience most of the development while the rest are largely unaffected.

Fragmented state A **state** whose territory consists of several separated parts, not a **contiguous** whole. The individual parts may be isolated from each other by the land area of other states or by international waters.

Functional region A **region** marked less by its sameness than by its dynamic internal structure; it is most commonly forged by a structured, urban-centered **system** of interaction.

Gender imbalance The demographic imbalance of males outnumbering females resulting from selective birth control. In China, this is an outcome of the **One-Child Policy**.

Gentrification The upgrading of an older residential area through private reinvestment, usually in the downtown area of a central city. Frequently, this involves the displacement of established lower-income residents, who cannot afford the heightened costs of living, and conflicts are not uncommon as such neighborhood change takes place.

Geographic realm The basic spatial unit in our world regionalization scheme. Each realm is defined in terms of a synthesis of its total human geography—a composite of its leading cultural, economic, historical, political, and appropriate environmental features.

Geopolitics Political relations among states or regions that are strongly influenced by their geographical setting, including proximity, **accessibility**, sovereign boundaries, natural resources, **population distribution**, and the like.

Global climate change The shift in the characteristics and spatial distribution of Earth's climates in response to a long-term trend in atmospheric warming.

Global core The constellation of countries with the most highly developed and influential economies (anchored by the United States, the 28 member-countries of the European Union, China, and Japan). Although the **global core** (mapped in Fig. G-12) is not a contiguous geographic region, its countries as a whole dominate international trade and investment flows.

Globalization The gradual reduction of regional differences at the world **scale**, resulting from increasing international cultural, economic, and political exchanges.

Golden Triangle The triangular-shaped region that encompasses northwestern Laos, northernmost Thailand, and (especially) part of eastern Myanmar (see Fig. 10-14); the second-biggest opium (poppy) producing region in the world after Afghanistan.

Green Revolution The successful recent development of higher-yield, fast-growing varieties of rice and other cereals in certain developing countries.

Growth Triangle An increasingly popular economic **development** concept along the western **Pacific Rim**, especially in Southeast Asia. It involves the linking of production in growth centers of three countries to achieve benefits for all.

Growth-pole concept An urban center with a number of attributes that, if augmented by investment support, will stimulate regional economic **development** in its **hinterland**.

Hacienda Literally, a large estate in a Spanish-speaking country. Sometimes equated with the **plantation**, but there are important differences between these two types of agricultural enterprise.

Hanification Imparting a cultural imprint by the ethnic Chinese (the "people of Han"). Within China often refers to the steadily increasing migration of Han Chinese into the country's **periphery**, especially Xinjiang and Xizang (Tibet). **Overseas Chinese** imprints, more generally referred to as **Sinicization**, have been significant as well, most importantly in the Southeast Asian realm.

High island Cultures associated with volcanic islands of the Pacific Realm that are high enough in elevation to wrest substantial moisture from the tropical ocean air (see **orographic precipitation**). They tend to be well watered, their volcanic soils enable productive agriculture, and they support larger populations than **low islands.**

High seas Areas of the oceans away from land, beyond national jurisdiction, open and free for all to use.

High-value-added goods Products of improved net worth.

Hindutva "Hinduness" as expressed through Hindu nationalism, Hindu heritage, and/or Hindu patriotism. The cornerstone of a fundamentalist movement that has been gaining strength since the late twentieth century that seeks to remake India as a society dominated by Hindu principles. It has been the guiding agenda of the Bharatiya Janata Party (BJP), which has emerged as a powerful force in national politics and in big States like Maharashtra, Gujarat, and Madhya Pradesh.

Hinterland Literally, "country behind," a term that applies to a surrounding area served by an urban center. That center is the focus of goods and services produced for its hinterland and is its dominant urban influence as well.

Holocene The current *interglacial* epoch (the warm period of glacial contraction between the glacial expansions of an **ice age**); extends from 10,000 years ago to the present. Also known as the *Recent Epoch*.

Hukou system A longstanding system used in China whereby all inhabitants must obtain and carry with them residency permits that indicate where an individual is from and where they may exercise particular rights such as education, healthcare, housing, and the like.

Human evolution Long-term biological maturation of the human species. Geographically, all evidence points toward East Africa as the source of humankind. Our species, *Homo sapiens*, emigrated from this hearth to eventually populate the rest of the **ecumene**.

Hurricane Alley The most frequent pathway followed by tropical storms and **hurricanes** over the past 150 years in their generally westward movement across the Caribbean Basin. Historically, hurricane tracks have bundled most tightly in the center of this route, most often affecting the Lesser Antilles between Antigua and the Virgin Islands, Puerto Rico, Hispaniola (Haiti/Dominican Republic), Jamaica, Cuba, southernmost Florida, Mexico's Yucatán, and the Gulf of Mexico.

Hydraulic civilization theory The theory that cities which managed to control **irrigated** farming over large **hinterlands** held political power over other cities. Particularly applies to early Asian civilizations based in such river valleys as the Chang (Yangzi), the Indus, and those of Mesopotamia.

Inclusive development The extent of equal economic (and social) development opportunities for different population groups, especially minorities and the poor.

Indigenous Aboriginal or native; an example would be the pre-Columbian inhabitants of the Americas.

Indirect rule British colonial practice that kept indigenous power structures in place, co-opting individual rulers. The purpose was to minimize armed conflict and maximize profits.

Indo-European languages The major world language family that dominates the European **geographic realm.** This language family is also the most widely dispersed globally (Fig. G-8), and about half of humankind speaks one of its languages.

Indochina The colonial territory that constituted the former French Empire in Southeast Asia, consisting of present-day Vietnam, Cambodia, and Laos.

Industrial Revolution The term applied to the social and economic changes in agriculture, commerce, and especially manufacturing and urbanization that resulted from technological innovations and greater specialization in late-eighteenth-century Europe.

Informal economy or Informal sector The part of a national economy that is not registered with the government, and for which reliable statistics are rarely available. The complement to a country's formal economy."

Insurgent state Territorial embodiment of a successful guerrilla movement. The establishment by antigovernment insurgents of a territorial base in which they exercise full control; thus a **state** within a state.

Irredentism A policy of cultural extension and potential political expansion by a **state** aimed at a community of its nationals living in a neighboring state.

Islamic Front The southern border of the African Transition Zone that marks the religious **frontier** of the **Muslim** faith in its southward penetration of Subsaharan Africa (see Fig. 6-20).

Jihad A doctrine within Islam. Commonly translated as *holy war*, it entails a personal or collective struggle on the part of **Muslims** to live up to the religious standards prescribed by the *Quran* (Koran).

Land alienation One society or culture group taking land from another.

Land bridge A narrow **isthmian** link between two large landmasses. They are temporary features—at least when measured in geologic time—subject to appearance and disappearance as the land or sea level rises and falls.

Land hemisphere The half of the globe containing the greatest amount of land surface, centered on Western Europe.

Land tenure The way people own, occupy, and use land.

Landlocked location An interior **state** surrounded by land. Without coasts, such a country is disadvantaged in terms of **accessibility** to international trade routes, and in the scramble for possession of areas of the **continental shelf** and control of the **exclusive economic zone** beyond.

Liberation theology A powerful religious movement that arose in South America during the 1950s, and subsequently gained followers all over the global **periphery**. At its heart is a belief system, based on a blend of Christian faith and socialist thinking, that interprets the teachings of Christ as a quest to liberate the impoverished masses from oppression.

Local functional specialization A hallmark of Europe's **economic geography** that later spread to many other parts of the world, whereby particular people in particular places concentrate on the production of particular goods and services.

Low island Cultures associated with low-lying coral islands of the Pacific Realm that cannot wrest sufficient moisture from the tropical maritime air to avoid chronic drought. Thus productive agriculture is impossible, and their modest populations must rely on fishing and the coconut palm for survival.

Maquiladoras The term given to modern industrial plants in Mexico's U.S. border zone. These foreign-owned factories assemble imported components and/or raw materials, and then export finished manufactures, mainly to the United States. Import duties have disappeared under **NAFTA**, bringing jobs to Mexico and the advantages of low wage rates to the foreign entrepreneurs.

Marine geography The geographic study of oceans and seas. Its practitioners investigate both the physical (e.g., coral-reef **biogeography**, ocean–**atmosphere** interactions, coastal **geomorphology**) as well as human (e.g., **maritime boundary**-making, fisheries, beachside development) aspects of oceanic environments.

Maritime boundary An international boundary that lies in the ocean. Like all boundaries, it is a vertical plane, extending from the seafloor to the upper limit of the air space in the atmosphere above the water.

Median-line boundary An international **maritime boundary** drawn where the width of a sea is less than 400 **nautical miles**. Because the **states** on either side of that sea claim **exclusive economic zones** of 200 nautical miles, it is necessary to reduce those claims to a (median) distance equidistant from each shoreline. **Delimitation** on the map almost always appears as a set of straight-line segments that reflect the configurations of the coastlines involved.

Medical geography The study of health and disease within a geographic context and from a spatial perspective. Among other things, this geographic field examines the sources, **diffusion** routes, and distributions of diseases.

Megacity Informal term referring to the world's most heavily populated cities; in this book, the term refers to a **metropolis** containing a population of greater than 10 million.

Melting pot Traditional characterization of American society as a blend of numerous **immigrant ethnic** groups that over time were assimilated into a single societal mainstream. This notion always had its challengers among social scientists, and is now increasingly difficult to sustain given the increasing complexity and sheer scale of the U.S. ethnic mosaic in the twenty-first century.

Mental maps Maps that individuals carry around in their minds that reflect their constantly evolving perception of how geographic space (ranging from their everyday activity space to the entire world) is organized around them.

Mestizo Derived from the Latin word for *mixed*, refers to a person of mixed European (white) and Amerindian ancestry.

Micro-credit Small loans extended to poverty-stricken borrowers who would not otherwise qualify for them. The aim is to help combat poverty, encourage entrepreneurship, and empower poor communities—especially their women.

Microstate A sovereign **state** that contains a minuscule land area and population. They do not have the attributes of "complete" states, but are on the map as tiny yet independent entities nonetheless.

Migration A change in residence intended to be permanent.

Monocentric geographic realm A **realm** dominated by a single **core area** or **state**.

Monsoon Refers to the seasonal reversal of wind and moisture flows in certain parts of the subtropics and lower-middle latitudes. The *dry monsoon* occurs during the cool season when dry offshore winds prevail. The *wet monsoon* occurs in the hot summer months, which produce onshore winds that bring large amounts of rainfall. The air-pressure differential over land and sea is the triggering mechanism, with windflows always moving from areas of relatively higher pressure toward areas of relatively lower pressure. Monsoons make their greatest regional impact in the coastal and near-coastal zones of South Asia, Southeast Asia, and East Asia.

Mulatto A person of mixed African (black) and European (white) ancestry.

Multilingualism A society marked by a mosaic of local languages. Constitutes a **centrifugal force** because it impedes communication within the larger population. Often a *lingua franca* is used as a "common language," as in many countries of Subsaharan Africa.

NAFTA The North American Free Trade Association. The **free-trade area** launched in 1994 involving United States, Canada, and Mexico.

Nation Legally a term encompassing all the citizens of a **state**, it also has other connotations. Most definitions now tend to refer to a group of tightly knit people possessing bonds of language, **ethnicity**, religion, and other shared **cultural** attributes. Such homogeneity actually prevails within very few states.

Nation-state A country whose population possesses a substantial degree of **cultural** homogeneity and unity. The ideal form to which most **nations** and **states** aspire—a political unit wherein the territorial state coincides with the area settled by a certain national group or people.

Natural landscape The array of landforms that constitutes the Earth's surface (mountains, hills, plains, and plateaus) and the physical features that mark them (such as water bodies, soils, and vegetation). Each **geographic realm** has its distinctive combination of natural landscapes.

Near Abroad The 14 former Soviet republics that, in combination with the dominant Russian Republic, constituted the USSR. Since the 1991 breakup of the Soviet Union, Russia has asserted a sphere of influence in these now-independent countries, based on its proclaimed right to protect the interests of ethnic Russians who were settled there in substantial numbers during Soviet times. These 14 countries include Armenia, Azerbaijan, Belarus, Estonia, Georgia, Kazakhstan, Kyrgyzstan, Latvia, Lithuania, Moldova, Tajikistan, Turkmenistan, Ukraine, and Uzbekistan.

Neoliberal policies Government policies aimed at supporting private entrepreneurship and free markets.

Neoliberalism An ideology or development strategy that advocates the privatization of state-run companies, lowering of international trade tariffs, reduction of government subsidies, cutting of corporate taxes, and overall deregulation of business activity.

New Silk Road The Chinese initiative to construct an overland routeway (railroads and highways), extending from western China via Central Asia to Europe, to connect markets, allow rapid movement of goods, and build Chinese influence.

Nine-dashed-line map The Chinese government's map that, through the use of nine dashes, delineates China's maritime territorial claims that cover most of the South China Sea (see Fig. 10-10).

Node A center that functions as a point of **connectivity** within a regional **network** or **system**. All urban settlements possess this function, and the higher the position of a settlement in its **urban system** or **hierarchy**, the greater its nodality.

Non-governmental organization (NGO) Legitimate organizations that operate independently from any form of government and do not function as for-profit businesses. They mostly seek to improve social conditions, but are not affiliated with political organizations.

Northeast Passage The high-latitude sea route of the Arctic Ocean that follows the entire north coast of Eurasia from northern Norway in the west to the northeasternmost corner of Russia where it meets the Bering Strait. Increased seasonal melting of the Arctic ice cap in recent years has begun to open up this waterway as a summer route for shipping between Europe and East Asia.

Northwest Passage The high-latitude, Arctic Ocean sea route around North America extending from Alaska's Bering Strait in the west to the Davis Strait between Canada and Greenland in the east. Heightened summertime melting of the Arctic ice cap in recent years has increased the likelihood of opening up this waterway for shipping between Asia's Pacific Rim and the Atlantic seaboard of the Americas.

Offshore banking Term referring to financial havens for foreign companies and individuals, who channel their earnings to accounts in such a country (usually an "offshore" island-state) to avoid paying taxes in their home countries.

One-Child Policy Chinese population control policy initiated in the late 1970s that proscribed (and enforced) a limit of one child per family of most population groups (mainly urban, **Han** populations). This policy has become increasingly controversial over the past decade.

Outback The name given by Australians to the vast, peripheral, sparsely settled interior of their country.

Outer city The non-central-city portion of the American **metropolis**; no longer "sub" to the "urb," this outer ring was transformed into a full-fledged city during the late twentieth century.

Overseas Chinese The more than 50 million ethnic Chinese who live outside China. About two-thirds live in Southeast Asia, and many have become quite successful. A large number maintain links to China and as investors played a major economic role in stimulating the growth of **SEZs** and Open Cities in China's **Pacific Rim**.

Pacific Rim A far-flung group of countries and components of countries (extending clockwise on the map from New Zealand to Chile) sharing the following criteria: they face the Pacific Ocean; they exhibit relatively high levels of economic development, industrialization, and urbanization; their imports and exports mainly move across Pacific waters.

Pacific Ring of Fire Zone of crustal instability along **tectonic plate** boundaries, marked by earthquakes and volcanic activity, that ring the Pacific Ocean Basin.

Pandemic An outbreak of a disease that spreads worldwide.

Partition The subdivision of the British Indian Empire into India and Pakistan at the end of colonial rule on August 15, 1947.

Peripheral development Spatial pattern in which a country's or region's development (and population) is most heavily concentrated along its outer edges rather than in its interior.

Permafrost Permanently frozen water in the near-surface soil and bedrock of cold environments, producing the effect of completely frozen ground. Surface can thaw during brief warm season.

Physiographic region A **region** within which there prevails substantial **natural-landscape** homogeneity, expressed by a certain degree of uniformity in surface **relief, climate**, vegetation, and soils.

Physiologic density The number of people per unit area of **arable** land.

Plantation A large estate owned by an individual, family, or corporation and organized to produce a cash crop. Almost all plantations were established within the tropics; in recent decades, many have been divided into smaller holdings or reorganized as cooperatives.

Population density The number of people per unit area. Also see **arithmetic density** and **physiologic density** measures.

Population distribution The way people have arranged themselves in geographic space. One of human geography's most essential expressions because it represents the sum total of the adjustments that a population has made to its natural, cultural, and economic environments. A population distribution map is included in every chapter in this book.

Population geography The field of geography that focuses on the spatial aspects of **demography** and the influences of demographic change on particular countries and regions.

Population implosion The opposite of **population explosion**; refers to the declining populations of many European countries and Russia in which the **death rate** exceeds the **birth rate** and **immigration** rate.

Population pyramid Graphic representation or *profile* of a national population according to age and gender. Such a diagram of age-sex structure typically displays the percentage of each age group (commonly in five-year increments) as a horizontal bar, whose length represents its relationship to the total population.

Primary sector Activities engaged in the direct extraction of natural resources from the environment such as mining, fishing, lumbering, and especially agriculture.

Primate city A country's largest city—ranking atop its urban **hierarchy**—most expressive of the national culture and usually (but not in every case) the capital city as well.

Protruded state Territorial shape of a **state** that exhibits a narrow, elongated land extension (or *protrusion*) leading away from the main body of territory.

Rain shadow effect The relative dryness in areas downwind of mountain ranges resulting from *orographic precipitation*, wherein moist air masses are forced to deposit most of their water content as they cross the highlands.

Region A commonly used term and a geographic concept of paramount importance. An **area** on the Earth's surface marked by specific criteria, which are discussed in the Introduction.

Regional complementarity Exists when two regions, through an exchange of raw materials and/or finished products, can specifically satisfy each other's demands.

Relative location The regional position or **situation** of a place relative to the position of other places. Distance, **accessibility**, and **connectivity** affect relative location.

Religious revivalism Religious movement whose objectives are to return to the foundations of that faith and to influence state policy. Often called *religious fundamentalism*; but in the case of Islam, **Muslims** prefer the term *revivalism*.

Relocation diffusion Sequential **diffusion process** in which the items being diffused are transmitted by their carrier agents as they relocate to new areas. The most common form of relocation diffusion involves the spreading of innovations by a **migrating** population.

Remittances Money earned by **emigrants** that is sent back to family and friends in their home country, mostly in cash; forms an important part of the economy in poorer countries.

Revanchism In **geopolitics**, a retaliatory government policy aimed at recovering a **state's** lost territory.

Rift valley The trough or trench that forms when a thinning strip of the Earth's crust sinks between two parallel faults (surface fractures).

Rising sea level One of the expected major impacts of global warming on the world ocean resulting from the large-scale melting of Arctic and Antarctic ice. Low-lying coastal settlements and human activities are at greatest risk of inundation. Worldwide, huge urban and rural population concentrations are threatened; in South Asia, southern Bangladesh and the Maldives **archipelago** are most vulnerable.

Rural-to-urban migration The dominant **migration** flow from countryside to city that continues to transform the world's population, most notably in the less advantaged geographic realms.

Russification Demographic resettlement policies pursued by the central planners of the Soviet Empire (1924–1991), whereby ethnic Russians were encouraged to **emigrate** from the Russian Republic to the 14 non-Russian republics of the USSR.

Satellite state The countries of Eastern Europe under Soviet **hegemony** between 1945 and 1989. This tier of countries—the "satellites" captured in Moscow's "orbit" following World War II—was bordered on the west by the Iron Curtain and on the east by the USSR. Using the names then in force, they included Bulgaria, Czechoslovakia, East Germany, Hungary, Poland, and Romania.

Scale Representation of a real-world phenomenon at a certain level of reduction or generalization. In **cartography**, the ratio of map distance to ground distance; indicated on a map as a bar graph, representative fraction, and/or verbal statement. *Macroscale* refers to a large area of national proportions; *microscale* refers to a local area no bigger than a county.

Sharia **law** The strict criminal code based in Islamic law that prescribes corporal punishment, amputations, stonings, and lashings for both major and minor offenses. Its occurrence today is associated with the spread of **religious revivalism** in **Muslim** societies.

Shatter belt Region caught between stronger, colliding external cultural-political forces, under persistent stress, and often fragmented by aggressive rivals.

Shi'ite Islam The smaller of the two main Islamic sects, comprising about 10 percent of Muslims overall, but in the majority in both Iran and Iraq. The origin of Shi'ism dates back to the death of the Prophet Muhammad, whose adherents believe that only a blood relative of Muhammad (his cousin) could be considered his legitimate successor.

Sinicization Giving a Chinese cultural imprint; Chinese **acculturation**. See also **Hanification**.

Site The internal locational attributes of an urban center, including its local spatial organization and physical setting.

Situation The external locational attributes of an urban center; its **relative location** or regional position with reference to other non-local places.

Small-island developing economies The additional disadvantages faced by lower-income island-states because of their often small territorial size and populations as well

as overland **inaccessibility**. Limited resources require expensive importing of many goods and services; the cost of government operations per capita are higher; and local production is unable to benefit from **economies of scale**.

Social stratification In a layered or stratified society, the population is divided into a **hierarchy** of social classes. In an industrialized society, the working class is at the lower end; **elites** that possess capital and control the means of production are at the upper level.

Southern Ocean The ocean that surrounds Antarctica.

Sovereignty Controlling power and influence over a territory, especially by the government of an autonomous state over the people it rules.

Spatial diffusion The spatial spreading or dissemination of a **culture** element (such as a technological innovation) or some other phenomenon (e.g., a disease outbreak).

Spatial perspective Broadly, the geographic dimension or expression of any phenomenon; more specifically, anything related to the organization of space on the Earth's surface.

Spatial system The components and interactions of a **functional region**, which is defined by the areal extent of those interactions.

Special Economic Zone (SEZ) Manufacturing and export center in China, created since 1980 to attract foreign investment and technology transfers. Seven SEZs—all located on China's Pacific coast—currently operate: Shenzhen, adjacent to Hong Kong; Zhuhai; Shantou; Xiamen; Hainan Island, in the far south; Pudong, across the river from Shanghai; and Binhai New Area, next to the port of Tianjin.

State A politically organized territory that is administered by a sovereign government and is recognized by a significant portion of the international community. A state must also contain a permanent resident population, an organized economy, and a functioning internal circulation system.

State capitalism Government-controlled corporations competing under free-market conditions, usually in a tightly regimented society.

State formation The creation of a **state**, exemplifying traditions of human **territoriality** that go back thousands of years.

State territorial configuration The geographical shape of a country, which has implications for its political geography and governmental functioning, both domestically and in the international sphere.

Stateless nation A national group that aspires to become a **nation-state** but lacks the territorial means to do so.

Subsistence agriculture Farmers who eke out a living on a small plot of land on which they are only able to grow enough food to support their families or at best a small community.

Subtropical Convergence A narrow marine **transition zone**, girdling the globe at approximately latitude 40°S, that marks the equatorward limit of the frigid **Southern Ocean** and the poleward limits of the warmer Atlantic, Pacific, and Indian oceans to the north.

Sunbelt The popular name given to the southern tier of the United States, which is anchored by the mega-States of California, Texas, and Florida. Its warmer climate, superior recreational opportunities, and other amenities have been attracting large numbers of relocating people and activities since the 1960s; broader definitions of the Sunbelt also include much of the western United States, even Colorado and the coastal Pacific Northwest.

Sunni Islam The larger of Islam's two main sects (encompassing) roughly 90 percent of all Muslims) who adhere to the conviction that any devout follower of the Prophet Muhammad is eligible to be his legitimate Successor.

Supranationalism A venture involving three or more **states**—political, economic, and/or cultural cooperation to promote shared objectives.

Sustainable development Viable, long-term development that does not deplete resources and/or cause significantly negative side-effects.

Taiga The subarctic, mostly **coniferous** snowforest that blankets northern Russia and Canada south of the **tundra** that lines the Arctic shore. Known as the **boreal forest** in North America.

Taliban The term means "students" or "seekers of religion." Specifically, refers to the Islamist militia group that emerged from *madrassas* in Pakistan and ruled neighboring Afghanistan between 1996 and 2001; it has been trying to regain control of that country in its continuing conflict with U.S.-led NATO troops. Taliban rule, in adherence with an extremist interpretation of *Sharia* law, was marked by one of the most virulent forms of militant Islam ever seen.

Tar sands The main source of oil from non-liquid petroleum reserves. The oil is mixed with sand and requires massive open-pit mining as well as a costly, complicated process to extract it. The largest known deposits are located in the northeast of Canada's province of Alberta, and by most estimates these Athabasca Tar Sands constitute one of the largest oil reserves in the world. The high oil prices of recent years have led to greatly expanded production here, but the accompanying environmental degradation caused by strip-mining and waste disposal has triggered a widening protest movement that may limit the exploitation of this resource.

Technopole A planned techno-industrial complex (such as California's Silicon Valley) that innovates, promotes, and manufactures the products of the **postindustrial** information economy.

Tectonic plates The slabs of heavier rock on which the lighter rocks of the continents rest. The plates are in motion, propelled by gigantic circulation cells in the red-hot, molten rock below. Most earthquakes and volcanic eruptions are associated with collisions of the mobile plates, as is the building of mountain chains.

Territorial sea Zone of seawater adjacent to a country's coast, held to be part of the national territory and treated as a component of the sovereign **state**.

Transculturation Cultural borrowing and two-way exchanges that occur when different **cultures** of approximately equal complexity and technological level come into close contact.

Transferability The capacity to move a good from one place to another at a bearable cost; the ease with which a commodity may be transported.

Transition zone An area of spatial change where the **peripheries** of two adjacent **realms** or **regions** join; marked by a gradual shift (rather than a sharp break) in the characteristics that distinguish these neighboring geographic entities from one another.

Transmigration The now-ended policy of the Indonesian government to induce residents of the overcrowded, **core-area** island of Jawa to move to the country's other islands.

Triple Frontier The turbulent and chaotic area in southern South America that surrounds the convergence of Brazil, Argentina, and Paraguay. Lawlessness pervades this haven for criminal elements, which is notorious for money laundering, arms and other smuggling, drug trafficking, and links to terrorist organizations, including money flows to the Middle East.

Tropical deforestation The clearing and destruction of tropical rainforests in order to make way for expanding settlement frontiers and the exploitation of new economic opportunities.

Tsunami A seismic (earthquake-generated) sea wave that can attain gigantic proportions and cause coastal devastation. The tsunami of December 26, 2004, centered in the Indian Ocean near the Indonesian island of Sumatera (Sumatra), produced the first great natural disaster of the twenty-first century. Our new century's second major tsunami disaster occurred along the coast of Japan's northeastern Honshu Island on March 11, 2011.

Tundra The treeless plain that lies along the Arctic shore in northernmost Russia and Canada, whose vegetation consists of mosses, lichens, and certain hardy grasses.

Turkish Model In the wake of the regime changes in the North Africa/Southwest Asia realm brought about by the "Arab Spring" of 2011, moderates have cited Turkey as the best model of democratic governance for this part of the world. Specifically, this involves a multiparty democracy that has a place for, but is not dominated by, Islamic political parties.

Uneven development The notion that economic development varies spatially, a central tenet of **core-periphery relationships** in realms, regions, and lesser geographic entities.

Unitary state A **nation-state** that has a centralized government and administration that exercises power equally over all parts of the **state**.

Unity of place The great German natural scientist Alexander von Humboldt's notion that in a particular locale or region intricate connections exist among climate, geology, biology, and human cultures. This laid the foundation for modern geography as an *integrative discipline* marked by a spatial perspective.

Urban system A **hierarchical** network or grouping of urban areas within a country; the spatial distribution and interrelationships among cities of various sizes within this network.

Urbanization A term with a variety of connotations. The proportion of a country's population living in urban places is its level of urbanization. The **process** of urbanization involves the movement to, and the clustering of, people in towns and cities—a major force in every geographic realm today. Another kind of urbanization occurs when an expanding city absorbs rural countryside and transforms it into suburbs; in the case of cities in disadvantaged countries, this also generates peripheral **shantytowns**.

Wallace's Line The zoogeographical boundary proposed by Alfred Russel Wallace that separates the marsupial fauna of Australia and New Guinea from the non-marsupial fauna of Indonesia (see Figure 11-4).

West Wind Drift The clockwise movement of water as a current that circles around Antarctica in the **Southern Ocean**.

World-city A large city with particularly significant international (economic) linkages that also has a high ranking in the global urban system. Leading world-cities include London, New York, Tokyo, Shanghai, Singapore, and Paris.

Xenophobia Extreme dislike and fear of foreigners, sometimes fueled by populist politicians.

INDEX

A page number followed by "f" indicates the entry is within a figure. A page number followed by "t" indicates the entry is within a table.

APPENDIX A

METRIC (STANDARD INTERNATIONAL [SI]) AND CUSTOMARY UNITS AND THEIR CONVERSIONS

LENGTH

Metric Measure

1 kilometer (km) = 1000 meters (m)

1 meter (m) = 100 centimeters (cm)

1 centimeter (cm) = 10 millimeters (mm)

Nonmetric Measure

1 mile (mi) = 5280 feet (ft)

= 1760 yards (yd)

1 yard (yd) = 3 feet (ft)

1 foot (ft) = 12 inches (in)

1 fathom (fath) = 6 feet (ft)

Conversions

1 kilometer (km) = 0.6214 mile (mi)

1 meter (m) = 3.281 feet (ft)

= 1.094 yards (yd)

1 centimeter (cm) = 0.3937 inch (in)

1 millimeter (mm) = 0.0394 inch (in)

1 mile (mi) = 1.609 kilometers (km)

1 foot (ft) = 0.3048 meter (m)

1 inch (in) = 2.54 centimeters (cm)

= 25.4 millimeters (mm)

AREA

Metric Measure

1 square kilometer (km^2) = 1,000,000 square meters (m^2)

= 100 hectares (ha)

1 square meter (m^2) = 10,000 square centimeters (cm^2)

1 hectare (ha) = 10,000 square meters (m^2)

Nonmetric Measure

1 square mile (mi^2) = 640 acres (ac)

1 acre (ac) = 4840 square yards (yd^2)

1 square foot (ft^2) = 144 square inches (in^2)

Conversions

1 square kilometer (km^2) = 0.386 square mile (mi^2)

1 hectare (ha) = 2.471 acres (ac)

1 square meter (m^2) = 10.764 square feet (ft^2)

= 1.196 square yards (yd^2)

1 square centimeter (cm^2) = 0.155 square inch (in^2)

1 square mile (mi^2) = 2.59 square kilometers (km^2)

1 acre (ac) = 0.4047 hectare (ha)

1 square foot (ft^2) = 0.0929 square meter (m^2)

1 square inch (in^2) = 6.4516 square centimeters (cm^2)

TEMPERATURE

To change from Fahrenheit (F) to Celsius (C)

$$°C = \frac{°F - 32}{1.8}$$

To change from Celsius (C) to Fahrenheit (F)

$$°F = °C \times 1.8 + 32$$